DEVELOPMENTS IN **PETROLEUM SCIENCE 36**

THE PRACTICE OF
RESERVOIR ENGINEERING
(REVISED EDITION)

T0348837

DEVELOPMENTS IN **PETROLEUM SCIENCE** 36

Volumes 1-7, 9-18, 19b, 20-29, 31, 34, 35, 37-39 are out of print.

DEVELOPMENTS IN **PETROLEUM SCIENCE 36**

THE PRACTICE OF RESERVOIR ENGINEERING (REVISED EDITION)

L.P. DAKE [†]

ELSEVIER

Amsterdam – Boston – Heidelberg – London – New York – Oxford
Paris – San Diego – San Francisco – Singapore – Sydney – Tokyo

Elsevier
Radarweg 29, PO Box 211, 1000 AE Amsterdam, The Netherlands
The Boulevard, Langford Lane, Kidlington, Oxford OX5 1GB, UK

First edition 1994
Second edition 1997
Revised edition 2001 (Hardbound & Paperback)
Reprinted 2004, 2006, 2007, 2008

British Library Cataloguing in Publication Data
A catalogue record for this book is available from the British Library

Library of Congress Cataloging-in-Publication Data
A catalog record for this book is available from the Library of Congress

ISBN: 978-0-444-50671-9

For information on all Elsevier publications
visit our website at books.elsevier.com

Printed and bound in the United Kingdom

Transferred to Digital Printing, 2010

Working together to grow
libraries in developing countries

www.elsevier.com | www.bookaid.org | www.sabre.org

ELSEVIER   BOOK AID International   Sabre Foundation

to Grace

FOREWORD TO THE REVISED EDITION

This revised edition presents a series of small text improvements throughout the book and a certain revision of the text of chapter 4 which was required to enable a better understanding of some physical explanations.

A more important change was carried out in subchapter 5.9 in relation to "the examination of water drive performance", where an excellent demonstration for a new procedure was developed for two real field cases. All elements of design, such as injection pressure, oil rate, and recovery prediction are explained in detail and illustrated with two field examples: one in the North Sea and another one in East Texas.

The philosophy introduced by Laurie Dake in chapter 5.9 concerns the key to understanding the reservoir fractional flow technique by the appreciation that the Buckley-Leverett theory is dimensionless and thus represents the simplest statement of the material balance for water drive.

In this book, containing the basic material and modifications prepared by the author Laurie Dake, any Petroleum Engineer will find the essential basis not only for understanding a gas or oil field, but also for predicting the future behaviour of a reservoir. It represents one of the most precious heritages of one of the most brilliant minds who dedicated his life to the advancement of Petroleum Science and Engineering.

Prof. Dr. T.D. van Golf Racht
Petroleum Department
Trondheim University

PREFACE

The *Practice of Reservoir Engineering* has been written for those in the oil industry requiring a working knowledge of how the complex subject of hydrocarbon reservoir engineering can be applied in the field in a practical manner. The book is a simple statement of how to do the job and is particularly suitable for the hard pressed reservoir/production engineers in its advice, illustrated with 27 examples and exercises based mainly on actual field developments. It should also be useful for those associated with this central subject of hydrocarbon recovery, from geoscientists and petrophysicists to those involved in the management of oil and gas fields.

Reservoir engineering is a complex subject for two reasons. In the first place, we never see enough of the reservoirs we are trying to describe. Therefore, it is difficult to define the physics of the system and, therefore, select the correct mathematics to describe the physics with any degree of certainty. The second problem is that even having selected a sensible mathematical model there are never enough equations to solve for the number of unknowns involved. The latter problem extends across the broad spectrum of the subject, from material balance application to well test interpretation and leads to an inevitable lack of uniqueness in describing reservoirs. Given these basic limitations, the only approach to the subject must be one of simplicity and such is the theme of the book. In fact, the basic tenet of science: Occam's Razor, applies to reservoir engineering to a greater extent than for most physical sciences — that if there are two ways to account for a physical phenomenon, it is the simpler that is the more useful.

Chapter 1, *Introduction to Reservoir Engineering*, is a description of the subject and of the main themes of the book. It was inspired on reading the excellent history of the industry *The Prize (The Quest for Oil, Money and Power)* written by Daniel Yergin and published by Simon and Schuster in 1991. Rather surprisingly, in the extensive index to this 880 page treatise the word "Reservoir" does not appear and, therefore, neither does "Reservoir Engineering", practitioners of which will appreciate is the most important subject in the whole industry. Chapter 1, therefore, attempts to redress the balance with a statement of the meaning of reservoir engineering. It includes a description of the main activities and the role of engineers, particularly in offshore developments, which is a topic that has received little attention in the literature. The history and future of reservoir engineering are considered and the chapter ends with a description of the basic physical principles involved in its application.

Chapter 2, *The Appraisal of Oil and Gas Fields*, focuses on the appraisal stage of field development, which is particularly relevant to offshore projects. Subjects covered include: PVT-fluid properties, with particular emphasis on sample collection and the correction of laboratory data to match field conditions, the estimation of hydrocarbons in place and the contentious issue of equity determination. The chapter concludes with a description of pressure–depth relations and, in particular, RFT-interpretation, the purpose and practice of appraisal well testing (DSTs) and the design of extended well tests (EWTs). The chapter is written in such a manner as to serve as a useful introduction to field appraisal for all disciplines involved in this activity.

Chapter 3, *Material Balance Applied to Oilfields*, is on the application of material balance to fields influenced by a variety of different drive mechanisms. In the author's opinion, the subject has become as dead as the Dodo in recent years, the general belief being that it has been superseded by the more sophisticated technique of numerical simulation modelling. Nothing could be further from the truth, however, since material balance is the fundamental physical statement of reservoir engineering, not only explaining the mechanics of reservoir behaviour but also being the basic principle of the mechanics of fluid displacement (Buckley-Leverett). The chapter points out that material balance and simulation should not be regarded as competitive modes of describing field performance but must instead be fully supportive. The former is the ideal tool for history matching field performance, the results of which are used to construct a simulation model for the purpose of prediction. Such has been the neglect of material balance, however, that younger engineers have neither experience nor confidence in its application. To overcome this, the chapter contains six fully worked examples of material balance application to real field developments.

Chapter 4, *Oilwell Testing*, is devoted to the examination of the purpose, practice and interpretation of well tests in both appraisal and development wells. Since the early 1980's, the subject has been dominated by the philosophy of attempting to solve the *inverse problem*: using mathematics to define the physical state of a system. With the exception of developments at the forefront of physical science, this is an unconventional approach in practical physical/engineering disciplines and amounts to little more than curve fitting, which suffers from a severe lack of uniqueness. The chapter attempts to persuade the engineer that the only rational approach to test interpretation is to first define the physical state of the system under test by comprehensive observation of all relevant reservoir/mechanical data and then reach for the appropriate mathematical model (if it exists) to analyse the test. This is a much more difficult approach but, bearing in mind the importance of the field development decisions based on well test interpretation, is one that is mandatory.

In examining the history of well testing, the author has had cause to revise some of the earlier, simplifying assumptions that have dominated the subject. The most pervasive is that of *transience* (infinite acting behaviour) which, on account of its mathematical simplicity, has long prevailed in the subject and is still enshrined in many modern texts and computer software for test interpretation. Removal of this assumption on fifteen occasions, from the conventional presentation of the subject, confronts the engineer with a completely different perspective on test interpretation; in some respects more restrictive, in others more liberating but always more realistic.

In spite of the burgeoning use of log–log pressure plotting since the early 1980's, by far the most popular means of pressure buildup interpretation remains the

Horner semi-log plot (1951). Yet the most widespread error in the whole subject lies in its interpretation: where should the straight line be drawn and what does it mean? To overcome this, the author has resurrected and extended application of the simpler form of buildup analysis technique of Miller, Dyes and Hutchinson (MDH-1950) and demonstrated that it is capable of matching anything that Horner analysis can do — and a little bit more, in a simpler and less error prone manner. Use of the technique suggests that perhaps we waste too much time and money indulging in lengthy pressure buildups, when a few hours of closure is all that is ever required. Examining Horner and MDH time derivative plots in conjunction is presented as a *guaranteed* method for defining the correct straight line on semi-log buildup plots.

Chapter 5, *Waterdrive*, describes the most widespread form of secondary recovery technique: engineered waterdrive. Some of the description relates to the development of North Sea fields, the majority of which operate under this condition. This is not chauvinistic because the argument is made that the North Sea has been the biggest laboratory ever for the study of waterdrive.

The chapter starts with a description of the practicalities of waterdrive with particular emphasis on matching the capacities of surface facilities for injection/ production of liquids to the reservoir performance. Next, the basic theory of waterdrive (Buckley-Leverett) and its components are examined in detail. These consist of relative permeabilities and the concept of the fractional flow of water. It is argued that the former have little relevance in themselves and it is the fractional flow relationship that predominates in the subject. In fact, it is the main purpose in writing the chapter to try and re-assert the importance of fractional flow, which, like material balance, has practically disappeared from reservoir engineering in recent years simply because the concept has never (or only recently) been incorporated in the construction of numerical simulation models — and, therefore, it has ceased to exist. Data requirements and their interpretation and incorporation in the calculation of vertical sweep efficiency in heterogeneous reservoir sections are described, all using pseudo fractional flow functions in Welge calculations. The chapter finishes with an account of methods for history matching and predicting the performance of *difficult* waterdrive fields, which sometimes defy the use of numerical simulation due to their sheer complexity.

Chapter 6, *Gas Reservoir Engineering*, covers three aspects of gas reservoir engineering: material balance, immiscible gas drive and dry gas recycling in retrograde gas condensate reservoirs. Gas material balance is probably the simplest subject in reservoir engineering, yet the universal use of p/Z-plots in isolation leads to some alarming errors in overestimating the GIIP, the worst example noted by the author being an excess of 107%. Surely we can do better than that — and indeed we can. A more rational and sensitive approach to material balance application, to be used in conjunction with p/Z-plots, is suggested and its use illustrated.

The mobility ratio for immiscible gas–oil displacement is very unfavourable, making the process intrinsically unstable, unless the gravity term in the gas fractional flow is dominant. The section concentrates on the vetting of reservoirs for their suitability for gas drive and provides an example of gas drive efficiency calculations.

In considering dry gas recycling, much of the analysis in the literature is focused on compositional effects and what is overlooked is that the process is basically unstable. The section describes the influence of heterogeneity and gravity on the efficiency of recycling, which is illustrated with an example.

Acknowledgements. I should like to thank all those who have helped me during the course of writing this book. Foremost are all those engineers with whom I have worked or who have attended my lecture courses in reservoir engineering. I am particularly indebted to Scottish Enterprise (formerly the Scottish Development Agency) for their support and Enterprise Oil of London for their continual assistance. My thanks are also due to the staff of the Department of Energy, London (now the Department of Trade and Industry) and also to members of the staff of The Danish Energy Agency in Copenhagen. Particular thanks also to my colleague Professor Th. van Golf Racht of Paris for his advice.

Laurie Dake

IN MEMORIAM: LAURENCE P. DAKE

In the family of reservoir and petroleum engineers it was always so natural and rewarding to talk about "Laurie" (the name he preferred to his official one, Laurence Patrick Dake) about his point of view, and about his acceptance of, or opposition to, certain ideas or procedures. Today, sitting in front of a blank sheet of paper, I understand for the first time how difficult, how sad, and how impossible it is for any of his friends to talk about Laurie in *memoriam*. The only way to proceed is by remembering Laurie's life and his contribution to our petroleum engineering profession, and in evoking his exceptional creative spirit.

I remember the unforgettable conversations during the long winter nights of 1985 in my Norsk Hydro Oslo Office, when Laurie elaborated on the key objective of reservoir engineering: The capacity to turn the time-mirror around, so that a coherent image of the future prediction of an oil field can in return give us valuable insight into today's understanding of the same field, in order to ensure that every statement about the future behaviour of the reservoir is not accompanied by a long series of "ifs", "buts" and an avalanche of "maybes".

It was during this period that Laurie began using this approach to lay the foundations for the book "Practice of Reservoir Engineering".

Laurence Dake was born *11 March 1941* on the Isle of Man. He received his education at King Williams College and graduated in Natural Philosophy at the University of Glasgow in *1964*.

Recruited by Shell in *1964*, he joined Shell International as a Petroleum Engineer. Following a thorough training program at the Shell Training Center in The Hague, he participated as Petroleum Engineer in a variety of field operations in Australia, Brunei, Turkey and Australia until 1971, when he was once again called back to the Shell Training Center in The Hague. For seven years, from *1971 until 1978*, he taught the subject of Reservoir Engineering to Shell graduates.

In *1978* Laurie Dake left Shell after 14 years of service, at which time he made two significant steps which would determine his further professional career:

(1) He joined the newly established State Oil Company BNOC (British National Oil Cooperation) as Chief, Reservoir Engineering. In this function he participated in the discovery, development and deciphering of the secrets of the large North Sea reservoirs. His contribution during the early days of the UK offshore industry was so significant that in *1987* he received the OBE recognition for his Reservoir Engineering services to the UK industry. In these days this recognition not only honoured him for his exceptional work, but also indirectly honoured the reservoir engineering profession for its potential to influence the results of the oil and gas industry.

(2) In *1978* Laurie Dake published his first book with Elsevier on reservoir engineering under the title *"The Fundamentals of Reservoir Engineering"*. In this work he introduced a modern vision on Reservoir Engineering based on the

synthesis between rigorous physics and applied science, necessary in any field operative work. The exceptional success of this book with the entire petroleum world resulted from:

- its utility for Petroleum Engineers in applying simplified procedures to complex problems of hydrocarbon reservoirs;
- its utility as fundamental text for students at almost every University where the scientific basis of the reservoir discipline is combined with a large amount of field applications and examples.

In *1982* Laurie Dake left BNOC at the time of its privatisation and started as an independent consultant, based in Edinburgh. His comprehensive activities were divided among:

- a "direct consulting activity" with medium and large companies where Laurie made a substantial contribution to the appraisal and development of over 150 world wide oil and gas fields, *between 1982 and 1994*. He became one of the most appreciated international petroleum consultants, and was consulted by very large companies (BP, Agip, Norsk Hydro, Statoil, etc.) and banks (Bank of Scotland – Edinburgh, BankWest Perth, Australia, etc.);
- an important collaboration with the Petroleum Department of the Heriot—Watt University, where he started initially (*after 1978*) as an external examiner and where he later became a "Honorary Professor";
- the elaboration of his second book " The Practice of Reservoir Engineering", published by Elsevier in *1994*. In addition to many field operative concepts, the text included specific procedures and analyses developed by Laurie and proven successful in various fields studied by him.

In the middle of these exceptional activities, his real help to the entire petroleum engineering family through his books and courses, his consulting activities and his advice to the Financial World and Petroleum Companies, Laurie Dake's death on *July 19, 1999* left us disoriented. All of us who appreciated him, who admired his work and loved him for his exceptional qualities and distinction suddenly felt impoverished.

However, if we now look back to the horizon opened by Laurie, knowing that there exists an accepted horizon — visible but sterile, and another ... an imaginative and creative one, we may change our point of view. Knowing that the creative horizon in a sense defines the boundaries between spirit and matter, between resources and platitude, we start to understand the role played by Laurie Dake — who disregarded the customary procedure and fought to grasp the real meaning of reservoir behaviour.

He has been able with his intelligence to enlarge the opened horizon by combining the will of creativity with the knowledge of reality versus the size of possibility ... , all of which we find in the solutions proposed by him.

It is this enlarged horizon which Laurie left to all of us as a splendid heritage ...

Prof. Dr. T.D. van Golf Racht
Petroleum Department, Trondheim University

CONTENTS

NOMENCLATURE

A	area (sq.ft)
B_g	gas formation volume factor: gas (rcf/scf)
	: oil (rb/scf)
B_o	oil formation volume factor (rb/stb)
B_w	water formation volume factor (rb/stb)
c	isothermal compressibility (1/psi)
c_f	pore compressibility (1/psi)
c_o	oil compressibility (1/psi)
c_w	water compressibility (1/psi)
C_A	Dietz shape factor (dimensionless)
C'_A	Larsen shape factor (dimensionless)
D	vertical depth (ft.ss)
e	exponential
ei	exponential integral function
E	gas expansion factor (scf/rcf)
$E_{f,w}$	term in the material balance equation accounting for connate water expansion and pore compaction (rb/stb)
E_g	term in the material balance equation accounting for the expansion of the gas cap (rb/stb)
E_o	term in the material balance equation accounting for the expansion of the oil and its originally dissolved gas (rb/stb)
f_g	fractional flow of gas (dimensionless)
f_w	fractional flow of water (dimensionless)
F	underground withdrawal of fluids (rb)
G	gas initially in place (scf)
G	gravity number in the factional flow equation (dimensionless)
G_p	cumulative gas production (scf)
h	formation thickness (ft)
k	permeability (mD)
k_r	relative permeability (dimensionless)
k'_r	end point relative permeability (dimensionless)
$\overline{k}_r$	pseudo relative permeability (dimensionless)
l	length (ft)
L	length (ft)
m	ratio of the initial hydrocarbon pore volume of the gascap to that of the oil column (material balance) (dimensionless)
m	slope of the early linear section of semi-log plots for well test interpretation psi/log cycle

$m(p)$ pseudo pressure (psi^2/cp)

M end point mobility ratio (dimensionless)

n number of lb moles

N stock tank oil initially in place (stb)

N_p cumulative oil recovery (stb)

N_{pd} dimensionless cumulative oil recovery (PV)

N_{pD} dimensionless cumulative oil recovery (HCPV)

p pressure (psia)

p_b bubble point pressure (psia)

p_D dimensionless pressure

p_e pressure at the external boundary (psia)

p_i initial pressure (psia)

p_{sc} pressure at standard conditions (psia)

p_{wf} wellbore flowing pressure (psia)

p_{ws} wellbore static pressure (psia)

p_{wsl} static pressure on the early linear trend of a semi-log buildup or two rate drawdown plot (psia)

$\bar{p}$ average pressure (psia)

p^* specific value of p_{wsl} at infinite closed-in time (Horner plot) (psia). (*See also* Z^*.)

q liquid production rate (stb/d)

q_{wi} water injection rate (b/d)

Q gas production rate (MMscf/d)

r radial distance (ft)

r_e external boundary radius (ft)

r_{eD} aquifer/reservoir radius ratio (dimensionless)

r_o reservoir radius (ft)

r_w wellbore radius (ft)

R producing (instantaneous) gas oil ratio (scf/stb)

R universal gas constant

R_p cumulative gas oil ratio (scf/stb)

R_s solution or dissolved gas oil ratio (scf/stb)

S skin factor (dimensionless)

S_g gas saturation (PV)

S_{gr} residual gas saturation (PV)

S_o oil saturation (PV)

S_{or} residual oil saturation (PV)

S_w water saturation (PV)

S_{wbt} breakthrough water saturation (PV)

S_{wc} connate water saturation (PV)

S_{wf} shock front water saturation (PV)

$\bar{S}_g$ thickness averaged gas saturation (PV)

$\bar{S}_{gd}$ thickness averaged dry gas saturation (PV)

$\bar{S}_w$ average water saturation (PV)

t	time hours to years, as appropriate
t_D	dimensionless time
t_{DA}	dimensionless time $(t_D r_w^2 / A)$
Δt	closed-in time (pressure buildup) (hours)
Δt_s	closed-in time during a buildup at which p_{wsl} extrapolates to p_i or $\overline{p}$ on an MDH plot
T	abslolute temperature (degrees Rankin)
U	aquifer constant (bbl/psi)
v	velocity (ft/d)
V	volume (cu.ft)
W	width (ft)
W_D	dimensionless cumulative water influx
W_e	cumulative water influx (bbl)
W_i	cumulative water injected (bbl)
W_{id}	dimensionless cumulative water injected (PV)
W_{iD}	dimensionless cumulative water injected (HCPV)
W_p	cumulative water produced (bbl)
Z	Z-factor (dimensionless)
Z^*	alternative symbol to replace p^* in Horner buildup analysis (psia)

GREEK

α	volumetric sweep (dimensionless)
γ	specific gravity (liquids: relative to water = 1, gas: relative to air = 1, at standard conditions)
γ	exponent of Euler's constant (1.781)
Δ	difference
θ	dip angle (degrees)
μ	viscosity (cp)
ρ	density (lb/cu.ft)
σ	coefficient in well testing equations = $7.08 \times 10^{-3}\, kh/q\mu B_o$

SUBSCRIPTS

b	bubble point
bt	breakthrough
d	differential (PVT analysis)
d	dimensionless (PV)
d	displacing phase
D	dimensionless (pressure, radius, time)
D	dimensionless (HCPV)
DA	dimensionless (time)
e	production end of a system
f	flash separation

f	flood front
g	gas
i	cumulative injection
i	initial
o	oil
p	cumulative production
r	relative
r	residual
s	solution gas
sc	standard conditions
t	total
w	water
wf	wellbore, flowing
ws	wellbore, static

ABBREVIATIONS

CTP	constant terminal pressure
CTR	constant terminal rate
CVD	constant volume depletion
DST	drill stem test
EOR	enhanced oil recovery
EOS	equation of state
EWT	extended well test
FIT	formation interval tester
FVF	formation volume factor
GDT	gas down to
GOC	gas oil contact
GOR	gas oil ratio
HCPV	hydrocarbon pore volume
IARF	infinite acting radial flow
JOA	joint operating agreement
KB	kelly bushing
MDT	modular formation dynamic tester
MGV	movable gas volume
MOV	movable oil volume
OWC	oil water contact
PV	pore volume
PVT	pressure, volume, temperature
RF	recovery factor
RFT	repeat formation tester
SCAL	special core analysis
SIP	sequential inflow performance
WUT	water up to

Chapter 1

INTRODUCTION TO RESERVOIR ENGINEERING

1.1. ACTIVITIES IN RESERVOIR ENGINEERING

Reservoir engineering shares the distinction with geology in being one of the great "underground sciences" of the oil industry, attempting to describe what occurs in the wide open spaces of the reservoir between the sparse points of observation — the wells. In applying the subject, it is possible to define four major activities, which are:

- Observations
- *Assumptions*
- Calculations
- Development decisions

and these can be described as follows.

(a) Observations

These include the geological model, the drilling of wells and the data acquired in each: cores, logs, tests, fluid samples. Following the start of field production, the oil, gas and water rates must be continuously and accurately monitored together with any injection of water and gas. Frequent pressure and production logging surveys should also be conducted throughout the lifetime of the project. The importance of data collection is a subject that is frequently referred to throughout the book and is perhaps best illustrated in Chapter 3 when describing material balance which is so fundamental to reservoir engineering. It is pointed out that the material balance equation can easily contain eight "unknowns" and failure to collect the essential rate, pressure and pressure–volume–temperature (PVT) data to give oneself a "sporting chance" of attaining a meaningful solution of the equation can make the application of quantitative engineering impossible. The author has seen many fields in which the data collection was so inadequate that it would have been dangerous to attempt to apply any quantitative reservoir engineering technique and the alternative: guesswork, although occasionally unavoidable, is something to be shunned as much as possible. Thorough data acquisition is expensive (very expensive) and it is the duty of practising reservoir engineers to convince those who hold the purse strings of the necessity of the exercise. To do this, it is important that the engineer knows exactly to what use the data will be put and in this respect it is hoped this book will prove helpful in advising on the matter.

Once the data have been collected and verified, the engineer must interpret them very carefully and collate them from well-to-well throughout the reservoir and adjoining aquifer. This is a most delicate phase of the whole business of understanding reservoirs, in which it can prove dangerous to rely too much on automated techniques. It is fine to use computer programs to interpret well tests, for instance, but to apply sophisticated numerical methods to generalise about formation properties by such means as generating statistical correlations and use of regression analysis should be reduced to a minimum. Reservoir performance is so often dominated by some particular physical feature (the bit in the corner), which may be a weakness or strength, that can be completely overlooked through the smudging effect inherent in the application of number crunching methods. In this respect, three examples of the dangers of "statistical smearing", which adversely affect the calculation of sweep efficiency in waterdrive or gas-drive projects, as described in Chapters 5 and 6, respectively are.

– The evaluation of formation heterogeneity using probability distributions of permeability. This totally neglects gravity and therefore disregards Newton's second law of motion.

– Application of convoluted petrophysical transforms to generate permeability distributions across formations. Considering the expensive errors this leads to, it is much cheaper to core "everything".

– Plotting core permeabilities on a logarithmic scale when viewing their distribution across a reservoir section. Darcy's flow law specifically states that the velocity of advance of a fluid is proportional to the permeability of a layer: not the logarithm of its permeability.

These are not small, fastidious points. Together they have cost international operators billions of dollars in lost and deferred production in secondary recovery projects worldwide. Most of the difficulties mentioned would be overcome by adherence to the basic laws of physics in the interpretation and collation of reservoir data.

The most valuable reservoir engineers are those who *see* the clearest and the most and who know what they are looking for.

The last comment implies the need for experience, which may be a bit discouraging to newcomers to the subject but there has and always will be a very large element of deja vu associated with reservoir engineering.

(b) Assumptions

Now the main difficulty arises: having thoroughly examined and collated all the available data, the engineer is usually obliged to make a set of assumptions concerning the physical state of the "system" for which an appropriate mathematical description must be sought. For instance:

– The oil or gas reservoir is or is not affected by natural water influx from an adjoining aquifer.

– There will or will not be complete pressure equilibrium across the reservoir section under depletion or waterdrive conditions.

– The late-time upward curvature of points in a pressure buildup survey results from: the presence of faults, dual porosity behaviour or the breakout of free gas around the wellbore.

This is the crucial step in practical engineering, so much so that the word *assumptions* is highlighted (page 1) to signify that once they have been made that is effectively the end of the reservoir engineering. The third activity, calculations, is entirely dependent on the nature of the assumptions made as is the fourth, development decisions, which rely on the results of the calculations. It is therefore necessary to be extremely cautious when making physical assumptions and the most convincing reservoir studies are those containing the least number. In this respect, the application of material balance is strongly recommended (oil, Chapter 3; gas, Chapter 6), in history matching field performance, as being one of the safest techniques in the business for the simple reason that it requires fewer (far fewer) assumptions than the alternative technique, numerical simulation modelling, which should therefore be used as a logical follow-up (Chapter 3, section 3.5). Assumptions can usually be confirmed only by observation, not calculation, but this can be problematical. In the first two examples listed above, for instance, the observations necessary to confirm the influence of an aquifer or the degree of pressure equilibrium across a reservoir can only be made some time after the start of continuous field production, which causes difficulties in offshore developments, as described in section 1.2d. Finally, considering the importance attached to assumptions, it is mandatory that they be listed and rationalized right at the start of any study report — if for no other reason than to permit the reader the choice of whether to explore the text further or not.

(c) Calculations

While the above comments may seem to denigrate the importance of calculations, that is only meant to be relative to the physical assumptions upon which any mathematical model selected for use is entirely dependent. Once a physical condition has been defined (assumed) then calculations are an absolute must and it should be remembered at all times that we reservoir engineers receive our salaries to perform calculations rather than merely express opinions.

Since the 1960's mathematics has been awarded a much more exalted status in the business than ever before for the simple reason that now, with our sophisticated computers, we can do it, whereas previously we could not, or only with great difficulty. With each passing year and each improved mathematical model, there is a great danger that numerical modelling is becoming a subject in its own right in which there is a tendency for it to evolve in a direction that removes it ever further from the requirements of reservoir engineering. Hence the fairly new title in the Industry of "Simulation Engineers". There should be no such thing, only reservoir engineers who happen to have simulation packages at their disposal for use, amongst other tools, as and when required.

Engineers, and particularly those entering the subject, should be aware of this pitfall and not become too mesmerised by mathematical models — no matter how user-friendly. It is not entirely unusual, for instance, to find a group of engineers slaving away in the back room for months with their computers only to emerge with the pronouncement that — "the model is great but there seems to be something wrong with the reservoir." The majority of a reservoir engineer's time should be spent looking at things very carefully and only then deciding on the correct approach to the problem rather than bombarding it with different mathematical models in the hope that one of them might fit. The approach to avoid is that noted recently by the author in which up to 600 computer runs were budgeted for in a study to history match the performance of a field. This sort of planning makes you wonder whether it was a reservoir engineering study or a lottery.

The requirement of the engineer is to assess in advance, through sound physical judgement, which of the runs is liable to be correct — and why. If mathematics is used carefully and correctly then we should have a great advantage over our predecessors in this subject but if it is abused by relying on mathematics to define physics, then reservoir engineering is itself in danger.

(d) Development decisions

Every action contemplated, planned and executed by reservoir engineers must lead to some form of development decision — otherwise it should not be undertaken in the first place.

1.2. BASIC THEMES OF THE TEXT

(a) Simplicity

Even if we could see in complete detail all the complications of the reservoir geology and its fluid contents, trying to describe it mathematically would be a daunting task. But since we hardly see anything of the systems we are attempting to describe and even when we formulate equations they contain far too many unknowns to provide a unique solution, then the situation begins to look grim. Under these circumstances, this book advocates that all problems be approached in a simple fashion, in fact, there seems to be an inverse law applicable to reservoir engineering that the more complex the system, the more appropriate is the attempt at simplicity and the more convincing. It is virtually impossible, for instance, to construct a reliable simulation model for waterdrive in a complex delta top reservoir, with criss-crossing channels that do not seem to correlate from well to well, for the simple reason that you cannot model what you cannot see — although it does not prevent people from trying.

The alternative approach (as illustrated in Chapter 5, section 5.9) is to apply the simple concept of material balance to match waterdrive history and make a reasonable attempt at prediction. This will provide an understanding of the physics of the system which enables sound operating judgements to be made, even though it may not yield exactly correct results which, by the very nature of the subject, are usually unattainable. It is, therefore, frequently stressed in the text to adopt the simplest physical method possible in tackling all forms of reservoir engineering problems, although this must not be overdone. Albert Einstein appears to have summed up the situation in a concise manner:

"Let us do everything as simply as possible, but then, not more simply than that."

(b) What works and what does not — and why?

Some time ago the author was confronted by a university geologist who declared that of all the subjects he had ever encountered, reservoir engineering was the one that stood in greatest need of an "Academic Audit". In other words, it's about time that somebody stood up and made a clear pronouncement about what works and what does not amongst the myriad of theoretical papers and books on the subject. Unfortunately, this is easier said than done because there is nobody in the business who knows, since we can never get down there in the reservoir and examine the in situ consequences of our elegant theories.

This author is certainly not in a position to attempt an academic audit but, nevertheless, it is possible to consider a damage-limitation exercise, merely by doing a "trace-back" through the history of the subject. Although the physics and mathematics applied date from earlier centuries and are, therefore, well founded, their application to reservoir engineering is comparatively new. It is only since the early 1930's, with the introduction of such basic concepts as material balance (Chapter 3) and the Buckley-Leverett theory of immiscible displacement (waterdrive, Chapter 5; gas drive, Chapter 6) that field engineers were first provided with sound quantitative methods for describing reservoir performance. Therefore, it would be unwise to believe that this young subject is well established, in its present form, such that we can build upon it with confidence — far from it. One of the earlier difficulties was that our predecessors, armed only with log-tables and slide-rules, were trying to tackle the same mathematical problems that today we feed into our mega computers. Understandably, they were forced into making simplifying assumptions and tended to sweep inconvenient pieces under the carpet, where many of them remain to this day. In researching and writing this book, the author has delved into the history of the subject to determine the effect of removing these unnecessary assumptions and dragging out and inspecting some of the hidden detritus.

The effect is perhaps most noticeable in the subject of well testing (Chapter 4) in which the removal of the persistent and all pervasive assumption of transience (that every reservoir exhibits infinite acting behaviour on test) seems to work wonders in clarifying and simplifying the subject and even suggests the abandonment of some of our traditional testing practices, such as the indulgence in lengthy pressure buildups. Similarly, in Chapter 5, on waterdrive, the meaning and use of relative permeability functions (which have been a part of the folklore of the subject for the past 50 years: ever since they first knocked Buckley-Leverett off balance) is seriously questioned. It appears that, except in the most extreme case (the failure event), continuous relative permeability functions can be dispensed with altogether when describing how one fluid displaces another in practical flooding situations and this tends to greatly simplify the subject.

Therefore, while not comprising a technical audit, the book at least attempts to look at the subject in a straightforward manner, discarding those elements from the past that it should be no longer necessary to preserve. But is it right? Well, everything appearing in the text has been validated in the field. The chapter on waterdrive, for instance, is not just some theoretical treatise on the subject, it has been checked-out in the biggest waterdrive laboratory of all time — the North Sea (and elsewhere), where it appears to work quite well.

(c) Analytical methods

Nowadays people tend to draw a distinction between "classical" and "modern" methods in reservoir engineering. By the former is meant using the analytical solutions of linear diffcrential equations, whereas the latter refers to the use of finite difference, multi-cell, numerical simulation models. As pointed out previously, however, no special status should be awarded to simulation, it's simply another tool for attempting to solve the same old problems.

Of the two approaches, the use of analytical methods must be regarded as the more specialised and difficult to apply, since it requires considerable knowledge and judgement in the subject to use a particular equation to describe a physical situation in a meaningful fashion. On the other hand, provided the engineer gets the input to a simulator correct and assigned to the right boxes then, applying the finite difference analogue for solving the basic differential equations for mass conservation, volume change and inter-grid block flow, in principle the simulator should be able to provide a solution to any problem irrespective of its complexity. Because of this, simulation would appear to be the obvious way forward in this subject and, indeed, today it is used to tackle just about every reservoir engineering problem in the business. The unfortunate aspect of this trend is that using simulators exclusively is the worst means of ever learning and coming to a more mature understanding of the subject. Data are fed in at one end, results emerge at the other, and in-between the black box does little to reveal what is actually occurring that would broaden the engineer's physical understanding of the reservoir mechanics. This state of affairs is revealed in modern reservoir engineering reports which merely describe the input and what the computer produced as output. Gone are the equations that used to be an integral part of reporting and there is a distinct lack of any conclusions concerning the physical occurrences that produce a particular result. Therefore, there is a concentration on the application of simple analytical techniques in the text (lest we forget) to increase the engineers awareness of the subject. On so many occasions, practising reservoir engineers are obliged to think on their feet, there is not always time for simulation to solve all problems. This extends from middle of the night operational decisions to meetings, both internal in a company and with partners. All too frequently, engineers at meetings are unable to address fairly obvious questions without resorting to construction of a simulation model. It will do your career prospects no harm at all if you can apply sound intuition in these situations founded on an understanding of the basic theory which itself proceeds from complete familiarity with analytical methods.

The ideal situation would be that prior to any lengthy numerical simulation engineers should spend some time trying to progress the study as far as possible using analytical techniques to gain an understanding of the most sensitive factors influencing the outcome. It is amazing the extent to which this step usually leads to a simplification in major studies and in some cases it is a mandatory opening move, such as in the history matching of field performance (Chapter 3, section 3.5), to define reservoir drive mechanisms and volumes of hydrocarbons in place, before detailed model construction. Numerical simulation receives quite a lot of attention in the text but the description is focused more on the need and use of models, the processing of input data and interpreting the output of studies rather than the intricate mathematical details which have been described already in several excellent textbooks on the subject [1–4].

(d) Offshore versus onshore developments

There seems to be a deficiency in the literature in that, to date, there has been no textbook and few technical papers that describe the difference between reservoir engineering for offshore and onshore fields; this, in spite of the fact that there are now decades of relevant experience in offshore developments in such major areas as the U.S. Gulf Coast, the Middle East and the North Sea, to name but a few. The present book attempts to fill this obvious gap.

A field is a field whether located beneath land or water and the basic physics and mathematics required in its description is naturally the same. Where the main difference lies in the application of reservoir engineering to field development is in decision making: the nature, magnitude and timing of decisions being quite different in the offshore environment, as depicted in Fig. 1.1, which contrasts the production profiles of typical onshore and offshore fields.

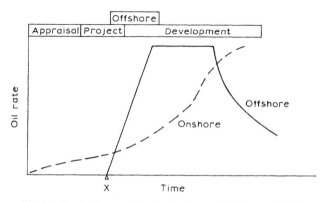

Fig. 1.1. Production profiles for onshore and offshore oilfields.

Onshore: Governmental regulations permitting and provided there are production facilities in the locality, then as soon as the explorationists release a discovery well to the production department, it should be tied back to the nearest block station and produced at high rate on a continuous basis. An obvious advantage in this manoeuvre is that it provides a positive cash flow from day one of the project but of greater benefit is that it permits the reservoirs to viewed under dynamic conditions from the earliest possible date. Continuous withdrawal of fluids creates a pressure sink at the location of the discovery well that, with time, will radiate both areally and vertically throughout the producing formations. When each subsequent appraisal/development well is drilled, the conducting of drillstem tests (DSTs) or, more significantly, repeat formation tester (RFT) surveys (Chapter 2, sections 2.7 and 8) will convey to the engineer the degree of lateral and vertical pressure communication: data that are indispensable in the planning of a successful secondary recovery flood for water or gas injection. Even if little or no pressure communication is observed, this at least warns the operator that wide scale flooding may not be feasible, which could prevent an expensive mistake being made. Therefore, as each new appraisal well is drilled it is placed on continuous production to help propagate the pressure sink. During the early years of the project, the systematic collection of rate and pressure data together with the PVT fluid properties permit the estimation of the strength of any natural energising mechanisms such as water influx from an aquifer or gas cap expansion and also lead to the estimation of hydrocarbons in place. The determination of these significant factors is by application of the concept of material balance (Chapter 3) which leads to the construction of meaningful numerical simulation models, whose main purpose is in predicting the areal movement of fluids in the reservoir and, therefore, decides such important matters as the number of wells required for successful development, their locations and the timing of their drilling.

If there is little natural energy (pressure support) in the reservoir then a pilot scheme for water or gas injection can be implemented and its efficacy tested. Altogether, it may take years before the optimum development plan evolves, after carefully piecing together the observations resulting from natural drive mechanisms and field experiments. If conducted properly, it is a disciplined and thoughtful process in which decisions are based on observations and should follow one another in a logical sequence.

Offshore: In this environment the sequence of events in field developments is much more compartmentalised than onshore (Fig. 1.1). To begin with, following the discovery well on an accumulation a series of appraisal wells is drilled to determine the volume of hydrocarbons in place and assess the ease with which they can be produced: two obvious requirements in deciding upon the commercial viability of the project. Unfortunately, the appraisal wells, which may range in number from one or two on a small accumulation to twenty or more on a large, cannot usually be produced on a continuous basis from the time of their drilling, since the offshore production and hydrocarbon transportation facilities are not in existence at this stage of the development. Consequently, the engineer is faced with the difficulty that, although there may be a perfectly adequate amount of data collected in each appraisal well, using the latest and greatest high-tech equipment, they are nevertheless the lowest quality data that will ever be acquired during the lifetime of the project because they are collected under purely static conditions. The most dynamic event that occurs at this stage is the conducting of DSTs in which perhaps a few thousand barrels of oil are withdrawn from each interval tested but this is usually insufficient to cause a significant pressure drop that can be detected in appraisal wells drilled subsequently. Extended well tests (EWTs, Chapter 2, section 2.10) can provide a partial solution to this problem but these are of quite rare occurrence and are never as conclusive as continuous production that is possible in land developments.

Therefore, even at the very end of the appraisal stage the reservoir engineer is confronted with the dilemma of not knowing precisely, or sometimes even approximately, the degree of pressure communication both areally and vertically in the reservoirs that have been appraised. Since, for the reasons explained in Chapter 5, section 5.2, large offshore projects tend to be developed by water or gas injection and since knowledge of the areal and vertical communication is crucial in the successful planning of such recovery schemes, the engineer has a problem: can injection and production wells be located in layers that are contiguous or, in the extreme, is the reservoir so fragmented by faulting that pressure cannot be maintained in the system as a whole. To exacerbate matters, all the important decisions relating to platform design: the number of wells required, capacities of surface equipment to inject and produce fluids commensurate with the required oil production profile, must be made up-front and therefore be based on the poor quality static data. Furthermore, the decisions made at this stage can on occasions be irrevocable and, of course, in offshore projects it is usually necessary to start multiplying onshore capital expenditure by orders of magnitude which adds further spice to the decision making.

Reservoir studies at the end of the appraisal stage are necessarily of a purely predictive nature, there being no production or pressure history to match. Under these circumstances, the only sensible tool to use in making predictions is numerical simulation modelling because simpler techniques such as material balance tend to break down (Chapter 3, section 3.5) because the main use of this tool is in history matching performance. Nevertheless, sound analytical techniques are presented in Chapters 5 and 6 for predicting the performance of water and gas drive projects and these should be used in conjunction with simulation modelling as an aid in understanding the processes. To be candid, however, at the end of the appraisal stage, irrespective of the technique employed, the quantification of predictions of field performance is anybody's guess and is therefore a highly subjective activity. Simulation models do not necessarily inform on the nature of the physical forces dictating reservoir performance (as some believe) they merely reflect the consequences of the input assumptions made by the engineer. Therefore, assumptions should be kept to an absolute minimum and carefully justified and every pertinent scrap of reservoir data collected during appraisal drilling must be thoroughly examined and used in any form of study. If there are neighbouring fields producing from similar reservoir sections, their histories should be carefully scrutinised for lessons that might be learned and incorporated in the model. For instance, the rate of watercut or gas/oil ratio (GOR) development in the real fields should be compared with the model predictions and suitable adjustments made to the latter if necessary. Unfortunately, by tradition, there is often a veil of secrecy over the transfer of this type of information between companies especially if the performance of the real fields has been disappointing. Some details of this type of prediction study, which is the very basis of offshore platform and facilities design, and the consequences of making errors at this stage are described in connection with the development of some of the major North Sea waterdrive projects in Chapter 5. Having completed the study and passed on the results to the project engineers, the reservoir engineer can take a back seat for a few years during the "project" phase when the real engineers of this world are out there with their hammers and cranes constructing the platform and installing all the production equipment and pipelines.

Sooner or later (usually later) the project phase comes to an end (point X, Fig. 1.1), all the equipment has been hooked-up and the development phase commences. In terms of career advice, this is as good a time as any for the reservoir engineer to transfer from one company to another, preferably a long way away, for now the truth is about to be revealed. The early development drilling phase is a period of frantic activity for the geologists, petrophysicists, reservoir and production engineers, as operators endeavour to set new drilling records in getting wells completed and on production as rapidly as possible to obtain a healthy, positive cash flow after years of heavy, up-front expenditure. The difficulty is that new geological and reservoir data are being collected with such haste from the wells that the experienced team hardly has time to assimilate all the facts in order to make decisions about the locations of new wells and when and exactly how should secondary recovery flooding be initiated. Adding to the complications is the fact that the reservoirs are now being viewed under dynamic conditions for the first time which can often result in drastic revisions being made to the initial development plan, formulated with only a static view of the reservoir (Chapter 5, section 5.2). Matters can become so extreme at this stage that it is not unknown for operators to hold a moratorium on drilling activity, sometimes lasting for months (an anathema to drilling departments), to give the experts time to revise their earlier plans. There are only a limited number of well slots on an offshore platform and they must not be wasted. For the reservoir engineer in particular, events happen with such rapidity that they must have the ability to think on their feet, there being too little time between decisions to permit the construction and use of numerical simulation models for guidance.

Eventually the plateau production rate is achieved (but not always). In waterdrive fields this usually results from a balance between additional oil from new producers and the decline in production of wells in which the injected water has broken through and started to be produced in significant quantities. By the end of the plateau, the excessive water production gains the upper hand and, producing through surface facilities of fixed capacity, the production decline period commences. In many respects, this should be the most active period for reservoir engineers and the one in which they should be at their most useful. The reason is because so much dynamic data has been collected from the field that numerical simulation models can be constructed and history matched in a meaningful manner making them reliable as predictive tools. In particular, in secondary recovery operations, the pressure is usually maintained constant by injection so model calibration must be affected by matching the timing of water or gas breakthrough in wells and their subsequent watercut or GOR developments. As described in Chapter 5, this is the most useful role for large simulation model studies in trying to determine how to prolong the "end game" prior to field abandonment for as long as possible by the means described in section 1.4.

The above scenario for an offshore development relates to the type of project encountered in the early days of the North Sea, with platforms standing in 500–600 ft water depths and each carrying the billion dollar plus price tag. There are many variations on this theme, however, such as in the Danish Sector of the North Sea, or offshore Indonesia, where the sea is shallower which relieves many of the tensions associated with deep water developments. Under these circumstances a lightweight platform can be installed and the field tested under dynamic conditions, perhaps for several years, to be followed by a second water or gas injection platform if required, which makes the project closer in nature to that described for onshore developments, although usually somewhat more expensive.

Alternatively, there can be onshore fields whose development resembles those in the offshore environment. One such field in the (aptly named) Empty Quarter of the Yemen was so remote that, by the time an oil production pipeline had been laid, all the appraisal and development wells had been drilled permitting only a static view of the reservoirs.

The main difference between offshore and onshore developments is therefore largely associated with decision making and the fact that offshore there are usually two distinct phases: *appraisal*, under static conditions, followed by the dynamic *development*. For onshore fields there is usually little or no distinction between these phases, they tend to get rolled into one and viewing reservoirs is invariably under dynamic conditions. In this book, there is a leaning towards the description of offshore fields but this should not deter the reader who is totally committed to land developments. For a start, you never know when you might be transferred and secondly, what has been learned about the application of reservoir engineering under the more stringent conditions prevailing offshore can only be of benefit to those operating on terra firma.

1.3. THE ROLE OF RESERVOIR ENGINEERS

There has been a considerable change in emphasis in the role of reservoir engineers since the early 1970's, so that their position within a multi-disciplinary team planning the development of a new field, particularly if it is offshore, is as depicted in Fig. 1.2. In this the reservoir engineers are seen occupying a location at the centre of the Universe — a comment not always appreciated by the other specialists involved. The fact is, however, that the reservoir engineers fulfil a

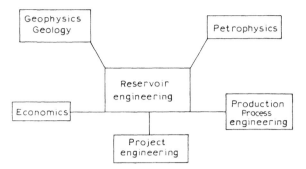

Fig. 1.2. The position of reservoir engineers within a development team.

central coordinating role in which data are received from others, processed and the interpreted results communicated to interested parties. That is, structural contour and other maps are obtained from the geophysicists and geologists together with formation properties from the petrophysicists: net pay thicknesses, porosities and fluid saturations. The reservoir engineers assimilate these with their own data (core permeability distributions, PVT fluid properties and interpreted DST results) to construct some form of model, simple or complex, dependent on the quantity and quality of the data available, which is used to predict the likely performance of the field under a variety of possible development scenarios. In this there is a close contact with the production technologists and process engineers responsible for ensuring that the fluids reach the surface and pass through the production facilities in the most efficient manner.

At the end of the day, the initial development study results, in terms of production profiles of oil, gas and water and injection profiles for water and/or gas, are presented to the project engineers charged with the responsibility of constructing the platform, surface production and injection facilities and pipework to cater for the predicted fluid rates. The essential, close liaison between reservoir and project engineers is a relatively new requirement in the business and has evolved with the growth of offshore developments. In the past, for instance, if unanticipated high water production occurred far ahead of schedule in an onshore waterdrive field, due to the unforseen effect of reservoir heterogeneity, there was no real problem. The reservoir engineer simply picked up the phone and ordered a couple of new injection pumps and additional separator capacity, or whatever was necessary to maintain the oil production profile at the required level. The equipment was duly delivered, offloaded and installed, usually without having to be concerned about the availability of space on which to site the new facilities. In such a case the oversight was never regarded as a reservoir engineering mistake but merely as a phase in progressing on the overall learning curve. Consequently, there was no great need for close links between reservoir and project engineers.

Offshore, however, the situation is markedly different. If severe water production occurs prematurely, so that the oil production rate cannot not be maintained through the installed facilities, then it is not always possible to upgrade their capacities to the desired level. The difficulty encountered, of course, is one of space constraint: often there simply is not the deck area available to install any additional facilities. As a consequence, the oil rate declines and so too does the profitability of the project. This particular problem is focused upon in detail in Chapter 5, the examples being taken from some of the deep water, North Sea projects which are particularly vulnerable because you cannot install a second platform in 500 ft of water just to overcome an initial mistake on the part of the reservoir engineers. The word "mistake" is used advisedly and in offshore activities it must be recognised as such and euphemisms such as "learning curve" are quite unacceptable in this altogether harsher environment. Failure to size the surface facilities to match the actual gross fluid rates is very common and has affected most offshore operators, furthermore, it can be very expensive in terms of lost and deferred production.

How does the reservoir engineer avoid this potential hazard? Well, the onus is on engineers to get the fluid production and injection profiles correct in the first place, in the initial development study. But, as pointed out in section 1.2d, the input data to such studies is of low quality because it is collected under static conditions whereas to assess the efficiency of a waterflood requires its viewing under dynamic conditions. Therefore, evaluating offshore fields is difficult but far from impossible. What seems to give rise to most of the errors has nothing to do with the sophistication of the computer model used but instead relates to failure to observe, understand and take full account of all the data so expensively acquired during the appraisal drilling programme. An attempt is made throughout Chapter 5, on waterdrive, to explain how best to view these data so as to reduce the likelihood of serious error.

The situation demands very close linkage, throughout the lifetime of an offshore field, between the reservoir and project engineers. There must necessarily be a two way flow of information because, as explained in Chapter 5, section 5.3, there are two basic equations involved. One is the reservoir material balance dictating its performance and the second is a platform facilities balance and unless these can be married into one, there may be a danger of some failure occurring. It follows that reservoir engineers involved in offshore developments cannot afford the luxury of confining their thoughts merely to what happens underground: everything ceases at the wellbore. Instead, they must broaden their perspective to consider everything in the well, at the surface and even the availability of facilities that might be shared with neighbouring platforms, since all these factors may affect the way in which the field is developed. If, for instance, there is no possibility of disposing of large volumes of produced gas, which is often the case offshore, then the reservoir pressure must not be allowed to fall below the bubble point leading to the production of excessive and largely unpredictable volumes of gas. This constraint would imply that the reservoir pressure must be maintained by either water or gas injection. If it is anticipated that in the latter years of a waterdrive project there will be insufficient gas production to provide gaslift for high watercut wells, then plans must be laid to raise the reservoir pressure to a level at which wells can produce by natural flow or, if this is not an option, make-up gas from elsewhere must be imported to the platform to defer field abandonment. These are but two of the myriad examples of how the reservoir engineer must consider the broader aspects

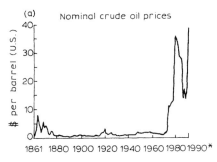 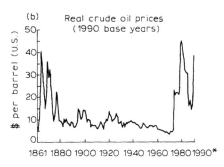

Fig. 1.3. Historic oil prices: (a) money-of-the-day terms, (b) based on the 1990 dollar value [5].

of offshore field developments. Above all, however, is the golden rule that all the expensively installed equipment must be used to the fullest at all times.

Prior to the extraordinary rise in the price of oil in 1973 (Fig. 1.3), resulting from severe pressure from the newly emerging OPEC member states, the suggestion that the reservoir engineer should be rightfully located at the centre of the universe in development studies would have been laughed at as fanciful. Before that time, with the price of oil at $3 per barrel, or less, it was quite definitely the economists (occupying a position on the left wing in Fig. 1.2) who were well entrenched at the centre of the action and all development decisions were made by them based on the low but stable price that had prevailed for so many years. In fact, it was this very stability that lent great conviction to the whole subject of economics and the validity of its long-term predictions — until the early seventies. Figure 1.3 can therefore be regarded as almost representing a history of the evolution of reservoir engineering and its importance within the Industry. Before the price rise, if an engineer had suggested that it might be a good idea to inject water into a reservoir to improve recovery but that the cost might be in excess of, say, $2.5 per barrel of incremental recovery, then the scheme would be quietly shelved. Consequently, before 1973, secondary recovery methods of water or gas injection were of rare occurrence and only practised in fields where the size and economics made it justifiable to do so. Instead, primary recovery was the order of the day, which consists of drilling holes in the ground, producing, crossing one's fingers and hoping for the best. If there was an element of natural pressure support from water or gas influx into the oil column recoveries could be respectable but if straightforward depletion occurred, resulting in pressure falling below the bubble point, triggering off the wasteful solution gas drive process (Chapter 3, section 3.7b), then recoveries would be much lower. As a consequence, of all the oil the Industry has discovered and attempted to recover since the drilling of Drake's first well in Titusville, Pennsylvania in 1859, more than 70% of it remains underground and, for the reasons stated in section 1.4, is likely to remain there. Since one of the prime aims of the Oil Industry in general (and reservoir engineering in particular) is the recovery of oil from fields then, although it might well be referred to as "The Biggest Business" [6] it would be difficult to credit it with being successful in meeting its stated aim.

This is not the time to delve into the reasons for the maintenance of what must be regarded as an artificially low oil price for so many years leading up to the early seventies. The history of competition between the major companies and between themselves and governments is both intricate and bewildering [5–7] and while it may be tempting to blame the "ugly face of capitalism" for the abandonment of so much of a wasting asset merely for short-term gain, to be fair, the alternative must also be considered since with the total absence of economic competition the former Soviet Union finished up with an even worse record for oil recovery.

The harsh price rise starting in 1973 presented severe difficulties to the western world economy, precipitating what has been referred to as the "inflationary spiral" but on the other hand, the event heralded the start of a period from 1973 to 1986 which was the heyday for reservoir engineering. Earlier, there were few (outside research institutes) who pursued it as a career subject since it was not an up-front decision making part of the business. Therefore, engineers with ambition did their statutory two years in reservoir engineering (just enough to acquire dangerous notions about the subject) before graduating towards economics which was the obvious route to management, the chauffeur driven car and all the accoutrements associated with power. During the seventies and early eighties, however, reservoir engineering flourished as a meaningful and decision making career subject. This was particularly noticeable in areas like the North Sea which came of age during that period. Not only that, engineers suddenly found themselves endowed with quantitative powers never dreamed of before, for it was in the early seventies that commercial versions of numerical simulation models hit the street for the first time. Everything seemed to be progressing in a most satisfactory manner until the price collapse in 1986.

While, as mentioned above, there is little point now in criticising the reasons for the inhibition of a reasonable rise in the price of oil in the decades leading up to the seventies, which would have stimulated reservoir engineering and oil recovery; what happened after 1973, however, does warrant some critical examination. According to one economic commentator [8] — "Much of the [price] increase can only be defined as a self inflicted wound on the fabric of western society arising from policies which reflected the unsubstantiated belief in an inevitable scarcity of oil." This is reflected in the nature of the extraordinary oil price predictions (Fig. 1.4a) by august economic institutes, some of whom had oil approaching $100 per barrel early in the twenty first century. The inevitable result of such optimism was that a host of entrepreneurial exploration companies was spawned and more and more oil was discovered and produced without much regard to the simple law of supply and demand. The high prices naturally stimulated conservation measures in the use of oil and its substitution by other sources of energy, where possible, which simply exacerbated the difference between newly discovered reserves and requirement (Fig. 1.4b), which by the end of the eighties stood at an all-time high so that the present situation is that [8] "the world is running into oil, not out of it." Could it not have been foreseen that this situation would inevitably lead to the price collapse that in March 1986 saw the oil price plummeting temporarily to $10 per barrel? But following the price collapse (or the restoration of normality — as many saw

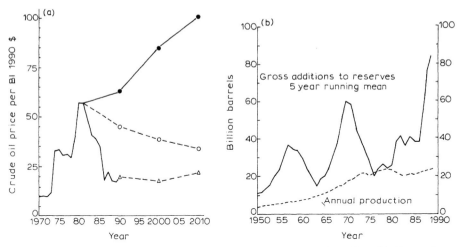

Fig. 1.4. (a) Oil price forecasts: wild optimism leading to economic reality. (b) Comparison of incremental reserves to production.

it) the harsh economic lesson did little to assuage the enthusiasm for finding more and more oil. Even at the time of writing (1992) the Industry is poised to enter Russia and the independent states of the former Soviet Union to participate in the discovery and development of their apparently vast reserves, this being regarded as one of the last, great and largely untapped oil areas in the world.

Economic considerations apart, what has been the effect of the sudden reduction in prices on reservoir engineering since, what is good for industry and the populace (low oil prices, low petrol pump prices) is anathema to those whose task is to recover as much oil from fields as (reasonably) possible. The immediate effect of the 1986 price collapse was an inevitable panic response. Companies cut back on technical staff, the experienced being put out to early retirement, university (college) petroleum engineering courses were severely undersubscribed, in fact, the usual knee-jerk reaction that has happened before although never as severely as in the mid-eighties. The current problem is that a vacuum has been created in which there are too few experienced engineering staff to provide guidance to young recruits and it is hoped that this book will, in some small part, assist in this essential training.

The prospects for reservoir engineers still looks very promising. We cannot now regress to the "solution gas drive scenario" of the past, in fact, producing below the bubble point pressure is quite rightly frowned upon by an increasing number of governments worldwide so that the job of oil extraction has to be done in a considered manner or not at all. Besides, it might be asked — who makes the all important development decisions these days? It can never revert to being the sole prerogative of the economists because, since the loss of price stability of oil in the seventies, they have lost faith in the reliability of their main decision making tool — the discounted cash flow. Nowadays, in making a 20 year economic forecast for a project, they are confronted with even greater uncertainties than those facing the underground scientists: the explorationists and reservoir engineers, which makes economists worthy of a considerable amount of sympathy. Any long-term cash flow requires, as input, predictions of the variation in oil price, inflation rates and the dollar rate of exchange (amongst others) and, considering the last, for instance, it is difficult to guess this 24 hours ahead, never mind 20 years. Nowadays, decision making on project viability is shared between the explorers, reservoir engineers and the economists. The first must present a reasonably sound figure for the hydrocarbons in place but this in itself is somewhat arbitrary unless the reservoir engineer is prepared to sign a piece of paper attesting to the ease with which the hydrocarbons can be recovered, the ultimate recovery, and how long it will take. Finally, the economists step in with their comments on the likelihood of the venture being profitable in the long term (plus or minus). In this shared activity, however, it is still the responsibility of the reservoir engineer to occupy the central, coordinating role, as depicted in Fig. 1.2 and there is no reason why this position should be changed.

Looking to the future, there must also be, eventually, a change in the emphasis and balance between exploration and production in the Industry which will accentuate the need for sound reservoir engineering. The main, speculative interest in the business has always been the exciting, romantic activity of discovery, while hydrocarbon production has been overshadowed in comparison as somewhat of a drudge. Even now, in a mature producing area like the North Sea, the fact that some entrepreneurial outfit has found a minor oil accumulation grabs the newspaper headlines, whereas the fact that an operator may have performed a series of successful, innovative workovers or modified a water injection project, which recovers twice as much oil as contained in the minor discovery, is a dull statistic that remains buried in the filing cabinet. Sooner or later it must be appreciated that we have discovered quite enough oil, the world is awash with it, and concentrate more on attaining a higher recovery of what has already been found than ever aimed at or attained in the past. Oil gained from improved recovery and production methods is usually cheaper than discovering more and producing it in a wasteful manner. Such a change in attitude would be of long-term benefit to both the Industry and consumers in attempting to conserve this wasting asset and those who merely aim at more and more discovery, without serious commitment to optimising recovery, are neither helping themselves, the Industry or reservoir engineering, in particular. While most things seem to happen later rather than sooner, there are some encouraging signs that this necessary shift in attitude is gaining momentum.

1.4. TECHNICAL RESPONSIBILITIES OF RESERVOIR ENGINEERS

Four distinct technical responsibilities can be defined:
– contributing, with the geologists and petrophysicists, to the estimation of hydrocarbons in place

– determining the fraction of discovered hydrocarbons that can be (reasonably) recovered
    – attaching a time scale to the recovery
    – day-to-day operational reservoir engineering throughout the lifetime of the project.

The first of these is described in Chapter 2 and is a shared activity. The second is the estimation of the recovery factor or recovery efficiency and is the sole prerogative of the reservoir engineer. Formerly, the recovery factor used to have the adjective "economic" attached but on account of the difficulties in making long-term economic predictions, described in the previous section, the word "reasonable" has been substituted in its place. The third activity, the attachment of a time scale, is the development of a production profile, while the fourth is an ongoing operational activity described throughout the book. The four technical duties are spread during the appraisal and development stages of the field, as described below.

(a) Appraisal

The reservoir engineer is involved in the development from the day of discovery of the accumulation and is (or should be) responsible for the collection, collation and interpretation of the following data in the discovery and subsequent appraisal wells.

– Well tests (DSTs)
    – Pressure–depth relationships
    – Fluid sampling
    – Coring

Although most of the attention in the literature is focused on the pressure analysis aspects of well testing, during the appraisal stage, and especially offshore, by far the most important information gathered from a well test (DST) is the type of hydrocarbon present and whether the formation capable of producing at what may be judged as a commercial rate. If not, then that is not only the end of the appraisal, it is also the end of the entire development! It is not uncommon, for instance, to find explorers and management becoming excited about the discovery of hundreds of millions of barrels of oil, two hundred miles offshore in 500 ft of water at a depth of 13,000 ft with permeability of 10 mD. Such a reservoir will probably only produce a few hundred barrels of oil per day on test, making its development a commercial impossibility in such an environment. Further attention is focused on testing at the appraisal stage in Chapter 2, section 2.9 and Chapter 4, section 4.4a.

Pressure–depth relationships and fluid sampling are also described in detail in Chapter 2, sections 2.6/7 and 2.2b, respectively. These are both closely related to testing: the manner of testing and the techniques employed. One of the main results determined from pressure-depth surveys, best obtained using the RFT, is the location of fluid contacts (gas–oil, oil–water, gas–water contacts) which are required in the estimation of hydrocarbons in place. The determination of fluid pressure gradients required in such work depends on the collection of reliable fluid samples and their accurate PVT analysis in the laboratory. Also dependent on the PVT analysis is any form of recovery calculation performed (Chapter 3) and the design of surface facilities to handle/separate and dispose of the produced fluids, the oil, gas and water.

Core data is described in Chapters 5 and 6 as one of the most important requirements from reservoirs in which it is intended that secondary recovery (water–oil or gas–oil) displacement will be the elected means of development. What matters in viewing core data is the all important permeability distribution across the producing formations; it is this, more than anything else, that dictates the efficiency of displacement processes.

(b) End of appraisal

Usually, the timing of this tends to be gauged by the explorers, when they feel satisfied that they have acquired sufficient data to define within a tolerable range of certainty whether they have proven sufficient stock tank oil initially in place/gas initially in place for a viable field development. But this must not be regarded as the sole criterion upon which development decisions are made. The reservoir engineer must be also heavily involved concerning when the appraisal phase should be terminated, based on whether enough meaningful production orientated data have been acquired to perform long-term project development calculations to evaluate:

– Production profiles of oil, gas and water
– Injection profiles of water and/or gas
– Production/injection well requirements
– Surface topsides facilities design.

If the data collection has been inadequate to attach meaningful figures to the above and the engineer can clearly define the deficiencies in the data and how these may be resolved by the drilling of one or more additional appraisal wells, then pressure must be brought to bear on those responsible for appraisal decision making to have these vital wells drilled (often easier said than done).

The importance and difficulty in developing production/injection profiles at the end of the appraisal stage, especially in offshore developments, has already been stressed in section 1.2d of this chapter. It remains, above all, the responsibility of the reservoir engineer to preserve excellent communication with the explorers and petrophysicists from whom information is received and the project engineers and economists who will react to the profiles determined and their implications for surface facilities design and economic viability of the project.

(c) Development

This is regarded as the period extending from the time that continuous hydrocarbon production commences until the time of field abandonment. Throughout this time, some of the day-to-day activities of the field reservoir engineer are as follows.

New well locations: These are selected in conjunction with the geophysicists and geologists but, as those who have been involved with this particular activity will appreciate, everybody seems to want to get involved in the act, from the managing director downwards. From the engineer's point of view, it is a matter of determining those parts of the reservoir that may have been unswept or poorly drained and trying to locate new wells in such areas, with geological advice.

Well completion intervals: Precisely where to perforate production and injection wells which penetrate massive sand/limestone sections is largely the responsibility of the reservoir engineer. One of the main factors affecting decision making is whether there is a reasonable degree of pressure communication across the section or are there impermeable barriers causing the isolation of intervals within the total formation. Also of importance is the areal continuity of either the total section or isolated zones within the interval. Unfortunately, objective decision making on completion intervals (perforating policy) cannot be made until after the start of continuous field production when dynamic pressure surveys can be run across the section using the RFT (Chapter 5). Therefore, when deciding upon a policy at the start of the development, based on only a static view of the pressure-depth trend (which usually reveals apparent equilibrium), it is wise to be cautious and preferably under-perforate rather than over-perforate, because it is easier to add extra holes in the casing than to isolate unwanted perforations.

Well recompletions/sidetracking: Once oil production from one perforated interval is adversely affected by excess water/gas production, the reservoir engineer is responsible for making the decision on when to cease production and recomplete the well within the same casing string upon a different horizon or by milling through the casing and sidetracking (re-drilling) the well to a promising horizon at a different areal location.

Regular well surveys: These include:

- The initial RFT survey
- Regular pressure surveys
- Production logging surveys

The initial RFT survey is mandatory in each new development well that is drilled after the start of continuous field production. Since it is an open-hole tool run prior to setting the production casing, it can usually be run only once during the lifetime of a well, providing a unique opportunity to acquire a dynamic pressure–depth profile across the formations (Chapter 2, section 2.8 and Chapter 5, section 5.2d). The dynamic profiles obtained from such surveys prove invaluable in calibrating (history matching) layered numerical simulation models.

Pressure surveys, usually pressure buildups, are conducted at regular intervals throughout the project lifetime, the frequency being dependent on the importance of the pressure data acquired which, in a depletion type field, will be of greater significance than if pressure is being maintained by secondary recovery water or gas injection. The main purpose in collecting pressure data is to develop a history matched model of the reservoir under study: matching pressures to obtain a reliable predictive tool. There are circumstances, such as in layered sections with limited cross-flow, in which pressure buildup surveys provide unreliable results (Chapter 4, section 4.20c) and production logging surveys must be relied upon instead to determine, in this case, individual layer pressures and the contribution to the total flow from each separate layer.

Areal fluid movement: This is one of the main objectives in running large, areal simulation models, as described in Chapter 5, section 5.5a. In a secondary recovery water injection project, for instance, the model is run to match the timing of water breakthrough in individual wells and their subsequent rates of watercut development. To be successful in such modelling requires that the vertical description, as determined from logs, cores, DSTs and production logging surveys, should be correctly incorporated in the model in great detail. Otherwise any attempt to steer the injected water around the reservoir by changing areal rock properties (k, ϕ) or introducing areal discontinuities such as faults can be made to produce a history match which may look convincing but will be unrealistic and could lead to damaging development decisions being made about such matters as the location of additional wells in the reservoir.

Production profiles: There are many production profiles required throughout the lifetime of a project and their generation is entirely the responsibility of the reservoir engineer. So great is the requirement for profiles, mainly to satisfy the needs of economists, that the task consumes a great deal of the engineers time and, on occasions can become a pain in the neck. Profiles are required at the end of the appraisal stage over the full project lifetime (20+ years) followed by continually updated, more detailed five year profiles for company medium term budgeting. A detailed, annual profile is usually required for operational planning, monthly profiles for regulatory authorities and even weekly profiles, possibly necessary for the scheduling of oil transportation facilities such as tankers. Superimposed on all these are the seemingly endless sensitivities requested to variation in reservoir/production parameters. Obviously, to keep abreast of all these requirements which, of course, are extremely important within a company, the engineer needs an automated, accurate method for profile generation. Usually, there is not the time available to apply numerical simulation and the cost could be prohibitive. Alternatively, methods are presented in this text for obtaining profiles by simple analytical means and, in particular, how to extend simulation results using analytical methods. In waterdrive fields, for instance, as described in Chapter 5, section 5.3, once the detailed numerical simulation model has been run to establish a "base case" and the overall reservoir watercut development, the results can be used with a simple material balance equation to generate alternative profiles in a rapid fashion. Obviously, there are limitations to this practice. If, for instance, a sensitivity is required including a different number of wells, their locations and completion intervals, all factors which would affect the watercut development, then additional simulation would be required. Nevertheless, the engineer should take every opportunity of applying the useful technique of using simulation followed-up by simple analytical procedures for saving time, without sacrificing accuracy, in generating production profiles.

Decline policy: In a depletion type field the onset of this most significant phase of production decline occurs shortly after the average reservoir pressure falls below the bubble point, releasing free gas in the reservoir thus precipitating the (usually) harmful solution gas drive process. In secondary recovery operations, such as engineered waterdrive, the decline from peak or plateau production (Fig. 1.1) occurs when the rate of increase of the produced injection water becomes greater than the rate of increase of new oil production from additional producers. In either case, the decline period should count as the most active and meaningful phase of the development for the reservoir engineer, the main aim being to arrest the decline in production for as long as possible by identifying problems and seeking practical solutions. Amongst these are the drilling of new wells, sidetracks of existing wells, recompletions and stimulation workovers: in fact, trying anything that proves to work in improving the reservoir performance.

On account of the large amount of production/pressure data available at this stage of development, the application of numerical simulation modelling should be on its soundest footing in the aspect of history matching to provide a reliable predictive tool. Unfortunately, however, sometimes the sheer amount of available data complicates the attainment of the "perfect" history matched model and in many cases it proves more cost effective to spend the time studying detailed, individual well histories in an effort to determine the most appropriate treatments to achieve their optimum productivity. Although not the most glamorous phase of overall field development, it often turns out to be the most interesting and rewarding for the reservoir engineer, doing a sleuthing job using all the available well data.

Enhanced oil recovery: Writing on enhanced oil recovery (EOR) in the early nineteen nineties is not exactly "getting the timing right" — or has there ever really been a right time for this subject? The difficulty that has and always will plague EOR is economics, as described below. EOR ("End Of the Road" — 1993 interpretation) defies precise definition but in general (and as used in this book) three categories of oil recovery are distinguished as follows.

Primary: Drilling wells, producing and hoping for the best: no attempt being made to enhance recovery by injection of fluids.

Secondary: Adopting a policy of water or gas injection, with the aim of complete or partial pressure maintenance and accelerated development through the positive displacement of oil towards the producing wells.

EOR: Injection of anything that will increase the recovery attained by the above methods.

It should be noted that there seems to be no internationally recognised definition of these recovery categories and this is not simply a matter of semantics but rather taxation rules and regulations which can and do vary from country to country for recovery by the different processes.

Perhaps the most common and best understood application of EOR is in an attempt to recover the oil remaining in the reservoir following the cessation of secondary recovery operations. In a secondary flood, the displacing fluid and oil are immiscible meaning, that on account of a finite surface tension between the two, they do not physically mix and therefore at the end of the flood there is a finite volume of oil trapped by surface tension forces in each pore space contacted. At the start of a waterflood the pore space is filled with oil and connate water which, in a water wet system, is the phase that preferentially adheres to the rock particles. This water is expressed as a saturation, S_{wc} (PV), which is the fraction of the pore space occupied by the water. By the time the flood has finished, all the oil has been displaced from an individual pore space except the residual volume expressed as a "residual oil saturation" S_{or} (PV), which is trapped in a minimum energy configuration by the finite surface tension forces. Therefore, the theoretical amount of oil that can be removed in a waterflood (or gas flood) can be expressed as:

$$MOV = PV(1 - S_{or} - S_{wc}) \qquad (PV) \qquad (1.1)$$

which is the movable oil volume (MOV). Typical values of the two saturations are $S_{wc} = 0.20$ PV, $S_{or} = 0.30$ PV, giving an MOV of 0.50 PV. If the residual oil saturation could be somehow reduced to near zero, the MOV would increase by 60% to a value of 0.80 PV and this is the main target for EOR application.

A fluid must be injected, following the waterflood, that is miscible (physically mixes) with the residual oil, meaning that the surface tension between the injected fluid and oil is reduced to zero so that the residual oil is mobilised and produced. While there may be 100% recovery of the oil in a controlled laboratory EOR flood in a thin core plug, in the reservoir matters are somewhat different and due to degradation of the chemical properties of the EOR fluids in their movement over large distances in the reservoir, together with the combined influence of heterogeneity and gravity (not a feature in core flooding experiments), the fraction of the residual oil that is actually recovered can be considerably diminished.

The EOR fluid is selected in terms of the physical/chemical requirements and the availability and cost of the fluid. Flooding agents comprise hydrocarbon and other gases, such as carbon dioxide, which at the appropriate pressure and temperature are miscible with the oil. All manner of chemicals and surfactants which are judged to be miscible are also used and flood control agents, such as polymers, that improve the sweep efficiency by slowing and stabilising the frontal advance of the displacing fluid.

The basic physical (if not chemical) principles of EOR have been appreciated for a great many years but at the low oil prices prevailing prior to 1973, there was little incentive to attempt such expensive recovery methods in practice. If, as mentioned in section 1.3, before the seventies price rise even water injection was considered as something of a luxury — then what chance for injecting very costly surfactants? Following the remarkable oil price increase, everybody seemed to jump on the EOR bandwagon. Brilliant theoretical scientists and engineers were recruited to the Industry and laboratories opened in which they could pursue the research in the subject that had been neglected in the past. Much valuable insight was gained into the complex theory of EOR flooding and the displacement processes, primarily on the scale of one dimensional core flooding experiments. There were field trials but these were few and far between and largely conducted in depleted US fields onshore, that were usually shallow therefore having relatively low pressure and temperature, which helped in selecting EOR flooding agents that were chemically stable under these conditions. Even in assessing the results of these field trials, however, an interesting debate raged (and still does) concerning whether the incremental recovery achieved resulted from the EOR flooding or the fact that field application required the drilling of many infill production and injection wells which naturally led to an enhancement in oil recovery.

Progress and interest remained high, as attested by the number of technical papers on the subject appearing in the literature since 1973. Unfortunately, the oil price collapse in 1986 claimed EOR as its first major victim. Most of the laboratories were closed, some of them overnight, and there were many redundancies amongst those who had struggled to try and rationalize the subject over the previous decade. In addition, many of the field trials that had been scheduled were cancelled, altogether a most disillusioning experience for those theoreticians who had been wooed into the Industry in the expectation of long and fulfilling careers in significantly increasing the overall recovery of hydrocarbons. The situation gives rise to speculation about whether if the oil price were, for some reason, to "shoot through the roof" — would it be possible for the oil industry to ever again recruit the calibre of technical experts that they attracted during the 1970's?

The mid-eighties recession even affected the most successful, proven method of EOR application: thermal recovery. This differs from the type of EOR defined above which was focused on the recovery of residual oil following secondary recovery flooding. Instead, thermal methods aim primarily at reducing the *in situ* viscosity of heavy oils through the application of heat. So successful is the method of steam injection that it is claimed that over 90% of the total incremental EOR recovery worldwide can be attributed to it. What was most affected by the slump in prices were the large projects planned to recover oil from some of the major tar sands in Athabaska (Canada) and the Orinoco tar belt (Venezuela) which were severely reduced in scale.

What is clear from the unfortunate history of EOR theory and application to date can be phrased as another piece of career advice, that if you intend to select reservoir engineering as a "career", then you should steer clear of the more esoteric subjects such as EOR flooding and the recovery of highly viscous oils. Instead, devote your talents to recovery of "easy" and "popular" hydrocarbons: light (low viscosity) oil and natural gas. These are the easiest pickings for which there is greatest demand. While EOR may present the more satisfying intellectual challenge, there is always the risk that it may also lead prematurely to the dole queue.

EOR is not specifically referred to in this text simply because the title of the book is "The *Practice* of Reservoir Engineering" and, at the time of writing, it would be rather difficult to give unbiased, practical advice on what works and what does not work in the field because there has been insufficient experience. Nevertheless, the book implicitly contains much advice that is of great importance in considering EOR floods and particularly in catering for the interdependent effects of reservoir heterogeneity and gravity. In studying the copious literature on EOR flooding there is very seldom any reference to these effects because most of the papers emanate from academic institutes where the reality of flooding in hillsides rather than in core plugs is not always appreciated. For instance, although a one dimensional core flooding experiment of gas drive may indicate complete miscibility with the oil if, in the reservoir, the higher permeabilities are towards the top of the section there may be such severe override of the gas that miscibility is only achieved in a thin interval at the top of the reservoir, and when the flood is viewed in its entirety it is hardly miscible at all. The field engineer must be particularly careful when viewing what may appear to be very impressive experimental results of floods in core plugs and chemical/compositional numerical simulations of such floods. Sometimes the simulation models can be so "top-heavy" in dealing with the complex physical/ chemical phenomena that they simply cannot handle also the details of reservoir heterogeneity which, in so many cases, if considered properly can overwhelm the beneficial effects of the flood viewed only on the microscopic scale. Therefore, if the engineer is fortunate enough to be involved with an EOR project (for they are of rare occurrence), it would prove beneficial to read how heterogeneity and gravity are accounted for in displacement efficiency calculations, as described in Chapter 5 for waterdrive and Chapter 6 for gas drive

What then are the prospects for the successful application of EOR in the future of the Industry? The answer to this question is primarily influenced by the trend in oil prices and also on such factors as environment and timing. The dependence on price is obvious but concerning the latter two, comparison is once again drawn between onshore and offshore developments. Following the 1986 collapse in the oil price, thousands of onshore oil producing stripper wells in Texas and Oklahoma were abandoned as uneconomic, in fact, according to statistics at that time the average oil rate of producers in the latter state was only 4 stb/d. Yet is "abandonment" the correct status to apply to these wells — possibly not. Assuming a future scenario of a shortage of oil supply and significant increase in its price, this onshore oil is at least accessible to further development by EOR flooding although, since most of the gas has been stripped from the oil by the solution gas drive process, its viscosity will have risen considerably making its eventual recovery more difficult and expensive. Offshore, however, the situation can be quite different. Once a North Sea field has been declared uneconomic and the billion+ dollar platform has been removed, then the field can be well and truly described as abandoned. In this respect it would appear that EOR will not find broad application in an area such as the North Sea. It should have started years ago but the low oil prices have precluded its widescale application.

1.5. THE PHYSICAL PRINCIPLES OF RESERVOIR ENGINEERING

To describe 90% of reservoir engineering (excluding thermal effects) requires application of the following physical principles:

– Conservation of mass
– Darcy's law
– Isothermal compressibility
– Newton's second and third laws of motion

The first of these is one of the basic tenets of physical science, while Darcy's flow law is an empirical relationship dating from the middle of the nineteenth century and was the first statement of the nature of fluid flow through a porous medium. It has subsequently been ratified in theoretical terms as derivable from the Navier-Stokes [9] equation of motion of a viscous fluid. For one dimensional, linear, horizontal flow (core flooding experiment), Darcy's law can be expressed in absolute units as

$$q = -\frac{kA}{\mu}\frac{\partial p}{\partial l} \tag{1.2}$$

or in field units as

$$q = -1.127 \times 10^{-3}\frac{kA}{\mu B_o}\frac{\partial p}{\partial l} \quad \text{(stb/d)} \tag{1.3}$$

For linear flow, the pressure gradient $\partial p/\partial l$ is intrinsically negative hence the minus sign to make the rate q positive.

Isothermal compressibility accounts for the change in volume of fluids, and the pore volume itself, as the pressure varies at constant temperature. It is defined as:

$$c = -\frac{1}{V}\frac{\partial V}{\partial p} \quad \text{(/psi)} \tag{1.4}$$

Again, since the rate of change of volume with respect to pressure is negative, the minus sign is required to assure that the compressibility is positive. The assumption of the condition of isothermal depletion (constant temperature) is one that prevails in conventional reservoir engineering and implies that the infinite heat sources of the cap and base rock maintain the temperature in the reservoir constant. The most common expression of equation 1.4 is:

$$dV = cV\Delta p \tag{1.5}$$

in which $\Delta p = p_i - p$ is the pressure drop from the initial value p_i to the current average reservoir pressure p. This is an expression for the material balance of a depletion type reservoir completely unaffected by influx of fluids, water or gas.

Specific inclusion of Newton's second and third laws of motion [10] is, in this author's experience, an innovation in the extensive literature on reservoir engineering but is necessitated by the current trends in the subject. The second law defines force as equal to mass times acceleration and, in particular, defines the force resulting from the attraction of gravity. When considering the displacement of one fluid by another with different density, it is imperative that the effect of gravity be accounted for in displacement efficiency calculations in macroscopic reservoir sections. There is a dearth of description of the influence of gravity in the literature and yet it is the most significant force active. This seems to result from the fact that the majority of technical papers on the subject relate to the physics of displacement in one dimensional core flooding experiments, in which gravity plays no part. The realism of water displacing oil or gas displacing oil in the field is that the process occurs in hillsides rather than core plugs and in this domain gravity predominates. The inclusion of gravity is emphasised in Chapter 5 for waterdrive and in Chapter 6 for gas drive.

Newton's third law of motion states that "action and reaction are equal and opposite". Its direct consideration in reservoir engineering in the past was unnecessary because, since the law was first published on July 5, 1686, the progress in physical science, using analytical methods, naturally incorporated the third law. For instance, Welge's method [11] of describing the basic theory of immiscible displacement of Buckley-Leverett [12], relying on the concept of the fractional flow of the displacing fluid, naturally incorporates the third law of motion but the more recent technique of numerical simulation does not necessarily. As described throughout Chapter 5, the neglect of the concept of fractional flow in "modern" reservoir engineering is a backward step that does necessitate the re-consideration of Newton's third law of motion — if we wish to model the physics of displacement correctly.

The reader may feel that the physical laws governing the subject of reservoir engineering are sparse. Where, for instance is the, by now, well established "Murphy's Law"? Since the first realisation of this law in the 1940's — "that things go from bad to worse", the meaning of entropy as established by the second law of thermodynamics has become a lot clearer to practical engineers. Murphy's law is implicitly incorporated in reservoir engineering, although we do not readily acknowledge it.

The author has always believed that there should be a place in reservoir engineering for the very basic theory of physics which is (perhaps unfortunately) the Heisenberg "Uncertainty Principle" of quantum mechanics. This is not an original thought in the subject because as long ago as 1949 the ultimate reservoir engineer, Morris Muskat [13], had flirted with the same idea but concluded that:

"In its operational sense the principle of uncertainty, which is usually considered as limited to the realm of microscopic physics, constitutes the very essence of applied reservoir engineering as a science."

An excellent thought — but what can be done about it?

Nevertheless, the subject is vulnerable to change, the latest approach being the adoption of "Chaos Theory". This would seem to be a convenient concept to hide behind in reservoir engineering but at the time of writing is still in its infancy — thank goodness.

REFERENCES

[1] Peaceman, D.W.: Fundamentals of Numerical Simulation, Elsevier, Amsterdam, 1978.

[2] Aziz, K. and Settari, A.: Petroleum Reservoir Simulation, Applied Science Publishers, London, 1979.

[3] Thomas, G.W.: Principles of Hydrocarbon Reservoir Simulation, IHRDC Publishers, Boston, Mass., 1982.

[4] Various authors: Reservoir Simulation, SPE Monograph Series, 1989.

[5] Yergin, D.: The Prize — The Epic Quest for Oil, Money, and Power, Simon and Schuster, New York, 1991.

[6] Tugendhat, C.: Oil, The Biggest Business, Eyre and Spottiswoode, London, 1968.

[7] Sampson, A.: The Seven Sisters" The Viking Press, New York, 1975.

[8] Odell, P.: Odell's Parting Shots Stress Oil Supply Fictions, Offshore Engineer, June 1991.

[9] King Hubbert, M.: Darcy's Law and the Field Equations of the Flow of Underground Fluids, Trans. AIME, 1956.

[10] Newton, J.S.: Philosophiæ Naturalis Principia Mathematica, S. Pepys, London, 1686.

[11] Welge, H.J.: A Simplified Method for Computing Oil Recovery by Gas or Water Drive, Trans. AIME, 216 (271), 1952.

[12] Buckley, S.E. and Leverett, M.C.: Mechanism of Fluid Displacement in Sands, Trans. AIME, 146 (107), 1942.

[13] Muskat, M.: Physical Principles of Oil Production, McGraw-Hill, Inc., New York, N.Y., 1949.

Chapter 2

THE APPRAISAL OF OIL AND GAS FIELDS

2.1. INTRODUCTION

As described in Chapter 1, the difference between field appraisal and subsequent development is more pronounced in offshore projects than onshore. In the latter environment, there is often the opportunity to tie in the exploration and early delineation wells to producing facilities early in the lifetime of the project. In this way the engineer is able to view the reservoir under dynamic production conditions from the outset and make sensible decisions based on the observed performance. By contrast, during the exploration and appraisal of offshore fields the reservoirs are viewed under purely static conditions which is a severe disadvantage in the "up-front" planning that is a feature of offshore developments.

The present chapter is therefore devoted to a description of the necessary data that must be collected during the static appraisal of oil and gas fields for use in reservoir development studies. The subjects covered are: PVT properties of hydrocarbons, the reservoir engineering aspects of the estimation of hydrocarbons in place and equity determination. Pressure–depth plotting is also considered together with a general description of the purpose and practice of conventional and extended well tests. It is hoped that the chapter will be read by and prove useful to a broad spectrum of professionals associated with field developments.

2.2. PRESSURE–VOLUME–TEMPERATURE FLUID PROPERTIES FOR OIL

Inspection and rationalisation of the pressure-volume–temperature (PVT) fluid properties must be the opening move in the study of any oilfield since the PVT functions, which relate surface to reservoir volumes, are required in practically every aspect of reservoir engineering: calculation of hydrocarbons in place, pressure–depth regimes, any form of recovery calculations and to assure correct design of surface facilities.

(a) Basic PVT parameters

For an oil reservoir these are:

B_o: oil formation volume factor (FVF), which is the number of reservoir barrels of oil and dissolved gas that must be produced to obtain one stock tank barrel of stable oil at the surface (rb/stb).

R_s: solution or dissolved gas/oil ratio (GOR), which is the number of standard cubic feet of gas produced with each stock tank barrel of oil that was dissolved in the oil in the reservoir (scf/stb).

B_g: gas formation volume factor, which is the volume in barrels that one standard cubic foot of gas at the surface occupies as free gas in the reservoir (rb/scf).

All three are strictly functions of pressure at the (assumed) constant reservoir temperature.

The first two allow for the fact that oil in the reservoir, at high pressure and temperature, contains dissolved gas that is released as it is produced to the surface. For example, if $B_o = 1.45$ rb/stb and $R_s = 500$ scf/stb, then 1.45 rb of oil and dissolved gas must be produced to obtain 1 stb of oil, releasing 500 scf of gas at the surface. As mentioned above, all three functions are required to relate reservoir volumes to those measured at the surface. The engineer works physically in terms of reservoir volumes yet must use practical surface measured volumes in equations. Thus, in setting-up the basic equation of waterdrive in terms of: injection rate = gross production rate (at constant pressure), it becomes:

$$q_{wi} = q_o B_o + q_{wp} B_w \quad \text{(rb/d)} \tag{2.1}$$

in which all the rates, q, are surface measured values (stb/d) but multiplication by the FVFs (rb/stb), establishes the equation as an "underground" balance (rb/d), in which $q_o B_o$ is the flow rate of the single phase oil containing R_s (scf/stb) of dissolved gas while $q_{wp} B_w$ is the reservoir flow rate of the produced water which may contain a small amount of dissolved gas, hence the need for the water FVF, B_w, which is usually close to unity. The injected water is invariably stripped of any gas prior to injection, therefore the surface and reservoir rate, q_{wi}, is the same. The significance of equation 2.1 in waterdrive calculations is illustrated in Chapter 5, section 5.3c.

Another simple example relates to the use of the basic definition of isothermal compressibility, introduced in Chapter 1, section 1.5. Expressed as the material balance for an undersaturated, depletion type oil reservoir: above the bubble point, no free gas, no fluid influx, it is:

$$dV = cV\Delta p \quad \text{(rb)} \tag{1.5}$$

But while this is conceptually useful: production (expansion) = compressibility × original volume × pressure drop, it is not very informative in a practical sense, for which it must be re-written as:

$$N_p B_o = N B_{oi} c \Delta p \quad \text{(rb)} \tag{2.2}$$

in which N_p is the number of stb of oil produced for the pressure drop Δp and $N_p B_o$ is therefore its reservoir volume including the dissolved gas, $N_p R_s$ scf. N is the original oil volume (STOIIP-stb) and $N B_{oi}$ the corresponding volume of this oil plus its $N R_{si}$ scf of dissolved gas (the suffix "i" relates to the initial condition).

All three PVT parameters are measured in laboratory experiments [1] conducted at the constant reservoir temperature. The results are then dependent on the

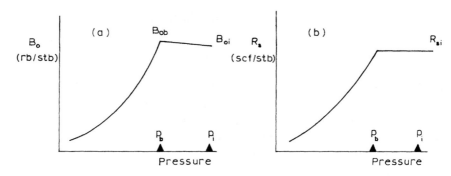

Fig. 2.1. PVT functions for a moderate volatility oil. (a) FVF; (b) solution GOR.

declining pressure alone. Typical shapes of the B_o and R_s functions for a moderate volatility oil are shown in Fig. 2.1. At the initial pressure, p_i, the oil is described as "undersaturated", meaning that if there was free gas available it would be dissolved — but there is none. The oil is therefore in the single phase state containing dissolved gas. Eventually, through continued depletion, the pressure drops to the bubble point, p_b, at which the oil becomes gas saturated. Below this pressure gas is liberated from the oil giving rise to the presence of two phases: oil + dissolved gas and free gas. Above the bubble point, as the pressure declines, the oil must expand hence B_o increases. Since the oil compressibility is usually fairly small, however, the extent of the expansion is slight. Also, above the bubble point each stb of oil produced contains the constant R_{si} scf of dissolved gas, since none has been liberated in the reservoir.

Below the bubble point, matters are more complicated. Gas is freed from the oil *in situ*, consequently B_o decreases continuously since the dissolved gas content is reduced. The decrease in oil volume from its maximum value at the bubble point is referred to as the "shrinkage", $B_{ob} - B_o$, at which point the amount of gas released is $R_{si} - R_s$ scf/stb. Producing below the bubble point is invariably "messy", because the gas viscosity is typically fifty times lower than that of oil, therefore, according to Darcy's law, equation 1.3, it is capable of travelling much faster. This leads to recovery under the so-called "solution gas drive" condition, which is described in Chapter 3, section 3.7b. It would be wrong to state that production below the bubble point necessarily results in a low oil recovery but on most occasions it does. There is an uncontrollable element associated with the process due to the completely different mobilities of the oil and free gas, which leads to the tendency to strip the gas from the reservoir prematurely at the expense of the less mobile oil. In doing so, it removes the highest energy component from the system and this can be viewed by comparing the compressibilities of everything that can change volume during depletion and contribute to the production, typically:

c_o = 10 × 10^{-6}/psi: oil
c_w = 3 × 10^{-6}/psi: water
c_f = 5 × 10^{-6}/psi: pore space (Chapter 3, section 3.10)

$c_g \approx 1/p = 200 \times 10^{-6}/\text{psi} \ (p = 5000 \ \text{psi})$ (Chapter 6, section 6.2d)

As can be seen, the gas compressibility is more than an order of magnitude greater than anything else which gives it the greatest drive energy, as can be appreciated by consideration of equation 1.5.

Because of the possible lack of control in depleting below the bubble point, there has been an increasing tendency in recent years, both on the part of governments and operators, to overcome this by not letting the pressure fall to this level in the first place, which can be done by modifying the material balance for an undersaturated system to the following form

$$N_p B_o = N B_{oi} c \Delta p + W_i \quad (\text{rb}) \tag{2.3}$$

in which W_i is the cumulative volume of water injected. This is an expression for the cumulative material balance for an "engineered" waterdrive (equation 2.1 is the differential form expressed in terms of rates) which is a secondary recovery operation in which water is injected to maintain the reservoir pressure (energy) at a high level while accelerating the oil recovery through its positive displacement towards the producing wells. Most significantly, it keeps the pressure above the bubble point leading to a more controlled development in which now the two phases in the reservoir, water and oil have similar mobilities (if the flood is to be successful). The process is described in Chapter 5. Gas injection is also a form of secondary recovery (Chapter 6) but is intrinsically unstable on account of the difference in mobilities of the oil and injected gas. It can only be controlled if gravity is active in stabilising the frontal advance of the gas.

In volatile oil reservoirs, in which the oil contains a large volume of dissolved gas that will be released at the surface (typically: $B_{oi} = 2.8$ rb/stb, $R_{si} = 3000$ scf/stb) the variation in FVF with pressure is as shown in Fig. 2.2. Even above the bubble point, the high dissolved gas content in the oil means that the change in volume with pressure is very significant. This is reflected in the oil compressibility which might typically have a value of $60-80 \times 10^{-6}$ /psi. Problems start, however, immediately below the bubble point where the shrinkage, $B_{ob} - B_o$, is very dramatic and, as the oil shrinks *in situ*, it gives the gas greater freedom to move through the formation and be produced in excess quantities. The situation can be so extreme that what was accepted as an oil field one day can appear to behave like a gas field the next, simply by passing through the bubble point pressure. This is reflected in the highly

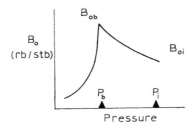

Fig. 2.2. FVF of a volatile oil.

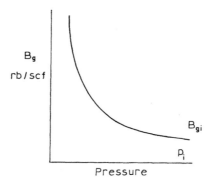

Fig. 2.3. Gas FVF as a function of pressure.

unfavourable oil relative permeability which inhibits its flow (Chapter 3, section 3.7b). Therefore, in such volatile oil reservoirs it is obviously preferable to aim at operating in the undersaturated pressure range:just above the bubble point (to take advantage of the natural oil expansion) through the injection of either water or gas to provide pressure maintenance.

The third PVT parameter, the gas FVF, B_g (rb/scf), is only required in initially undersaturated reservoirs once gas has been liberated below the bubble point pressure. It then relates free gas volumes in the reservoir (rb) to the surface volume (scf). Alternatively, if there is a gas cap at initial conditions, B_g would be required from the start. On account of its units (rb/scf) the FVF has a very small value at high pressure, typically, 0.0007 rb/scf. While this should not be of concern to us these days, as a historical note, it is worth mentioning the problems it posed our predecessors in the subject: armed computationally with only slide rules, with which the difficulty was trying to figure out where the decimal point should be located! The B_g function has a rather unfortunate, near hyperbolic shape (Fig. 2.3), causing some difficulties in interpolation/extrapolation. When working in the low-pressure range, the engineer should be careful to choose very small pressure increments when entering the B_g function to computer programs to assure accuracy in its use.

(b) Sampling reservoir fluids

By far the main responsibility of the practising reservoir engineer in this matter is the collection of valid fluid samples for transfer to the laboratory where the basic PVT experiments are performed. That is, irrespective of how the fluids are collected, the laboratory must somehow be furnished with (or be capable of recombining) oil and gas samples in their correct proportion of:

1 stb Oil + R_{si} scf Gas

which is often a lot easier said than done.

Sampling is usually conducted during testing at the appraisal stage of field development, when advantage is taken of the fact that the system is at initial, static conditions. Under these circumstances the source of origin of the fluid sample is well defined as the centre of the tested interval. If sampling is conducted after the start of continuous production the samples could have emanated from elsewhere in the reservoir and been transported to the tested interval. Therefore, if there are variations in PVT properties areally or with depth, these will not be defined correctly. Integral to the collection of samples is the accurate measurement of the initial pressure and temperature of the reservoir since the experiments are all conducted in the laboratory at the assumed constant temperature. This is measured on thermometers run with each logging tool but perhaps the most reliable value is that measured during the DST itself, since the tools are in the hole for longer.

The most common techniques for sampling appraisal wells are:

– downhole sampling using the RFT or MDT tools
– downhole sampling during a DST
– direct surface sampling of oil that is still undersaturated at the wellhead
– recombination of oil and gas samples after their separation at the surface

Although the Repeat Formation Tester (RFT) was originally intended to be used primarily for fluid sampling, it never quite fulfilled this role being much more usefully and universally employed for measuring pressure–depth profiles across reservoirs, as described in section 2.7 and in several other chapters in the book. Its more sophisticated successor the Modular Formation Dynamics Tester (MDT), is claimed to provide much more reliable downhole samples because it has a "pump-through" facility, meaning that the flow can be continued and monitored until it is evident that the chamber is filled with a valid sample of oil and dissolved gas (at the time of writing, the MDT tool was undergoing initial field trials). Nevertheless, a draw-back in this type downhole sampling is that it provides only relatively small volumes of the reservoir fluids.

The most common method of collecting downhole samples is by running wireline sampling chambers during DSTs. If the oil is undersaturated in the wellbore then reliable samples can be collected, if not, there will be inevitably uncertainty in the validity that the oil and gas have been collected in the required proportions, as described below.

If the oil is undersaturated by a significant amount, then flowing the well at a low rate may result in the wellhead pressure being above the bubble point. Then it is possible to collect samples directly from the flow string at the surface in which the gas is still dissolved in the oil in the ratio R_{si} scf/stb. This is the simplest form of sampling which was appropriate during the appraisal stage in many of the fields in the North Sea.

If the oil falls below the bubble point either in the reservoir or at any depth in the production string, then it is appropriate to collect separate surface samples of oil and gas for recombination in the laboratory [1]. In this event, the well is flowed until there is stability in the GOR. Gas samples are collected at the separator and oil from the stock tank. These together with the operating pressures and temperatures of the separator and stock tank are provided to the laboratory to permit the correct recombination of the oil and gas, which is the first stage of the experimental procedure. So important is the determination of reliable PVT relations, that operators will try to sample a well using all the above methods, if appropriate, and check the different samples for consistency.

No matter how carefully the sampling is conducted, however, it is often very difficult to obtain the oil and gas mixed in the required ratio. One of the main reasons for this is if the wellbore pressure falls below the bubble point which can occur if:

- the reservoir has low permeability
- the reservoir contains a free gas cap, or its initial pressure is close to the bubble point

In the former case, to produce the well on test at commercial rates means that the reservoir pressure in the vicinity of the wellbore may fall below the bubble point. When this occurs gas is liberated from the oil (Fig. 2.4a) and, having a very much lower viscosity (and therefore higher velocity, equation 1.2), the oil and gas can be collected in disproportionate amounts. To begin with, the liberated gas will not flow until it reaches the so-called critical gas saturation, S_{gc} (typically 5% PV). This phenomenon is common between any two immiscible fluids (that do not physically mix) and between which there is, therefore, a finite surface tension. The gas will form in discrete bubbles in each separate pore space until the critical saturation is exceeded when gas in neighbouring pores will unite and it begins to move collectively. Until this happens, the fluids sampled in the wellbore may be deficient in gas but thereafter there is liable to be an excess of gas in the fluid samples. In fractured reservoirs, however, it is often observed that the effect of the critical gas saturation is negligible and gas is free to move in the fractures as soon as it is liberated. But whether there is too much or too little gas in the samples collected cannot be distinguished by the engineer when testing a new reservoir.

In the event that the reservoir has a free gas cap, it means that the oil at the gas–oil contact is fully saturated with gas and is therefore at the saturation or bubble point pressure ($p_i = p_b$). That is, if the oil could dissolve more gas it would because it is freely available but the fact that it does not implies its saturation. Consequently,

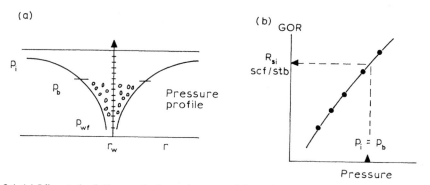

Fig. 2.4. (a) Liberated solution gas in the region around the well in which the pressure falls below the bubble point. (b) Determination of R_{si} for a gascap reservoir.

as soon as the reservoir is tested, the pressure in the region of drawdown around the wellbore must necessarily fall below the bubble point liberating solution gas and giving rise to the same uncertainty in the reliability of the GOR in samples collected in the wellbore as described above for low-permeability reservoirs.

A lot of the difficulty lies in the timing of sampling in appraisal wells whose flowing pressure may fall below the bubble point. The advice usually proffered in the literature and adhered to by most operators is to sample late in the test: following an initial clean-up period and one or more flow periods at high production rate. Following these the well is then closed in for a pressure buildup after which sampling is conducted flowing at modest rates to ensure a low-pressure drawdown, under the belief that during the buildup any gas that has been liberated around the wellbore will have been redissolved in the oil as the pressure rises. But this is not necessarily the case because often there is a disproportionately large volume of gas in this region: too much for the oil to dissolve. Therefore, the oil produced during the sampling periods may still be accompanied by production of free gas. This is not really a point to theorise about. Suffice to say that at the end of the high rate flow periods there is a "mess" in the fluid distribution around the well which is not necessarily going to be rectified by a pressure buildup. Sampling early in the test may not appear to be a sanitary practice because the well is cleaning-up but, nevertheless, the oil collected is much more likely to have the correct GOR than samples that are collected later in the test sequence. The early samples should be compared with those collected subsequently. If the GORs do not correspond it probably means that the well has produced below the bubble point at some stage thus casting doubt on the reliability of samples collected late in the test and the early ones should be accepted in preference.

An example of collecting samples too late during a test is illustrated in Fig. 2.5. The formation was a tight fractured chalk that was initially flowed at low rate, *A–B*, during which the GOR remained constant at 1100 scf/stb. After point *B* the GOR rose significantly accompanied by a sudden drop in the flowing pressure, both indicating that the pressure in the immediate vicinity of the wellbore had fallen below the bubble point, leading to excess gas production through the fractures. At point *C*

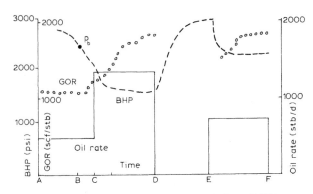

Fig. 2.5. Collection of samples for PVT analysis too late during a DST in an appraisal well.

the oil rate was increased to 1400 stb/d leading to a further drop in pressure and increase in the GOR; the well was then closed in for a 12 hour pressure buildup, *D–E*. Following this, it was produced at a reduced rate of 700 stb/d, *E–F*, for the purpose of collecting fluid samples both downhole and by surface recombination. As can be seen, however, during this final flow, the wellbore pressure was marginally below the bubble point causing the GOR to rise rapidly to 1800 scf/stb, before it appeared to stabilize. It was the samples collected during this last flow period that were accepted as representative for the formation and subjected to full PVT analysis. The results revealed a bubble point pressure of 3600 psia compared to the correct value of 2450 psia that was evident from the initial flow period. This represents a serious error in assessing the fluid properties of the reservoir implying that all subsequent attempts at applying reservoir engineering become reduced in meaning. This type of error can be overcome by carefully observing the rate–pressure–GOR behaviour during the test, especially in tight reservoirs, such as described, in which the wellbore pressure can fall below the bubble point. Therefore, in all appraisal tests, it is recommended that the well be sampled very early in the sequence of events when flowing at a restricted rate since it is never certain in such wells precisely how close the bubble point is to the initial pressure.

In sampling gas cap reservoirs, defining the correct GOR is usually less problematical than for tight formations. Samples of separator gas and stock tank oil are dispatched to the laboratory and mixed with different GORs believed to be in the vicinity of the correct value. The bubble point pressure of each sample is determined by expanding an enclosed volume (constant composition expansion) from high pressure, at reservoir temperature, until a sudden large change in volume during a pressure decline step reveals the evolution of free gas and hence defines the bubble point. Plotting the GORs of the different recombined samples versus their bubble points (Fig. 2.4b) allows definition of the correct saturation pressure when $p_i = p_b$ and permits the oil and gas to be combined with the corresponding value of R_{si}.

(c) Laboratory experiments

In sending fluid samples to the laboratory, the engineer must specify the type of experiments that should be performed. A routine PVT analysis on an initially undersaturated oil consists of the following:

 I. Constant composition expansion of the undersaturated oil from initial pressure to the bubble point.
 II. Differential vaporization: expansion below the bubble point with varying fluid composition.
 III. Separator flash expansion experiments.
 IV. Compositional analysis of the reservoir fluid.
 V. Measurement of oil and gas viscosities as functions of pressure.

All these experiments are performed under isothermal conditions at the reservoir temperature.

(I) A volume of undersaturated oil is expanded in a PV-cell in which all of the fluid is contained within the cell during a series of pressure decrements: usually from the initial pressure (or above) to the bubble point. After each pressure step, the oil volume is measured and reported as a relative volume (RV):

$$RV = \frac{V}{V_b} \quad (rb/rb_b) \qquad (2.4)$$

in which V_b is the volume at the bubble point. From inspection of Fig. 2.1a, it will be apparent that the RV is necessarily less than unity and decreases with increasing pressure. Oil density and compressibility are also reported. The bubble point pressure is manifest, as described above, by the sudden large change in volume for a small change in pressure, due to the evolution of high compressibility gas.

(II) The differential vaporization experiment defines the changing FVF and solution GOR below the bubble point pressure. Gas saturated oil is charged to a PV-cell and the pressure is reduced in steps below the bubble point; after each, the evolved gas is removed from the cell and its volume measured along with the volume of gas saturated oil remaining in the cell. Once the pressure has been reduced to atmospheric, the temperature is decreased from the reservoir value to 60ºF, meaning that the final state is the stock tank condition (14.7 psia, 60ºF) at which the remaining oil in the cell is referred to as the "residual" volume (stb), to which all oil volumes are related at any stage of depletion.

From the volumetric measurements of the oil and gas it is possible to calculate [1]:

B_{od} = FVF of the oil at each stage of depletion (rb/stb).

R_{sd} = corresponding solution GOR (scf/stb).

The subscript "d" relates to the differential experiment and the initial values of these parameters at the bubble point are B_{obd} (rb$_b$/stb) and R_{sid} (scf/stb) respectively.

(III) The series of separator tests is required to correct the results of the differential vaporization, which may be regarded as an absolute experiment, for the conditions of gas–oil surface separation applied in the field. Saturated oil is charged to a PV-cell at the bubble point and allowed to expand through separators (single, two, three stage, or more, each operating at different pressures and temperatures) to stock tank conditions. These are referred to as flash expansions to which the subscript "f" is applied. Each experiment will provide different values of:

B_{obf} = flash FVF of the bubble point oil (rb$_b$/stb).

R_{sif} = solution GOR of the bubble point oil (scf/stb).

the latter being the sum of the GOR's measured for each stage of separation. The values of these parameters will vary dependent on the number of separator stages applied in each experiment and their operating conditions. Usually, two stage separation will provide a smaller value of B_{obf} than one and three a smaller value than two, although one cannot generalise about this matter. Obviously, the smaller the value the better since it means that a stable, stock tank barrel is obtained from a smaller volume of reservoir oil and dissolved gas and therefore, as described in section 2.3, the smaller the value of the FVF, the larger the STOIIP. Furthermore, the value of B_{obf} for any separator condition is smaller than B_{obd}. This is because during each of the multi-stages of the differential vaporization experiment, the lighter gas is removed from physical contact with the oil which encourages the escape of gas molecules in subsequent stages. This means that the stock tank barrel eventually achieved at the end of the experiment must emanate from a larger volume of bubble point oil than in the flash expansion experiments in which the oil and gas remain confined, temporarily, within each of the separators, which inhibits the release of gas [1].

Below the bubble point, the absolute results of the differential vaporization require modification to cater for the effect of the conditions of separation. The required FVF then becomes:

$$B_o = B_{od} \left(\frac{B_{obf}}{B_{obd}} \right) < B_{od} \quad \text{(rb/stb)} \tag{2.5}$$

That is, the separator corrected FVF is shifted from the differential vaporization function by a constant factor at any pressure, the former being the smaller, and so too must be the GOR; such that at any pressure, the amount of gas released from one barrel of oil at the bubble point must be the same in both types of experiment. This implies that:

$$\frac{R_{sid} - R_{sd}}{B_{obd}} = \frac{R_{sif} - R_s}{B_{obf}} \quad \text{(scf/rb}_b\text{)} \tag{2.6}$$

From which the required GOR can be evaluated as:

$$R_s = R_{sif} - (R_{sid} - R_{sd}) \left(\frac{B_{obf}}{B_{obd}} \right) \quad \text{(scf/stb)} \tag{2.7}$$

Alternatively, above the bubble point, applying equation 2.4,

$$B_o = \frac{V}{V_b} B_{obf} \quad \text{(rb/stb)} \tag{2.8}$$

and:

$$R_s = R_{sif} = \text{constant} \quad \text{(scf/stb)} \tag{2.9}$$

These are the values of the FVF and GOR, as functions of pressure, that must be used in all reservoir studies and, in particular, to maximise the STOIIP (equation 2.12), rather than employing the differential vaporization results directly. Maximising the STOIIP requires the determination of the optimum conditions of separation for which B_{obf} and R_{sif} both have minimum values. It would be both time consuming and expensive to attempt to achieve this optimisation by performing repeated separator flash experiments in the laboratory. Instead, optimisation can best be established by calculation, as described below.

(IV) The compositional analysis defines the mole fractions of all the paraffin series (C_nH_{2n+2}) plus the inert gases: N_2, O_2, H_2S and CO_2. This analysis is

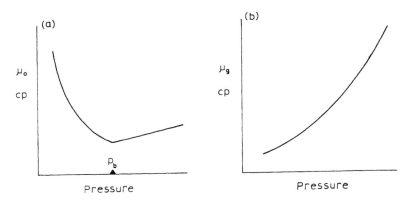

Fig. 2.6. Variation in viscosity with pressure of (a) oil, and (b) gas.

performed for the undersaturated oil and at each stage of depletion during the differential vaporization experiment. With these data, it is possible to calculate the optimum conditions of surface separation applying sophisticated computer packages that rely upon the use of thermodynamic equations of state (EOS). These, in turn, contain both constants and exponents meaning that they require calibration before use. This is achieved by successfully matching the results of the PVT experiments and, in particular, the various separator tests performed. Once this has been done, the package can be applied with some confidence to calculate the optimum separator conditions. The analysis is a sophisticated technical activity and in high-cost areas such as the North Sea, where optimisation in separation can bring significant financial rewards, the work is usually performed by specialised process engineers. It is the responsibility of the reservoir engineer, however, to make sure that reliable fluid samples are collected in the first place and subjected to meaningful, basic PVT analysis. Otherwise the whole exercise of optimisation can be invalidated.

(V) The final stage of the conventional PVT analysis is the determination of the oil and gas viscosities, as a function of declining pressure, and at reservoir temperature. Typical shapes of these functions are depicted in Fig. 2.6. The viscosities are required in all forms of "dynamic" reservoir engineering in which fluid movement is catered for: waterdrive (Chapter 5), gas drive (Chapter 6) and in the analysis of well tests (Chapter 4).

(d) Comparison of laboratory and field PVT data

No matter how careful the engineer may be in collecting fluid samples for laboratory PVT analysis, it seldom happens that the resulting functions match exactly the PVT relations demonstrated by the field itself, once it has been brought on continuous production. There can be many reasons for the disparities, such as deferring the sampling until too late in the test, as described in section 2.2b, or possibly because the separators installed are not quite as theoretically planned.

But perhaps the main reason is simply that if a well is sampled at the appraisal stage during the course of a DST, it must be appreciated that it is producing in a non-equilibrium state, especially with respect to conditions in the vicinity of the wellbore and the collection of completely representative samples in a few "small bottles" is too much to be expected. Yet, surprisingly, many engineers seem quite content to rely upon the laboratory results long after the field has started production and has revealed the actual PVT relations to be somewhat different.

Although laboratory techniques are very sophisticated, the experimental PVT results are only as reliable as the fluid samples collected and should never be regarded as more than a good approximation to the real fluid properties. Laboratory PVT results must be calibrated and, as is often necessary, corrected to match field observations after the start of production. At this time, the engineer should monitor the production data very carefully and, in particular, the producing GOR through the separator configuration used in the field and seek any evidence that the reservoir has passed through the bubble point pressure, which will occur in depletion type fields. The typical GOR behaviour of fractured and homogeneous acting reservoirs is depicted in Fig. 2.7a. In the former, there is usually little or no development of a critical gas saturation, S_{gc}, below the bubble point, which must be exceeded before the liberated gas becomes mobile (section 2.2b). Instead the free gas tends to flow towards the wellbore through the fracture system as soon as it is released. In this case the saturation pressure can be directly related to the prevailing reservoir pressure at which the GOR begins to rise sharply above the value of, R_{si}, defined during the initial phase of production at undersaturated conditions (Fig. 2.7a). In the case of a more homogeneous reservoir, the development of a critical gas saturation is usually observed below the bubble point and is manifest as a slight decrease in the producing GOR as the small volume of gas is trapped in the reservoir rather than produced to the surface (Fig. 2.7a). The bubble point can then be defined as the pressure at which the decline in GOR first occurs, although this is often a more subtle observation than for fractured reservoirs. Other ways of confirming that the reservoir pressure has fallen below the bubble point are by observation of:

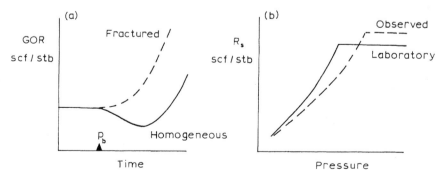

Fig. 2.7. (a) GOR behaviour of fractured and homogeneous reservoirs. (b) Correction of laboratory measured GOR for field conditions.

 – an increase in the API gravity of the oil
 – noting any sudden reduction in the productivity of wells

The first of these is evident in more volatile oil reservoirs (GOR > 1500 scf/stb). As described in section 2.2e, the gas liberated from such oil in the reservoir deposits gas-liquids at the surface which when fed into the liquid oil stream lessens the specific gravity, γ_o, or, alternatively, raises its API gravity.

$$\text{API}^\text{o} = \frac{141.5}{\gamma_o} - 131.5 \qquad\qquad (2.10)$$

The drop in productivity results from the liberation of gas close to the wellbore and consequent reduction in the relative permeability to oil (Chapter 3, section 3.7b) which inhibits its flow into the well.

If it is determined, by direct observation of reservoir performance, that the initial GOR is different from that determined in the PVT analysis, then it becomes necessary to correct the laboratory PVT functions. If, as shown in Fig. 2.7b, the field GOR exceeds that determined by experiment and the elevated bubble point pressure has also been observed, then the R_s trend below the bubble point can simply be "eye-balled" in the knowledge that at atmospheric pressure $R_s \to 0$. A more scientific approach might be to use standard correlations [2] to predict the decline or even more sophisticated EOS thermodynamic software packages. If the bubble point has not been directly observed the latter techniques must be applied to determine it and, whether the bubble point has been observed or not, they need to be used to correct the oil FVF both above and below the bubble point pressure.

The engineer must also bear in mind that the PVT functions will not necessarily remain constant throughout the lifetime of the project but are liable to alter commensurate with changes that might be made to the conditions of surface separation. A good example of this is provided in the development of fields in the East Shetland Basin which has been the main producing area in the North Sea. Although the oil in most of these fields is of moderate or low volatility ($R_{si} < 1000$ scf/stb), it is still possible to squeeze additional quantities of gas-liquids from the gas by suitable processing at the surface. The separator gas (most platforms were equipped with three stages of separation) is refrigerated and compressed which produces liquid hydrocarbons that are fed into the export liquid line, the process being known as "spike-back". The physical reason for the recovery of this extra volume of liquid hydrocarbons is illustrated in Fig. 2.8. This is a phase diagram for a hydrocarbon system (as more comprehensively described in Chapter 6, section 6.2). Within the two phase envelope, oil and gas can coexist in the reservoir. To the left of the critical point (CP) and above the envelope, the hydrocarbons are in the liquid state whereas, to the right and below the envelope the hydrocarbons are in the gaseous state. If the separator gas is at point *A* on the diagram, it can be seen that reduction in temperature and increase in pressure will move the gas within the two phase envelope in which some liquid hydrocarbons will be condensed. The effect of the additional liquid hydrocarbon recovery is to reduce the effective FVF of the oil. To begin with, in the development of these fields, the complex processing

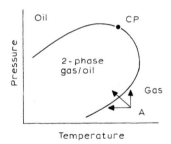

Fig. 2.8. Phase diagram.

equipment required for spike-back was not installed and a fairly high level of flaring of the separator gas was tolerated. Even following installation, however, the complex processing equipment proved to be "temperamental" in operation and there were periods when it had to be totally by-passed or only partially functioned, causing day-to-day variations in the oil FVF. Many of the fields were brought on production during the late 1970's and it was not until some years later that a major oil terminal at Sullom Voe in the Shetland Islands was commissioned. This permitted operators to export oil (and dissolved gas) at higher pressure to take advantage of more efficient separation onshore which, again had the effect of reducing the FVF further. This example is simply to indicate that the PVT properties of reservoir fluids can be somewhat more complex than determined in the laboratory by performing experiments on samples collected at the appraisal stage of field development and can vary throughout the lifetime of a project. Consequently, the reservoir engineer should take advice from specialised process engineers before finalising the input to studies, whether analytical or using numerical simulation.

(e) PVT for volatile oil systems

The PVT described above is appropriate for moderate–low volatility "black oil" which, when produced to the surface, is separated into stable crude from which the solution gas has been released. For more volatile oils, however, the situation is more complex in that the solution gas liberated in the reservoir below the bubble point is rich in condensate which is condensed at the surface and fed into the oil export stream. If:

q_o = black oil rate (stb/d)
q'_o = enhanced rate including the condensate (stb/d)
R' = total measured GOR (scf/stb: oil + condensate)
R_s = solution GOR (scf/stb: oil)
r_s = condensate yield (stb/scf) (the yield is determined by the constant volume depletion experiment, as described in Chapter 6, section 6.2).

Then:

$$q_o = q'_o - (q'_o R' - q_o R_s) r_s$$

and therefore:

$$q_o = q_o' \frac{1 - R'r_s}{1 - R_s r_s} \tag{2.11}$$

This expression permits the cumulative "black oil", N_p, to be calculated and if G_p is the cumulative gas production then $R_p = G_p/N_p$ (scf/stb), which is the cumulative GOR. When used in the material balance equation (Chapter 3, section 3.2), the left hand side of equation 3.5 can then be evaluated in conventional terms of N_p and R_p using black oil PVT properties while still accommodating the additional condensate production. This technique is applied in exercise 3.6.

2.3. CALCULATION OF THE STOCK TANK OIL INITIALLY IN PLACE

The stock tank oil initially in place (STOIIP) equation can be expressed in symbolic form as:

$$\text{STOIIP} = N = \frac{V\phi(1 - S_{wc})}{B_{oi}} \quad \text{(stb)} \tag{2.12}$$

in which V is the net rock volume of the reservoir, ϕ the average porosity and S_{wc} the average connate water saturation. In a more exact manner it is evaluated as:

$$\text{STOIIP} = N = C \times \sum_j V_j \phi_j (1 - S_{wcj})/B_{oij} \quad \text{(stb)} \tag{2.13}$$

implying the use of a gridded mapping package in which the parameters in the equation can vary from cell to cell and the total STOIIP is the sum of the individual values calculated for each. The constant, C, is required to convert the units of volume to barrels. The equation also caters for the possible variation in B_{oi} with areal position or depth if this has been established.

Two volumes appearing in equation 2.12 are used frequently in reservoir engineering calculations and these are:

$$\text{pore volume (PV)} = V\phi = \frac{NB_{oi}}{1 - S_{wc}} \quad \text{(rb)} \tag{2.14}$$

$$\text{hydrocarbon pore volume (HCPV)} = V\phi(1 - S_{wc}) = NB_{oi} \quad \text{(rb)} \tag{2.15}$$

The latter being the actual reservoir volume filled with hydrocarbons. By convention, HCPVs tend to be used in the subject of Material Balance (Chapter 3), whereas in Oilwell Testing (Chapter 4), the use of PVs is preferred.

If the detailed approach suggested by equation 2.13 is not applied an alternative and popular method of evaluating the numerator of equation 2.12 is by resorting to statistical methods. That is, some form of probability distributions for V, ϕ and S_{wc} are evaluated, their definition being commensurate with the state of knowledge of the parameters. The numerator is then usually evaluated by Monte Carlo analysis, in which values of the three parameters are selected randomly and multiplied and this step is repeated perhaps several thousand times to define a probability distribution of the HCPV. From this, a mean or expectation value can be determined and the range of uncertainty attached. Application of this method is perfectly appropriate for the numerator of equation 2.12 because it is known that there are "exact" values of V, ϕ and S_{wc} knowledge of which should be improved with the acquisition of data from each new well so that the range in uncertainty is reduced.

The same approach should not be applied to the oil FVF in the denominator of equation 2.12, required to convert the HCPV, which is a reservoir volume, to stb. The reason is because this is a "engineering number" whose value can be influenced by the amount of time and money the operator is prepared to expend in optimising the conditions of surface separation. As explained in sections 2.2c and d, it is normal to try and establish, either by direct experiment or detailed calculation with thermodynamic EOS packages, the gas-oil separator conditions that will minimize B_{oi} and so maximize the volume of stabilized crude oil recovered in the stock tank. In this respect, there is nothing "random" about the magnitude of the FVF, it should be under strict engineering control.

2.4. FIELD UNITIZATION/EQUITY DETERMINATION

If an exploration well drilled on an accumulation by Company A discovers oil (Fig. 2.9), it is quite evident from a glance at the early seismic and geological inter-pretations that the structure straddles the licence boundary between the concessions of Company A and B. Appreciating this, both companies (and each licence block may contain several companies each with an equity interest) do the sensible thing and agree that the field should be unitized. The basic aim in such an agreement is an excellent one, in that nothing should be done in the field development by either company that would compromise the aim of attaining the maximum amount of (economically) recoverable oil. In this respect a Joint Operating Agreement (JOA) is formulated between the companies headed by a management committee that elects an operator of the field: usually the company with the greatest share, provided they have sufficient operating experience. It also supervises the activities

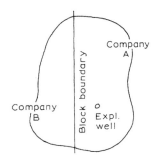

Fig. 2.9. Oilfield straddling two licence blocks.

of various technical subcommittees, including one devoted to the application of reservoir engineering. In applying the operating agreement, if there are differences of opinion on development decisions these are usually settled by vote, applying the voting rights stipulated in the agreement itself.

So far, so good, and field development usually proceeds in an orderly and considered manner. The problems arise, however, when it comes to the matter of deciding who owns what in the accumulation, that is — Equity Determination. If, for instance, Company A owns 65% of the oil, then it collects 65% of the revenue from its sale but is obliged to pay the same percentage of the capital expenditure for the project and its operating costs — and this is where difficulties can and do arise in the calculations that lead to the division of equity.

There is an initial determination, based on the data collected in appraisal wells and prior to continuous field production. During the production phase, however, there can be a sequence of re-determinations triggered by the acquisition of the additional reservoir data from the drilling of development wells. If the partners on one side of the licence boundary believe that the new data could be used to their advantage in increasing their equity share, then they are liable to "ring the bell" (according to the JOA rules) causing the exercise to begin. An Equity Management Committee is formed to supervise the activities of the subordinate technical committees: Geophysical, Geological, Petrophysical and Reservoir Engineering in which agreement must be reached amongst the members from different sides of the licence line on the interpretation of the data collected that might affect the outcome of the determination. Strict rules are laid down in a set of Equity Procedures, agreed by all and signed and counter-signed by their legal representatives which must be adhered to by all parties. These include all the technical details of calculations for the different disciplines such as, for instance, velocity–depth transforms for Geophysicists, and water saturation calculation procedures for Petrophysicists. The equations defining these relations contain empirical constants and exponents that will vary with greater knowledge of the field and, therefore, the participants are obliged to reach agreement on the appropriateness of the techniques enshrined in the "procedures" before the new determination commences, and this in itself can be a lengthy exercise. To add to the difficulties, decision making in most equity determinations is not by vote but must be unanimous.

Obviously, the most important technical matter to decide upon is on what principle the equity division should be made: oil in place, STOIIP, recoverable reserves or movable oil. That is, suppose division by STOIIP was the elected method then the oil revenue returns and expenditure outgoings would be dictated in accordance with the percentage of STOIIP on each side of the licence boundary line, the figure often being stated to six places of decimal! Some of the pros and cons in attempting to apply the different methods of determining equity are described below.

(a) Oil initially in place (OIIP)

The symbolic equation representing this method is:

$$\text{OIIP} = V\phi(1 - S_{wc}) \quad \text{(rb)} \tag{2.16}$$

but in reality in this form of exercise the calculation is performed using a fine gridded model containing all the areal and depth variation in shape and formation properties. In this description of the use of different equations for determination, only their symbolic form will be used for simplicity.

One reason for favouring the OIIP method is that usually it involves the smallest number of technical committees because the oil FVF required in the STOIIP calculation is absent and, therefore, so too the reservoir engineers who are free to devote themselves to the more serious matter of field development studies. Those who have been involved in this type of exercise will appreciate the amount of "horse trading" that occurs and therefore the fewer people concerned the simpler the process becomes and the more rapidly it can be conclusively dispatched.

(b) Stock tank oil initially in place (STOIIP)

As seen already, the relevant equation for this method is:

$$\text{STOIIP} = \frac{V\phi(1 - S_{wc})}{B_{oi}} \quad \text{(stb)} \tag{2.12}$$

and in its evaluation the reservoir engineers are dragged in to settle matters concerning the FVF. In particular, what is necessary to quantify is whether there is any variation in the value of B_{oi} either areally or with depth that could be to the advantage of one of the partnerships and have the reverse effect on the other. Areal variation can occur due to oil migrating into a faulted accumulation from different sources and directions. What is more common, however, is variation of the FVF with depth, especially in thick formations. There is usually a tendency for the oil to become less volatile with depth, meaning its dissolved gas content diminishes and so too does the value of B_{oi}. Such a variation is shown in Fig. 2.10a and the impact of this on an equity determination for the field shown in Fig. 2.10b is patently obvious. Those reservoir engineers representing companies to the east would swear that no variation in FVF with depth could be detected and those claiming such needed their eyesight tested, since the higher value of B_{oi} at the crest would diminish their

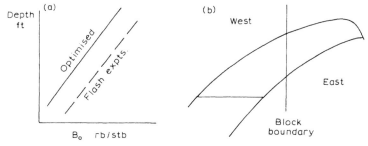

Fig. 2.10. (a) Variation of FVF with depth. (b) Type of field that could be seriously affected by such a variation.

share of the total oil (equation 2.12). Alternatively those representing the favoured downdip partnership in the west would have an array of technical arguments to counter this.

If an engineer believes that there is a trend such as shown in Fig. 2.10a and wishes to prove the point then great care must be exercised in sampling appraisal wells and the laboratory determination of B_{oi} (usually samples collected after the start of continuous production are not considered valid because their depth of origin is uncertain). Under the confrontational circumstances encountered in many equity determinations, it is often surprising how little faith is placed in high technology. Complex EOS packages are banned, since each side of the line will be represented by different ones containing the latest and greatest equations of state. Therefore, what must be considered carefully is both uniformity and simplicity in the experimental determination of B_{oi}. Oil samples from different depths should be sent to the same laboratory and single stage flash expansion experiments performed on each using a standard separator functioning at precisely the same pressure and temperature for each experiment. This will provide a set of values of B_{obf} which can be directly compared to check the validity of any depth variation. The values will not be optimised (section 2.2c) but that is of no concern because equity determination is all about division of the STOIIP, not its absolute value. Therefore, whether the optimised or unoptimised trends shown in Fig. 2.10a are accepted and used will not affect the result.

One of the strangest re-determinations witnessed by the author concerned a field in which the FVF, although included in the calculations was, according to the equity rules, to be regarded as a constant; in which case it could have been left out of the proceedings altogether. Since the previous determination exercise the STOIIP had been considerably reduced due to the drilling of some disappointing development wells and the resulting re-mapping. One of the smaller partners involved felt that they might be seriously financially embarrassed when this news was eventually reported in the press and therefore the remaining partners unanimously agreed to keep the FVF constant, as required, but at the reduced level of $B_{oi} = 0.80$ rb/stb (!!) which demonstrates that the spirit of cooperation can still exist even in such a highly competitive exercise.

(c) Recoverable reserves

The equation describing this method of applying equity is:

$$\text{recoverable oil} = \frac{V\phi(1 - S_{wc})}{B_{oi}} \times \text{RF} \tag{2.17}$$

in which RF is the recovery factor. Now the involvement of reservoir engineers becomes massive. The separate partnerships will construct separate numerical simulation models of the field and run them in "competitive mode". Not only does this greatly extend the lifetime of the determination but it can get down to such gritty arguments as — which model has the best numerical solver, degree of implicitness, etc.

Much more fundamental than this, however, is that equity based on recoverable reserves can be against the basic spirit of field unitization, which in itself is an excellent concept. In the field depicted in Fig. 2.9, for instance, the partners on the western flank may claim reservoir is of poorer quality and therefore requires a greater well density than in the east and while this may be in the equity interest of the western partners, it may not be in the best interest of securing maximum economic recovery from the field as a whole, which is the aim of unitization.

If one side of a licence boundary genuinely does have poorer quality reservoir than the other, then it would seem unfair for those on the better side of the line to have to settle the equity on, say, STOIIP. What would appear to be the fairer option of deciding the issue on recoverable reserves, however, can become so intractable and therefore lengthy that it is preferable for both sides to seek some compromise option such as settling matters basically on STOIIP and applying some agreed discount factors to each side of the line in acknowledgement of different reservoir qualities. To attempt to resolve matters in terms of recoverable reserves can lead to a complete stalemate and there are examples of such fields in the U.K. North Sea where equity determination that commenced in the late-sixties, using this method, have not yet been resolved, which is an unsatisfactory state of affairs for all involved.

(d) Movable oil

This was defined in Chapter 1 (equation 1.1) as that oil that can be physically moved during a secondary recovery flood by either water or gas injection. The equity equation can be expressed as:

$$\text{movable oil} = \frac{V\phi(1 - S_{\text{or}} - S_{\text{wc}})}{B_{\text{oi}}} \tag{2.18}$$

This is a fairly rare form of organising a determination and gives rise to the complication of measuring and mapping the residual oil saturation throughout the reservoirs. It requires heavy involvement from the reservoir engineers since the measurement of residual oil saturation in waterflooding experiments on thin core plugs is largely regarded as their domain (Chapter 5. section 5.4g). The difficulty with the method is that it leans towards determination by recoverable reserves described above. Consider, for instance, the reservoir depicted in Fig. 2.11a in which there is a definite coarsening downwards in the permeability. During a waterflood the majority of the water will enter the base of the reservoir and being heavier than the oil it remains there. Consequently, at the end of the flood the water distribution might be as shown in Fig. 2.11b, in which the upper part of the reservoir has not experienced any waterflooding at all.

Therefore, a core flooding experiment performed on a plug from the top of the reservoir will be quite irrelevant to the determination exercise. What is required instead is the determination of some average value of S_{or} for the reservoir section as a whole and it is in this respect that consideration of recoverable reserves becomes necessary to apply the method correctly.

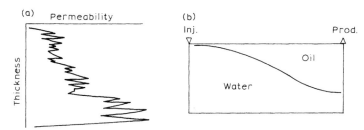

Fig. 2.11. (a) Permeability distribution displaying coarsening downward characteristic. (b) Corresponding water distribution at the abandonment of a waterflood.

Considering the subject of equity determination as a whole, it is an extremely time consuming exercise and at the end of the day proves to be non-wealth generating in that it merely transfers money from one side of the line to the other. It would be preferable if all the time devoted to equity by geophysicists, geologists, petrophysicists and reservoir engineers could instead be channelled into field development studies which would be to the financial benefit of all parties concerned.

2.5. CALCULATION OF GAS INITIALLY IN PLACE

PVT for gas is described at the beginning of Chapter 6 (section 6.2). The important parameter required in the calculation of the gas initially in place (GIIP) is the gas expansion factor, which at initial conditions is:

$$E_i = 35.37 \frac{p_i}{Z_i T} \quad \text{(scf/rcf)} \tag{6.2}$$

in which Z_i is the dimensionless Z-factor accounting for molecular scale effects which must be accounted for in the PVT for gas at reservoir conditions. T is the absolute temperature in degrees Rankine ($= {}^{\circ}F + 460$). The gas expansion factor relates surface to reservoir volumes of gas and is related to the gas formation volume factor as:

$$E = \frac{1}{5.615 \, B_g} \quad \text{(scf/rcf)}$$

The GIIP can then be calculated as:

$$\text{GIIP} = V\phi(1 - S_{wc})E_i \quad \text{(scf)} \tag{2.19}$$

All the comments relating to equity determination in oilfields in section 2.4 (with the exception of determination by movable oil volume) apply equally well to gas fields.

2.6. PRESSURE–DEPTH PLOTTING

This subject was described in reference 1 but is worth revisiting because of the advent of tools such as the RFT and its upgraded sister tool, the MDT. The former has been in operation since the mid-1970's and has revolutionised the subject of pressure–depth relationships in hydrocarbon columns and aquifers. The involvement of reservoir engineers in this subject is in the location of fluid contacts in the formations to enable the calculation of the net rock volume V appearing in equations 2.12 (STOIIP) and 2.19 (GIIP). The situation depicted in Fig. 2.12a is straightforward in that in such a massive horst-like structure, all wells will directly penetrate the oil–water contact (OWC) which will be detected on both cores and logs. Layered reservoirs separated by impermeable shales (Fig. 2.12b), however, represent a much greater challenge to the engineer in establishing the fluid contacts. In this complex situation, pressures are controlled by the common aquifer pressure but the hydrocarbon contents in any of the individual layers is dictated by migration paths, reservoir rock properties, etc.

The basic principle in pressure–depth plotting is illustrated in Fig. 2.13a for a reservoir which has an oil column and free gas cap, Fig. 2.13b. Well-1 in the gas is tested determining the pressure at a particular depth and a gas sample is collected from which the gas PVT properties can be ascertained by laboratory analysis. If the measured gas gravity is γ_g (Air = 1), then its density at standard conditions is $\rho_{gsc} = 0.0763\gamma_g$ (0.0763 lb/ft^3 is the density of air at standard conditions) and the gas density in the reservoir can then be calculated by consideration of mass conservation as:

$$\rho_{gr} = \rho_{gsc}\frac{V_{sc}}{V_r} = 0.0763\gamma_g E \quad (\text{lb/ft}^3) \tag{2.20}$$

and the gas gradient as:

$$\frac{dp_g}{dD} = \frac{0.0763\gamma_g E}{144} \quad (\text{psi/ft}) \tag{2.21}$$

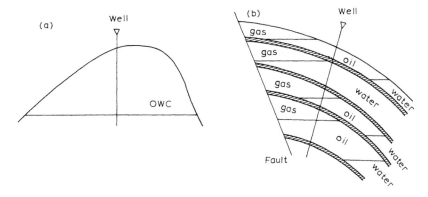

Fig. 2.12. (a) Massive horst structure. (b) Multi-layered reservoir system.

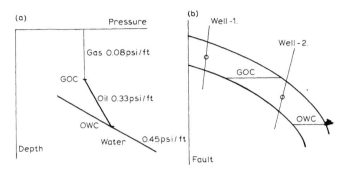

Fig. 2.13. (a) Pressure–depth plot for a reservoir with gas and oil columns (b).

in which E is evaluated at the relevant reservoir pressure. The combination of one pressure point in the gas in well-1 together with the gas gradient permit the gas pressure–depth trend to be constructed (Fig. 2.13a).

The procedure is the same for well-2 which penetrates the oil column. A test is made from which a single pressure is determined and an oil sample obtained. Usually the *in situ* density of the oil and its dissolved gas, ρ_{or}, is provided in the constant composition experiment but if not it can be calculated by application of mass conservation [1] using the surface densities of the oil and gas together with the PVT properties:

$$\rho_{or} = \frac{(\rho_{osc} \times 5.615) + (R_s \times \rho_{gsc})}{(B_o \times 5.615)} \quad \text{(lb/ft}^3) \tag{2.22}$$

following which the oil pressure gradient can be calculated as $\rho_{or}/144$ psi/ft. The combination of the single oil pressure point and the gradient permit the construction of the oil pressure–depth line whose intersection with gas line locates the position of the gas-oil contact.

One of the most important things the reservoir engineer must ascertain in a new area is the pressure–depth trend in the aquifer. No opportunity should be missed to measure pressures in water-bearing sands to establish this relationship and determine whether the aquifer is at normal hydrostatic pressure or is overpressured. The intersection of the oil line (Fig. 2.13a) with the water trend determines the depth of the oil–water contact and so both contacts can be established although neither was seen in the well itself.

Two potential uncertainties encountered in pressure–depth plotting are illustrated in Figs. 2.14a and b. In the first of these the well penetrates an oil column but poses the question of whether there might be a gas cap updip in the reservoir. If p_o is the pressure measured in the oil column and p_b the bubble point pressure, then the depth increment to the possible gas–oil contact (GOC), ΔD, can be determined from the relationship:

$$\Delta D = \frac{p_o - p_b}{\mathrm{d}p/\mathrm{d}D} \quad \text{(ft)} \tag{2.23}$$

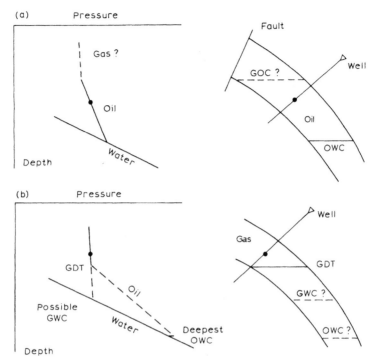

Fig. 2.14. Uncertainties in pressure–depth plotting: (a) the possibility of updip gas; (b) the possibility of downdip oil.

If the calculated value of ΔD locates the GOC within the reservoir, then there very well may be a free gas cap but this is not a certainty. In equation 2.23 it is assumed that the pressure gradient in the oil, dp/dD, is a constant but in some cases, especially in thick reservoirs, the PVT properties and therefore the gradient vary with depth which distorts the calculation of the GOC. The only way to be certain about the presence of updip gas is to drill a crestal well.

Fig. 2.14b illustrates a similar uncertainty associated with the appraisal of gas fields. Gas has only been seen down to the gas-down-to (GDT) level but this allows the possibility of a downdip oil rim, the maximum extension being indicated. The effect of this type of uncertainty in development planning is illustrated in Exercise 2.1.

Exercise 2.1: Gas field appraisal

Introduction

The offshore gas field shown in Fig. 2.15 has been appraised with discovery well, *A1*, and two later downdip wells, *A2* and *A3*. These penetrated the two thin gas-bearing reservoirs, *X* and *Y* as shown in Fig. 2.16 and a sequence of deeper oil reservoirs (not shown). DSTs conducted in the three wells are listed in Table 2.1.

The oil-bearing sands tested beneath the gas reservoirs were found to be over-

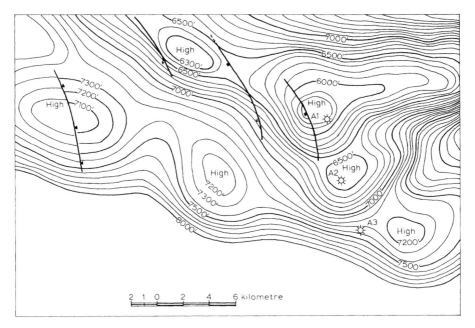

Fig. 2.15. Depth contour map of offshore gas field (Exercise 2.1).

pressured by increasing amounts with depth this being attributed to the extensive shale intervals between each reservoir which removed hydrostatic equilibrium during burial.

Question
• Have these reservoirs been adequately appraised by the three wells, sufficient to permit field development planning to commence and if not what further appraisal is required.

Solution
   It is first necessary to construct a pressure–depth diagram for the two reservoirs using the test data provided. For the X reservoir, DST 9 in well A3 was in a water-bearing sand and the pressure gradient from the surface is $3040/6928 = 0.439$ psi/ft, which implies normal, hydrostatic pressuring. Therefore, the water pressure line can be drawn through the test pressure point as shown in Fig. 2.17 (on this plot the tests are labelled as — well number/test number, e.g: *A3/9*). To plot the gas pressure–depth trends it is necessary to calculate the pressure gradients from the PVT data listed in Table 2.1. This requires first the application of equation 6.2 to calculate the gas expansion factor, E, and its inclusion in equation 2.21 to give the gradients, the values being listed in Table 2.2.
   For the X reservoir the test pressure (*A2/8*) is plotted on Fig. 2.17 and the gradient line of 0.057 psi/ft drawn through it. The gas-down-to (GDT) level in this well is at 5993 ft.ss but deeper than this, the presence of gas can only be inferred.

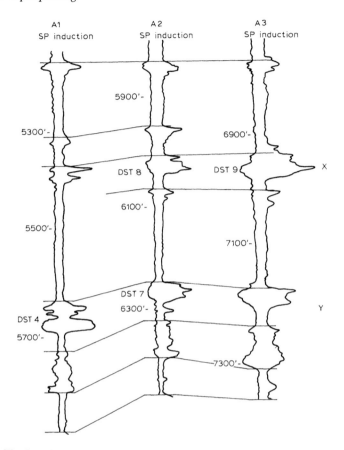

Fig. 2.16. Logs/DSTs across the two gas-bearing reservoirs (Exercise 2.1).

TABLE 2.1

Gas well test results

Test No.	Well	Sand	Initial pressure (psia)	Depth (ft.ss)	Z	γ_g (Air = 1)	Temperature (ºF)	Fluid
8	A2	X	2797	5993	0.91	0.69	242	Gas
9	A3	X	3040	6928			254	Water
4	A1	Y	3100	5544	1.00	0.69	232	Gas
7	A2	Y	3112	6212	0.91	0.69	244	Gas

If it is just gas, then the gas–water contact (GWC) would be at 6430 ft.ss but there is also the possibility of a downdip oil accumulation. The maximum extent of this is indicated by the dashed line in Fig. 2.18. It is dictated by the fact that the water-up-to (WUT) level in the X reservoir is at 6928 ft.ss but it is possible, in the most optimistic case that there could be oil immediately above this depth. The oil

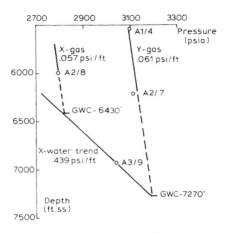

Fig. 2.17. Pressure–depth (Exercise 2.1) assuming gas alone.

TABLE 2.2

Calculated gas gradients

Test No.	Well	Sand	E (scf/rcf)	Gradient (psi/ft)
8	A2	X	155	0.057
4	A1	Y	158	0.058
7	A2	Y	172	0.063

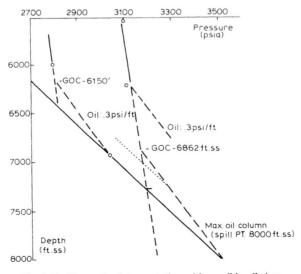

Fig. 2.18. Alternative interpretation with possible oil rims.

gradient in the deeper reservoirs is 0.3 psi/ft and using this figure, as shown in Fig. 2.18, would lead to a possible GOC at 6150 ft.ss, implying a maximum downdip oil column of 780 ft.

In the deeper Y reservoir there are two tests in the gas. Joining the pressures with a straight line would be incorrect because as can be seen from the pressures in Table 2.1, the resulting pressure gradient would be $(3112 - 3100)/(6212 - 5544) = 0.018$ psi/ft which is a physically unrealistic value. The two gas tests in this reservoir were conducted using different pressure gauges and there is therefore no validity in connecting the points. Considering that the average gradient for the two tests calculated from the PVT is 0.061 psi/ft then the error between the gauges amounts to some 29 psi which is considerable. Under the circumstances, the best policy is to draw the average gradient line (0.061 psi/ft) between the points as shown in Fig. 2.17. If it is assumed that the reservoir contains gas alone, extrapolation of the pressure trend below the GDT at 6212 ft.ss would imply a possible GWC at 7270 ft.ss (Fig. 2.17). There are, however, two uncertainties associated with this determination:

– there may be a downdip oil accumulation
– it is not known whether the water pressure trend established for the X reservoir applies also to the Y reservoir.

The maximum oil column can be estimated by drawing a pressure trend with gradient 0.3 psi/ft immediately below the GDT at 6212 ft.ss (Fig. 2.18). It can be calculated that this would give an OWC at below 9000 ft.ss which is considerably deeper than the spill point of the accumulation which is at ±8000 ft.ss. Assuming that the latter could be the deepest OWC, the elevation of the GOC above this level can be calculated as:

$$\Delta D = \frac{p_{\mathrm{o}} - p_{\mathrm{g}}}{\mathrm{d}p_{\mathrm{o}}/\mathrm{d}D - \mathrm{d}p_{\mathrm{g}}/\mathrm{d}D} \quad \text{(ft)} \tag{2.24}$$

in which the pressures are evaluated at 8000 ft.ss (Fig. 2.18) as $p_{\mathrm{o}} = 3512$ psia (at the possible OWC) and $p_{\mathrm{g}} = 3240$ psia (by extrapolation) and using the oil and gas gradients of 0.3 and 0.061 psi/ft then $\Delta D = 1138$ ft. which locates the shallowest GOC at 6862 ft.ss.

The second uncertainty associated with the Y reservoir results from the oversight of not measuring a water pressure in the downdip well $A3$. The log interpretation clearly indicated the sand to be wet and the operator decided to forgo a test. [It should be noted that this appraisal programme was conducted in the early 1970's before the advent of the RFT tool when the testing of the sand would be a lengthy and expensive process. Nowadays the RFT (section 2.7) would be used in a routine fashion to measure pressures in all water-bearing sands and this sort of error is unlikely to be made.] Since the aquifers in the deeper oil-bearing sands were systematically overpressured, there is no reason why the Y aquifer might not be overpressured with respect to the X on account of the 200 ft thick shale between the sands. Suppose the Y aquifer were overpressured by 100 psi, as indicated by the dotted line in Fig. 2.18. It can be seen that the effect would be to truncate the possible gas column by 260 ft and the maximum possible oil column by 720 ft.

Altogether, in spite of the drilling and testing of three wells, it is simply impossible to commence any sensible development planning for this field because it is not even certain whether it is predominantly an oil accumulation with a gas cap or simply a gas field. Since it is located offshore important decisions must be made up-front concerning the project engineering. If there is a downdip oil accumulation it could be a very significant volume because of the areal extent of the structure. Development wells would be drilled downdip to tap the oil while keeping the high-compressibility gas in the reservoirs to let it expand and displace the oil. The platform itself would require facilities for the production of oil gas and water and an oil pipeline to shore.

Alternatively, if it proves to be a gas field, the wells could be drilled in a fairly tight cluster at the crest, since the permeability is good, which keeps them safely away from any natural water influx. Gas processing facilities would be required on the platform and a gas pipeline to shore. Furthermore, a gas market would have to be identified before the project could proceed, on account of the difficulties of transporting gas compared to oil.

In the event, the operator elected to drill two additional appraisal wells one to the east and one to the west of the single line of the first three wells which through their geometry only provided a two dimensional, cross sectional perspective of the reservoirs. The responsibility of the reservoir engineer is to point to the deficiency in the initial appraisal programme in that a well, or wells, are required to penetrate the structure at a depth between wells A2 and A3, to prove or disprove the existence of a downdip oil column. The geophysicists and geologists are then largely responsible in choosing the best locations. The two additional wells proved that the simple situation depicted in Fig. 2.17 was the correct interpretation and that while there was possibly a slight oil rim in both reservoirs it was too small to consider its separate development. Planning therefore commenced for the development of the accumulation as a gas field.

2.7. APPLICATION OF THE REPEAT FORMATION TESTER

The RFT was introduced in the mid seventies. Its main advantage over its predecessor, the FIT (formation interval tester), was that it could measure an unlimited number of spot pressures in one trip in the hole, whereas the FIT was restricted to one. It was originally considered that the main application of the RFT would be for fluid sampling but it did not take the Industry long to appreciate that its greatest value was in providing pressure–depth profiles across reservoir sections. This proves particularly useful during the development drilling programme [3], as described in Chapter 5, section 5.2d, when a detailed survey run in each new well after the start of continuous field production permits the pressure–depth profiles across the reservoirs to be viewed under dynamic conditions. This reveals the degree of areal and vertical communication which is of great assistance in planning secondary recovery flooding (water or gas injection).

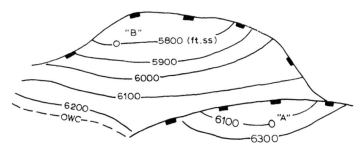

Fig. 2.19. Structural contour map showing locations of wells "A" and "B".

At the appraisal stage of development, RFT surveys provide the best quality pressure data and are routinely run to establish fluid contacts. The surveys are usually straightforward to interpret compared to DSTs because no complex buildup analysis is required to determine the pressure nor are there any extensive depth corrections to be applied since the gauge depth is practically coincidental with that of the RFT probe.

As an illustration of the use of the RFT in field appraisal, the contour map shown in Fig. 2.19 is of the top of a 50 foot thick oil reservoir and aquifer discovered in a region remote from any other developments. The first well drilled, A, was downdip in the aquifer but the operator had the foresight to measure water pressures over a 160 ft interval to ascertain the water pressure regime in this new area.

The second well, B, drilled several kilometres to the west was more fortunate and discovered a 50 ft thick oil-bearing reservoir with good porosity and permeability. Six RFT pressures were measured across this interval. The pressures recorded in the two wells are listed below.

Well A (water)		Well B (oil)	
Depth (ft.ss)	Pressure (psia)	Depth (ft.ss)	Pressure (psia)
6075	2662	5771	2602
6091	2669	5778	2604
6108	2677	5785	2608
6220	2725	5800	2610
6232	2731	5806	2612
		5813	2614

It should be noted that all depths must be converted to sub-sea values to facilitate direct comparison of the pressures. Similarly pressures should all be stated in psia. Earlier mechanical pressure gauges and strain gauges used to record in psig (gauge-pressure) which are relative to atmospheric pressure: psia = psig + 14.7. Engineers must be quite specific in defining the pressures being used. It is not uncommon to find pressures referred to as being measured simply in "psi", which is not an absolute pressure unit but represents a pressure difference.

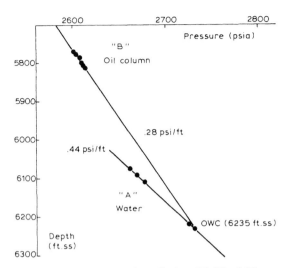

Fig. 2.20. RFT-surveys in wells A and B (Fig. 2.19).

The pressure measurements in the oil column and aquifer are plotted in Fig. 2.20. When making such plots, the most expanded pressure scale possible should be employed, across the range of interest, to facilitate accurate determination of the fluid contacts. The pressure–depth trend in the water has a gradient of 0.44 psi/ft, which is in agreement with its salinity. Furthermore, division of the individual water pressures by their depths in the above table reveals this same figure for each indicating a normal hydrostatic pressure regime to the surface.

The specific gravity of the oil and its dissolved gas at initial reservoir conditions, as determined in the PVT analysis, was 0.646 relative to water. Therefore, since the density of pure water is 62.43 lb/ft^3, the oil pressure gradient is $0.646 \times 62.43/144 = 0.28$ psi/ft. Such a gradient line fits through the measured pressure points in a very convincing manner and extrapolating this gives an OWC at 6235 ft.ss: 422 ft below the oil-down-to (ODT) level at 5813 ft.ss. The surveys also demonstrate that the fault to the north of well A is sealing.

This is a simple example of the interpretation of RFT surveys in two different wells, illustrating the superiority of the tool over earlier equipment. In the past, running DSTs, the operator would probably not have conducted a test in well A in the aquifer on account of the expense and a test in the oil column would only have yielded one pressure point, whereas there are six RFT pressures in the oil, whose gradient is confirmed by that calculated from the PVT analysis. Accuracy in pressure measurement is all important in this type of work and to improve matters the operator standardised by using the same high-resolution gauge in all RFT surveys and DSTs throughout the appraisal program in the area. Thus the pressures listed in the above table were all measured with the same gauge. The advantage of this is that even if the gauge had a systematic, absolute error of 10 psi, the OWC would still be determined at the same depth. Alternatively, if different gauges had been used and there was an absolute error difference of 10 psi between them, it can be calculated applying equation 2.24 for oil/water instead of gas/water, that the error would represent a uncertainty in the OWC of 62.5 ft and being at the base of the oil column this would represent a significant uncertainty in the STOIIP. The accurate location of the OWC resulting from this RFT interpretation, together with the geological modelling, permit the correct evaluation of the net rock volume V in equation 2.12 for the calculation of the STOIIP.

Unfortunately, not all RFT surveys are as straightforward to analyse as the one just described and the author has noted a widespread error in the basic interpretion technique which can lead to a severe distortion in the way reservoirs are viewed and fluid contacts located. This arises from adopting the approach shown in Fig. 2.21a of plotting the pressure points as a function of depth and then simply "forcing" the best looking straight lines through them without regard to the magnitude of the fluid pressure gradients. As can be seen in the plot this produces a gradient of 0.355 psi/ft in the oil column and the highly improbable figure of 0.577 psi/ft in the aquifer.

Interpreting RFTs is a bit more subtle than that and the correct analysis is shown in Fig. 2.21b. In this the gradient of 0.27 psi/ft determined in the PVT analysis has been honoured for the oil and 0.450 psi/ft for the water, the latter having been well established in the area. When these gradient lines (plotted in the top right of Fig. 2.21b) are fitted through the pressure points then a completely different picture emerges. The pressures are completely consistent with the fluid gradients and reveal that a non-equilibrium situation pertains across the reservoir and aquifer: there being slight perturbations in pressure of about 5 psi between separate layers. In the present example there was a tight interval close to the water in which RFT pressures could not be recorded nor could the logs be interpreted to determine the OWC, only values of ODT and WUT, which meant that RFT and log contacts could not

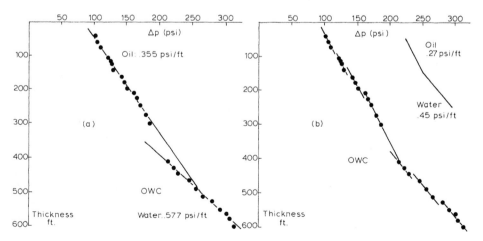

Fig. 2.21. RFT-survey interpretation: (a) Incorrect, (b) correct in honouring the PVT fluid gradients.

be compared. As can be seen the difference in OWCs between Figs. 2.21a and b is 90 ft: the incorrect interpretation giving the optimistic result.

There are several reasons why such a lack of equilibrium should occur:

Appraisal wells:
 - A quite natural difference of pressures between sands separated by impermeable barriers, with no external influence.
 - Pressure interference through the aquifer from neighbouring fields that have been on production for some time.

Development wells:
 - The fact that there is a lack of vertical pressure equilibrium across the reservoir/ aquifer under dynamic producing conditions.

The detection of interference effects in appraisal wells is not uncommon and is to be looked for. In provinces such as the northern or central areas of the U.K. North Sea, for instance, some of the major fields have been on production for 15 years (at the time of writing) and it is hardly surprising that if an exploration/appraisal well is drilled in one of these areas it should detect pressure perturbations across the sand section. Pressure in the producing fields is allowed to drop initially by perhaps 1000–2000 psi before this is arrested by water injection and it is this initial pulse that causes the interference effects at distant locations years afterwards.

The difficulty in offshore appraisal, as described in Chapter 1, section 1.2d, is that observation of the reservoirs invariably occurs under static conditions. Therefore, interference effects, such as shown in Fig. 2.21b, are of great benefit to the engineer in providing a "slightly" dynamic view of the reservoirs prior to continuous field production. In the first place the interference implies that there is a continuous aquifer and even if the intention is to inject water, the results give some encouragement that the injection wells can be located safely in the aquifer to conduct a peripheral waterflood which is usually the most satisfactory way of proceeding. If the pressure perturbations are large, then it might prove possible to perform interference calculations between the producing field that caused the effect and the exploration/ appraisal well location. This type of calculation is illustrated in Chapter 4, section 4.17 and can, if successful lead to a definition of aquifer properties. In even greater detail, a numerical simulation model can be set up to determine what sort of reduction in vertical permeabilities between the separate layers is required to cause the observed pressure perturbations across the section. This can be of great benefit in assessing the vertical sweep efficiency of waterdrive, as described in Chapter 5, sections 5.6 and 7. In this respect, to view matters properly, the core permeability distribution across the reservoir and aquifer (if available) should be plotted alongside the correct RFT interpretation (Fig. 2.21b). If the survey was in a development well, drilled after the start of continuous production, then the same comments apply. The data prove invaluable for calibrating (history matching) numerical simulation models to ensure that the vertical sweep is correctly modelled. In fact, such RFT surveys, which can only be run once in a well, prior to setting the production casing, provide the highest quality pressure data that are acquired during its lifetime.

The error shown in Fig. 2.21a is, unfortunately, of very common occurrence and engineers attribute the slight scatter in pressures as due to all sorts of different reasons connected both with the equipment and formation. The resolution in modern pressure gauges is such, however, that surveys are deliberately run to detect pressure differences of a few psi and provided the formation is of reasonable quality, the scatter in pressures implied by Fig. 2.21a would be quite intolerable. It is amazing how many apparently anomalous RFT surveys can be rationalised by adopting the obvious interpretive technique illustrated in Fig. 2.21b.

The source of this error appears to be in the predilection of engineers for computer graphics and least square fitting of lines through points. It is very simple to program the type of interpretation shown in Fig. 2.21a but not quite so simple that shown in Fig. 2.21b. To be fair, however, this type of work is often done on the wellsite in the middle of the night and, if it is an exploration well, the PVT data will not be available. Nevertheless, the engineer should be guided by the use of "reasonable" pressure gradients. In one case, for instance, the author noted a water pressure gradient of 0.76 psi/ft. This should have caused some alarm bells to ring, having possibly penetrated the first "heavy water" aquifer in the world.

2.8. PULSE TESTING USING THE REPEAT FORMATION TESTER

As mentioned previously, the best the most useful application of the RFT is in running surveys in each new production well, prior to running the production casing, when dynamic pressure profiles across the reservoir can be viewed due to the continuous production of the previously drilled wells. Application of such survey results in field development studies of waterdrive is described in Chapter 5, sections 5.6 and 7. Apart from this, RFT-pulse surveys can be extremely useful in making immediate operational decisions in fields.

The technique relies on running an RFT survey at the start of a logging job, deliberately causing some perturbation in the field's normal operating condition, then running a second survey at the end of the logging job to try and detect the influence of the disturbance. The double survey is naturally very popular with the service companies.

Two examples of the application of this technique are presented in reference 4, describing vertical and lateral pulse tests in what is probably the largest single reservoir in the entire North Sea. This is at the base of the Middle Jurassic, Brent Sand section in the major oilfields of the East Shetland Basin in the U.K. and Norwegian sectors. The permeability distribution across the reservoir section is shown in Fig. 2.22. It consists of two sands that have quite different depositional environments resulting in their marked contrast in flow characteristics. The high-permeability Etive and the underlying lower-permeability Rannoch (all the sands in the Brent section are named after Scottish Lochs) act as a single reservoir unit in that RFT surveys run under undisturbed producing conditions reveal a state of apparent hydrostatic equilibrium across the section. Therefore, in spite of the severe contrast in average permeabilities and the presence of a correlatable tight interval at the top of the

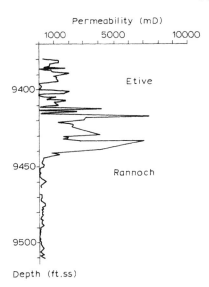

Fig. 2.22. Typical permeability distribution across the Etive–Rannoch reservoir; Middle Jurassic Brent section.

Rannoch (just below 9450 ft, Fig. 2.22), which acts as a partial barrier to vertical fluid movement and exacerbates the contrast in permeabilities, the two sands are hydraulically linked — which causes a production problem.

All of the fields containing this massive reservoir are under secondary recovery conditions, mainly by waterdrive, and it can be imagined what happens when water is injected across the entire sand section. 90% of the water enters the upper, Etive reservoir at the injection wellbore and, on account of the partial barrier at the top of the Rannoch, it stays there. This causes premature breakthrough of water in the production wells and a very poor and retarded flood of the lower Rannoch sand. The flooding condition is described in greater detail in Chapter 5, section 5.10b.

In the operational situation being considered, a well was drilled into the extreme southern part of fault block *TW*, Fig. 2.23a, with the intention of completing it as an injector on the high-permeability Etive sand alone to provide urgently required injection support to production wells in the north of the block. The well penetrated what was believed to be a sealing fault separating blocks *TW* from *TE* at location *A*. Neither the logs nor the initial RFT survey, which on this occasion did not exhibit fluid potential equilibrium across the Etive–Rannoch (Fig. 2.23c), were conclusive in pinning down the exact location of the fault in the wellbore.

Therefore, immediately after running the initial RFT survey, an injection well, *I*, in block *TW* (Fig. 2.23a) was closed-in simultaneously with a production well, *P*, in block *TE*. Both wells were completed exclusively on the Etive sand. Twenty four hours after the closure of these wells, a second RFT survey was run in well *A*, at the conclusion of the logging survey in the well, with the result shown in Fig. 2.23c. Closing-in injector *I* at a distance of 1060 ft from well *A* propagated a large negative pressure pulse of over 200 psi directly through the Etive sand whereas, closing-in

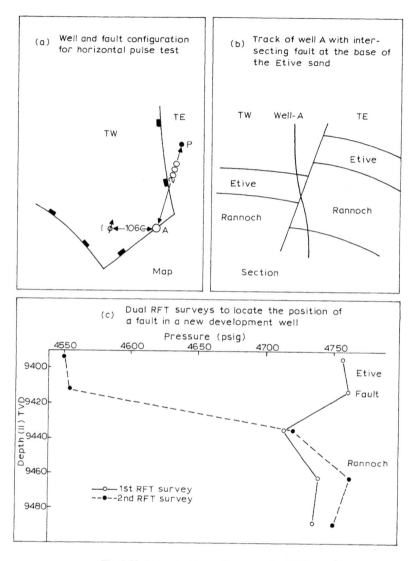

Fig. 2.23. Lateral pulse testing using the RFT.

the producer *P* located 2000 ft from well *A* gave rise to a smaller positive pulse of maximum value 25 psi, transmitted through the Etive and downwards into the Rannoch.

The difference in RFT survey results indicated that the position of the fault in the wellbore was near the base of the Etive sand, as depicted in Fig. 2.23b, and that in spite of the sand-to-sand contact the fault was an effective pressure seal between blocks *TW* and *TE*. The well was completed on the Etive sand as an injector. The dual RFT survey saved a costly and unnecessary sidetrack of the well back to a safe location in the Etive in block *TW*. To add to the pleasure of the service companies in running the dual RFT surveys, pressure measurements made in the first survey, above and below the interval in which the production perturbation occurred, were repeated in the second to calibrate the pressure gauge. In the present example the initial and repeat pressures in the unaffected sands were exact, thus confirming that the pressure differences between the first and second surveys across the interval of interest were genuine. It should be appreciated that this type of pressure pulse test with the RFT can only be applied in sands of moderate to high permeability in which the transmission of pressure responses is rapid.

An even more important example of this type of test is described in Chapter 5, section 5.10b, in which a vertical pulse test was conducted across the tight section between the Etive and Rannoch to establish the degree of vertical communication between the sands.

2.9. APPRAISAL WELL TESTING

Rather surprisingly, after years of international experience in developing offshore fields, there is still a dearth of papers on the purpose of the extremely expensive testing conducted in offshore appraisal wells. This owes to the fact that in earlier onshore developments, which predominated during the years when the subject of well testing was founded, there was little distinction between appraisal and development in that appraisal wells were usually tied-in as producers as soon as possible and testing was of the routine type for such wells, aimed at the determination of pressure and skin factor.

It is in the offshore environment, however, that the distinction between the static appraisal stage of a field and the dynamic production phase that follows is apparent, as described in Chapter 1, section 1.2d. While the aims in testing will vary slightly from one well to another, the main sequence in priorities in appraisal well testing should be as follows:

- determination of the production rate (q)
- calculation of the skin factor (S)
- collection of fluid samples for PVT analysis
- evaluation of formation characteristics (permeability, fractures, layering)
- influence of boundary conditions (faults, depletion)
- measurement of pressure

This list is presented and described in detail in Chapter 4, section 4.4a, but the points of general interest beyond the specialist subject of reservoir engineering will be described here. It is the first two steps that are of paramount importance in testing wells at the appraisal stage; between them they determine the number of wells required for the development which is the most important factor of all in making the "go" or "no-go" decision for an offshore project. A reservoir with a billion barrels of STOIIP at 13,000 ft and a water depth 500 ft whose wells produce only a few hundred stb/d on test would hardly be considered as economic, requiring perhaps hundreds of wells to develop, and that would be the end of the project, based on the results of the initial well tests. The number of wells required for a field development is assessed through determination of the productivity index (PI) of an "ideal" development well completed without any skin factor, resulting from formation damage caused by drilling (refer Chapter 4, section 4.4a). The ideal PI is evaluated as:

$$\text{PI}_{\text{IDEAL}} = \frac{q}{p_i - p_{wf} - \Delta p_{skin}} \tag{2.25}$$

in which p_i is the initial reservoir pressure, p_{wf}, the final flowing pressure and Δp_{skin} is the additional pressure drop across the damaged zone close to the wellbore. The latter is quantified by a dimensionless number S which is defined through the expression:

$$\Delta p_{skin} = 141.2 \frac{q\mu B_o}{kh} S \quad \text{(psi)} \tag{2.26}$$

Subtracting this from the observed pressure drawdown, $p_i - p_{wf}$, in equation 2.25 makes the assumption that while appraisal wells may be drilled with a heavy (safe) mud that gives rise to a large skin factor, development wells will be drilled through the production zones with a refined completion fluid such that, on average, their skin factors will be zero.

Estimation of the number of wells required to develop an offshore accumulation is no small matter since based on the number is the size of the well deck required to accommodate all the well heads and flow lines and this dictates the very size of the platform itself. If determination of the ideal PI is to be regarded as the most important result achieved in appraisal well testing, then it follows that:

APPRAISAL WELLS SHOULD BE PERFORATED JUST AS IF THEY WERE DEVELOPMENT WELLS

otherwise the aim of anticipating the performance of an average development well is unattainable. Therefore, even at the earliest stage of appraisal, the engineer must make an assessment of how the reservoir section would be perforated in a development well and do likewise in an appraisal well, if not appraisal well testing can be rendered meaningless.

In far too many instances, offshore appraisal well tests are conducted by partially perforating reservoirs, as in the three test sequence shown in Fig. 2.24. Essentially the tests were conducted in this exploration well to investigate the flow performance of different permeability layers in the section: high in DST-1, low in DST-3 and practically no permeability at all in DST-2. To consider the upper reservoir first, tests such as DSTs 2 and 3 cannot be condemned, the operator may wish to determine whether such low-permeability intervals will contribute to production or not. But the conclusion of testing the upper reservoir should be to perforate the entire interval, as would be done in development wells, to evaluate the PI of the entire formation.

DST-1 on the much better quality lower formation illustrates other aspects of test design and the reservoir configuration is depicted in Fig. 2.25.

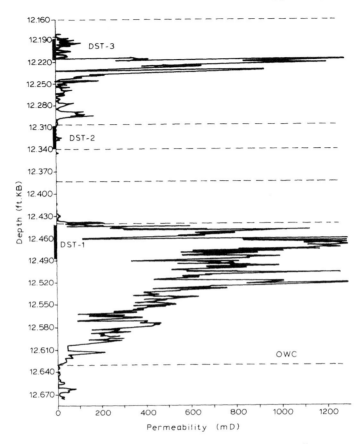

Fig. 2.24. DST sequence in an offshore exploration well.

The exploration well proved most successful in clipping the basal shale coincidentally with the depth of the OWC (location *A*). Thus, not only did it discover oil in the excellent-quality lower reservoir, it also delineated the limits of the accumulation. One of the aims in perforating only the upper, high-permeability interval was to investigate the formation remote from the wellbore to determine the existence of any sealing faults. The main concern in restricting the perforations to the top of the sand, however, was to avoid producing any basal water. It does not look good in a press statement to read that the well produced with a 20% watercut, which might have happened if the entire section had been perforated. In this respect, DST-1 can be regarded as a "management test".

While the exploration well was ideally located, a producer would not be drilled in this position but much further updip at location *B* and it would be undoubtedly perforated across the entire formation. Therefore, anticipating that wells *A* and *B* will have the same formation characteristics, the former should have been perforated across the whole formation, irrespective of the presence of the basal water to obtain the total PI of the formation. The problem with partial perforation, as in well *A*, is that it is not known whether all or only a part of the sand is contributing to production. The pressure gradients at test rates in such an excellent formation are small and therefore thin, tight intervals might well act as barriers to vertical communication. Consequently, it is not certain whether the PI measured is representative of just the perforated interval or the total formation. Similarly with the permeability, which is not determined in isolation in test analysis but rather as a *"kh"*-product (Chapter 4, section 4.12). In partially perforated wells if there is an uncertainty in the effective thickness contributing to flow then there will be an associated ambiguity in the value of the permeability. Most advanced forms of well test interpretation demand an explicit knowledge of the permeability to evaluate the dimensionless time:

$$t_\mathrm{D} = 0.000264 \frac{kt}{\phi\mu c r_\mathrm{w}^2} \quad (t, \text{ hours}) \tag{2.27}$$

for which dimensionless pressure functions characterising the formation under test are defined (Chapter 4, section 4.7). If the permeability cannot be uniquely determined, however, then any attempt to use these functions in sophisticated analysis to locate fault positions, characterise dual porosity systems, etc., can be totally invalidated.

In further consideration of the reservoir configuration in Fig. 2.25, if the well had been drilled at locations *C*, penetrating the OWC, or even fully in the aquifer at point *D*, a brief test to establish its productivity or injectivity and obtain a water sample would not be amiss. All too often operators cease with the formation evaluation once they discover that the well has penetrated the aquifer. Such downdip wells should be fully cored through the formation in the expectation that they will be similar in wells drilled further updip and to observe the rock properties in the aquifer itself. Since it is the aim in many offshore projects to inject water in the aquifer to apply pressure maintenance, then it is extremely important to check on the feasibility of doing so at the appraisal stage. It is not uncommon for diagenetic effects to diminish the aquifer permeability to the extent that injection is impracticable. If this occurred, it would require relocating the injection wells in the

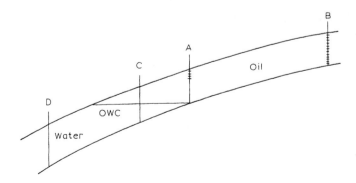

Fig. 2.25. Reservoir/aquifer configuration for DST-1 (Fig. 2.24)

oil column with a resulting loss in recovery of "edge oil" and the requirement for a greater number of injectors because of the added resistance to water injection in the oil column (Chapter 5, section 5.4b).

The exact requirements in appraisal well testing will vary from one reservoir to the next and with the nature of the appraisal well itself. Some wells are drilled and abandoned which means that they can be "tested to destruction". Others, if drilled in favourable parts of the reservoir may be suspended to be tied-back at the start of the development and used as producers. With these the perforating policy may have to be a little more circumspect but nevertheless the same basic rule should apply: to try and anticipate how development wells will be perforated and do likewise with the appraisal wells.

2.10. EXTENDED WELL TESTING

Extended well testing (EWT) is becoming increasingly popular in appraising more marginal offshore developments and is described in greater technical detail in Chapter 4, section 4.19h. The main aims in performing such tests, which can last for weeks and in more ambitious cases for months, are the estimation of hydrocarbons in place and assessment of the nature and strength of the drive mechanism before committing to a full scale development. Since the oil is usually collected and marketed another reasonable aim is to make a profit from the test or, at least, try and break even financially.

There are, however, some basic difficulties confronting the engineer in interpreting such tests. The most significant of these arises from the attempted solution of the volumetric material balance equation to calculate the hydrocarbons in place. Consider the case of performing an EWT in an undersaturated oil reservoir possibly affected by natural water influx. The material balance for such a system (Chapter 3, section 3.8) is:

$$N_p B_o + W_p B_w = N B_{oi} c_{eff} \Delta p + W_e B_w \qquad (2.28)$$

production = STOIIP × unit expansion + water influx

While the cumulative oil and water production (N_p and W_p) on the left hand side will have been monitored and are therefore known quantities, the right hand side of the equation contains two major unknowns: the STOIIP (N) and any water influx (W_e). As such the equation poses the problem of lacking mathematical uniqueness and it is at this point that the engineer must be particularly careful of making unsubstantiated assumptions, one of the favourites being that there is no water influx ($W_e = 0$) which, if incorrect, will lead to an overestimate of the STOIIP.

More potential difficulties arise with the interpretation and use of the pressure data. Apart from appearing explicitly in the Δp term in equation 2.28, the oil and water FVFs are functions of the declining pressure which must therefore be measured accurately during the test. Quite often operators opt for a single, lengthy pressure buildup test to measure the average reservoir pressure at the end of the extended flow period giving themselves only one pressure point with which to evaluate equation 2.28. What would be preferable, however, is to plan for several closed-in periods to acquire sufficient average pressure measurements to apply the material balance method of Havlena and Odeh [5]. This, which is regarded as the most accurate way of solving equation 2.28, is described for a waterdrive field in Chapter 3, section 3.8b and illustrated in Exercise 3.4. The periods of closure need not be for long, usually a few hours will suffice. It is a repeated argument throughout Chapter 4, on well testing, that pressure buildups as generally conducted are unnecessarily long and if their duration could be restricted, the profitability of the EWT may not be affected. It is also demonstrated in Chapter 4, section 4.20a that, theoretically, it should be possible to determine the average reservoir pressure simply by changing the rate during the main flow period which avoids the necessity of well closure for conventional pressure buildup surveys.

The interpretation of buildups is itself a traditional source of error. In spite of innovations in analysis techniques, particularly since the early 1980's, the most popular test interpretation technique is to apply the buildup plot of Horner which dates from 1951 [6]. This is a plot of the static pressure p_{ws} versus the logarithmic time function $\log(t + \Delta t)/\Delta t$, in which t is the flowing time and Δt the closed-in time. Yet there is a misapprehension, which is perhaps the most common error in test analysis, that *linear* extrapolation of the late time trend in points on a Horner plot to infinite closed-in time ($\Delta t \to \infty$, $\log(t + \Delta t)/\Delta t \to 0$) somehow identifies the reservoir pressure — it doesn't. If there has been depletion during the EWT, then theoretically the pressure points on a buildup must necessarily curve downwards and eventually flatten. Consequently, the conventional linear extrapolation is liable to determine an incorrect pressure, which is too high, and that leads to an overestimation of the STOIIP since Δp in equation 2.28 is too small. There is also an inherent weakness in interpreting tests in which there has been an element of depletion in that the area of the hydrocarbon accumulation, its shape and the position of the test well with respect to the boundaries must be "known" which in turn implies that the STOIIP must also be known. This reduces the efficacy of test interpretation under depletion conditions. It should be appreciated that the difficulties described above are simply basic limitations in reservoir engineering technique: the mathematics is always perfect but the physical situation to which it is being applied is unseen and, therefore, uncertain. Many believe that numerical simulation should offer the means of overcoming these difficulties but the technique is so assumption laden (Chapter 3, section 3.5) that it adds nothing to the solution of the problem, furthermore, the basic simulation technique of history matching reservoir performance to determine hydrocarbons in place and define the drive mechanism is flawed, as described in Chapter 3, section 3.8c. The engineer must therefore appreciate that interpretation of tests and particularly EWTs is inevitably a subjective matter that will give, at best, approximate results.

The most effective way of conducting an EWT to establish the economic viability of a project is to "do something else" to enhance the value of the test. By this it is meant that other appraisal/development wells should be drilled, say, six or twelve months after the EWT and pressures measured across their entire reservoir/

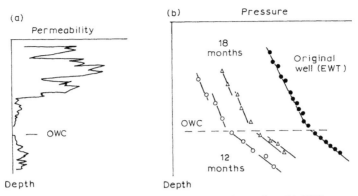

Fig. 2.26. (a) Permeability distribution across a reservoir/aquifer section. (b) RFT surveys illustrating delayed interference effects in wells drilled subsequent to an EWT.

aquifer sections using the RFT. Figure 2.26 illustrates the significance of such a test sequence. The reservoir being tested with the initial EWT was very heterogeneous in its permeability variation. It was not known whether the oil in the basal section could be recovered by cross-flow into the upper high-permeability sands (dual porosity behaviour) or if the basal aquifer would be effective in providing pressure support. The EWT was perfectly conducted but interpretation left an ambiguity in estimating the STOIIP between 60 and 100 MMstb. Two appraisal wells were drilled 12 and 18 months after the EWT and RFT-surveys in each are shown in Fig. 2.26b. These revealed a reasonable degree of pressure communication across both the basal part of the reservoir and the aquifer, confirming both dual porosity behaviour and pressure support. The value of such information is that it permits the construction and calibration of a numerical simulation model in which the RFT pressures can be matched layer by layer to produce a credible model which in this case helped in defining the STOIIP in the formation and led to a positive development decision.

REFERENCES

[1] Dake, L.P.: Fundamentals of Reservoir Engineering, Elsevier, Amsterdam, 1978.
[2] Hewlett Packard: Fluid Pac.
[3] Bishlawi, M. and Moore, R.L.: Montrose Field Reservoir Management, SPE Europec Conference, London (EUR 166), October 1980.
[4] Dake, L.P.: Application of the Repeat Formation Tester in Vertical and Horizontal Pulse Testing in the Middle Jurassic Brent Sands, SPE Europec Conference, London, 1982 (EUR 272).
[5] Havlena, D and Odeh, A.S.: The Material Balance as the Equation of a Straight Line, JPT, August 1964.
[6] Horner, D.R.: Pressure Buildup in Wells, Proc., Third World Petroleum Congress, Leiden, 1951.

Chapter 3

MATERIAL BALANCE APPLIED TO OILFIELDS

3.1. INTRODUCTION

It seems no longer fashionable to apply the concept of material balance to oilfields, the belief being that it has now been superseded by the application of the more modern technique of numerical simulation modelling. Acceptance of this idea has been a tragedy and has robbed engineers of their most powerful tool for investigating reservoirs and understanding their performance rather than imposing their wills upon them, as is often the case when applying numerical simulation directly in history matching.

As demonstrated in this chapter, by defining an average pressure decline trend for a reservoir, which is *always* possible, irrespective of any lack of pressure equilibrium, then material balance can be applied using simply the production and pressure histories together with the fluid PVT properties. No geometrical considerations (geological models) are involved, hence the material balance can be used to calculate the hydrocarbons in place and define the drive mechanisms. In this respect, it is the safest technique in the business since it is the minimum assumption route through the subject of reservoir engineering. Conversely, the mere act of construction of a simulation model, using the geological maps and petrophysically determined formation properties implies that the STOIIP is "known". Therefore, history matching by simulation can hardly be regarded as an investigative technique but one that merely reflects the input assumptions of the engineer performing the study.

There should be no competition between material balance and simulation, instead they must be supportive of one another: the former defining the system which is then used as input to the model. Material balance is excellent at history matching production performance but has considerable disadvantages when it comes to prediction, which is the domain of numerical simulation modelling.

Because engineers have drifted away from oilfield material balance in recent years, the unfamiliarity breeds a lack of confidence in its meaningfullness and, indeed, how to use it properly. To counter this, the chapter provides a comprehensive description of various methods of application of the technique and includes six fully worked exercises illustrating the history matching of oilfields. It is perhaps worth commenting that in none of these fields had the operators attempted to apply material balance, which denied them vital information concerning the basic understanding of the physics of reservoir performance.

3.2. DERIVATION OF THE CUMULATIVE MATERIAL BALANCE FOR OIL RESERVOIRS

The derivation of the volumetric material balance equation was first presented in 1936 by Schilthuis [1]. Since then it has appeared in many papers and text books, including the author's earlier volume [2]. It will, nevertheless, be repeated here for the sake of completeness and because a thorough understanding of the equation's derivation is fundamental to its meaningful application.

To appreciate the full range of complexity in derivation, consider the case of depletion of the reservoir depicted in Fig. 3.1, which has an active aquifer and a gas cap.

If N is the STOIIP in the oil column and

$$m = \frac{\text{HCPV of the gascap}}{\text{HCPV of the oil column}} \tag{3.1}$$

at initial conditions, which is therefore a constant, then applying the defining expression for the STOIIP, equation 2.12:

$$\text{HCPV}_{\text{OIL}} = N B_{\text{oi}} = V \phi (1 - S_{\text{wc}}) \quad \text{(rb)}$$

and

$$\text{HCPV}_{\text{GAS}} = m N B_{\text{oi}} \quad \text{(rb)}$$

giving a total HCPV of $(1 + m) N B_{\text{oi}}$ (rb). Suppose that at a given time after the start of production the pressure has dropped from its initial value at datum, p_i (psia), to some current average value, p (psia), as the result of fluid withdrawal from the reservoir. Then, if during this pressure drop, Δp, the entire reservoir system is allowed to expand (in an artificial sense) underground, the total expansion plus any natural water influx must equal the volume of fluids expelled from the reservoir as production. The volumetric material balance can then be expressed in reservoir barrels as:

Underground withdrawal = Expansion of the system + Cumulative water influx

Perhaps the simplest way to visualise the balance is that if a measured surface volume of production (oil/gas/water) were returned to the reservoir at the reduced pressure, p, it must fit exactly into the volume which is the sum of the total expansion and the water influx. The material balance equation can then be evaluated as follows.

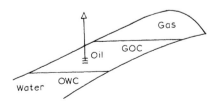

Fig. 3.1. Reservoir considered in the derivation of the fully expanded material balance equation.

(a) Left-hand side (underground withdrawal — rb)

Cumulative oil production. The reservoir depicted in Fig. 3.1 is particularly complicated in that it contains a gas cap. Consequently, the initial pressure in the oil column must also be equal to the bubble point pressure, $p_i = p_b$, since the oil is gas saturated; therefore, even a slight drop in the pressure will cause the liberation of solution gas in the reservoir. The cumulative volumes of produced fluids measured at the surface are:

N_p = cumulative oil (stb)
W_p = cumulative water (stb)
G_p = cumulative gas (scf)

and:

$$R_p = \frac{\text{cumulative gas (scf)}}{\text{cumulative oil (stb)}} = \frac{G_p}{N_p} \quad \text{(scf/stb)}$$

which is the cumulative or average GOR since the start of production.

Returning the produced fluids to the reservoir at the reduced pressure, p, at which the FVF's and solution GOR are: B_o(rb/stb), B_w (rb/stb), B_g (rb/scf) and R_s (scf/stb) gives:

oil + dissolved gas = $N_p B_o$ (rb)
water   = $W_p B_w$ (rb)
free gas   = $N_p(R_p - R_s)B_g$ (rb)

The first of these contains $N_p R_s$ scf of dissolved gas, consequently the free gas to be returned is $N_p R_p - N_p R_s$ (scf). To complicate matters, some of the free gas will be that evolved from the oil column in producing below the bubble point while there may also be a component of gas cap gas which is eventually produced; the difficulty is distinguishing between the two. The cumulative underground withdrawal is therefore:

$$N_p[B_o + (R_p - R_s)B_g] + W_p B_w \quad \text{(rb)} \tag{3.2}$$

(b) Right-hand side (expansion plus water influx)

In allowing the total system to expand as the pressure drops from p_i to p, consideration must be given to everything that can change volume during the depletion: the oil and it's originally dissolved gas, the gas cap, the connate water and the reservoir pore volume; added to which is the cumulative water influx. The separate components may be accounted for as follows.

Oil plus originally dissolved gas. The original HCPV of the oil column is $N B_{oi}$ (rb). Since the reservoir under study, Fig. 3.1, is initially at the bubble point, reducing the pressure will result in the release of gas and the shrinkage of oil such that its change in volume is:

$$N(B_o - B_{oi}) \quad \text{(rb)} \tag{3.3}$$

The initial volume of gas in the oil column is NR_{si} (scf) and the amount still dissolved at the reduced pressure, p, is NR_s (scf). Therefore, the amount liberated is:

$$N(R_{si} - R_s)B_g \quad \text{(rb)} \tag{3.4}$$

and the total change in volume (expansion) of the oil column is:

$$N[(B_o - B_{oi}) + (R_{si} - R_s)B_g] \quad \text{(rb)} \tag{3.5}$$

Gas cap. The original HCPV of the gas cap is mNB_{oi} (rb) which expressed as a surface volume is:

$$mN\frac{B_{oi}}{B_{gi}} \quad \text{(scf)}$$

and, at the reduced pressure, p, it occupies a reservoir volume of:

$$mNB_{oi}\frac{B_g}{B_{gi}} \quad \text{(rb)}$$

The gas cap expansion is therefore:

$$mNB_{oi}\left(\frac{B_g}{B_{gi}} - 1\right) \quad \text{(rb)} \tag{3.6}$$

Connate water. Although of low compressibility, the volume of connate water in the reservoir is large and therefore its expansion must be catered for during depletion. If S_{wc} is the average connate water saturation, determined as described in section 3.4, then its expansion can be calculated applying the definition of isothermal compressibility to the water as:

$$dV_w = c_w V_w \Delta p$$

where V_w is the total volume of water:

$$V_w = PVS_{wc} = \frac{\text{HCPV}}{1 - S_{wc}}S_{wc}$$

The total HCPV, as defined above, is $(1 + m)NB_{oi}$ (rb), therefore:

$$dV_w = \frac{(1 + m)NB_{oi}c_w S_{wc}\Delta p}{1 - S_{wc}} \quad \text{(rb)} \tag{3.7}$$

Pore compaction. As fluids are produced and the pressure declines the entire reservoir pore volume is reduced (compaction) and the change in volume, although negative, expels an equal volume of fluid as production and is therefore additive to the expansion terms. Again, applying the definition of isothermal compressibility to the entire pore volume:

$$d(\text{PV}) = c_f(\text{PV})\Delta p = \frac{(1+m)NB_{oi}c_f\Delta p}{1 - S_{wc}} \quad \text{(rb)} \tag{3.8}$$

Water influx. If the reservoir is connected to an active aquifer then, once the pressure drop is communicated either wholly or partially throughout its volume, the water will expand resulting in an influx of W_e (stb) or $W_e B_w$ (rb).

Adding equations 3.5, 3.6, 3.7 and 3.8, together with the influx and equating the sum to the underground withdrawal, equation 3.2, gives an expression for the Fully Expanded Material Balance as:

$$N_p[B_o + (R_p - R_s)B_g] + W_p B_w = N[(B_o - B_{oi}) + (R_{si} - R_s)B_g]$$
$$+ mNB_{oi}\left(\frac{B_g}{B_{gi}} - 1\right) + \frac{(1+m)NB_{oi}(c_w S_w + c_f)\Delta p}{1 - S_{wc}} + W_e B_w \tag{3.9}$$

In this general form there are so many unknowns in the equation (section 3.4) that the attainment of an exact or unique solution is virtually impossible. Fortunately, most reservoirs are simpler than that depicted in Fig. 3.1, leading to more straightforward formulations of the equation. In terms of importance it is as powerful in reservoir engineering as Einstein's $E = mc^2$ in nuclear physics but unfortunately is a little more cumbersome in form.

If there is water or gas injection, or both, into the reservoir, the cumulative surface volumes: W_i (stb) and G_i(scf) would represent equivalent reservoir volumes at pressure p of W_i (rb) and $G_i B_{gI}$ (rb), in which B_{gI} is the FVF of the lean injected gas and it is assumed that the water is injected free of gas. The cumulative injected volumes can be either subtracted from the left-hand side of the material balance, resulting in a net underground withdrawal, or added to the expansion terms on the right, as appropriate in the particular application.

In an attempt to condense the material balance into a more tractable form, the nomenclature of Havlena and Odeh [3,4] will be employed throughout the chapter. This reduces equation 3.9 to the form:

$$F = N(E_o + mE_g + E_{fw}) + W_e B_w \tag{3.10}$$

in which:

$$F = N_p[B_o + (R_p - R_s)B_g] + W_p B_w \quad \text{(rb)} \tag{3.11}$$

$$E_o = (B_o - B_{oi}) + (R_{si} - R_s)B_g \quad \text{(rb/stb)} \tag{3.12}$$

$$E_g = B_{oi}\left(\frac{B_g}{B_{gi}} - 1\right) \quad \text{(rb/stb)} \tag{3.13}$$

$$E_{\text{fw}} = (1 + m) B_{\text{oi}} \frac{(c_w S_{\text{wc}} + c_f) \Delta p}{1 - S_{\text{wc}}} \quad \text{(rb/stb)} \tag{3.14}$$

The final three terms, having units of rb/stb, represent underground expansions per unit volume of stock tank oil.

3.3. NECESSARY CONDITIONS FOR APPLICATION OF MATERIAL BALANCE

There are no "sufficient" conditions for the meaningful application of material balance to a reservoir, a statement that applies across the broad spectrum of reservoir engineering activity, but there are two "necessary" conditions that must be satisfied. In the first place, there should be adequate data collection (production/pressure/PVT), both in quantity and quality, otherwise the attempted application of the technique can become quite meaningless. The second condition is that it must be possible to define an average pressure decline trend for the system under study. In a great many reservoirs (or parts of reservoirs) the simple condition prevails that they display "tank-like" behaviour, that is, the pressures when referred to a common datum plane exhibit uniformity in decline. The speed with which pressure disturbances are propagated throughout a reservoir so that equilibrium is attained is dependent on the magnitude of the hydraulic diffusivity constant, $k/\phi\mu c$: the larger its value, the more rapid the equilibration of pressure (Chapter 4, section 4.5). But rather than relying on the numerical value of this parametric group, the obvious way of checking on the degree of pressure communication in an accumulation is by plotting the individual well pressures as a function of time, as shown in Fig. 3.2. The pressures, p, are the average values within the drainage area of each well referred to a selected datum plane in the reservoir. The means of analysing routine well tests to determine such pressures is described in Chapter 4, section 4.19. If a uniform pressure decline can be defined for the system under study then in evaluating the material balance, equation 3.9, the pressure dependent PVT parameters and Δp are evaluated as a function of time using this decline trend.

It is commonly believed that rapid pressure equilibration is a prerequisite for successful application of material balance but this is not the case; the necessary condition is that an average pressure decline trend can be defined, which is possible

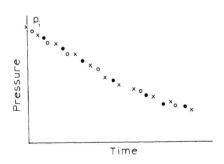

Fig. 3.2. Individual well pressure declines displaying equilibrium in the reservoir.

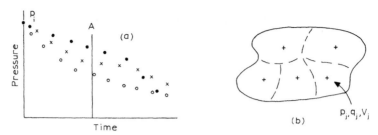

Fig. 3.3. (a) Non-equilibrium pressure decline in a reservoir. (b) Well positions and drainage boundaries

even if there are large pressure differentials across the accumulation under normal producing conditions. All that is necessary is to devise some means of averaging individual well pressure declines to determine a uniform trend for the reservoir as a whole. Consider the reservoir depicted in Fig. 3.3b, if its permeability is low so that there are significant pressure differentials across it under normal producing conditions then each well will have its own distinct pressure decline, as shown in Fig. 3.3a.

The average pressure decline can be determined by the volume weighting of pressures within the drainage area of each well. If p_j and V_j represent the pressure and volume drained by the jth well, then:

$$\overline{p} = \sum_j p_j V_j / \sum_j V_j \tag{3.15}$$

In this expression, however, the individual volumes are intractable to unique determination and their estimation tends to be subjective. An alternative means of evaluating equation 3.15 is by considering the time derivative of the isothermal compressibility (equation 1.5) which, expressed in terms of a reservoir rate is:

$$\frac{dV_j}{dt} = q_j = cV_j \frac{dp_j}{dt} = cV_j p'_j \tag{3.16}$$

which indicates that for a reasonably constant compressibility at the time of measurement:

$$V_j \propto q_j / p'_j \tag{3.17}$$

Therefore, equation 3.15 can be more conveniently expressed as:

$$\overline{p} = \frac{\displaystyle\sum_j p_j q_j / p'_j}{\displaystyle\sum_j q_j / p'_j} \tag{3.18}$$

Material balance is usually applied at regular intervals of, say, every six months throughout the lifetime of the field. If, during such a period, the change in underground withdrawal (UW) of the jth well is ΔUW_j and of pressure Δp_j, then equation 3.18 can be expressed as:

$$\overline{p} = \frac{\sum\limits_{j} p_j \Delta UW_j / \Delta p_j}{\sum\limits_{j} UW_j / \Delta p_j} \tag{3.19}$$

and, whereas the direct evaluation of the volumes in equation 3.15 obviously lacks uniqueness, the underground withdrawal and pressure changes in equation 3.19 are direct observations. Numerical simulation models evaluate equation 3.15 at the end of each time step, in which the V_j are the well defined individual grid block volumes but equation 3.19 usually replicates simulation model average pressure trends very accurately. During the most important period for applying material balance, which is early in the producing lifetime of a reservoir, before the bubble point pressure has been reached and prior to any water production, the full change in underground withdrawal (left-hand side of equation 3.9) is reduced to $\Delta N_p B_o$ and quite often use of ΔN_p alone is sufficient to determine an average pressure decline, provided pressure differentials across the reservoir are not too large.

Application of equation 3.16 implicitly assumes that volumetric depletion conditions prevail. If there is a weak or moderate waterdrive, however, equation 3.19 will still provide a reasonably accurate average pressure decline, especially towards the start of the development, when it is important. Alternatively, if the water influx is strong, it implies that the reservoir has high permeability and will most likely display tank-like behaviour (Fig. 3.2) removing the need for applying the averaging technique.

In this manner, the pressure declines of individual wells can be averaged at any time (such as point at A in Fig. 3.3a) to give the decline for the reservoir as a whole, irrespective of the lack of equilibrium throughout the system. Consequently, most reservoirs may be regarded as suitable candidates for the application of material balance. The method is similar although more rigorous than that presented by Matthews et al. [5], which relies on a uniform decline rate of pressures (semi-steady state condition), as described in Chapter 4, section 4.6.

The reader may wonder at the necessity of going to such lengths in determining the average pressure on a well-by-well basis: surely it would be preferable to apply numerical simulation modelling from the outset which treats the reservoir as cellular and would automatically cater for non-equilibrium conditions in the system. The (tedious) requirement still exists, however, since, as described in section 3.5, material balance and numerical simulation modelling serve quite different purposes when applied to the history matching of reservoir production /pressure records. There are often compelling reasons for applying material balance as an investigative tool in advance of structuring a more complex simulation model.

3.4. SOLVING THE MATERIAL BALANCE (KNOWNS AND UNKNOWNS)

Considering the fully expanded material balance equation, 3.9, the parameters involved can be divided into the following categories:

Should be "known"	Potential unknowns
N_p	N
R_p	W_e
W_p	p
c_w	B_o, R_s, B_g
S_{wc}	m
B_w	c_f
Total 6	Total 8

The score of 6 knowns and 8 potential unknowns reveals the perennial difficulty not just with material balance but throughout the subject of reservoir engineering as a whole: there are far too few equations for the number of unknowns for which they must be solved. Numerical simulation does not alleviate the condition, it merely adds to the potential unknowns appearing in the sparse equations: geometry, ϕ, k, k_r, etc. But rather than writing off reservoir engineering as a serious subject, the uncertainty simply adds to its allure, throwing emphasis on observation and judgement rather than numeracy for there can be rarely a unique mathematical solution to problems. Considering further the "knowns" and "unknowns."

Knowns: For strictly commercial reasons, N_p (cumulative oil production) is the best "known" of all but, frustratingly, in many older fields or fields that are in remote locations where produced gas cannot be utilised, R_p (cumulative or average GOR = G_p/N_p) and W_p (cumulative water production) are not necessarily measured. This moves them into the unknown category and exacerbates the difficulty in making sense of the material balance, if not making it an impossibility. It is assumed that the petrophysical evaluation is "always correct", in which case the average connate water saturation for the reservoir is evaluated as:

$$S_{wc} = \sum_{wells} \left(\sum h_i \, \phi_i \, S_{wci} / \sum h_i \, \phi_i \right) \tag{3.20}$$

which is averaged over all the wells and the subscript "i" relates to the different layers selected within each.

Unknowns: From early in the appraisal stage of a field there is always a volumetric estimate of the STOIIP, N, available but as soon as there is production/pressure history for a reservoir this figure must be disregarded and an attempt made to establish it through solution of the material balance, as demonstrated in exercises 3.1 and 2. The value so determined is the "effective" or "active" STOIIP: that oil contributing to the production/pressure history, which is usually somewhat different from the volumetric estimate (usually being smaller) due to oil being trapped in undrained fault compartments or low-permeability regions of the reservoir.

Perhaps the greatest unknown of all is whether there has been any cumulative water influx, W_e, or not. If the waterdrive is from the base of the reservoir its advance can be monitored directly by time-lapsed logging in selected wells but if the influx is largely from the edge of the accumulation it may not be observed at all until individual wells begin to water-out. The gas cap size, defined by m, may fall in the semi-known category because, unless it areally covers a substantial part of the accumulation, there is a tendency to avoid it in drilling development wells.

The average pressure decline trend is placed in the unknown category because interpreting tests to determine the average pressure within the drainage area of a well at the time of the survey is something of a "black art" (Chapter 4, section 4.19) and is likely to be subjective in nature. Therefore pressures in individual wells may need refining before including in the overall pressure decline trend and moved into the "known" category of parameters. The same applies to the PVT functions which invariably require correction for field operating conditions before being regarded as known data (Chapter 2, sections 2.2d).

Concerning the pore compressibility, c_f, there is a tendency to consider this as being small and constant but in a significant number of cases it has turned out to be both large and variable- and surprising! If it is large, then the resulting compaction drive can supply a significant contribution to the overall hydrocarbon recovery. The unfortunate aspect of the mechanism is that the compaction of the reservoir inevitably leads to subsidence at the surface. If the field is located in the middle of a desert, this is of little consequence but if there happens to be a free water level at the surface (offshore and shoreline locations) then the effects of subsidence can be very expensive. The subject is described in a more detailed manner in section 3.10.

Because of the many unknowns, there is no such thing as a conventional solution of the material balance equation. Often it is solved for the fractional oil recovery at any stage of depletion, N_p/N, which at the abandonment condition becomes the recovery factor or recovery efficiency. But, as demonstrated in later sections of the chapter there are many other solutions that can be sought: for the STOIIP, natural water influx, gas cap size, pore compressibility and even the pressure. As a result of this diversity in solution, it is very difficult to structure the all purpose computer program for solving the material balance. It is preferable, if felt necessary, to write special spread sheet programs for each separate application. The calculations always prove trivial, however, in comparison to the effort that must be expended in examining, validating and collating the input data for calculations and physically justifying the application of material balance.

3.5. COMPARISON BETWEEN MATERIAL BALANCE AND NUMERICAL SIMULATION MODELLING

It is a quite widely held belief that since the advent of numerical simulation modelling in the 1960's, material balance can now be discarded as an approximate technique of historical interest only. Nothing could be further from the truth, however, and anyone who embarks on a numerical simulation study without first

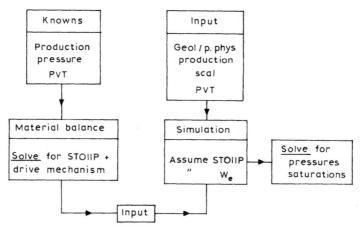

Fig. 3.4. Contrast between material balance and numerical simulation in history matching field performance.

applying material balance to the observed production/pressure history of a field is running the risk of error.

Consider an undersaturated oil reservoir in which the principle unknowns are the STOIIP, N, and whether there has been a finite water influx, W_e, or not. The difference between material balance and simulation in matching field performance for such a reservoir is illustrated in Fig. 3.4.

Provided the PVT and pressure decline trend have been rationalised then they are input to material balance calculations along with the production data as known quantities, nothing else is required. Then by "adept manipulation" — as described in later sections of the chapter — an attempt is made to solve the material balance equation to determine the two principle unknowns, N and W_e, the latter representing the drive mechanism. The main advantage in this approach is that provided an average pressure decline can be defined for the system, as described in section 3.3, then no geometrical model is required. The accumulation is seen simply as an underground "blob" of hydrocarbons. In this respect the model is zero dimensional with the average pressure determined at a point (datum plane) which is representative of the reservoir as a whole. It is this unique property that permits the attempted solution of the equation to determine the hydrocarbons in place and define the drive mechanism.

The alternative approach of employing numerical simulation modelling to match production performance is also illustrated in Fig. 3.4. Now the input must contain a full geological and petrophysical description of the reservoir necessary to give physical structure to the model. But if these are defined then so too is the STOIIP which is therefore input as a known quantity. Similarly a part of the physical description is the aquifer model, one is either (mathematically) attached to the reservoir or it is not, at the discretion of the engineer. Consequently, the drive mechanism is also being fed in as an input assumption or partial assumption (i.e. if the aquifer is defined through history matching with the simulator then its properties are dependent on the assumed STOIIP). Therefore, since the required answer, the STOIIP, must necessarily be specified from the outset, then numerical simulation applied to the history matching of field performance can hardly be regarded as an investigative technique: the results achieved being simply a reflection of the engineer's input assumptions. The remainder of the simulation input consists of the production and PVT data together with SCAL results most notably the relative permeability functions required to shift fluids from one grid block to the next (Chapter 5, sections 5.4). Strangely enough the pressures are not input to the model as known quantities, instead individual grid block pressures and fluid saturations comprise the "solution" achieved at the end of each time step by simultaneously solving large numbers of second order differential equations expressed in finite difference form. In this respect the simulator tackles the problem backwards: "unknowns" such as the STOIIP are fed in to determine the "knowns" such as the pressure.

Using the material balance to attempt to make an estimate of the STOIIP after the start of continuous field production used to be a part of the culture of reservoir engineering that has been rejected in recent times, at least when studying oil fields. For gas fields, however, the tradition still survives, for the reasons explained in Chapter 6, section 6.3a, although the common method used in gas field history matching leaves a lot to be desired. The important fact that material balance needs no shape is synonymous to saying "who needs geologists"! In fact, things are not quite as bad as that and in the past this used to be a valuable point of contact between engineers an geologists. If the material balance STOIIP turned out to be, say, 10% lower than the volumetric estimate they would get together to try and figure out why this disparity existed: where was the missing oil possibly trapped in the reservoir and was there any practical means of recovering it?

Nowadays, however, reservoir engineers have abdicated and seem to have ac- knowledged that after fifty or more years of near open warfare between the "underground" disciplines, the geologists had it right after all. This is necessarily implicit in the acceptance of their maps as unchecked input for simulation studies. Material balance provides a means of checking the validity of the mapping, as illustrated in Exercises 3.1 and 2. The obvious approach to history matching the per- formance of a field should be first of all to check carefully the production, pressure and PVT data and use them in the material balance to define the STOIIP and drive mechanism. Because these are the sole input requirements, this may be regarded as the "minimum assumption" route through the whole subject of understanding reservoirs. Then, following consultation with the geologists, a realistic simulation model can be structured (Fig. 3.4), based on the findings. If the application of ma- terial balance leads to ambiguity, then beware, simulation will yield the same.And if material balance doesn't work it is most likely that the reason is because the data collection has been inadequate or careless, under which circumstance no technique will provide sensible answers to reservoir engineering problems.

3.6. THE OPENING MOVE IN APPLYING MATERIAL BALANCE

The first thing to do in applying material balance is to check whether the reservoir is of the volumetric depletion type or is being energised by some drive mechanism such as natural water influx or gas cap expansion. Consider again the type of reservoir described in the previous section which has no gas cap but might be influenced by an adjoining aquifer. In this case, the simplified form of the material balance, equation 3.10, can be expressed as:

$$F = N(E_o + E_{fw}) + W_e B_w \quad \text{(rb)} \tag{3.21}$$

and dividing both sides by $E_o + E_{fw}$, gives:

$$\frac{F}{E_o + E_{fw}} = N + \frac{W_e B_w}{E_o + E_{fw}} \quad \text{(stb)} \tag{3.22}$$

While the right-hand side of the equation contains two unknowns, N and W_e, making it very difficult to deal with, the left-hand side should be readily calculable from the production/pressure/PVT data, provided the field performance has been carefully monitored. Evaluating this for regular time steps of, say, six monthly intervals, a plot is made of $F/(E_o + E_{fw})$ versus cumulative production, N_p, or time or the pressure drop, Δp, simply to display its variance throughout the history matching period:the result being used qualitatively rather than quantitatively. The plot can assume various shapes, as shown in Fig. 3.5. If the points lie on a horizontal straight line (*A*) it implies from equation 3.22 that $W_e = 0$ and the pore compressibility is constant. This defines a purely depletion drive reservoir whose energy derives solely from the expansion of the oil and its originally dissolved gas plus a regular component of compaction drive. Furthermore, the ordinate value of the plateau determines the STOIIP, N. Alternatively, if the points rise (plots *B* or *C*) it indicates that the reservoir has been energised by "something else": water influx, abnormal pore compaction or any subtle combination of these two. Plot *B* might be for a strong waterdrive field in which the aquifer is displaying infinite acting behaviour, whereas *C* represents an aquifer whose outer boundary had been felt and the aquifer is depleting in unison with the reservoir itself, hence the downward

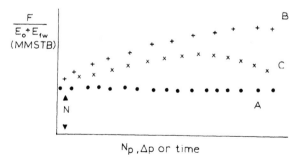

Fig. 3.5. The "opening move" in material balance application

trend in points as time progresses denotes a diminishing degree of energising by the aquifer. In principle the backward extrapolation of either plots B or C to the ordinate should yield the value of N. But there is no reason why this extrapolation should be linear.

Caution should be exercised, however, in not attempting to read too much into the shape of the plot. In waterdrive fields, for instance, the shape is highly rate dependent: if the rate is increased the oil is evacuated before the water can catch-up and the points will dip downwards revealing a lack of energising by the aquifer, whereas, if the rate is decreased the reverse happens and the points are elevated. Because of this it is helpful in interpretation to plot the rate of underground withdrawal of reservoir fluids beneath Fig. 3.5. Ideally, to facilitate the backward extrapolation to estimate the STOIIP two things are necessary:

– The field should be produced at a steady rate, especially at the start of its producing life.

– There should be a high frequency of pressure surveys in wells early in the development.

3.7. VOLUMETRIC DEPLETION FIELDS

In considering the performance of oilfields characterised by different forms of predominant drive mechanism, the first to be studied will be the strictly depletion type field in which there is negligible water influx and the reservoir contains no gas cap. Above the bubble point, the drive energy is due to the expansion of the undersaturated, single phase oil, the connate water expansion and the pore compaction, while below, the complex solution gas drive process is activated once gas has been liberated from the oil.

(a) Depletion above the bubble point

For such a reservoir, $m = 0$, $W_e = 0$, and since all the produced gas is dissolved in the oil in the reservoir, then $R_s = R_{si} = R_p$ and inserting these in the material balance, equation 3.9, reduces it to the form:

$$N_p B_o + W_p B_w = N B_{oi} \left(\frac{B_o - B_{oi}}{B_{oi}} + \frac{c_w S_{wc} + c_f}{1 - S_{wc}} \right) \Delta p \qquad (3.23)$$

But, from the definition of isothermal compressibility (equation 1.4), it will be recognised that:

$$c_o = \frac{B_o - B_{oi}}{B_{oi} \Delta p} \qquad (3.24)$$

and since, for the two phase system, $S_o = 1 - S_{wc}$, then equation 3.23 can be more concisely expressed as:

$$N_p B_o + W_p B_w = N B_{oi} c_{eff} \Delta p \qquad (3.25)$$

in which:

$$c_{\text{eff}} = \frac{c_o S_o + c_w S_{wc} + c_f}{1 - S_{wc}} \qquad (3.26)$$

which is the effective compressibility of the undersaturated system, catering for the expansion of the oil and water and reduction in the pore volume. As mentioned in Chapter 2, section 2.2a, the three compressibilities in the above equation are small and using the values quoted ($c_o = 10$, $c_w = 3$, $c_f = 5 \times 10^{-6}$/psi) for $S_{wc} = 0.20$ PV gives an effective compressibility of $c_{\text{eff}} = 17 \times 10^{-6}$/psi. If, $W_p = 0$, $B_{oi} = 1.430$ rb/stb and the pressure drop from initial to bubble point pressure is $\Delta p = 1500$ psi at which $B_{ob} = 1.452$ rb/stb, then the fractional oil recovery by undersaturated depletion is $N_p/N = 0.025$. That is, depletion of 1500 psi only recovers 2.5% of the STOIIP in this intrinsically low-energy system.

Exercise 3.1: Material balance applied to an undersaturated volatile oilfield

Introduction

This exercise illustrates the simplest possible application of material balance to a relatively small but complex reservoir being produced by depletion above the bubble point pressure. It contains a volatile oil and has a severe degree of heterogeneity, although the average permeability is low being less than 10 mD. Material balance is used to determine the effective STOIIP contributing to the production/pressure history.

Question

The reservoir to be studied can be defined as follows:

STOIIP = 95.0 MMstb, volumetric estimate
p_i = 4250 psia, at a datum depth of 9400 ft.ss.
p_b = 2900 psia, bubble point pressure at datum
B_{oi} = 2.654 rb/stb
B_w = 1.020 rb/stb
c_w = 3.0×10^{-6}/psi
c_f = 5.0×10^{-6}/psi
S_{wc} = 0.30 PV, average connate water saturation.

The variation in oil FVF with pressure is plotted in Fig. 3.6. As can be seen, on account of the high oil volatility, the slope and therefore the compressibility is high, even in this undersaturated pressure range. It is interesting to note that in calculating the volumetric STOIIP, no petrophysical cut-offs on porosity, permeability or water saturation were imposed, as is usually the case. Thus, all the oil saturated rock is included in the total net rock volume. Under normal producing conditions, the low permeability meant that there was a poor degree of pressure equilibrium in the reservoir, each well having its own distinct pressure decline. On one occasion, however, this offshore field had to be closed-in for the better part of two months for repair and extension to the sub-sea oil gathering system in the area. The production

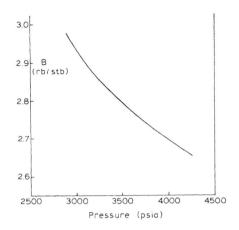

Fig. 3.6. FVF for the volatile oilfield (Exercise 3.1).

TABLE 3.1

Pressure and cumulative production statistics

Well No.	Datum pressure (psia)	Δp (psi)	Cumulative oil (stb)	Produced water (stb)
1	3511	739	239830	2730
2			15140	150
3	3546	704	242330	2780
4	3522	728	140900	1740
5			17170	1630
6	3507	743	204310	2100
7	3520	730	277730	2500
Totals (stb)			1137410	13630

department quite wisely took advantage of the situation by running pressure gauges in five of the seven producing wells in the field to record extended pressure buildup surveys, the results of which are listed in Table 3.1

Using the production-pressure data:
• Calculate the effective STOIIP contributing to oil recovery from the reservoir. No account need be taken of any natural water influx since this is a lens type accumulation completely devoid of any aquifer support.

Solution
   The pressure data listed in Table 3.1 demonstrate that during the period of field closure, a reasonable degree of pressure equilibrium was attained in the reservoir, the range of individual well pressure declines being between 704 and 743 psi. Whether the representative reservoir pressure is evaluated as a straightforward average or using the weighting procedures described in section 3.3 makes no

difference to the result, the average reservoir pressure at the end of the buildup being 3521 psia, giving a pressure drop of $\Delta p = 729$ psi. The material balance for this simple system depleting at undersaturated conditions is:

$$N_p B_o + W_p B_w = N B_{oi} c_{eff} \Delta p$$

At the average pressure of 3521 psia, the FVF can be read from Fig. 3.6 as $B_o = 2.790$ rb/stb and the oil compressibility over the pressure range of interest can be calculated as:

$$c_o = \frac{B_o - B_{oi}}{B_{oi} \Delta p} = \frac{2.790 - 2.654}{2.654 \times 729} = 70.29 \times 10^{-6}/\text{psi}$$

the high value resulting from volatility of the oil. The effective compressibility can next be calculated (equation 3.26) as:

$$c_{eff} = \frac{(70.29 \times 0.7 + 3 \times 0.3 + 5)}{0.7} \times 10^{-6} = 78.72 \times 10^{-6}/\text{psi}$$

The material balance is then solved to determine the initial HCPV as:

$$[N B_{oi}]_{MB} = \frac{N_p B_o + W_p B_w}{c_{eff} \Delta p} = \frac{1137410 \times 2.79 + 13630 \times 1.02}{78.72 \times 10^{-6} \times 729} = 55.54 \text{ MMrb}$$

whereas the corresponding volume obtained from the volumetric estimate is:

$$[N B_{oi}]_{VOL} = 95 \times 2.654 = 252.13 \text{ MMrb}$$

The ratio of the material balance to volumetric estimate is therefore:

$$\frac{[N B_{oi}]_{MB}}{[N B_{oi}]_{VOL}} = \frac{55.54}{252.13} = 0.22$$

which reveals that only 22% of the total mapped oil in place was contributing to the observed production-pressure history at the time of closure for the field wide pressure survey. In fact, the closure provided the sole opportunity in the history of the field to apply material balance directly for a unique pressure drop in the reservoir, although the alternative but more laborious method of averaging pressures described in section 3.3 could have been applied at any time during depletion.

The reason for the severe disparity in oil in place figures arises because of the severe degree of vertical heterogeneity in the reservoir as revealed from a core permeability distribution across the sand section in one of the centrally located development wells, which is plotted in Fig. 3.7. As can be seen, although the average permeability is low, there is a relatively high-permeability "tunnel" ($k_{max} \sim 50$ mD) at the centre of the reservoir. This same feature is correlatable in all wells: at some locations higher in the reservoir, at others lower but always present. Not surprisingly, volumetric calculations indicated that the high-permeability interval contains about 20% of the oil in place.

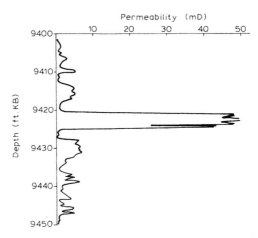

Fig. 3.7. Core measured permeability distribution of a typical well (Exercise 3.1).

Conclusion

This field example illustrates in a rather extreme manner the difference that can exist between the total volumetric oil in place and that referred to by Schilthuis [1] as the "active" volume contributing to production. The type of formation revealed by the permeability distribution (Fig. 3.7) is a typical example of what is referred to as a "dual porosity " system. This is defined such that fluid flow into the wellbore can only occur through the high-permeability channels and not from the tighter rock, from which production is only obtained by cross-flow into the high-permeability conduits through which the oil is produced. (The behaviour of naturally fractured reservoirs is the same. Flow into the wellbore occurs exclusively through the fractures into which the tight matrix rock produces). The mechanics of dual porosity behavoir can best be appreciated by consideration of Darcy's law:

$$q = -\frac{kA}{\mu}\frac{\partial p}{\partial l} \tag{1.2}$$

Liquid production from the tight sands directly into the well is limited by both the low permeability and area for flow which is proportional to r_w^2 and is therefore very small. Alternatively, for cross flow in the reservoir from the tight sands into the high-permeability production channels, although the vertical permeability is usually lower than the horizontal, this is more than compensated by the very large area for flow which is that of the entire reservoir itself. The original development plan for the field envisaged oil recovery by water injection but obviously such a scheme requires careful consideration. The injected water will maintain the reservoir pressure thus inhibiting the potential for oil recovery by cross flow. This is the type of reservoir in which straightforward depletion would appear to offer advantages since a high pressure differential would be encouraged between the two types of sand which, again referring to Darcy's equation, would promote the cross flow of fluids. There is a problem associated with this approach, however, in that if the pressure is allowed to fall below the bubble point at 2900 psia, the degree of shrinkage of the high-volatility oil will be so severe that the gas production will become excessive, as described in Chapter 2, section 2.2. In the event, the pressure was allowed to fall to almost 2000 psia before a low-pressure waterflood was initiated to clear out the oil remaining in the high-permeability layer.

The exercise illustrates an extreme case of the disparity between the volumetric and realistic value of hydrocarbons in place that is exaggerated by the fact that petrophysical "cut-offs" were not applied in the first place which would have led to the exclusion of some of the oil in the tighter rock. It can be argued, however, that in an ideal world, there should be no such thing as cut-offs applied to reservoirs. Instead, the total oil in place should be calculated including all oil saturations and distinction between the oil that can and cannot flow is then determined as the result of reservoir engineering calculations and reflected in the recovery factor. Application of petrophysical cut-offs is a somewhat arbitrary exercise and is highly subjective in that the practitioner in his office makes decisions about fluid movement in reservoirs without the slightest resort to any quantitative technique other than the inevitable regression analysis. The appropriateness of elimination of net rock volume on account of low porosity or oil saturation is exclusively in the petrophysical domain but its exclusion because of low permeability is not. To announce, for instance, that fluid will not be recovered from rock with a permeability of less than 0.5 mD is not a meaningful physical statement in itself. Permeability is defined through Darcy's equation 1.2, and is not an isolated variable such as porosity or saturation. Consequently, it is not valid to exclude rock for low permeability without taking account of the area, fluid viscosity and pressure gradient. Imposing a pressure differential of over 1000 psi across a low-permeability interval, as happened in the reservoir described above, may well cause a significant amount of fluid flow through it.

One further point about the application of cut-offs is the way they are reported. Invariably, what is quoted is simply the gross section thickness and the net, once the excluded rock has been removed from the section. This is quite inadequate for the reservoir engineer who also must know precisely where in the section the non-reservoir rock has been removed. This is particularly the case in evaluating reservoir performance in which one fluid displaces another of different density, as in water or gas drive. Unless the location of the excluded intervals is detailed in the reporting of the results of cut-off exercises: top, bottom or middle of the reservoir, errors can be made in correctly incorporating the influence of gravity in displacement efficiency calculations, as described in Chapters 5 and 6 for water and gas drive respectively. Having made that comment, it is appreciated that this is not an ideal world and therefore petrophysical cut-offs will probably survive and thrive.

Exercise 3.2: Identification of the drive mechanism and calculation of the STOIIP for a depletion type reservoir

Introduction

In this example, a field with six years of production/pressure history is subjected to material balance analysis in an attempt to define the STOIIP and establish the main drive mechanism. It is a tight, naturally fractured chalk reservoir with a much greater intensity of fractures towards the crest of the accumulation. The field was developed without pressure support by water injection because it was feared that the water would react in an unfavourable manner with the chalk. Furthermore, there is always the risk that the injected water will move almost exclusively through the fracture system leading to its premature breakthrough in the producing wells. The accompanying pressure maintenance can also inhibit the flow of oil from the tight matrix blocks into the fractures through which it is produced in such a dual porosity system. To compensate, there is always the possibility of capillary imbibition of water into the matrix to displace the oil but the extent to which this will occur is very difficult to predict. Under these circumstances, it is preferable to examine the field's production performance under depletion conditions prior to finalizing any plans for secondary recovery flooding, as in the present case. The exercise applies material balance both to the undersaturated depletion and also to the performance below the bubble point pressure, using the "opening move" technique described in section 3.6.

Question

The following data are available for the reservoir under study:

STOIIP = 650 MMstb, latest volumetric estimate
p_i = 7150 psia, initial pressure at datum depth
p_b = 4500 psia, bubble point pressure
B_{oi} = 1.743 rb/stb, initial oil FVF
B_{ob} = 1.850 rb/stb, FVF at the bubble point
c_w = 3.0×10^{-6}/psi
c_f = 3.3×10^{-6}/psi
S_{wc} = 0.43 PV, high average connate water saturation in the tight chalk reservoir
N_{pb} = 43.473 MMstb, oil recovery at the bubble point pressure
W_p = 0, negligible water production
m = 0, no initial gas cap.

The field's production history data are listed in Table 3.2 and the production rate and GOR development plotted in Fig. 3.8. Above the bubble point pressure, which was reached after 21 months of production, the oil rate peaked at almost 80,000 stb/d thereafter a significant reduction occurred, which is typical in solution gas drive fields, resulting from the removal of high-compressibility gas from the reservoir together with a reduction in the relative permeability to oil flow into the producing wells which destroys their PI's. At the end of the six year period under review, the oil rate had fallen below 14,000 stb/d.

TABLE 3.2

Production data (Exercise 3.2)

Time (months)	Oil rate (stb/d)	Gas rate (MMscf/d)	Cumulative oil (MMstb)	Cumulative gas (MMMscf)	R (scf/stb)	R_p (scf/stb)
6	44230	64.110	8.072	11.7	1449	1450
12	79326	115.616	22.549	32.8	1457	1455
18	75726	110.192	36.369	52.91	1455	1455
24	70208	134.685	49.182	77.49	1918	1576
30	50416	147.414	58.383	104.393	2924	1788
36	35227	135.282	64.812	129.082	3840	1992
42	26027	115.277	69.562	150.120	4429	2158
48	27452	151.167	74.572	177.708	5507	2383
54	20975	141.326	78.400	203.500	6738	2596
60	15753	125.107	81.275	226.332	7942	2785
66	14268	116.970	83.879	247.679	8198	2953
72	13819	111.792	86.401	268.081	8090	3103

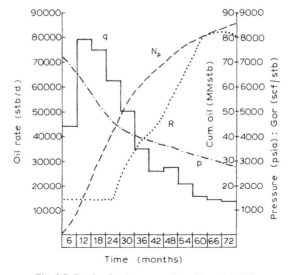

Fig. 3.8. Production/pressure data (Exercise 3.2).

The pressure decline history of individual wells is plotted in Fig. 3.9, in which the interpreted pressures have all been referred to a common datum depth. Considering the generally low permeability of this tight chalk reservoir ($k \sim 5$ mD), it is surprising that there is such uniformity in pressures across the fairly large structure. The reason is because of the rapid communication through the extensive and well connected fracture system and then locally into the tighter matrix blocks. Above the bubble point, the pressure decline is severe in the low-compressibility system but below, even though liberated gas is being removed from the reservoir, a significant volume is being retained and its high compressibility tends to alleviate the rate

TABLE 3.3

Material balance estimate of STOIIP and the drive mechanism (Exercise 3.2)

Months	Pressure (psia)	B_o (rb/stb)	R_s (scf/stb)	B_g (rb/scf)	N_p (MMstb)	R_p (scf/stb)	F (MMrb)	E_o (rb/stb)	E_{fw} (rb/stb)	$F/E_o + E_{fw}$ (MMstb)
0	7150	1.743	1450			1450				
6	6600	1.76	1450		8.072	1450	14.20672	0.017	0.00772	574.7102
12	5800	1.796	1450		22.549	1455	40.498	0.053	0.018949	562.8741
18	4950	1.83	1450		36.369	1455	66.55527	0.087	0.030879	564.6057
21	4500	1.85	1450		43.473	1447	80.42505	0.107	0.037195	557.752
24	4350	1.775	1323	0.000797	49.182	1576	97.21516	0.133219	0.039301	563.5015
30	4060	1.67	1143	0.00084	58.383	1788	129.1315	0.18488	0.043371	565.7429
36	3840	1.611	1037	0.000881	64.812	1992	158.942	0.231853	0.046459	571.0927
42	3660	1.566	958	0.000916	69.562	2158	185.3966	0.273672	0.048986	574.5924
48	3480	1.523	882	0.000959	74.572	2383	220.9165	0.324712	0.051512	587.1939
54	3260	1.474	791	0.001015	78.4	2596	259.1963	0.399885	0.0546	570.3076
60	3100	1.44	734	0.001065	81.275	2785	294.5662	0.45954	0.056846	570.4382
66	2940	1.409	682	0.001121	83.879	2953	331.7239	0.526928	0.059092	566.0629
72	2800	1.382	637	0.00117	86.401	3103	368.6921	0.59021	0.061057	566.1154
Average										568.8453

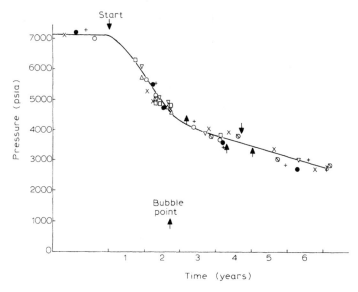

Fig. 3.9. Pressure decline history of individual wells (Exercise 3.2).

of pressure decline. The greater scatter in the individual well pressures below the bubble point compared to above results mainly from the added complexity in interpreting well tests to determine average well pressures in a two phase (oil/gas) system [2]. A further indicator of the degree of pressure uniformity in the reservoir is that all the wells passed through the bubble point pressure simultaneously: within a few weeks of each other. This was manifest by the increase in individual well GOR's (not detailed) above the undersaturated value of 1450 scf/stb.

The average pressure decline is plotted as the solid line in Fig. 3.9 and pressures have been read from this at six monthly intervals and are listed in Table 3.3, together with the PVT fluid properties (B_o, R_s and B_g) evaluated at the same pressures. These data are corrected for the operating separator conditions.

For the field data provided:
- Estimate the maximum STOIIP by application of material balance above the bubble point pressure.
- Apply material balance throughout the entire six year production period to calculate the STOIIP and establish the drive mechanism.
- Determine the fraction of the liberated solution gas that has remained in the reservoir at the end of the six years of production.

Solution

Considering the high degree of pressure equilibrium observed directly (Fig. 3.9) and inferred from the simultaneous production of liberated gas in all wells throughout the field, it is appropriate to apply material balance directly to the entire accumulation using the average decline defined in Fig. 3.9.

Undersaturated performance. In depleting from the initial pressure to the bubble point the material balance is:

$$N_{pb} B_{ob} = N B_{oi} c_{eff} \Delta p + W_e B_w \quad (W_p = 0)$$

If it is initially assumed that there is no water influx ($W_e = 0$), then the STOIIP determined by solving this equation will be the maximum possible value. The oil compressibility can be calculated using equation 3.24 as:

$$c_o = \frac{B_{ob} - B_{oi}}{B_{oi} \Delta p} = \frac{1.850 - 1.743}{1.743 \times 2650} = 23.17 \times 10^{-6}/psi$$

and the effective compressibility applying equation 3.26 as:

$$c_{eff} = \frac{c_o S_o + c_w S_{wc} + c_f}{1 - S_{wc}} = \frac{23.17 \times 0.57 + 3.0 \times 0.43 + 3.3}{0.57} = 31.22 \times 10^{-6}/psi$$

Then for the given values of: $N_{pb} = 43.473$ MMstb, $B_{oi} = 1.743$ rb/stb, $B_{ob} = 1.850$ rb/stb and $\Delta p = 7150 - 4500 = 2650$ psi, the STOIIP can be calculated as $N = 558$ MMstb.

This maximum possible value of the STOIIP is some 14% lower than the volumetric estimate of 650 MMstb. Usually such a result would lead to discussions with the geologists about the likely reasons for the disparity but, in this case, the explanation is more likely to be sought from the petrophysicists. Evidently, not all of the net pay they are attributing to the reservoir is actually contributing to the production/pressure history of the field: presumably some of the oil in the very tight matrix rock is simply not flowing, at least, not under the pressure difference of 2650 psi imposed on the reservoir during depletion to the bubble point pressure.

Total performance. As described in section 3.6, the most useful opening move in studying the production/pressure history of a field is to make a plot of $F/(E_o + E_{fw})$, as a function of time, cumulative production or pressure decline, simply to examine its variance with continued depletion. Values in this ratio are defined by equations 3.11, 12 and 14. It will be appreciated that above the bubble point, the numerator is reduced to $F = N_p B_o$, since there is no free gas in the reservoir.

The data and calculations are listed in Table 3.3. With the exception of the values for 21 months, which is when the reservoir passed through the bubble point, six monthly time steps have been selected. The resulting values of $F/(E_o + E_{fw})$ are listed in the final column and demonstrate a remarkable degree of constancy over the full six years of production history which, as described in section 3.6, is characteristic of a strictly depletion type reservoir. Furthermore, the numerical value of the constant is the STOIIP (Fig. 3.5), the average being $N = 569$ MMstb, with a variance of only 24.3 MMstb (4.3%).

The final part of the exercise is to estimate how much of the liberated solution gas has been retained in the reservoir during the solution gas drive phase. At any stage of depletion below the bubble point, the liberated gas that has been produced is $N_p(R_p - R_s)$ scf, whereas the total gas liberated is $N(R_{si} - R_s)$ scf. Therefore, the fraction of gas remaining in the reservoir is:

$$\alpha = 1 - \frac{N_p(R_p - R_s)}{N(R_{si} - R_s)} \tag{3.27}$$

Values of α have been calculated below the bubble point using the data contained in Table 3.3 for a value of $N = 570$ MMstb and are listed below.

Time (months)	21	24	30	36	42	48	56	60	66	72
α		1.0	0.83	0.78	0.74	0.70	0.65	0.62	0.59	0.56	0.54

These reveal that by the end of the period considered, 54% of the total volume of gas released had, by some mechanism, been retained in the reservoir, which appears to be a most satisfactory result and it is for this reason that there occurs an alleviation of the harsh initial pressure decline below the bubble point pressure. This field is considered further in Exercise 3.3.

Conclusion

Although the exercise employs real field data, it must be appreciated that it represents an idealised example of the performance of a depletion type reservoir. The reasons why material balance works so well in this field are because of the high communication of pressures through the fracture system which make it exhibit "tank like" behaviour and also because of the high standard of data collection. In most cases when application of material balance proves ambiguous, it can usually be traced to poor monitoring of reservoir performance rather than any deficiencies in the technique itself.

The fact that material balance proves so effective in what is a highly complex, dual porosity reservoir is most encouraging. Although sophisticated numerical simulation models have been constructed to describe the intricacies of fractured reservoirs, their practical application often leaves a lot to be desired. One of the main problems is that the input requires specification of the tight matrix block geometry and dimensions, which has a profound effect on the recovery of oil from such systems. Consequently, the outcome of the simulation study is dictated by the input assumptions of the engineer because there is no way yet devised of assessing matrix block geometry throughout a reservoir as a whole. What is comforting about the application of material balance is that all this complexity is implicit in the production/pressure performance, no geometry being required in its solution. Engineers would be well advised to apply material balance to fractured reservoirs before attempting to use any more sophisticated technique, it is surprising how often it proves useful in illuminating the mechanics of oil recovery.

The exercise illustrates the point made in section 3.5, that material balance application should always precede the construction of a more detailed simulation model. In the present case, a few simple calculations determine the STOIIP and the fact that the reservoir has not been energised by any natural water influx. Furthermore, the constancy of values of $F/(E_o + E_{fw})$ indicates that while there has been a degree of reservoir compaction, it has been regular rather than abnormal, meaning that the pore compressibility is a constant rather than variable. Therefore, a simulation model can be meaningfully constructed incorporating these important features: STOIIP = 570 MMstb, $c_f = 3.3 \times 10^{-6}$/psi and no aquifer model.

(b) Depletion below the bubble point (solution gas drive)

Once the pressure falls below the bubble point solution gas is liberated from the oil leading, in many cases, to a chaotic and largely uncontrollable situation in the reservoir which is characteristic of what is referred to as the solution gas drive process. The problem arises on account of the widely different viscosities of the oil and gas (typically: $\mu_o = 1$ cP, $\mu_g = 0.025$ cP). Because of this disparity there is a tendency to strip the more mobile gas from the reservoir at the expense of the oil, unless physical conditions are favourable such that they promote the retention of liberated gas. What is required is high porosity/permeability and a steeply dipping formation that will encourage the segregation of the oil and gas under the influence of gravity. Unfortunately, such conditions are rare and the under more normal circumstances the removal of excessive quantities of the high-compressibility gas robs the reservoir of its highest energy component thus seriously reducing the oil recovery.

Perhaps the most straightforward method of predicting the performance of a solution gas drive reservoir is that presented by Morris Muskat [6], the engineer who during the 1940's and 50's did more than anyone else to evolve reservoir engineering technique in terms of well established physical principles and their supportive mathematics. Consider an initially gas saturated reservoir from which N_p stb of oil have been produced. Then the oil remaining in the reservoir at that stage of depletion is:

$$N_r = N - N_p = \frac{V S_o}{B_o} \quad \text{(stb)}$$

where V is the pore volume (rb). The change in this volume with pressure is:

$$\frac{dN_r}{dp} = V \frac{1}{B_o} \frac{dS_o}{dp} - V \frac{S_o}{B_o^2} \frac{dB_o}{dp} \qquad (3.28)$$

The total volume of dissolved and free gas in the reservoir is:

$$G_r = V \frac{S_o R_s}{B_o} + (1 - S_o - S_{wc}) \frac{V}{B_g} \quad \text{(scf)} \qquad (3.29)$$

and its change in volume with pressure:

$$\frac{dG_r}{dp} = V \left(\frac{S_o}{B_o} \frac{dR_s}{dp} + \frac{R_s}{B_o} \frac{dS_o}{dp} - \frac{R_s S_o}{B_o^2} \frac{dB_o}{dp} - \frac{1}{B_g} \frac{dS_o}{dp} - \frac{1 - S_o - S_{wc}}{B_g^2} \frac{dB_g}{dp} \right) \qquad (3.30)$$

The instantaneous GOR while producing at this stage of depletion can be obtained by the division of equation 3.30 by 3.28 to give:

$$R = \cfrac{\cfrac{S_o}{B_o}\cfrac{dR_s}{dp} + \cfrac{R_s}{B_o}\cfrac{dS_o}{dp} - \cfrac{R_sS_o}{B_o^2}\cfrac{dB_o}{dp} - \cfrac{1}{B_g}\cfrac{dS_o}{dp} - \cfrac{1-S_o-S_{wc}}{B_g^2}\cfrac{dB_g}{dp}}{\cfrac{1}{B_o}\cfrac{dS_o}{dp} - \cfrac{S_o}{B_o^2}\cfrac{dB_o}{dp}} \tag{3.31}$$

An alternative expression for the producing GOR can be obtained by applying Darcy's law for gas/oil flow in the reservoir as:

$$R = \frac{k_{rg}}{B_g}\frac{B_o}{k_{ro}}\frac{\mu_o}{\mu_g} + R_s \quad \text{(scf/stb)} \tag{3.32}$$

in which k_{rg} and k_{ro} are the relative permeabilities to oil and gas, as described below. Equations 3.31 and 32 can be equated and solved to give the oil saturation derivative with respect to pressure as:

$$\frac{dS_o}{dp} = \cfrac{\cfrac{S_oB_g}{B_o}\cfrac{dR_s}{dp} + \cfrac{S_o}{B_o}\cfrac{k_{rg}}{k_{ro}}\cfrac{\mu_o}{\mu_g}\cfrac{dB_o}{dp} - \cfrac{(1-S_o-S_{wc})}{B_g}\cfrac{dB_g}{dp}}{1 + \cfrac{k_{rg}}{k_{ro}}\cfrac{\mu_o}{\mu_g}}$$

which can be expressed in a more simple form as:

$$\Delta S_o = \Delta p \frac{\left(S_o A(p) + S_o B(p) k_{rg}/k_{ro} - (1 - S_o - S_{wc})C(p)\right)}{1 + \cfrac{k_{rg}}{k_{ro}}\cfrac{\mu_o}{\mu_g}} \tag{3.33}$$

in which:

$$A(p) = \frac{B_g}{B_o}\frac{dR_s}{dp} \tag{3.34}$$

$$B(p) = \frac{1}{B_o}\frac{dB_o}{dp}\frac{\mu_o}{\mu_g} \tag{3.35}$$

$$C(p) = \frac{1}{B_g}\frac{dB_g}{dp} \tag{3.36}$$

These are all functions of the pressure and are readily calculable from the PVT relations. It should be noticed that while $A(p)$ and $B(p)$ are intrinsically positive, the shape of the B_g function is such that $C(p)$ is negative (Fig. 2.3). At any stage of depletion, the oil saturation and recovery in an initially saturated reservoir are related as follows:

$$S_o = \frac{\text{oil remaining}}{\text{one PV}} = \frac{(N-N_p)B_o}{NB_{oi}}(1-S_{wc}) \tag{3.37}$$

giving:

$$\frac{N_p}{N} = 1 - \frac{S_o}{1-S_{wc}}\frac{B_{oi}}{B_o} \tag{3.38}$$

The derivation of equation 3.33 neglects the expansion of the connate water and compaction: these being regarded as negligible once high-compressibility gas has been liberated in the reservoir. If the reservoir is undersaturated at initial conditions, however, then these terms must be catered for in depletion above the bubble point. Then the total recovery can be evaluated as:

$$\frac{N_p}{N} = \frac{N_{pb} + N'_p}{N} = \frac{N_{pb}}{N} + \frac{N'_p}{N_b}\frac{N_b}{N} = \frac{N_{pb}}{N} + \frac{N'_p}{N_b}\left(1 - \frac{N_{pb}}{N}\right) \tag{3.39}$$

in which: N_{pb} = oil recovered by depletion to the bubble point, evaluated using equation 3.25; N'_p = oil recovered below the bubble point and N_b = stock tank oil remaining in the reservoir at the bubble point.

Therefore, for an initially undersaturated reservoir, equation 3.38 is used in slightly modified form to calculate the oil recovery below the bubble point, N'_p/N_b, and in its evaluation, B_{oi} is replaced by B_{ob}. Equation 3.39 is then used to calculate the total recovery. Applying the Muskat material balance, involves the following steps.

- Using the PVT data, calculate the functions $A(p)$, $B(p)$ and $C(p)$ across the full range of pressure depletion below the bubble point.
- Structure a table as shown below (refer Exercise 3.3 for a practical example).

I	II	III	IV	V	VI	VII	VIII
Pressure (psia)	$A(p)$ $B(p)$ $C(p)$ (/psi)	S_o (PV)	k_{rg}/k_{ro}	ΔS_o (PV)	S_o (PV)	N_p/N	GOR (scf/stb)

for which the columns can be defined as follows:

I.   Pressure decline in steps below the bubble point
II.   Table of values of A, B, and C: these are evaluated at the average pressure between successive table values
III. S_o prior to the pressure drop Δp
IV.   Relative permeability ratio evaluated at the last determined value of S_o
V. ΔS_o evaluated using equation 3.33
VI.   The new lower value of S_o at the reduced pressure level
VII.   The fractional oil recovery, using equations 3.38 or 3.39, as appropriate
VIII. The GOR evaluated applying equation 3.32 at the reduced pressure and using a k_{rg}/k_{ro} value determined at the reduced S_o level in column VI

For refinement of solution, several iterations of steps III to V are required. Having calculated an initial value of ΔS_o, the necessary iterations are performed using $S_o = S_o - \Delta S_o/2$ in equation 3.33, since the pressure dependent values, A, B and C are evaluated at the mid point of the pressure decrements. Usually two or three iterations are sufficient for convergence to determine a final value of ΔS_o (column V).

The solution gas drive process is highly dependent on the fluid PVT properties of the oil and gas [7]. Favourable recovery requires a moderate oil volatility and, in particular, a low oil viscosity. The factor that exerts greatest influence on recovery, however, is the gas/oil relative permeability ratio and associated with this the critical gas saturation, S_{gc}, at which the liberated gas first becomes mobile. This is a surface tension effect in that, once the bubble point pressure is reached, the liberated gas is trapped within each pore space in discrete bubbles. It is not until the gas saturation increases to a level at which bubbles in adjoining pores coalesce to exceed the critical saturation that the gas can start to move and be produced. The value of the critical gas saturation is extremely important in dictating the ultimate oil recovery by solution gas drive. Typically it is about 5% PV meaning that the gas expansion to this value and its retention in the reservoir expels an equal volume of oil as production. Sometimes, for reasons that are difficult to theorise about, this value can be as high as 15%[+] PV which gives a significant enhancement in production. The effect of the buildup of liberated gas towards the critical saturation means that while this is occurring there is a deficiency of gas production at the surface so that for a while the GOR $< R_{si}$. This behaviour is apparent in non-fractured reservoirs but to a lesser extent in fractured systems (Fig. 3.10a). In these, the gas liberated in the high-conductivity fractures becomes mobile immediately which is to the detriment of oil recovery.

Darcy's basic law applies to the flow of a fluid at 100% saturation but if two fluids are flowing simultaneously through the system the law requires modification so that, for oil and gas, the rates are described separately by the equations:

$$q_o = -\frac{kk_{ro}A}{\mu_o}\frac{\partial p}{\partial l}, \quad q_g = -\frac{kk_{rg}A}{\mu_g}\frac{\partial p}{\partial l}$$

in which k_{ro} and k_{rg} are the relative permeabilities to oil and gas respectively. They are expressed as fractions and are strictly functions of the increasing saturation of the displacing phase, the gas (Fig. 3.10b). It is the division of these equations, with the rates expressed at standard conditions, that produces the GOR equation, 3.32. Division also yields the so called end-point mobility ratio:

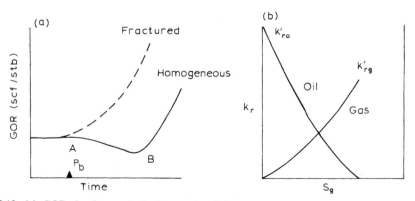

Fig. 3.10. (a) GOR developments in fractured and homogeneous reservoirs. (b) Gas–oil relative permeability functions.

$$M = \frac{k'_{rg}}{\mu_g} \bigg/ \frac{k'_{ro}}{\mu_o} \tag{3.40}$$

in which k'_{ro} and k'_{rg} are the maximum or end-point relative permeabilities to oil and gas. Because of their inclusion, the mobility ratio expresses the maximum velocity of gas flow to that of oil. Using typical figures ($k'_{ro} = 1.0$, $k'_{rg} = 0.5$, $\mu_o = 1.0$ cP, $\mu_g = 0.025$ cP), $M = 20$, meaning that under a given pressure differential the gas is capable of travelling twenty times faster than the oil it is displacing. This is the problem with having free gas in the reservoir whether it is evolved during depletion or is injected. It is the extremely low gas viscosity that is the dominant factor that leads to considerable instability in the manner that gas displaces oil, usually resulting in premature and excessive gas production. Control can only be achieved if gravity acts in a favourable manner to stabilize the frontal advance of the gas (Chapter 6, section 6.4).

Gas–oil relative permeabilities are usually measured in the laboratory using the viscous displacement technique in which gas or air is injected into an oil saturated core plug. The oil recovery, N_{pd} (PV's) is measured as a function of the cumulative gas injected, G_{id} (PV's) and the relative permeabilities are then evaluated applying the Welge equation (Chapter 5. section 5.4e). Unfortunately, while these functions are appropriate for oil recovery by gas injection, they have little relevance to the much more complex solution gas drive process, although they are frequently used in its description in equations 3.32 and 3. 33. They do not cater for the development of a critical gas saturation nor the dispersed rather than frontal displacement of the oil by gas. To model the process correctly would require using a core plug filled with gas saturated oil of the correct volatility. The pressure should be decreased in small decrements at first to establish the critical gas saturation and once the gas becomes mobile the relative permeabilities should then be measured applying a realistic pressure differential across the core plug. This type of complex experiment is hardly ever performed, however, and falls more in the category of research rather than a routine procedure. In lieu of relevant experimental results, a good starting point is to use the functions presented by Arps and Roberts [8]. These, which are included as Figs. 3.11a and b for sandstones and limestones, were compiled from both experimental results and direct observations of field performance. They are plots of the relative permeability ratio k_{rg}/k_{ro} on a log scale versus the total liquid saturation $S_L = S_o + S_{wc}$; therefore, the gas saturation increases from right to left on the abscissa. Their best use is in comparison with relative permeabilities derived from history matching field performance. A successful match can provide a means of extrapolating the field derived functions to higher gas saturations. On their plots, Arps and Roberts distinguish between Minimum, Average and Maximum functions and although these categories are not easy to define, the following description may serve as a guide

  – *Minimum:* Low-permeability sandstones or sections in which there are barriers to vertical fluid movement that preclude gravity segregation. In fractured reservoirs, a region of high fracture intensity usually corresponds to the minimum curve in Fig. 3.11b,in which it will be noticed that the critical gas saturation is extremely low.

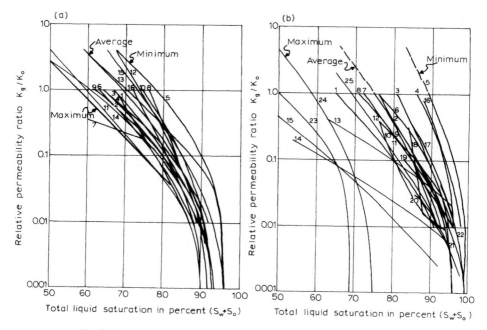

Fig. 3.11. Arps-Roberts k_{rg}/k_{ro}-functions: (a) sandstones; (b) limestones.

– *Maximum:* Reservoirs with good porosity and permeability and a reasonable dip angle: features that will promote segregation of the oil and gas and the retention of the latter in the reservoir.

– *Average:* Anything in between these extremes.

The paper of Arps and Roberts dates from 1955 and there are other articles on the subject of similar vintage [9] and from earlier [10]. This, of course, was the age in which depletion leading to solution gas drive was the predominant means of oil recovery, secondary recovery flooding being regarded as generally too expensive at the prevailing oil price. Therefore, anyone wishing to study the process would be well advised to dig deep into the archives and read the publications from the 1940's and 50's.

More recently, Honarpour et al. [11] have presented an empirical means of calculating separate oil and gas relative permeability functions. The relations for sandstones are as follows:

$$k_{ro} = 0.98372 \left(\frac{S_o}{1 - S_{wc}} \right)^4 \left(\frac{S_o - S_{or}}{1 - S_{or} - S_{wc}} \right)^2 \tag{3.41}$$

and:

$$k_{rg} = 1.1072 \left(\frac{S_g - S_{gc}}{1 - S_{wc}} \right)^2 k'_{rg} + 2.7794 \left(\frac{S_{or}(S_g - S_{gc})}{1 - S_{wc}} \right) k'_{rg} \tag{3.42}$$

It will be noted that the critical gas saturation S_{gc} is catered for in generating the gas relative permeability function. These relations were based on matching 133 sets of experimental data and are claimed to closely match field observations also.

The difficulty in using the above methods as anything more than guidelines when performing analytical material balance calculations lies in the lack of geometrical definition implicit in their use. The functions of Arps and Roberts and Honarpour, while making some claims at matching observed field performance, lean more towards matching experimental data in core flooding experiments. As such they are more suitable as input to numerical simulation models in each separate layer: assuming there are sufficient layers such that each has a thickness tending towards that of a core plug itself. What is required in applying material balance, however, is some form of "pseudo" relative permeabilities, as they are referred to, that will incorporate all the complexities of reservoir heterogeneity and the nature of the displacement mechanics, such as the degree of gravity segregation of the oil and gas. Paradoxically, the better the reservoir quality the less appropriate it becomes to use the correlations directly in material balance calculations. Therefore, thick, well connected sections in which segregation may be anticipated are poor candidates whereas a section comprising thin layers separated by shales which restrict vertical fluid movement can be realistically matched using, for instance, the Arps and Roberts "Minimum" function since the section acts as a set of discrete core plugs for which the use of direct experimental data is appropriate. As demonstrated in Exercise 3.3, the required pseudo-functions can only be generated by history matching the reservoir performance, using the above correlations for qualitative assessment of the efficiency of the solution gas drive mechanism (Minimum, Average, Maximum) and sometimes using them as a guide in extrapolating the k_{rg}/k_{ro}-function to higher gas saturations thus facilitating performance prediction. Alternatively, if a simulation model is constructed in an ideal manner, using a large number of layers with the correct transmissibility between each, then input of the above relative permeability correlations in each layer should lead to the correct prediction of such factors as gravity segregation. This is the main advantage of numerical simulation in its power to predict.

History matching of the production/pressure data is the most important function of material balance application. In this respect, the following techniques should be applied once the reservoir pressure has fallen below the bubble point:

- Examination of the GOR behaviour to define the bubble point pressure, evaluate the critical gas saturation and also assess the gas drive efficiency as illustrated in Exercise 3.2.
- Generation of a pseudo k_{rg}/k_{ro} versus S_L function for the reservoir as a whole.

The first step is necessary to compare the observed bubble point pressure in the field with that determined in laboratory experiments. If they differ, which is often the case, the laboratory PVT functions will require correction, as described in Chapter 2, section 2.2d. The most sensitive way of determining the bubble point pressure is by examination of the instantaneous and cumulative GOR's (R and R_p) of individual wells and of the reservoir as a whole. In the reservoir examined in

Exercise 3.2, identification of the bubble point was clear cut: the values of R and R_p rising sharply above $R_{si} = 1450$ scf/stb after 21 months of production at a pressure of 4500 psia (Table 3.2). The reason for this precision is because there was no development of a critical gas saturation in this fractured reservoir; if there is, then evaluation of the saturation pressure is more complex and can be less precise. As illustrated in Fig. 3.10a, it depends on identifying point A on the GOR trend at which R first decreases below R_{si} but the initial decrease in GOR can be quite subtle and it requires accurate oil and gas monitoring to locate the start of the downward trend. Once the GOR starts to increase (point B, Fig. 3.10a) it means that the critical gas saturation has been exceeded and the gas has become collectively mobile. If, at this time, the pressure and cumulative oil production are known, then equation 3.37 can be solved for the current oil saturation, S_o, using a value of the STOIIP determined preferably by application of material balance during the initial depletion at undersaturated conditions or, failing this, using the volumetric estimate. The critical gas saturation can then be evaluated as $S_{gc} = 1 - S_o - S_{wc}$.

The relative permeability ratio, k_{rg}/k_{ro}, which, as described above, is a "pseudo" function containing all the complexity of heterogeneity and the nature of the drive mechanism, can be derived from the production/pressure history by solution of equation 3.32. This requires knowledge of the instantaneous GOR (R), the average pressure and the PVT. The k_{rg}/k_{ro} ratio is generated as a function of the total liquid saturation $S_L = S_o + S_{wc}$ which can be evaluated by solving equation 3.37 for S_o. Extrapolation of the established trend permits the prediction of solution gas drive performance by using it in the Muskat material balance equation, 3.33. The technique is illustrated in Exercise 3.3.

The performance prediction provides the fractional oil recovery as a function of the declining pressure and increasing GOR. These results are incomplete, however, for they give no indication of how long it will take to recover the oil. To determine this requires the generation of a production profile which, in turn, necessitates the study of individual well performance. The prediction of productivity decline of wells under solution gas drive conditions has received considerable attention in the literature. In a summary paper on the subject [12], it was concluded that the relatively straightforward inflow equation of Fetkovich [13], provided as sound a prediction tool as any, particularly since it is based on matching field observations. The equation has the form:

$$q = \text{constant} \times \frac{kh}{\ln 0.47X} \frac{pk_{ro}}{\mu_o B_o} (1 - R^2)$$

in which the first term is described as the geometric factor: X is analogous to the Dietz Shape Factor in well testing (Chapter 4, section 4.19b). The second is the depletion stage factor and the third the drawdown factor in which $R = p_{wf}/p$. The equation is used in ratio form to calculate q/q_i, with q the rate at any stage of depletion and q_i the rate at the bubble point pressure. In this form, only the last two terms are important and the proportionality holds that:

$$q \propto \frac{k_{ro}}{\mu_o B_o} \frac{p^2 - p_{wf}^2}{p} \tag{3.43}$$

This, divided by its value at initial conditions, gives the rate decline ratio from which a time scale can be attached to the oil recovery by setting up a table as follows:

I	II			III	IV	V	VI	VII		
Pressure	B_o	μ_o	S_o	k_{ro}	q/q_i	q	N_p/N	ΔN_p	Δt	Time
(psia)	(rb/stb)	(cp)	(PV)			(stb/d)		(stb)	(days)	(years)

in which the columns can be described as follows:

I.   Pressure steps used in the material balance equation
II.   Evaluated from the PVT data
III. S_o determined from the Muskat material balance (equation 3.33) giving k_{ro}
IV.   Calculated using equation 3.43, divided by its evaluation at initial conditions
V.   Average rate during the pressure decrement
VI.   Oil recovery evaluated from the material balance: ΔN_p is the incremental oil recovery during the pressure decrement
VII. $\Delta t = \Delta N_p/q$ (days) which is the duration of the pressure step, the cumulative of which gives the total time.

The method can be applied on a well by well basis or to an entire reservoir being drained by a "super" well.

Exercise 3.3: Application of the Muskat material balance in history matching and prediction of solution gas drive

Introduction
   This exercise continues to study the chalk field producing under depletion conditions described in Exercise 3.2. The purpose is to history match its solution gas drive performance and predict the ultimate recovery applying the material balance method of Muskat.

Question
Using the six years of production/pressure history and the PVT for the field described in Exercise 3.2.
• Generate a k_{rg}/k_{ro} function.
• Extrapolate this and predict field performance to an abandonment pressure of 800 psi.

Solution
   The gas–oil relative permeability ratio can be evaluated by solving equation 3.32 expressed in the form:

TABLE 3.4a

PVT relationships (Exercise 3.3)

Pressure (psia)	B_o (rb/stb)	R_s (scf/stb)	B_g (rb/scf)	μ_o/μ_g
4500	1.8500	1450	0.00078	5.604
4400	1.7956	1356	0.00079	5.981
4200	1.7131	1224	0.000817	6.815
4000	1.6505	1108	0.00085	7.591
3600	1.5507	931	0.000928	9.350
3200	1.4600	766	0.00103	11.384
2800	1.3817	637	0.00117	13.544
2400	1.3224	515	0.00136	16.037
2000	1.2663	411	0.00163	19.045
1600	1.2152	319	0.00204	22.335
1200	1.1707	222	0.00276	26.688
800	1.1344	129	0.00419	32.568

TABLE 3.4b

Calculation of the gas/oil relative permeability ratio by history matching the field production performance

Pressure (psia)	N_p (MMstb)	N_p/N	N_p'/N_b	R (scf/stb)	$\log k_{rg}/k_{ro}$	S_o (PV)	S_{liq} (PV)
4500	43.473	0.0763		1450			
4400	47.279	0.0829	0.0072	2150	−1.2336	0.549	0.979
4200	53.941	0.0946	0.0199	2940	−0.9205	0.517	0.947
4000	60.136	0.1055	0.0316	3660	−0.7617	0.492	0.922
3600	71.232	0.1250	0.0527	5330	−0.5504	0.453	0.883
3200	79.478	0.1394	0.0684	7815	−0.3597	0.419	0.849
2800	86.401	0.1516	0.0815	8060	−0.3334	0.391	0.821

$$\frac{k_{rg}}{k_{ro}} = \frac{R - R_s}{\mu_o/\mu_g} \frac{B_g}{B_o} \tag{3.44}$$

in which values of the instantaneous GOR, R, are taken from Fig. 3.8 and the PVT parameters from Table 3.3. These data are presented in Tables 3.4a and b. Since the reservoir was undersaturated by 2650 psi at initial conditions the total fractional oil recovery at any stage of depletion below the bubble point is defined by equation 3.39 which, since the recovery at the bubble point is $N_{pb}/N = 0.0763$ ($N = 570$ MMstb), can be expressed as:

$$\frac{N_p}{N} = 0.0763 + 0.9237 \frac{N_p'}{N_b} \tag{3.45}$$

The corresponding oil saturation can be calculated using equation 3.38 in the form:

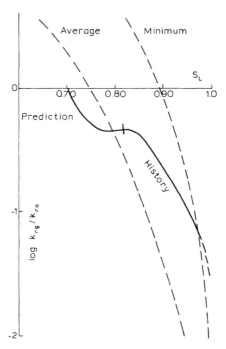

Fig. 3.12. Gas–oil relative permeability function (Exercise 3.3) compared to Arps-Roberts minimum and average curves.

$$S_o = \left(1 - \frac{N'_p}{N_b}\right) \frac{B_o}{B_{ob}} (1 - S_{wc}) \tag{3.46}$$

($B_{ob} = 1.850$ rb/stb: $S_{wc} = 0.43$ PV).

Using these relationships and the data supplied in Tables 3.2 and 3, the gas/oil relative permeability ratio has been calculated as a function of $S_L = S_o + S_w$, the data being listed in Table 3.4b and plotted in Fig. 3.12, in which the Minimum and Average functions of Arps an Roberts for limestone (Fig. 11b) have been plotted for comparison. The field's macroscopic (pseudo) function starts off worse than the Arps-Roberts Minimum function as the gas liberated just below the bubble point is immediately produced through the fracture system without the development of a critical gas saturation. Thereafter, a degree of gas segregation occurs as the liberated gas moves through the extensive fracture system to the crest of the structure to form a secondary gas cap and the k_{rg}/k_{ro} function swings sharply to the left towards the Arps-Roberts Average curve. This explains the result calculated in Exercise 3.2 that after the six years of production history 54% of the liberated gas had remained in the reservoir. Extrapolation of the function below $S_L = 0.82$ PV is speculative but based on the histories of similar fields in the area it has been extrapolated to $k_{rg}/k_{ro} = 1.0$ as shown in Fig. 3.12 and this is used in performance prediction.

The first step in applying the Muskat material balance technique is to calculate

TABLE 3.5

Muskat material balance calculations (Exercise 3.3)

Pressure (psia)	$A(p)$ (1/psi $\times 10^{-4}$)	$B(p)$ (1/psi $\times 10^{-4}$)	$C(p)$ (1/psi $\times 10^{-4}$)	S_o (PV)	k_{rg}/k_{ro}	ΔS_o (PV)	S_o (PV)	N_p/N	GOR (scf/stb)
4500				0.57			0.57	0.0763	1450
4400	4.048	17.287	−1.274	0.57	0.0407	0.0216	0.5484	0.0844	2140
4200	3.023	15.044	−1.68	0.5484	0.0832	0.0306	0.5178	0.0938	2790
4000	2.874	13.406	−1.98	0.5178	0.1413	0.0252	0.4926	0.1052	3610
3600	2.458	13.204	−2.193	0.4926	0.2239	0.0383	0.4543	0.1217	5330
3200	2.683	15.616	−2.605	0.4543	0.3548	0.0337	0.4206	0.1363	7420
2800	2.497	17.172	−3.182	0.4206	0.4571	0.0282	0.3924	0.1486	8030
2400	2.845	16.237	−3.755	0.3924	0.4571	0.0236	0.3688	0.1639	7660
2000	3.003	19.026	−4.515	0.3688	0.455	0.0232	0.3456	0.1818	7170
1600	3.402	21.303	−5.586	0.3456	0.4656	0.0217	0.3239	0.2009	6990
1200	4.879	22.858	−7.5	0.3239	0.5458	0.0205	0.3034	0.223	6890
800	7.01	23.329	−10.288	0.3034	0.6607	0.0183	0.2851	0.2465	6520

the three PVT functions A, B and C using equations 3.34–3.36 and the PVT data listed in Table 3.4a which includes the oil/gas viscosity ratio data.

The values are listed in Table 3.5 as functions of the declining pressure below the bubble point. Next the Muskat material balance calculations are performed applying equation 3.33 as described earlier in the section. These use the gas/oil relative permeability function shown in Fig. 3.12 both for the history match and prediction ($S_L < 0.82$ PV).

The results which are listed in Table 3.5 and plotted in Fig. 3.13 are most encouraging showing by depleting to 800 psi (which is quite possible since there is sufficient produced gas to "lift" the wells) that a recovery of over 24% STOIIP could be obtained at a GOR of 6500 scf/stb.

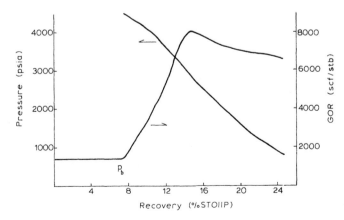

Fig. 3.13. Pressure decline and GOR development as functions of the oil recovery (Exercise 3.3).

3.8. WATER INFLUX CALCULATIONS

The absolute method of calculating natural water influx resulting from pressure depletion of the reservoir/aquifer system is that of van Everdingen and Hurst [14] dating from 1949. It relies on the use of the constant terminal pressure solution of the diffusivity equation. That is, the pressure at the original oil–water contact is dropped by a finite amount, Δp, and maintained at the lower level which gives rise to a finite influx of water from the aquifer. The pressure decline history is divided into a set of such discrete pressure steps and the cumulative influx at any time is the superposed sum of the separate influxes resulting from each.

In the past the method gave engineers computational difficulties because of the need to apply superposition. That is, in progressing from time step "n" to "$n + 1$" it is not possible to directly calculate the incremental water influx and add it to the previous value, instead it is necessary to go right back to the beginning and calculate the sum of all the separate influxes evaluated at the new time step. Trying to do this using a slide rule was a little tedious, to say the least, and therefore engineers devised methods in which the influx could be calculated directly in passing from one time step to the next without the historic involvement required in applying superposition. Nowadays, of course, superposition can be applied using a pocket calculator making Hurst and van Everdingen straightforward to use. The method has been described at some length in reference 2 and is also illustrated in Chapter 6, Exercise 6.1 to calculate water influx into a gas reservoir.

Of the "direct" methods, that of Fetkovich [15] was described in reference 2 but what appears to be the more popular, the routine of Carter and Tracy [16], was omitted. Since it is simple and accurate in application and has been coded into several commercial numerical simulators it is worthwhile including here for completeness and its application will be illustrated with a field example.

(a) Carter-Tracy water influx calculations

The Carter-Tracy water influx equation employs the constant terminal rate (CTR) solution of the diffusivity equation (Chapter 4, section 4.6), as opposed to the constant terminal pressure used in the Hurst-van Everdingen approach. Their equation is:

$$W_e(t_{Dj}) = W_e(t_{Dj-1}) + \left(\frac{U \Delta p(t_{Dj}) - W_e(t_{Dj-1}) P'(t_{Dj})}{P(t_{Dj}) - t_{Dj-1} P'(t_{Dj})} \right) (t_{Dj} - t_{Dj-1}) \qquad (3.47)$$

in which the subscript "j" refers to the present time step and "$j - 1$" the previous. The parameters in the equation are as follows:

U = aquifer constant = $1.119 f \phi h c r_o^2$ (bbl/psi) for radial geometry [2]
         [f = fractional encroachment angle: $c = c_w + c_f$ (/psi): r_o = reservoir radius (ft)]

Δp = total pressure drop, $p_i - p$, (psi)

W_e = cumulative water influx (bbl)

TABLE 3.6

Carter-Tracy influence functions regression coefficients for the constant terminal rate case [17]

r_{eD}	Regression coefficients			
	a_0	a_1	a_2	a_3
1.5	0.10371	1.66657	−0.04579	−0.01023
2.0	0.30210	0.68178	−0.01599	−0.01356
3.0	0.51243	0.29317	0.01534	−0.06732
4.0	0.63656	0.16101	0.15812	−0.09104
5.0	0.65106	0.10414	0.30953	−0.11258
6.0	0.63367	0.06940	0.41750	−0.11137
8.0	0.40132	0.04104	0.69592	−0.14350
10.0	0.14386	0.02649	0.89646	−0.15502
∞	0.82092	-3.68×10^{-4}	0.28908	0.02882

t_D = dimensionless time = $0.00634 \, (kt/\phi\mu cr_o^2) \, t$ in days: the constant is replaced by 2.309 if t is measured in years

$P(t_D)$ = dimensionless CTR solution of the diffusivity equation

$P'(t_D)$ = is its time derivative $dP(t_D)/dt_D$.

The $P(t_D)$ have been presented in tabular form by van Everdingen and Hurst [14] but a more convenient way of evaluating them has been presented by Fanchi [17] who matched the functions with the regression equation:

$$P(t_D) = a_o + a_1 t_D + a_2 ln t_D + a_3 (\ln t_D)^2 \qquad (3.48)$$

in which the regression coefficients are as listed in Table 3.6 for different values of the ratio $r_{eD} = r_e/r_o$, where r_e is the outer radius of the aquifer for radial geometry.

Application of equation 3.47 in water influx calculations with $P(t_D)$ functions and their time derivatives determined using equation 3.48 gives results that are very close to those obtained using the method of Hurst and van Everdingen.

(b) Aquifer "fitting" using the method of Havlena-Odeh

This has already been described in reference 2 and is the most accurate means of manipulating the historic production/pressure/PVT data of a reservoir to "fit" a suitable aquifer model while at the same time making an estimate of the STOIIP. Consider the case of an undersaturated oil reservoir affected by natural water influx. The most appropriate expression of the material balance to history match the performance is that of Havlena and Odeh [3,4] as expressed by equation 3.22. In principle, if the data monitoring has been satisfactory, the left-hand side of the equation should be determinate but the right-hand side contains two major unknowns: the STOIIP, N, and the cumulative water influx, W_e. With geological/petrophysical guidance the reservoir engineer constructs a suitable physical model of the aquifer, as described in the literature [2,3,14], based on the geology and estimated rock properties and applying a theoretical model, such as described by

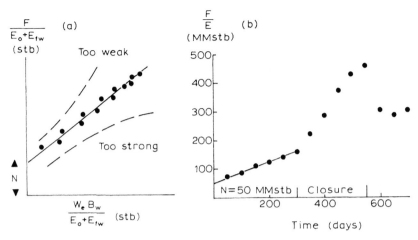

Fig. 3.14. (a) Havlena-Odeh aquifer fitting technique; (b) F/E versus W_e/E (Exercise 3.4).

equation 3.47, calculates the water influx to match the reservoir offtake. A plot is then made of $F/(E_o + E_{fw})$ versus $W_e/(E_o + E_{fw})$ as shown in Fig. 3.14a. If the correct model has been selected the points must lie on a 45° straight line whose intercept on the ordinate will provide an estimate of the STOIIP, N. If the aquifer model is inappropriate deviations from linearity will occur, as shown, and it must be modified until the required result is obtained.

Exercise 3.4: History matching using the Carter-Tracy aquifer model and the "fitting" technique of Havlena and Odeh

Introduction

A small offshore oilfield has the pressure history shown in Fig. 3.15 for the first 700 days of production. After 300 days the field was closed in for a period of 250 days and the response of the aquifer can be clearly seen in elevating the pressure by almost 560 psi. The closure was due to the extensive upgrading of the oil gathering/transportation system in the area and the immediate response of the water influx is taken as an indication of excellent pressure communication between the reservoir and aquifer which justifies the use of material balance. The pressure remained above the bubble point during the entire period and the field and aquifer are characterised by the following properties:

B_{oi} = 1.118 rb/stb at p_i = 4217 psia
B_w = 1.0 rb/stb
c_o = 7.51 × 10⁻⁶/psi, therefore:
B_o = 1.118 (1 + 7.51 × 10⁻⁶ Δp) rb/stb
c_f = 5.0 × 10⁻⁶/psi:
c_w = 3.0 × 10⁻⁶/psi

The aquifer is regarded as being radial with encroachment angle 140°. Therefore

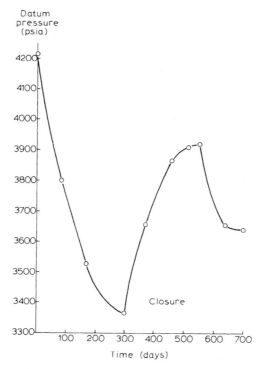

Fig. 3.15. Pressure decline history (Exercise 3.4).

f = 140/360 = 0.389.
ϕ = 0.19, aquifer porosity
h = 150 ft, aquifer thickness
c = $c_f + c_w = 8 \times 10^{-6}$/psi, effective aquifer compressibility
r_o = 6000 ft, reservoir radius
k = 15 mD, aquifer permeability
μ = 0.4 cP, aquifer water viscosity
S_{wc} = 0.248 PV
c_{eff} = $(c_o S_o + c_w S_{wc} + c_f)/(1 - S_{wc}) = 15.15 \times 10^{-6}$/psi

Question
History match the 700 days production/pressure history to determine the STOIIP and water influx.

Solution
Using these data, the aquifer constant and dimensionless time [2] can be evaluated as:

$$U = 1.119 f \phi h c r_o^2 = 1.119 \times 0.389 \times 0.19 \times 150 \times 8 \times 10^{-6} \times 6000^2 \approx 3600 \text{ b/psi)}$$

TABLE 3.7

Production/pressure data (Exercise 3.4)

Time (days)	N_p (stb)	W_p (stb)	p (psia)	B_o (rb/stb)	E (rb/stb)	F (rb)	F/E (MMstb)
0			4217	1.1180			
50	342704	–	3915	1.1205	0.00512	384000	75.00
100	635048	1211	3725	1.1221	0.00833	713798	85.69
150	1068981	26356	3570	1.1234	0.01096	1227249	111.97
200	1367882	70414	3450	1.1244	0.01299	1608461	123.82
250	1643973	122183	3390	1.1249	0.01401	1971488	140.72
300	1914646	173036	3365	1.1252	0.01443	2327396	161.29
350	1914646	173036	3600	1.1232	0.01045	2323566	222.35
400	1914646	173036	3740	1.1220	0.00808	2321269	287.29
450	1914646	173036	3850	1.1211	0.00622	2319546	372.92
500	1914646	173036	3900	1.1207	0.00537	2318780	431.80
550	1914646	173036	3920	1.1205	0.00503	2318397	460.91
600	2114019	203684	3720	1.1222	0.00842	2576036	305.94
650	2249213	250938	3650	1.1228	0.00960	2776354	289.20
700	2331328	316898	3640	1.1228	0.00977	2934513	300.36

$$t_D = 0.00634 \frac{kt}{\phi \mu c r_o^2} = \frac{0.00634 \times 15 \times t}{0.19 \times 0.4 \times 8 \times 10^{-6} \times 6000^2} = 0.00434\, t \qquad \text{(t-days)}$$

The cumulative production data at 50 day intervals are listed in Table 3.7 together with the average reservoir pressures read from Fig. 3.15.

In Table 3.7, $E = E_o + E_{fw} = B_{oi} c_{eff} \Delta p$ for undersaturated conditions and using the above data this can be evaluated as $E = 16.938 x 10^{-6} \Delta p$ (rb/stb). Values of the underground withdrawal are calculated as $F = N_p B_o + W_p$ (rb). The final column, F/E (MMstb), is plotted in Fig. 3.14b as a function of time. The backward extrapolation of the points to the ordinate suggests a STOIIP of about 50 MMstb although caution must be attached to this extrapolation, as explained in section 3.6. The rate dependence of F/E_o is clearly visible: during the 250 days of closure the aquifer energises the reservoir elevating the points far above the trend established while the field was on production. The basic technique for history matching waterdrive performance is to apply the Havlena-Odeh equation:

$$\frac{F}{E} = N + \frac{W_e}{E}$$

in which the water influx is calculated using the Carter-Tracy equation, 3.47, for the reservoir-aquifer data listed above. When this was attempted for $U = 3600$ bbl/psi, $t_D = 0.00434$ t and a finite aquifer radius ($r_{eD} = 2.0$), the resulting plot of F/E versus W_e/E produced the rather peculiar result shown in Fig. 3.16a. Initially the aquifer was too strong but then a strange effect occurred in that during the period of closure there was an efflux of water from the reservoir, which is hardly realistic. The only way to overcome this anomaly was to use an infinite acting aquifer. In addition it was found that the aquifer constant had to be reduced by a factor of two

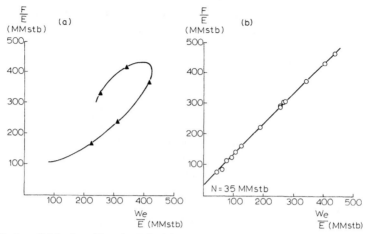

Fig. 3.16. Havlena-Odeh plots (Exercise 3.4). (a) U = 3600 b/psi, t_D = 0.00434t, r_{eD} = 2.0: incorrect. (b) U = 1800 b/psi, t_D = 0.00325t, r_{eD} = ∞: satisfactory history match.

to U = 1800 bbl/psi and the dimensionless time coefficient to t_D = 0.00325t, a 25% reduction of the original value. Concerning the number of "unknowns" contained in these parametric groups, the modifications were considered quite reasonable. Water influx calculations for the modified values of U and t_D are listed in Table 3.8, in which the P_D-functions have been evaluated for r_{eD} = ∞ (Table 3.6).

The resulting plot of F/E versus W_e/E is shown as Fig. 3.16b and, as can be seen, a fairly convincing straight line is obtained with the requisite slope of 45°. Extrapolation to the ordinate determines a value of the STOIIP of about 35 MMstb.

TABLE 3.8

Carter-Tracy water influx calculations

Time (days)	t_D	$P(t_D)$	$P'(t_D)$	Δp (psi)	W_e (MMrb)	W_e/E (MMstb)	N (MMstb)
50	0.163	0.3913	2.4006	302	0.2264	44.22	30.80
100	0.325	0.5323	0.8704	492	0.5121	61.48	24.21
150	0.488	0.6282	0.5883	647	0.8341	76.10	35.87
200	0.650	0.7015	0.4525	767	1.1722	90.24	33.58
250	0.813	0.7620	0.3712	827	1.5020	107.21	33.51
300	0.975	0.8133	0.3164	852	1.8103	125.45	35.84
350	1.138	0.8584	0.2767	617	1.9791	189.39	32.96
400	1.300	0.8983	0.2463	477	2.0764	256.98	30.31
450	1.463	0.9345	0.2221	367	2.1267	341.91	31.01
500	1.625	0.9675	0.2037	317	2.1599	402.22	29.58
550	1.788	0.9980	0.1871	297	2.1905	435.49	25.42
600	1.950	1.0261	0.1735	497	2.3069	273.98	31.96
650	2.113	1.0525	0.1620	567	2.4500	255.21	34.00
700	2.275	1.0772	0.1525	577	2.5927	265.37	35.00

Listed in the final column of Table 3.8 are values of the STOIIP, N, calculated as $F/E - W_e/E$. Excluding three spuriously low figures ($N < 30$ MMstb), the average of the remaining 11 values is $N = 33.17$ MMstb.

As is quite typical nowadays, it was believed that material balance was not a credible technique for history matching the field compared to numerical simulation. It was also argued that the mobilities of the oil and water were too similar to permit its application. This point of view is often expressed, especially in areas like the North Sea — where the condition does prevail but it is nevertheless irrelevant. The material balance does not take account of fluid dynamics, and therefore mobilities, but only distinguishes between relative volume changes as defined by the PVT properties of the fluids. But even for this low volatility oil reservoir ($B_{oi} = 1.118$ rb/stb), the difference is sufficient to permit a very convincing application of the technique.

Instead, a numerical simulation model was constructed based on poor quality seismic and little well control, with a volumetrically estimated STOIIP of 65 MMstb ($\sim 90\%$ too large) and this produced a perfectly reasonable history match, for the reasons explained in section 3.5. Using the data relevant for 700 days in Tables 3.7 and 8, it can be calculated that for $N = 65$ MMstb the water influx would be $W_e = 2.299$ MMrb instead of the value of 2.593 MMrb, which is only an 11% reduction. This illustrates the great sensitivity in applying not just material balance but any form of reservoir engineering technique in history matching field performance. The whole difficulty arises for the reasons stated already: that equation 3.22 contains two major unknowns, N and W_e (together with a host of lesser, dependent unknowns). Consequently, there is an infinite number of possible solutions and simulation provides only one, based on the assumed volumetric STOIIP.

(c) History matching with numerical simulation models

One of the primary means of history matching numerical simulation models is by comparison of the measured reservoir pressures with those calculated by the simulator, as shown in Fig. 3.17. Not to put too fine a point on it, the technique can only be described as "primitive" and is one that gives rise to considerable error in attempting to understand the reservoir being studied.

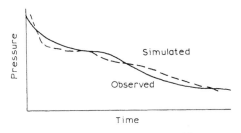

Fig. 3.17. Numerical simulation-history matching on pressure.

The necessity for applying "history matching on pressure" is because, as explained in section 3.5, simulators are structured in such a manner that the solutions they provide at the end of each time step are fluid saturations and pressures in each grid block. It is not possible to feed into a simulator five years of pressure history and ask it to match it exactly, they simply are not structured to do so. Therefore, the quality of the history match is subjective and will vary from engineer to engineer dependent on: personal fastidiousness, eyesight, sleeping patterns and general social behaviour — which is a most unfortunate circumstance in an otherwise serious engineering subject. In particular, it is at the start of production when small errors in pressure matching imply large errors in volumetrics where the history match must be perfect, otherwise it is wrong.

Considering that operating companies spend millions of dollars a year measuring pressures in development wells, the least that can be aimed for is that they should be honoured. Admittedly, test interpretation to determine average pressures within the drainage area of each well is not an exact science (Chapter 4, section 4.19) and also contains a subjective element but nevertheless some faith must be placed in the subject and the results. As pointed out in section 3.5, when applying material balance the situation is quite different and interpreted pressures are input as known quantities that must be honoured in the calculations. In this respect, from the point of view of correctness of technique, the method of history matching of Havlena and Odeh, illustrated in exercise 3.4 (Fig. 3.16b), must be regarded as superior to that applied using numerical simulation.

3.9. GASCAP DRIVE

If a reservoir has a gas cap the material balance can be stated as:

$$F = N(E_o + mE_g) + W_e B_w \tag{3.49}$$

in which it is assumed that, in the presence of high-compressibility free gas, the water and pore compressibility term E_{fw} is negligible (an assumption that must always be checked). The inclusion of the gas cap component raises the level of complication in trying to solve the equation by a considerable amount. Havlena and Odeh [3] acknowledged that they met with only "limited success" in its solution and stated that the reason for this was because:

> "Whenever a gas cap is to be solved for, an exceptional degree of accuracy of basic data, mainly pressures, is required."

The statement is perfectly correct but the problems are more acute than that. In the first place the right-hand side of equation 3.49 now contains three major, potential unknowns, N, W_e and m, which adds to the lack of mathematical uniqueness in its solution referred to in section 3.4. Secondly comes the difficulty mentioned by Havlena and Odeh of the great sensitivity to the quality of the input data, particularly the pressure. What they did not refer to, however, is that in "gassy" reservoirs the measurement of well pressures is extremely difficult. If there is a gas cap, the underlying oil is at the saturation pressure, or close to it. Therefore, the flowing pressures in production wellbores are liable to be below the bubble point which exacerbates the difficulty in conventional pressure buildup interpretation to determine the average reservoir pressure [2].

If there is no water influx ($W_e = 0$) and both N and m are unknown, the material balance is reduced to:

$$F = N(E_o + mE_g) \tag{3.50}$$

which can be also expressed as:

$$\frac{F}{E_o} = N + mN\frac{E_g}{E_o} \tag{3.51}$$

These equations offer two methods of solution for m and N. The first is to plot F as a function of $E_o + mE_g$ for an estimated value of m, as shown in Fig. 3.18a. If the value of m has been chosen correctly, the points will lie on a straight line with slope N that passes through the origin. The second method is to plot F/E_o versus E_g/E_o (Fig. 3.18b). The data points should plot on a straight line with slope mN whose intercept on the ordinate gives the STOIIP, N. Havlena and Odeh [3] suggest that the plotting method shown in Fig. 3.18a is the more powerful because the line is constrained to pass through the origin but suggest that both methods should be applied for checking purposes. The technique is illustrated with an example in reference 2.

In the event that there is a finite water influx, the technique recommended by Havlena and Odeh [3] is to differentiate equation 3.49 with respect to pressure. The resulting equation is then manipulated with equation 3.49 so as to eliminate m which results in the expression:

$$\frac{FE'_g - F'E_g}{E_oE'_g - E'_oE_g} = N + \frac{W_eE'_g - W'_eE_g}{E_oE'_g - E'_oE_g} \tag{3.52}$$

in which the primes denote derivatives with respect to pressure. A plot of the left-hand side of the equation versus the second term on the right for a selected

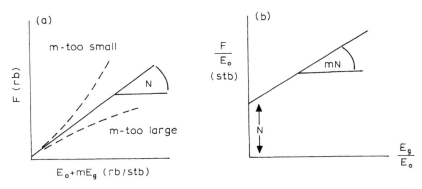

Fig. 3.18. Havlena-Odeh plots for gascap drive reservoirs: no water influx, m and N are unknowns.

aquifer model should, if the choice is correct, provide a straight line with unit slope whose intercept on the ordinate determines the STOIIP, N. This is the same form of plot as that described for waterdrive in section 3.8. Having correctly determined N and W_e, equation 3.49 can be solved directly for m.

Exercise 3.5: Application of material balance to the early production performance of a gascap drive field

Introduction

This exercise demonstrates the great sensitivity to the average pressure in solving the material balance for a gascap drive reservoir. The field is situated in a remote onshore location consequently, prior to installation of an oil pipeline, all 45 development wells had been drilled so that the required production plateau was attained as soon as the field was "switched-on". The configuration of the field is as depicted in Fig. 3.19. Since there was no gas market all the produced gas, except that used as fuel, was re-injected into the gas cap. The permeability in the reservoir is extremely high, justifying the application of material balance, but downdip in the aquifer the permeability is greatly reduced through diagenesis implying a negligible degree of natural water influx.

Question

The exercise considers two field wide pressure surveys in selected wells soon after the start of production. The details are as follows:

Survey 1. Conducted after a cumulative production of N_p = 5.684 MMstb, determined an average pressure drop of about 6–7 psi, although there was some uncertainty in this figure since different pressure gauges were used and the pressure drop is almost within the overall gauge accuracy. During this initial period there was no gas injection and the produced gas measurements failed to distinguish between R_p and R_{si}.

Survey 2. Conducted after N_p = 35.845 MMstb, G_p = 35.901 Bscf, W_p = 0, determined an average pressure drop of 30–40 psi, subject to the same uncertainty

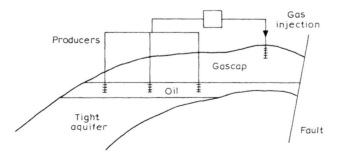

Fig. 3.19. Schematic cross-section of the gascap drive reservoir considered in Exercise 3.5.

as described for the first survey. By this stage of development the cumulative gas injection was $G_i = 17.923$ Bscf.

The field is defined as follows:

STOIIP (N) = 736 MMstb (volumetric estimate) } Reliable values based
GIIP (G) = 3200 Bscf (volumetric estimate) } on 45 drilled wells.
B_{oi} = 1.390 rb/stb
B_{gi} = 0.000919 rb/scf

From these data the HCPV's of the oil column and gas cap can be evaluated as:

HCPV_{oil} = $736 \times 1.39 = 1023.04$ MMrb
HCPV_{gas} = $3200000 \times 0.000919 = 2940.80$ MMrb

consequently $m \approx 2.9$.

$p_i = p_b = 2808.7$ psia
temperature = $157°F = 617°R$

PVT relations immediately below the initial/bubble point pressure are as follows:

$B_o = B_{oi}(1 - 2.07 \times 10^{-4}\Delta p)$, B_{oi} = 1.390 rb/stb
$R_s = R_{si}(1 - 5.27 \times 10^{-4}\Delta p)$, R_{si} = 755 scf/stb
$B_g = B_{gi}(1 + 4.13 \times 10^{-4}\Delta p)$, B_{gi} = 0.000919 rb/scf (produced gas)
$B_{gl} = B_{gli}(1 + 3.66 \times 10^{-4}\Delta p)$, B_{gli} = 0.000973 rb/scf (injected gas)

- Check whether the pressure drops recorded in the first and second surveys are consistent with the volumetric estimates of N and m. Following the second survey, determine the contributions to the total drive energy of the gas cap expansion and gas injection.

Solution
First survey. Since water influx is considered to be negligible, then during the brief initial period of production, the material balance can be expressed as:

$$F = N(E_o + mE_g)$$

which can be solved for m as:

$$m = \frac{F - NE_o}{NE_g} \qquad (3.53)$$

in which for the prevailing conditions:

$$F = N_p B_o \quad \text{(no water production: } R_p \sim R_s)$$

$$E_o = (B_o - B_{oi}) + (R_{si} - R_s)B_g, \quad E_g = B_{oi}\left(\frac{B_g}{B_{gi}} - 1\right)$$

TABLE 3.9

Estimation of the gas cap size: pressure survey 1

Δp (psi)	F (MMrb)	B_o (rb/stb)	R_s (scf/stb)	B_g (rb/scf)	E_o (rb/stb)	E_g (rb/stb)	m
5	7.8928	1.3886	753.0	0.0009209	0.000442	0.002874	3.58
6	7.8911	1.3883	752.6	0.0009213	0.000511	0.003479	2.93
7	7.8894	1.3880	752.2	0.0009217	0.000581	0.004084	2.48
8	7.8877	1.3877	751.8	0.0009220	0.000650	0.004538	2.22
9	7.8860	1.3874	751.4	0.0009224	0.000721	0.005143	1.94
10	7.8843	1.3871	751.0	0.0009228	0.000791	0.005748	1.73

Using values of $N_p = 5.684$ MMstb, at the time of the first pressure survey, $N = 736$ MMstb and the PVT correlations listed above, equation 3.53 can be solved for m as detailed in Table 3.9 for different assumed values of the average pressure drop in the reservoir. As can be seen, the results in this high-compressibility system are extremely sensitive to the interpreted pressure drop. Considering that the early gas measurements could not distinguish between R_p and R_s the values of F in Table 3.9 are likely to be slightly low and the values of m also. Therefore, the actual pressure drop of between 6–7 psi, measured in the first survey does appear to confirm the volumetric estimate of $m = 2.9$.

Second survey. By the time of this field wide survey the gas re-injection scheme had been commissioned so that the material balance can now be expressed as:

$$F = N(E_o + mE_g) + G_i B_{gi} \tag{3.54}$$

In view of the uncertainty in assessing the average pressure drop ($\Delta p \sim 30$–40 psi) and the great sensitivity to the pressure established in the first survey, the approach adopted in applying the material balance to the second survey results is to solve the equation for the pressure. That is, for selected average pressure drops between 10 and 60 psi, the left and right-hand sides of equation 3.54 are evaluated separately and when both are in "balance" the correct pressure should be determined.

Since reliable gas production measurements were possible prior to the second survey, the left-hand side of equation 3.54 can now be evaluated in its full form:

$$F = N_p[B_o + (R_p - R_s)B_g] \quad \text{(rb)} \quad (W_p = 0)$$

And since at the time of the second survey: $N_p = 35.845$ MMstb and $G_p = 35.901$ Bscf, then $R_p = 35901/35.845 = 1002$ scf/stb. Using the PVT relations presented above, values of F are evaluated in Table 3.10.

Using the volumetric determined value of $m = 2.9$ and the cumulative gas injection at the time of the survey of $G_i = 17.923$ Bscf, the right-hand side of equation 3.54 can be evaluated as listed in Table 3.11.

TABLE 3.10

Evaluation of the left-hand side of equation 3.54

Δp (psi)	B_o (rb/stb)	R_s (scf/stb)	B_g (rb/scf)	F (MMrb)
10	1.387	751	0.000923	58.021
20	1.384	747	0.000927	58.083
30	1.381	743	0.000930	58.136
40	1.378	739	0.000934	58.199
50	1.376	735	0.000938	58.300
60	1.373	731	0.000942	58.366

TABLE 3.11

Calculation of the right-hand side of equation 3.54

Δp (psi)	E_o (rb/stb)	mE_g (rb/stb)	$N(E_o + mE_g)$ (MMrb)	B_{gl} (rb/scf)	$G_i B_{gl}$ (MMrb)	RHS (MMrb)
10	0.00069	0.01755	13.425	0.000977	17.511	30.936
20	0.00142	0.03509	26.871	0.000980	17.565	44.436
30	0.00216	0.04825	37.102	0.000984	17.636	54.738
40	0.00294	0.06579	50.585	0.000987	17.690	68.275
50	0.00476	0.08334	64.842	0.000991	17.762	82.604
60	0.00561	0.10088	78.377	0.000994	17.815	96.192

The left- and right-hand sides of the material balance are plotted separately in Fig. 3.20 and their point of intersection occurs at a pressure drop of 32.5 psi at which $F = \text{RHS} \approx 58.25$ MMrb. The value of the average pressure drop corresponds to that interpreted from the second survey and again generally confirms the volumetric estimates of the STOIIP and GIIP.

For a pressure drop of $\Delta p = 32.5$ psi, the PVT parameters are:

$B_o = 1.3806$ rb/stb: $R_s = 742.1$ scf/stb
$B_g = 0.000931$ rb/scf: $B_{gl} = 0.000985$ rb/scf

Consequently, $E_o = 0.002610$ rb/stb and $mE_g = 0.05264$ rb/stb.

At the time of the second survey, the contributions of the various drive mechanisms to the total recovery are therefore:

Expansion of oil and originally dissolved gas:

$$= NE_o = 736 \times 0.002610 = 1.921 \text{ MMrb } (3.3\%)$$

Gas cap expansion:

$$= NmE_g = 736 \times 0.05264 = 38.743 \text{ MMrb } (66.4\%)$$

Gas injection:

$$= G_i B_{gl} = 17923 \times 0.000985 = 17.654 \text{ MMrb } (30.3\%)$$

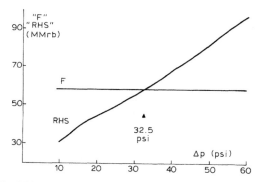

Fig. 3.20. Solution of the material balance for the pressure.

These figures illustrate the dominance of the large gas cap in supplying 66.4% of the total drive energy in the reservoir. At the time of the second pressure survey, the fractional oil recovery was $35.845/736 \approx 5\%$ of the STOIIP and the total gas drive efficiency was still high. That is, the volume of expanded gas cap and injected gas amounted to $38.743 + 17.654 = 56.397$ MMrb, yet the total volume of gas produced in excess of the solution gas was: $N_p(R_p - R_{si})B_g = 35.845(1002 - 755)0.000931 = 8.243$ MMrb. Consequently, only $8.243/56.397 = 14.6\%$ of the expanded gascap and injected gas had been produced. As the development continues, however, more and more of this gas will be produced and the eventual constraint becomes the capacity of the gas plant for processing and re-injecting the gas.

Predicting the performance of such a reservoir depends on assessing the gas–oil displacement efficiency, as described in Chapter 6, section 6.4. Fortunately, although the dip angle in the field in question was only a few degrees, the permeability in the upper parts of the reservoir was so high that the displacement efficiency was quite satisfactory.

To gain any understanding of such a gas drive reservoir it is essential that every effort be made to determine reservoir pressures with accuracy. As pointed out in Chapter 4, section 4.19, however, it is difficult enough to measure average pressures within the drainage areas of each well even in "hard" undersaturated systems and the presence of free gas makes matters worse. It is not just a reservoir difficulty but also relates to well completion practices. In such gassy fields, the degree of afterflow resulting from the surface closure of wells for routine pressure buildup surveys (Chapter 4, section 4.14b) can seriously affect the accuracy with which surveys can be analysed to determine average pressures. Furthermore, as depletion continues and more and more gas is produced, the situation can become so intolerable that test interpretation may be impossible.

Appreciating the difficulties, the operator of the field studied took the precaution of selecting ten wells in which regular pressure surveys would be conducted. These were completed with downhole mandrils that facilitated closure at the sand face at the start of buildup surveys, thus minimising (although not eliminating) the effects of afterflow. In addition all surveys were run with just three high-accuracy pressure gauges which were regularly calibrated against one another to improve the overall precision in determining the average pressure decline in the reservoir. Unless the engineer anticipates the difficulties in monitoring gas drive fields and takes such measures to improve pressure determination, history matching their performance, using either material balance or numerical simulation, becomes very problematical because there is another unknown in the equations: the pressure. In fact, it is in such fields where the numerical simulation practice of "history matching on pressure" (section 3.8c) is at its most vulnerable.

3.10. COMPACTION DRIVE

All reservoirs experience an element of compaction drive resulting from pressure depletion, the contribution to the total underground withdrawal being $(1 + m)NB_{oi}\,c_f\Delta p/(1 - S_{wc})$. It is the partial collapse of the reservoir that expels additional fluids as production. The factor regulating the degree of compaction is the *pore* compressibility c_f. This is frequently referred to in the literature as the "rock" compressibility which is a misnomer. The rock itself has a negligible compressibility and when a bulk rock sample is deformed in a uniaxial compaction experiment [2] by an amount Δh, what this represents is the change in its pore volume:

$$\Delta(\text{PV}) = c_f \times \text{(pore volume)} \times \Delta p$$

Whereas, what is physically measured in the experiment is:

$$\Delta(\text{BV}) = c_b \times \text{(bulk volume)} \times \Delta p$$

where BV is the bulk volume and c_b the bulk compressibility. But since the left-hand sides of the above expressions are equivalent then:

$$c_f = c_b \times \frac{\text{(bulk volume)}}{\text{(pore volume)}} = \frac{c_b}{\phi} = \frac{1}{\phi}\frac{\Delta h}{h}\frac{1}{\Delta p} \qquad (3.55)$$

It is worthwhile drawing attention to this point because inputting pore compressibility to reservoir engineering calculations is the sole means of accounting for reservoir compaction in studies. Therefore, to enter c_b (usually measured and reported in petrophysical SCAL experiments) instead of c_f would result in an underestimation of compaction by a factor of five for a reservoir with a twenty percent porosity.

In most reservoirs, the pore compressibility is small and remains almost constant during depletion, typical values being $c_f = 3$–6×10^{-6}/psi and under these circumstances compaction has only a marginal effect in enhancing the hydrocarbon recovery. There are occasions, however, when the compressibility may start off as small at initial conditions but increase to quite large values as pressure depletion continues. The rock mechanical theories of why such abnormal behaviour should occur are complex and often difficult to relate to actual reservoir performance but empirically it appears that two of the principal causes relate to high porosity and overpressuring, which are themselves interrelated. Whatever the cause of the overpressuring, the excess fluid pressure reduces the grain pressure between the rock particles [2] so that at initial conditions the pore system is in an artificial state for its depth of burial. Reducing the fluid pressure due to production causes a significant collapse in the pore volume which is reflected in an increase in the pore compressibility. Such behaviour was described in reference 2, relating to a well documented case history of abnormal compaction in the Bachaquero Field in Venezuela [18] but what was only briefly referred to was the practical difficulty and expense that confronted the operator of the field in overcoming the effects of surface subsidence.

Compaction of the reservoir leads to a related degree of subsidence at the surface, unless the overlying rock mantle is rigid. If the field is located at some remote surface location, the sinking of the surface is of little or no consequence because in the subsidence bowl above the field the settling is usually slow and fairly uniform, unlike the subsidence resulting from mining activities which can cause localised fracturing at the surface. Problems occur, however, if there is any body of free water at the surface in the vicinity of the field. In the case of Bachaquero, for instance, the field straddled the shoreline of Lake Maracaibo. Consequently, as the surface subsided the operator was obliged to keep on building a dyke around the lake to protect the surface facilities and oil camp that had grown up in the area. In fact, one of the main aims in performing predictive numerical simulation studies for the field was to estimate from the continued fluid withdrawal the degree of compaction and, therefore, surface subsidence, the ultimate aim being the prediction of how high the protective wall should be built as a function of time. The construction was a very costly business but fortunately most of the engineers involved were Dutch who, of necessity, have evolved as masters of the technique of building dykes. The reader may wonder why the operator of the field did not take the obvious step of inhibiting compaction/subsidence by injecting water to arrest the pressure decline. There are two reasons for this: in the first place the oil viscosity in parts of the field was extremely high making water–oil displacement very inefficient (Chapter 5, section 5.2c) and secondly, and much more significantly, the component of compaction drive contributed over 50% of the total drive energy of the field. Therefore, compaction drive was to be encouraged rather than inhibited and the dyke was built higher and higher.

A more spectacular and much more publicised case history of compaction occurred during the 1980's in the Ekofisk Field in the Norwegian sector of the North Sea. Rather than mention all the individual technical papers written on the reservoir/project engineering of the field, reference is made to a summary paper written in 1991 — "Ekofisk Field: The First Twenty Years " [19], in which twenty-three separate papers on the subject are referenced.

Ekofisk is one of the largest fields in the entire North Sea with a STOIIP of about 6000 MMstb. The initially undersaturated oil was contained in a massive fractured chalk section at a depth of 10,000 ft.ss and with a net oil column at the crest of over 1000 ft. The reservoir fluids were initially overpressured in excess of 2000 psi and the chalks exhibited exceptionally high porosities in the range of 25–48%, factors conducive to abnormal pore compressibility during depletion. Ekofisk was the first major oilfield discovered in the Norwegian North Sea, the original well being drilled in November 1969. Full scale field production commenced in May 1974, the basic recovery mechanism being volumetric depletion supplemented by the re-injection of produced gas (in excess of the sales gas contract) at the crest of the structure. A pilot water injection scheme was initiated in 1984 and since it proved successful was expanded in several phases thereafter, starting in 1987.

Seabed subsidence amounting to 10 ft was first noticed in 1984: a decade after the start of production. This had been unsuspected for two main reasons: abnormal compaction had never been reported in such a deep reservoir and reduction in well PI's, which is normally associated with reservoir compaction, was not observed. The latter appears to be because production was primarily through the, near vertical, extensive fracture system and such fractures do not necessarily close as a result of compaction, whereas in unfractured reservoirs reduction in PIs is normal. The subsidence was eventually noted from the direct observation that the production platforms were sinking rather than from any reservoir engineering consideration. Numerous, complex history matching simulation studies were conducted by all parties concerned in the development but, as commented in an editorial article (Petroleum Engineer International, November 1986):

"The models did not signal any alarms."

This is an extraordinary statement which seems to impute numerical simulation models with mystical powers. If, as input to such models, a pore compressibility function is used, based on laboratory experiments, that is too benign in its variation, then how is it possible that the simulator can be expected to inform the engineer that the input is wrong? Simulation studies invariably led to apparently satisfactory history matches over the initial 10 years of production, although these gave rise to inadequate predictions which had no permanence. Considering the equation that is implicitly being solved by the simulator for the initial period of Ekofisk production:

$$F = N(E_o + E_{fw}) + G_i B_{gi}$$

If the pore compressibility input in the E_{fw} term (equation 3.14) is too small, then in order to balance the left-hand side of the equation, something on the right-hand side must be increased. In the case of numerical simulation, there are so many degrees of freedom in attempting to match the history that temporary success can always be achieved: the STOIIP could be directly increased by altering the mapping or increasing the porosity; or the PVT altered to increase the oil volatility but these do not lead to valid history matches nor do they warn about irregular compaction/subsidence.

It is easy to be wise after an event such as occurred at Ekofisk but it is interesting to speculate what might have happened, say, in the 1950's if engineers were confronted with the same problem. They would have been obliged to apply material balance to history match the performance, which is possible with tolerable accuracy even in a field as large as Ekofisk. It would have then been noticed that using a small and constant compressibility in the above equation would have resulted in a lack of balance between both sides of the equation: the left-hand side exceeding the right, in a non-linear fashion with time, the ultimate disparity being ±500 MMrb (not an insubstantial amount) and this, in turn, may well have led to the suspicion that the pore compressibility was variable and increasing. This approach is illustrated in Exercise 3.6, in which the material balance is explicitly solved for the variable pore compressibility in a similar field situation. Prior to the Ekofisk "incident" most numerical simulation models could not cater for variable pore compressibility in their input but this situation subsequently changed rapidly.

Subsidence in offshore developments is a serious matter. It can mean that the lower decks of the platform become susceptible to damage by the "100 year wave", to the extent that the platform insurance and, therefore, operation is affected. Many suggestions were made to overcome the difficulty for the Ekofisk Field — some of them bizarre. One entrepreneur proposed, for instance, that forty moth-balled tankers, anchored in Norwegian Fjords, should be purchased and sunk close to the platforms on their windward sides to break up the threatening 100 year wave. Fortunately, someone mentioned to him that he had overlooked Murphy's Law and that such waves usually occurred every second week and came from the opposite direction! Finally, the operator embarked on one of the boldest engineering projects of all time in cutting the platform legs, jacking up the superstructures and inserting new leg sections of up to 20 ft in length. In addition a protective barrier had to be built around an offshore storage facility. The total expenditure for this exercise was in excess of $1 billion.

What has not been adequately stressed in connection with Ekofisk compaction are the positive aspects. In the first place, in spite of the cost of repairing the damage, the abnormal degree of reservoir compaction, caused by the pore compressibility increasing from 6×10^{-6}/psi to a maximum of $\sim 100 \times 10^{-6}$/psi ($\phi > 40\%$), had provided ±30% of the total drive energy by the mid-late 1980's and continued to do so which, although unanticipated, more than compensated for the costly repair. Secondly, the publicity attached to the incident has been most welcome in the Industry for it alerted operators to the potential dangers in developing offshore fields by depletion alone, without giving serious consideration to the phenomenon of compaction and its associated subsidence.

The great difficulty, however, is that even armed with the knowledge that abnormal compaction/subsidence may occur, how is it possible to quantify this in advance and build the platform 20 ft higher than required under more normal circumstances. This is expensive in itself and, if compaction is normal, could leave the operator with the tallest platform in the area. Unfortunately, at the appraisal stage, the planners are entirely reliant on the results of laboratory compaction experiments on small core plugs. These may be reliable on samples in which reservoir conditions are normal but in highly overpressured reservoirs with high porosity the mere act of hauling such distorted rock samples to the surface, where the excess fluid pressure is relaxed, means that the samples arrive in the laboratory in an entirely artificial state and due to hysteresis in tracing and re-tracing compaction trends [2], it is difficult to relate laboratory experimental results to in-situ performance in the reservoir. Furthermore, laboratory experiments on reservoir rock samples do not provide information on the rigidity of the overlying mantel which affects subsidence. In Ekofisk the mantel was originally assumed to be rigid but proved not to be so.Certainly, after the start of production material balance can be the most useful tool in defining abnormal compaction trends but by then, of course, it is just a little too late. The difficulty has not affected most fields in the large producing areas of the North Sea because pressure maintenance, mainly by water injection, has been the main recovery mechanism. It is for straightforward depletion of offshore fields that greater consideration must be given at the planning stage, although it must be admitted that this is one of the most difficult up-front decisions of all.

Exercise 3.6: Compaction drive

Introduction

The exercise studies the performance of a large offshore oilfield which experienced an abnormal degree of reservoir compaction. At the end of the fifteen year period of production history studied, the subsidence of the seabed amounted to 6 ft. The only way that the material balance can be made to "balance" to account for the production/pressure history is to allow for variable pore compressibility, in fact, the technique applied amounts to solving the material balance for the compressibility.

Question

The field being examined has extremely high porosity and is also overpressured, factors that have undoubtedly caused the abnormal degree of reservoir compaction and related surface subsidence. Initially, the reservoir contained an undersaturated moderately volatile oil, the system being defined as follows:

STOIIP ~ 3000 MMstb (reliable volumetric estimate)

p_i = 7200 psia @ 10,600 ft.ss. (overpressure = 2400 psia with respect to a water gradient of 0.45 psi/ft).

B_{oi} = 1.990 rb/stb

R_{si} = 1550 scf/stb

ϕ = 0.32

S_{wc} = 0.25 PV

c_w = 3.5 × 10^{-6}/psi

h = 230 ft average formation thickness.

Experience in the area implies that there should be no water influx the main reservoir ($W_e = 0$). The oil is contained in a tight, fractured rock and a pilot waterflood demonstrated the unsuitability of this as a practical recovery mechanism. Instead, some of the produced gas was re-injected in the hope that it would segregate in the fracture system to form a secondary gas cap. Unfortunately, premature gas breakthrough in several wells led to a reduction in the injection. The drive mechanism is therefore predominantly depletion with a diminishing amount of gas injection.

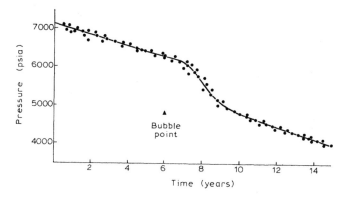

Fig. 3.21. Individual well pressures (Exercise 3.6).

Pressure decline. Individual well pressures are plotted as a function of time in Fig. 3.21. As can be seen, there is sufficient uniformity in the decline to justify application of material balance to the accumulation as a whole and this is confirmed by the production performance since practically all the wells on this large accumulation passed through the bubble point almost simultaneously. This could be directly observed by their rapidly rising GOR's, as to be expected in this fractured system (Fig. 3.10a). Furthermore, since the oil is fairly volatile, the API gravity of each wellstream rose due to the production of gas liquids from the excess gas once the bubble point had been reached, again confirming its value. The high degree of pressure equilibration in this large accumulation is attributed to transmittal of pressure directly through the extensive fracture system.

PVT fluid properties. These are listed in Table 3.12 as functions of the pressure at the end of each year, the pressures being taken from the trend shown in Fig. 3.21. On account of the moderate volatility of the oil, the production data require modification as described in Chapter 2, section 2.2e, to accommodate the gas-liquids condensed from the liberated solution gas.

Production/injection statistics. These are listed in Table 3.13.

- Calculate the initial pore compressibility above the bubble point pressure.
- Estimate the pore compressibility after 15 years of production based on the surface subsidence of 6 ft.
- Calculate the variable pore compressibility throughout the period of production history.

Solution

The oil compressibility above the bubble point pressure can be calculated from the PVT data in Table 3.12 as:

$$c_o = \frac{B_{ob} - B_{oi}}{B_{oi}\Delta p} = \frac{2.030 - 1.990}{1.990\,(7120 - 6300)} = 24.51 \times 10^{-6}/\text{psi}$$

TABLE 3.12

PVT functions (Exercise 3.6)

Year	Pressure (psia)	B_o (rb/stb)	R_s (scf/stb)	B_g (rb/scf)	B_{gI} (rb/scf)	r_s (stb/MMscf)
0	7120 (p_i)	1.990	1550			
6	6300 (p_b)	2.030	1550		0.00065	100
7	6175	2.000	1450	0.00061	0.00066	98
8	5625	1.825	1160	0.00065	0.00070	70
9	5075	1.675	940	0.00070	0.00075	53
10	4820	1.620	860	0.00073	0.00077	45
11	4620	1.565	790	0.00075	0.00080	40
12	4475	1.540	750	0.00076	0.00082	37
13	4310	1.515	700	0.00079	0.00085	33
14	4150	1.470	640	0.00081	0.00088	30
15	4000	1.455	610	0.00083	0.00090	28

B_{gI} = FVF of the injected dry gas (rb/scf)
r_s = condensate yield of the solution gas (stb/MMscf)

TABLE 3.13

Oil production/gas injection statistics

Year	q'_o (stb/d)	N'_p (MMstb)	G_p (MMscf)	R'_p (scf/stb)	Q_{inj} (MMscf/d)	G_i (MMscf)
1	4000	1.460	2368	1622		
2	21595	9.342	14497	1552		
3	20953	16.990	26482	1559		
4	18564	23.766	37019	1558		
5	134575	72.886	111843	1534	228	83050
6	210066	149.560	231096	1545	365	216180
7	194973	220.725	361270	1637	271	315144
8	193392	291.313	542541	1862	173	378415
9	159301	349.458	754957	2160	138	428667
10	124616	394.943	981541	2485	142	480625
11	99923	431.415	1192577	2764	99	516918
12	93348	465.487	1397891	3003	68	541809
13	99521	501.812	1638823	3266	166	602536
14	82301	531.852	1869561	3515	109	642170
15	64759	555.489	2076158	3738	150	696865

q'_o and N'_p relate to the total liquid hydrocarbon production: black oil + condensate.

Then applying the depletion material balance to the first six years of under-saturated production history listed in Table 3.13, including the two years of gas injection:

$$N_p B_o = N B_{oi} c_{eff} \Delta p + G_i B_{gI}$$

the effective compressibility can be evaluated as:

$$c_{eff} = \frac{149.56 \times 2.03 - 216180 \times 0.00065}{3000 \times 1.99 \times 820} = 33.31 \times 10^{-6}/\text{psi}$$

Therefore, the initial pore compressibility (equation 3.26) is:

$$
\begin{aligned}
c_f &= c_{eff}(1 - S_{wc}) - c_o S_o - c_w S_{wc} \\
&= [33.31(1 - 0.25) - 24.51 \times 0.75 - 3.5 \times 0.25] \times 10^{-6} \\
&= 5.73 \times 10^{-6}/\text{psi}
\end{aligned}
$$

If it is assumed that the six foot of seabed subsidence is roughly the same as the degree of compaction in the reservoir, then the pore compressibility after 15 years of production can be approximated as:

$$c_f = \frac{1}{\phi}\frac{\Delta h}{h}\frac{1}{\Delta p} = \frac{1}{0.32}\frac{6}{230}\frac{1}{3120} = 26.13 \times 10^{-6}/\text{psi}$$

Since the surface subsidence is usually less than the reservoir compaction, then this figure must be regarded as a low estimate of the increasing pore compressibility.

On account of the moderate volatility of the oil allowance must be made for the fact that the liberated gas produces liquid condensate at the surface. The means of doing this has already been presented in Chapter 2, section 2.2e. The "black oil" rate can be evaluated as:

$$q_o = q_o'\frac{1 - R'r_s/10^6}{1 - R_s r_s/10^6}$$

This function is evaluated in Table 3.14 using the average pressure, rate, GOR and condensate yield during each year and leads to the cumulative production statistics listed in Table 3.15.

TABLE 3.14

Conversion of production data to allow for condensate production from the liberated gas

Year	Pressure (psia)	q_o' (Mstb/yr)	R' (scf/stb)	R_s (scf/stb)	r_s (stb/MMscf)	q_o (Mstb/yr)
7	6237	71165	1829	1500	98	68475
8	5900	70588	2568	1305	84	62177
9	5350	58145	3653	1050	61	48281
10	4947	45485	4982	900	49	35967
11	4720	36472	5786	825	42	28600
12	4547	34072	6026	770	39	26872
13	4392	36325	6633	725	35	28618
14	4230	30040	7681	670	31	23373
15	4075	23637	8740	625	29	17972

q_o' = oil + condensate rate (Mstb/year)
R' = average GOR during the year (scf/stb oil + condensate)
r_s = average condensate yield (stb/MMscf)
q_o = black oil rate (Mstb/year)

TABLE 3.15

Calculation of the underground withdrawal F

Year	N_p (MMstb)	R_p (scf/stb)	B_o (rb/stb)	R_s (scf/stb)	B_g (rb/scf)	F (MMrb)
7	218.035	1657	2.000	1450	0.00061	463.601
8	280.212	1936	1.825	1160	0.00065	652.726
9	328.493	2298	1.675	940	0.00070	862.491
10	364.460	2693	1.620	860	0.00073	1078.105
11	393.060	3034	1.565	790	0.00075	1276.659
12	419.932	3329	1.540	750	0.00076	1469.779
13	448.550	3654	1.515	700	0.00079	1726.316
14	471.923	3962	1.470	640	0.00081	1963.587
15	489.895	4238	1.455	610	0.00083	2187.989

Since the cumulative gas production is: $G_p = N_p R_p = N'_p R'_p$, this modification to the PVT allows the left-hand side of the material balance, F, to be evaluated in terms of N_p and R_p using black oil PVT properties. Below the bubble point, the material balance can be expressed as:

$$F = N(E_o + E_{fw}) + G_i B_{gI} \tag{3.56}$$

in which:

$$F = N_p[B_o + (R_p - R_s)B_g] \quad (rb)$$

$$E_o = (B_o - B_{oi}) + (R_{si} - R_s)B_g \quad (rb/stb)$$

$$E_{fw} = B_{oi}\frac{(c_w S_{wc} + c_f)}{1 - S_{wc}}\Delta p \quad (rb/stb)$$

Values of the left-hand side of the material balance equation, F, are calculated in Table 3.15, while the right-hand side is evaluated in Table 3.16. If E_{fw} is calculated after 15 years using the value of $c_f = 5.73 \times 10^{-6}$/psi determined during the first six years of undersaturated production, its value would be $E_{fw} = 0.0547$ rb/stb. Inserting this in the material balance would lead to the evaluation of the right-hand side as 1526.87 MMrb which is 30% less than the value of $F = 2188$ MMrb at that time. This demonstrates the significance of the abnormal compaction and requires that the material balance equation should, in this case, be solved for the variable pore compressibility, c_f. That is, equation 3.56 is solved as:

$$E_{fw} = \frac{F - NE_o - GB_{gI}}{N}$$

and then c_f can be calculated as:

$$c_f = \frac{E_{fw}(1 - S_{wc})}{B_{oi}\Delta p} - c_w S_{wc} \tag{3.57}$$

TABLE 3.16

Calculation of the variable pore compressibility

Year	F (MMrb)	$G_i B_{gi}$ (MMrb)	E_o (rb/stb)	$N E_o$ (MMrb)	E_{fw} (rb/stb)	c_f (1/psi $\times 10^{-6}$)
7	463.601	207.995	0.0710	213.00	0.0142	4.79
8	652.726	264.891	0.0885	265.50	0.0408	9.41
9	862.491	321.500	0.1120	336.00	0.0683	11.71
10	1078.105	370.081	0.1337	401.10	0.1023	15.89
11	1276.659	413.534	0.1450	435.00	0.1427	20.64
12	1469.779	444.283	0.1580	474.00	0.1838	25.31
13	1726.316	512.156	0.1965	589.50	0.2082	27.05
14	1963.587	565.110	0.2171	651.30	0.2491	30.74
15	2187.989	627.179	0.2452	735.60	0.2751	32.36

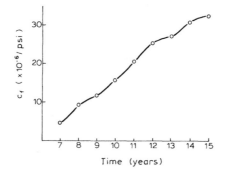

Fig. 3.22. Variable pore compressibility (Exercise 3.6).

These values are listed in the final two columns of Table 3.16 and c_f is plotted in Fig. 3.22. These reveal a sixfold increase from $c_f = 5$ to $30^+ \times 10^{-6}$/psi during the last 9 years of the production history. After 15 years of production, the contribution of the compaction term alone is:

$$\frac{N B_{oi} c_f \Delta p}{1 - S_{wc}} = \frac{3000 \times 1.99 \times 32.36 \times (7120 - 4000)}{0.75} = 803.7 \text{ MMrb}$$

which is 37% of the total underground withdrawal.

3.11. CONCLUSION

It is hoped that the chapter has demonstrated that volumetric material balance is not some anachronism that can now be dispensed with but is instead a vital tool in gaining an insight into reservoir mechanics and evaluating the hydrocarbons in place. By its very nature it is the antithesis of the technique of reservoir simulation which follows the current trend in reservoir engineering, that of "differentiation":

dividing a reservoir into discrete cells. Unfortunately, it is difficult to appreciate the *wholeness* of something by splintering it and examining the pieces. As soon as a reservoir is divided into cells problems naturally arise in assigning them individual rock properties and in accounting for fluid flow from one grid block to the next (relative permeabilities) which adds layers of assumptions in attempting to define the reservoir system.

Material balance, on the other hand, implies "integration", the reservoir being treated as a zero dimensional "black box" which may contain all manner of complications: fractures, severe heterogeneity, horizontal wells, etc. all of which are sublimated to the overall balance which, once an average pressure decline has been defined for the system, relies solely on the production, pressure and PVT data. This is the great strength of the technique that permits the definition of the STOIIP and drive mechanisms, as illustrated in the exercises in the chapter. It is not uncommon for engineers to attempt to add sophistication to their calculations by subdividing a reservoir into two or three grid blocks and applying material balance to each. This practice, however, should be treated with caution, because again it requires the handling of fluid fluxes between the blocks which raises the same problems as described above for numerical simulation. The motto of material balance application is "the bigger the better". The technique applies constraint on the engineer compared to simulation. In the latter, parametric values can be changed in each grid block to facilitate a match on the reservoir performance, whereas in applying material balance, this option simply isn't available. The limitation of material balance is in its ability to predict reservoir performance. Provided the future pressure decline in the reservoir can be guaranteed to be uniform (Fig. 3.2), predictive methods can be applied in simple systems but in cases where there is a lack of uniformity (Fig. 3.3a) the usefulness of material balance is greatly diminished.

Above all, numerical simulation and material balance must not be regarded as competitive techniques: we have too few tools in reservoir engineering to discard any of them. Material balance is for history matching whereas simulation is for prediction and in this respect they should be supportive.

REFERENCES

[1] Schilthuis, R.J.: Active Oil and Reservoir Energy, Trans., AIME 1936.
[2] Dake, L.P.: Fundamentals of Reservoir Engineering, Elsevier, Amsterdam, 1978.
[3] Havlena, D. and Odeh, A.S.: The Material Balance as the Equation of a Straight Line, JPT, August 1963.
[4] Havlena, D. and Odeh, A.S.: The Material Balance as the Equation of a Straight Line. Part II — Field Cases, JPT, July 1964.
[5] Matthews, C.S., Brons, F. and Hazebroek, P.: A Method for the Determination of Average Pressure in a Bounded Reservoir, Trans., AIME., 1954.
[6] Muskat, M.: The Production Histories of Oil Producing Gas-Drive Reservoirs, Journal of Applied Physics, 1945.
[7] Muskat, M.: Physical Principles of Oil Production, McGraw-Hill, New York, N.Y., 1949.
[8] Arps, J.J. and Roberts, T.G.: The Effect of Relative Permeability Ratio, the Oil Gravity and the Solution Gas–Oil Ratio on the Primary Recovery from a Depletion Type Reservoir, Trans., AIME, 1955.

[9] Wahl, W.L., Mullins, L.D. and Elfrink, E.B.: Estimation of Ultimate Recovery from Solution Gas Drive Reservoirs, Trans., AIME, 1958.

[10] Torcaso, A.M. and Wyllie, M.R.J.: A Comparison of Calculated k_{rg}/k_{ro} Ratios with a Correlation of Field Data, Trans., AIME 1936.

[11] Honarpour, M. et al.: Empirical Equations for Estimating Two Phase Relative Permeability Measurements, JPT, December 1982.

[12] Dias-Couto, L.E. and Golan, M.: General Inflow Performance Relationship for Solution–Gas Reservoir Wells, JPT, February 1982.

[13] Fetkovich, M.J.: The Isochronal Testing of Oil Wells, SPE 4529, Las Vegas, October 1973.

[14] van Everdingen, A.F. and Hurst, W.: The Application of the Laplace Transformation to Flow Problems in Reservoirs, Trans., AIME, 1949.

[15] Fetkovich, M.J.: A Simplified Approach to Water Influx Calculations — Finite Aquifer Systems, JPT, July, 1971.

[16] Carter, R.D. and Tracy, G.W.: An Improved Method for Calculating Water Influx, Trans., AIME, 1960.

[17] Fanchi, J.R.: Analytical Representation of the van Everdingen-Hurst Aquifer Influence Functions for Reservoir Simulation, SPE-Reservoir Engineering, June 1985.

[18] Merle, H.A. et al.: The Bachaquero Study — A Composite Analysis of the Behaviour of a Compaction/Solution Gas Drive Reservoir, JPT, September 1976.

[19] Sulak, R.M.: Ekofisk Field: The First 20 Years, JPT, October 1991.

Chapter 4

OILWELL TESTING

4.1. INTRODUCTION

Returning from his round-the-world cruise, the Victorian naturalist Charles Darwin was heard to mutter:

"If Mother Nature can, She will tell you a direct lie".

Considering Darwin's position of eminence in the history of science, such a blunt warning can be of little encouragement to the practising reservoir engineer for whom well testing happens to be the most direct interface with devious Mother Nature. Therefore, in an attempt to limit the possible harmful effects of either direct lies or even self delusion it is recommended that the whole subject of well testing: design, execution and analysis be kept as simple as practicably possible and such is the basic theme of this chapter.

Perhaps the most important part of the chapter is the description, right at the start, of the essential observations (the full data set) required for meaningful test interpretation. This is followed by an account of the purpose of well testing both in appraisal and development wells. The former, which is also referred to in Chapter 2, is worth considering in detail on account of the importance and expense of appraisal well testing, especially when conducted offshore. Concerning the theoretical description of test interpretation there are three main themes running through the chapter:

(1) The abandonment of the perennial assumption of transience.

(2) Concentration on the flowing performance of wells prior to closure for a buildup or a rate change.

(3) The extension of Miller, Dyes, Hutchinson (MDH) analysis to all forms of test interpretation.

The first of these is an attempt to purge the subject of the persistent and usually unjustified assumption of transience (infinite acting behaviour) which finds its way into so many technical papers and computer software packages. Historically there was an excuse for the assumption since the transient solution of the diffusivity equation is one of the most straightforward of all and our predecessors, armed computationally with only slide rules and log tables, were obliged to invoke such simplifying assumptions to make any progress at all in test interpretation. But it is quite inexcusable that this relic should persist in these days of high-tech interpretation. Throughout the chapter warnings and demonstrations of the fallibility and consequences of assuming transience are referred to on fifteen separate occasions;

often a more general approach renders familiar techniques redundant but in some cases it offers new scope in well test interpretation such as in the application of two-rate flow testing in which the traditional assumption of transience has limited its use. Above all, it is stressed that the condition of transience can never be assumed it must always be proven.

A very high proportion of tests conducted in the Industry are pressure buildup surveys and traditionally almost exclusive attention has been focused on the buildup in static pressure itself, with scant regard to what happened during the preceding pressure drawdown. Yet it is argued that the latter is the dynamic phase of the test, the buildup being merely is static reflection which is mathematically more ambiguous. Taking due regard of what occurred during the drawdown, even in a qualitative sense, can greatly assist in the analysis of the buildup and lead to a reduction in buildup times.

Horner plotting is still the most popular form of buildup interpretation yet it gives rise to serious errors if the technique is not fully understood. The trouble stems from the fact that the flowing time prior to well closure is required in the analysis and related to this is the need to extrapolate to infinite closed-in time; together these are the root cause of so many mistakes in interpretation. What is demonstrated in the chapter, however, is that *anything* that can be done with Horner plotting can usually be performed in a simpler, more elegant and less error prone manner using the much maligned MDH plot, which in the majority of applications does not require involvement of the flowing time nor the dangerous extrapolation to infinite closure. There is no attempt to perpetuate the competition between the methods that has persisted since the 1950's. Instead, both Horner and MDH plots (and their time derivatives) must be used in conjunction and their results cross checked for physical consistency.

The chapter concentrates on selected topics in oilwell testing, it being impossible in a general textbook on reservoir engineering to cover the entire spectrum of activities in this ever expanding subject. Gas well testing is not specifically referred to but it will be appreciated from chapter 8 of reference 3, that all the techniques described here for oilwell testing are equally appropriate provided that real gas pseudo-pressures are substituted for pressures and allowance is made for a component of rate dependent skin.

4.2. ESSENTIAL OBSERVATIONS IN WELL TESTING

It is the most important concern in writing this chapter that the engineer should appreciate the full range of observations necessary for complete test interpretation, as described below.

(a) Rate, pressure, time

The most common and practical method of testing wells, which is concentrated upon in this chapter, is the pressure buildup test for which the rate pressure,

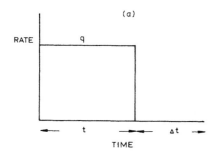
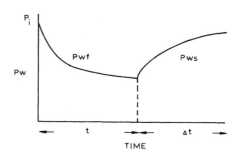

Fig. 4.1. (a) Rate and (b) pressure profile during a pressure buildup test.

time profiles are as shown in Fig. 4.1. Ideally, the well is produced at a constant rate, q (stb/d), for a flowing time, t (hours), after which it is closed-in for a pressure buildup. During the flowing period, the pressures recorded on a gauge in the wellbore are denoted as p_{wf} (psia-pressure, wellbore, flowing) and during the subsequent buildup, p_{ws} (psia-pressure, wellbore, static) which is measured as a function of the closed-in time Δt (hours). The data collected throughout the test are therefore the rate, pressure, time records which are analysed, as described in the following sections, to determine the reservoir pressure and formation characteristics. There has been a tendency, however, which seems to have been encouraged in recent years through the development of sophisticated software interpretation packages, to regard the matching of the rate–pressure–time records with mathematical functions (curve fitting) as the sole requirement in test analysis. It is not. There are also the seven observations listed below which are essential in gaining a comprehensive understanding of the full "message" of the test. Furthermore, even in restricting observations to rate, pressure and time, many have an apparent fixation with the buildup data (p_{ws}, Δt) almost to the exclusion of the drawdown response (p_{wf}, t), the latter being of vital importance in appreciating the subsequent buildup performance, as stressed frequently throughout the chapter.

(b) Core/log data

A fundamental assumption implicit in the majority of well test, interpretations (section 4.5a) is that the formation is "homogeneous acting". Before making such an assumption, however, it is necessary to check on the degree of reservoir heterogeneity by careful examination of the core and log data collected in the well and, in particular, the permeability distribution revealed by the former. Shown in Fig. 4.2 are two permeability distributions across reservoirs determined from routine core analysis. To inspect these data in true perspective it is essential, as emphasised in Chapter 5, section 5.5, to plot the permeabilities on a linear rather than the more conventional logarithmic scale. For a reservoir to act as homogeneous in the context of well testing, it is necessary that there should be complete pressure equilibrium across the formation throughout the flow and buildup periods thus implying a lack of barriers to vertical fluid movement. In a section such as depicted in Fig. 4.2a, it

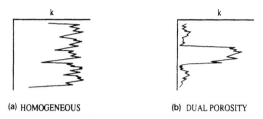

(a) HOMOGENEOUS   (b) DUAL POROSITY

Fig. 4.2. Schematic of permeability distributions across a reservoir section.

would be fairly safe to assume that with the random distribution of permeabilities across the clean sand the reservoir would display homogeneous behaviour and properties determined, such as permeability, would be the thickness averaged values.

The formation depicted in Fig. 4.2b, however, displays a severe contrast in permeability and represents the type of reservoir that displays dual porosity behaviour. The definition of such systems is that flow into the wellbore is strictly through the higher porosity/permeability layer or layers and the only contribution from the poorer quality rock is by cross-flow into the highly conductive channels through which the oil from the tighter intervals is produced. Not only does this definition apply to layered reservoirs with a large contrast in rock properties across the section but it also caters for naturally fractured reservoirs in which cross-flow occurs from the tight matrix blocks into the fracture system through which the oil is exclusively produced into the wellbore. The description of dual porosity systems is both different and more complex than for homogeneous acting reservoirs [1] and it is therefore necessary, prior to undertaking the interpretation, that the engineer be quite sure of the validity of the assumption concerning the homogeneity of the system under test — or lack of it. In this respect, every test interpretation report should *commence* with a plot of the core permeability distribution and composite log interpretation coupled with a clear statement concerning the assumptions made governing the manner in which the formation is to be modelled.

The description of dual porosity behaviour and means of recognising it and interpreting tests in such formations (using type-curve and time-derivative type-curve analysis techniques) has received much attention in the literature since the early 1980s. Yet hardly ever in the intensely theoretical papers written on the subject is the practical suggestion made that the engineer should carefully inspect the core (with expert geological advice — if necessary) and the permeability distribution, before making the all important decision on whether the formation is homogeneous acting or is likely to display dual porosity (fractured) behaviour on test. As an example of this neglect in basic observation — at a technical seminar in London in the late 1980s, several different consultancy groups marketing computer software were permitted to demonstrate their wares by interpreting several well tests for which only the rate–pressure–time records were made available. One of the tests exhibited a degree of upward curvature towards the end of the pressure buildup and half the consultancy groups interpreted the response as due to the presence of lateral boundaries (faults) while the other half decided it was the behaviour of a dual porosity system — since upward curvature in the buildup could manifest either reservoir condition. In fact, the right answer is that the physical condition was quite indeterminate, because the logs and core description necessary to make the decision were not available and none of the consultants should even have attempted the analysis. It merely reinforces the statement made in (a), above that modern test analysis is becoming more and more focused on the mathematical intricacies of "curve fitting" the rate–pressure–time data to the exclusion of more practical considerations based on observation.

(c) RFT, pressure–depth profiles

As mentioned in section 2.7 of Chapter 2, RFT pressure surveys conducted in exploration/appraisal wells are not usually of great interest because they invariably reveal the condition of apparent hydrostatic equilibrium established over geological time — almost irrespective of the nature of the reservoir section. The most useful application of the tool is in surveying newly drilled development wells, prior to running the production casing, when the reservoir can be viewed under dynamic conditions. That is, the pressure profile is directly influenced by the continuous production of previously drilled development wells and the resulting pressure sink propagated to the location of the new well. Inspection of such dynamic pressure profiles is of great assistance in deciding whether the formation is "homogeneous acting", which is the conventional assumption in test analysis, or whether the physical condition is more complex. In this respect the profiles must be viewed in conjunction with the core/log data described in (b), above. Dynamic surveys are illustrated in Fig. 4.3: the homogeneous reservoir (a), demonstrating a uniform pressure gradient across the reservoir compared to the heterogeneous case (b), indicating different degrees of pressure communication/depletion between the individual sands in the section. In the latter case, conventional pressure buildup testing is extremely difficult to interpret and running flowmeter surveys should be considered as an alternative, as described in section 4.20c.

Since its innovation in the mid-1970's, the RFT has been employed as a field development tool rather than one for "academic" circles. By that it is meant that it has received little attention in the universities and research institutes from which the majority of theoretical papers on well test analysis emanate. Considering that the majority of such papers rely on the basic assumption of formation homogeneity,

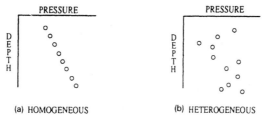

Fig. 4.3. Dynamic RFT surveys in development wells.

then it becomes the responsibility of the field engineer, based on the inspection of core/RFT data, to decide whether these theoretical papers can be applied in practice.

(d) Geological model

At the exploration/appraisal stage of field development there will naturally be considerable uncertainty in mapping the hydrocarbon accumulation: refinement of the model being attained with the drilling of each new appraisal well. Nevertheless, the engineer is obliged to keep in close contact with the exploration/production geologists, throughout the lifetime of the project, and take full account of their current structural interpretation (Fig. 4.4). Of particular interest for test analysis are the intensity and position of faults with respect to the well which will influence the drawdown/buildup pressure responses (section 4.16) and the proximity of any oil–water or gas–oil contacts. The last is of considerable importance for if there is a free gas cap the entire test will be affected from its design to the eventual analysis. The presence of free gas in the reservoir means that the underlying oil is invariably at its saturation or bubble point pressure at initial conditions and production of oil must inevitably lead to the liberation of free gas in the region of pressure drawdown close to the wellbore. Under these circumstances it is imperative to close-in the well downhole, at the start of the buildup, to avoid the damaging and largely uninterpretable effects of afterflow. Failure to appreciate the presence of a free gas cap can also lead to the invalid prediction of reservoir depletion on account of the retarded nature of the pressure buildup.

(e) Drive mechanism

In interpreting a test, it is necessary that the engineer appreciates the nature of the reservoir drive mechanism; whether it is primary: depletion (solution gas drive), waterdrive or gas drive, or secondary: water or gas injection. The complications associated with the liberation of free gas in the reservoir have been alluded to in (d), above and the situation must be recognised from the outset. Natural water or gas drive will provide an element of pressure support during the flow period, particularly

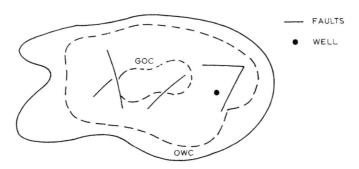

Fig. 4.4. Geological model including faults/fluid contacts.

the latter, that will influence the nature of the pressure buildup (sections 4.13 and 4.18) and which must be recognised when attempting to estimate the hydrocarbons in place in extended well tests (section 4.19h). Secondary recovery operations usually aim at pressure maintenance through water or gas injection which necessitates modifications to conventional test analysis, as described in section 4.19c.

(f) PVT fluid properties

For oilwell testing, the fluid properties specifically required are: the bubble point pressure, to assess whether the wellbore flowing pressure is below that level thus giving rise to the evolution of gas around the wellbore, the oil formation volume factor. B_o (rb/stb), oil viscosity, μ_o (cp), and compressibility, c_o. The latter can be calculated using equation 3.24, and together with the water and pore compressibilities is used to evaluate the total compressibility, c, using equation 4.2. These basic parameters are required in the calculation of the kh-product of the formation (equation 4.36) and the skin factor (equation 4.37).

In situ fluid densities (oil, gas water) are necessary to calculate pressure gradients both in the reservoir and the flow string. That is, for the interpretation of RFT surveys (Chapter 2, section 2.7) and the adjustment of DST pressures from their depth of measurement in the well to the selected datum level in the reservoir.

In the past there was a difficulty in that it often took months for the full, experimental PVT results to be received from the laboratory, necessitating a revision of the original test analysis performed using provisional PVT parameters derived possibly from standard correlations. Nowadays, however, instant and accurate PVT analysis is available on the wellsite [2] which alleviates the problem.

(g) Well completion

That is, whether the well has an open or cased hole completion. If the former, the entire net pay section exposed may be considered to contribute to production whereas for a cased hole test, especially in the event of partial perforation of the formation (Chapter 2, section 2.9), there is often uncertainty as to the extent of the total net section which is actually producing into the wellbore. Also included under this heading is whether the well has been stimulated prior to the test by acidisation or hydraulic fracturing.

(h) Equipment

Includes both the downhole and surface hardware and in the former case, whether the test is being conducted with DST equipment or with a conventional production packer and tubing. Of prime importance in this respect is whether the well is closed-in downhole or at the surface at the commencement of the pressure buildup, which affects the severity and duration of afterflow (section 4.14b). It is also necessary to know (specify) the types of pressure gauges used and their position in the flow string with respect to the reservoir datum depth.

At the surface, a record must be kept of the flowing wellhead pressure and temperature during the drawdown/buildup periods. This can prove of assistance in checking on the malfunction of downhole equipment, such as leaks during the buildup or variations in the fluid content of the string during the flowing period due, for instance, to water production. Also necessary is continual monitoring of the pressures/temperatures of the various stages of separation to enable reliable recombination of surface samples of oil and gas for PVT analysis [3]. This further requires the accurate measurement of the GOR during the production period which serves as the most significant indicator of whether the formation pressure has fallen below the bubble point during the drawdown period.

(i) Tests in neighbouring wells

If other wells have been tested on the same accumulation then their results and analyses should be compared with those from the current test. Consistency in interpretations is reassuring but contrasts can also prove significant. If, for instance, all previous tests demonstrated an element of pressure support which was absent in the well under consideration then the reason for this must be sought: is the well confined, or separated from direct contact with a gas cap, unlike the other wells.

Considering the number and complexity of the observations necessary for meaningful test interpretation, it will be appreciated that the modern trend in merely "curve fitting" the pressure buildup with mathematical functions does not do justice to the subject. It was promoted during the nineteen eighties as the technique of solving the "Inverse Problem" which, broadly speaking, means using mathematics to define the physical state of a system. Naturally the concept and its attempted application are fraught with difficulties, not least of which is that it happens to be putting the cart before the horse: when tackling physics/engineering problems, the conventional approach is to define the physical state of a system and then seek the appropriate mathematics to describe it. When dealing with unseen, underground hydrocarbon accumulations, however, there is an obvious difficulty in describing the physics first and consequently there has always been an element of mathematical pattern recognition in the subject. Nevertheless, the engineer must be continually on guard against the allure of matching test responses with what appears to be a convincing mathematical model for, as described in section 4.8, there is an inevitable lack of mathematical uniqueness associated with test analysis which on occasions can become complete (section 4.18).

By far the soundest approach to test interpretation is that the engineer must collate all the observations listed above (and any more that may seem relevant) in the analysis. Yet such advice is hardly ever proffered in the never ending stream of papers on the subject of testing for the simple reason that most of treatise emanate from universities and industrial research institutes in which the practical aspects of well testing are seldom if ever considered. Yet even in the very rare case of a paper that acknowledges that there might be more to testing than the solution of differential equations, the emphasis on priorities is often misplaced [57].

"The potential for nonuniqueness of an interpretation implies that one should consult *external* data before concluding that any particular model that matches the transient data provides the correct interpretation."

The external data referred to are the core and log data, the geological model and drive mechanism, etc., which far from being peripheral to the interpretation are the very guts of the physical problem; it is the mathematical model describing the observed physical state of the system under test that must be regarded as the "side show". Mathematics is, by its very nature, always correct, the perpetual difficulty is whether it can be made to match the remote, underground physics with any degree of uniqueness — a point which is frequently stressed throughout the chapter.

When all the above observations are duly accounted for it often happens that the test is far too complex to analyse with any of the analytical models available to the industry. If, for instance, a well is producing below the bubble point pressure in a fractured reservoir with a nearby fault and surface closure is affected at the start of the buildup, then it would not only be unwise to attempt a quantitative analysis but could also prove harmful if development decisions were based on a set of numbers emerging from any such interpretation. Failure to analyse tests in a quantitative manner is not a crime, the main aim should always be to gain a sound understanding of the physical state/performance of the reservoir which will permit sound judgements to be made concerning operational decisions.

4.3. WELL TESTING LITERATURE

Regarding the "popular" literature on oil and gas well testing the main emphasis over the years has been on the following topics:

(A) Analysis of tests in formations with low flow capacity (kh-product).
(B) Tests conducted in development wells in mature producing fields.
(C) The testing of reservoirs in which primary depletion is the main drive mechanism.

Consequently, testing under these circumstances has implicitly conditioned thinking in the subject. In many practical situations which confront the modern engineer, however, these three conditions require expanding to include the following:

(D) Analysis of tests conducted in moderate and high flow capacity reservoirs.
(E) Tests conducted in exploration and appraisal wells.
(F) The testing of reservoirs in which there is a degree of pressure maintenance provided either by nature or by engineered secondary recovery activities.

By "popular" literature it is meant texts such as the two excellent SPE monographs [4,5] (the Blue Books) published in 1967 and 1977 which occupy space on most Reservoir Engineers' bookshelves. Alternatively, papers do exist on testing under conditions D–F, above but these, while being readily available, form part of what might be described as the "specialist" literature which is frequently overlooked by the hard-pressed field engineer responsible for designing and interpreting well tests. An example of this is the comprehensive body of literature on the subject of testing reservoirs subjected to partial or full pressure maintenance [6,7,8] described in sections 4.18 and 19c.

Concerning conditions A–C; these have predominated in the literature mainly for historic/economic reasons and it must be remembered in this respect that many of the "classic" papers on well testing were written in the 1950's and 60's and related to onshore developments in the U.S.A., the birthplace not only of well testing but of Reservoir Engineering itself.

It is not suggested for a moment that every reservoir tested in the past in the United States was of low flow capacity as implied by A, above. Instead the condition is implicit in the persistent, simplifying assumption that the most common pressure response observed during any test is that of pure transience, during which pressures recorded on a gauge in the wellbore are unaffected by any boundary condition. The reservoir therefore appears to be infinite in extent which leads to a great simplification in the mathematics used in test analysis. Since this condition is mainly satisfied in the testing of low-permeability reservoirs it justifies the assertion made in condition A, above. The *ad hoc* assumption of transience prevailing at all times and in all reservoirs is one that has done a great deal of harm in the Industry and its validity is continually challenged throughout the chapter.

Conditions B and C have arisen primarily for historic reasons. During the development of the basic theoretical and practical approaches to well testing the main interest was in the routine tests conducted in development wells in the numerous mature producing fields in the United States. Besides, as described in Chapter 1, section 1.4, onshore the distinction between appraisal and development wells is rather vague anyway. Condition C results mainly from the suppression in the rise in oil prices during the twenty five years prior to 1973, also referred to in Chapter 1, section 1.3. At less than $2 per barrel, secondary recovery and the resulting pressure maintenance was a luxury only considered for a few selected fields. Therefore, unless nature provided pressure support by means of water influx from an aquifer or gas cap expansion, depletion drive was the order of the day. Consequently the majority of technical papers on testing relate to this recovery mechanism in which the reservoir pressure usually declines continually as a function of time.

This chapter not only describes testing in terms of its historical development, summarised in conditions A–C, but also takes account of the additional requirements necessary to conduct and interpret tests under the conditions suggested by D–F, above. These find particular application in the development of offshore fields although their relevance is not by any means restricted to that environment. Offshore, and particularly in deep water provinces such as the U.K. sector of the North Sea, the cost of installing permanent production platforms is so prohibitive that operators can only consider the development of reservoirs with moderate to high permeability. To do otherwise would prove to be uneconomic, hence the special requirement to cater for condition D. Also, in offshore developments appraisal wells tend to be treated as such and abandoned after full evaluation rather than used subsequently as development wells. This necessitates the inclusion of condition E which is elaborated upon in Chapter 2, section 2.9. Finally, since the rapid escalation of oil prices in the 1970's, pressure maintenance by the application of secondary recovery methods (engineered water and gas drive) has become almost standard practice. This is especially the case in major offshore oilfield developments where, as described in Chapter 5, section 5.2 there is a need to apply such techniques to increase and accelerate oil recovery and to insure against the failure event in which nature does not supply any reasonable degree of pressure support. This justifies inclusion of condition F above, but pressure maintenance is also frequently encountered when testing at the appraisal stage. In this case, it is nature, under a variety of circumstances described in section 4.6b, that supplies the energy.

Inclusion of conditions D–F is intended to broaden the reader's view of the subject of well testing and encourage further reading of the "specialist" literature on some of the subjects raised. By no means does it add complication to what may be already regarded as a difficult subject. For instance, if there is a high degree of pressure support (condition F), the steady-state solution of the diffusivity equation required to describe the condition is just about the simplest in the business. It is the appreciation, based on observation, of the need to apply this solution in preference to others that is of prime importance.

4.4. THE PURPOSE OF WELL TESTING

As noted in the previous section, the main concentration in theme in the extensive literature on well testing has been directed towards routine tests conducted in development wells in mature fields; little attention being focused on the testing of exploration and appraisal wells which will therefore be dealt with first in this section.

(a) Appraisal well testing

As described in Chapter 1, section 1.4, when developing onshore fields, there is little distinction between appraisal and development because, if possible, the appraisal wells are brought on continuous production from an early stage and integrated into the overall project. Exploration/appraisal wells (referred to in this section simply as appraisal wells) are largely, but not exclusively, an offshore phenomenon. They are drilled both to establish the hydrocarbons in place and the productivity of the reservoirs discovered, with the overall purpose of enabling development decisions to be made concerning the economic viability of the project. In general, the appraisal wells cannot be produced on a continuous basis once completed since the offshore production facilities and oil transportation system may not be installed for several years. Some appraisal wells are suspended upon drilling and "tied back" into the project once the facilities are available but in many cases they are simply abandoned once the formation evaluation is complete. The advantage with the latter is that they can be "tested to destruction": without regard to perforating policy and its compatibility to the long term field development. Since an underlying aim in appraisal well testing is to investigate potential difficulties that might adversely affect the long term production prospects of the accumulation, then testing wells which are to be abandoned gives the engineer a high degree of flexibility in accomplishing this aim.

Offshore appraisal well tests in such areas as the North Sea have been amongst the biggest and most expensive ever conducted by the Industry and therefore merit special attention that has not been afforded them in the literature before. The tests are invariably pressure buildups following a lengthy flow period in which DST equipment is used; the order of priority in data collection usually being:

(A) Measurement of the oil production rate (q, stb/d)
(B) Calculation of the skin factor (S, dimensionless)
(C) Collection of fluid samples
(D) Evaluation of formation characteristics (permeability, fractures, layering)
(E) Influence of boundary conditions (fault patterns, depletion)
(F) Determination of initial reservoir pressure (p_i, psia)

At first sight, this order may appear strange but can be justified item by item, as follows:

A. Rate determination

A new formation has been penetrated and naturally the prime interest of all concerned, for strictly commercial reasons, is — what fluid is produced and at what rate.

B. Skin factor

The skin factor, S, is a dimensionless number representing the degree of formation damage caused by the positive pressure differential between the wellbore and formation while drilling, which leads to an invasion of the latter by drilling mud whose solids particles are retained in the pores close to the wellbore thus reducing permeability in this restricted region. In an appraisal well the amount of damage done will be significant on account of the high (safe) mud weight used when drilling into reservoirs in which there is uncertainty in the pressure; whereas subsequent development wells will be drilled into the reservoir using a refined completion fluid with the intention of reducing the damage in an average well to a negligible level ($S = 0$). It is because of this difference that the skin factor must be carefully determined in appraisal wells so that the effect of its complete removal in development wells can be catered for in project planning.

The efficiency with which a well produced is defined in terms of its productivity index, PI, defined as

$$PI = \frac{\text{Oil rate}}{\text{Pressure drawdown}} = \underbrace{\frac{q}{p_i - p_{wf}}}_{\text{Observed}} : \underbrace{\frac{q}{p_i - p_{wf} - \Delta p_{skin}}}_{\text{Ideal } (S = 0)} \quad \text{(stb/d/psi)} \quad (4.1)$$

The larger the PI, the greater the oil rate for a given pressure drawdown and the smaller the number of wells required to develop the accumulation. The first of the above expressions is the directly observed PI during a test in which the total drawdown, $p_i - p_{wf}$, includes the additional component of pressure drop across the skin, Δp_{skin}, in the immediate vicinity of the wellbore, as defined in section 4.7 (equation 4.18). In an average development well, however, the skin factor and, therefore, the pressure drop across it are both reduced to zero. Consequently, to calculate the PI of an ideally completed well, PI_{ideal} (equation 4.1), requires the subtraction of the calculated value of Δp_{skin} in the appraisal well test from the observed pressure drawdown. Values of k and S required to calculate the component of pressure drop across the skin can be evaluated by conventional buildup analysis as described in sections 4.12 and 4.13.

Based on the evaluation of ideal PIs in appraisal wells, the engineer can make a reasonable assessment of the number of wells required to develop the accumulation. This is very important in offshore developments since it leads to an estimate of the size of the well-deck containing the slots through which the development wells will be drilled and their bulky wellheads and flow lines accommodated which, in turn, dictates the very size and strength of the offshore platform itself. It is therefore considered that accurate determinations of the rate and skin factor, which facilitate the calculation of ideal PIs, are the most important parameters to ascertain in appraisal well testing and, as described in Chapter 2, section 2.9, to ensure that valid measurements are made, it is essential to aim at perforating appraisal wells, on test, across the same intervals as anticipated for future development wells.

C. Fluid sampling

The importance of collecting reliable fluid samples for full PVT analysis has already been described in Chapter 2, section 2.2. The advantage in sampling wells under static conditions at the appraisal stage is that the exact source of the sample is known as the mid-point of the perforated interval. The aim is to collect oil and gas samples, whether downhole or at the surface, in the ratio: 1 stb oil + R_{si} scf gas. The PVT parameters resulting from laboratory analysis are used directly in the test analysis, in all manner of recovery calculations and in the design of surface topsides equipment.

D. Formation characteristics

The permeability of the formation can be calculated by identification of the early linear trend on a semi-log pressure buildup plot (sections 4.12 and 4.13) or using type-curve analysis (section 4.21). The value so determined is the average, effective permeability to oil in the presence of irreducible water. As noted in Chapter 5, on the subject of waterdrive, the value of the average permeability of a formation is not as important in reservoir engineering calculations as the permeability distribution across the reservoir, obtained from routine core analysis. Nevertheless, the permeability evaluated in tests finds a use in modelling well performance, using either analytical or numerical simulation techniques, for matching/predicting reservoir performance.

The test permeability should be compared with the average, absolute permeability determined from routine core analysis. Usually, the former is smaller because it is the "effective" value. Sometimes, however, it proves to be larger and this either means that the linear section of the pressure buildup has been incorrectly chosen and has too small a slope (sections 4.13 and 4.18) or else there are sections of the core with very high permeability that have not been recovered and are therefore not included in the thickness averaging.

Dual porosity or fractured drawdown/buildup behaviour is more complex to analyse than for homogeneous acting systems (section 4.2b) and requires lengthy tests to observe firstly the flow exclusively in the high-permeability layers or fractures, the transition period as the tighter matrix rock contributes to production and finally, flow from the entire system. It is also necessary to close-in the well downhole to observe the first of these buildup responses which may be of brief duration (although a counter argument to this practical point is forwarded in reference 1). This author has little to add the esoteric analysis techniques developed during the 1980's for dual porosity/fracture analysis other than to again stress that the most reliable method of ascertaining whether a reservoir is of this type is by careful inspection of the core data. A warning is also given in section 4.21b on the dependence of the shape of the time-derivative type curve, used to identify such complex systems, on the manner in which the pressure buildup is plotted: it is possible to diagnose a reservoir as displaying dual porosity behaviour when it is merely a mathematical hallucination.

It is often impossible and, indeed, meaningless to attempt to analyse conventional buildup tests performed in reservoirs which are layered in such a manner that there are impermeable barriers separating the individual productive sands. This is to be expected since as described in Chapter 5, section 5.7, if a well is fully perforated over such a sand section, which is effectively commingling a set of discrete reservoirs within a single wellbore, there is inevitably a loss of reservoir engineering control (not just in well test analysis but throughout the entire subject) which simply must be accepted. This is not to suggest that such commingled production constitutes malpractice, far from it, for when developing such fields with multiple, isolated thin sands it is the only means of completion which is economically viable. Perhaps the soundest approach to testing such formations, in which there may be cross-flow between sands while flowing and certainly upon well closure, is to dispense with conventional buildup tests altogether and instead perform production logging (flowmeter) surveys while flowing the well at a series of different rates. The method and analysis technique, which aims at determining the PI and pressure of each layer is described in section 4.20c. Although application of the technique is at its most useful in development well testing, it can also be used at the appraisal stage.

E. Boundary conditions

The basic problem in analysing tests to detect reservoir boundaries, or indeed to attempt to quantify anything occurring away from the wellbore, is that implicitly the interpreter is faced with the problem of solving second-order differential equations to determine the boundary conditions. This is a "back-to-front" procedure for solving such equations which is fraught with a lack of uniqueness in solution. Such tests, conducted at the appraisal stage, and their analysis is described in section

4.16. To locate boundaries (faults) requires a long flow test: sufficient to explore the reservoir to a depth of at least four times the distance to the fault, otherwise its presence may not be manifest during the subsequent pressure buildup. As such, the tests are expensive, especially if conducted offshore, and they should only be planned provided the engineer feels confident that a successful outcome, in terms of defining the position of boundaries, will lead to some useful development decision.

Otherwise, reliance should be placed on the geophysicists and geologists to delineate faults. The Industry remunerates these specialists in the most generous fashion to do just that and this engineer is prepared to rely on their judgement in such matters — in the majority of cases.

The observation of reservoir pressure depletion during an appraisal well test is a very serious matter. That is, after producing only a few thousand barrels of oil from a newly discovered reservoir, the pressure has fallen and does not appear to be building-up to the initial value, p_i. This would imply that a small volume of oil was being tested which may not prove commercially viable to develop, especially if located offshore. Unfortunately, other than in the most obvious cases it is very difficult to be certain whether the reservoir is depleting or not and in many cases depletion is diagnosed when it has not occurred at all. In this respect, the main source of error is the invalid assumption that the late time pressure trend of a Horner buildup plot (section 4.14) should extrapolate in a linear fashion to give the reservoir pressure at infinite closed-in time. There is nothing in the basic mathematics of well testing, however, to support this premise: other than in a few simple and well defined cases such as a well testing a truly infinite acting system. Normally, the late time pressure trend of a Horner plot is distinctly non-linear and, if the well could be closed-in for long enough would eventually rise towards the initial pressure. If there is genuine depletion, an attempt can be made to calculate the STOIIP being drained by application of material balance, as described in section 4.19h.

F. Initial pressure

The determination of the initial reservoir pressure in an appraisal well test is relegated in the order of priorities not because it is unimportant but rather because it is better measured by other means. Since the mid-1970's it has become standard practice in expensive appraisal wells to run the RFT (or its successor, the MDT) prior to setting the final production casing. This has the advantage of providing a pressure–depth relationship across the reservoir section (Chapter 2, section 2.7) and furthermore, the pressure is measured at the exact depth of the probe. In the case of pressures measured in a well test, however, there are two potential sources of error both of which are associated with extrapolation. In the first place, the pressure buildup plot requires some form of extrapolation to determine the initial pressure (sections 4.13 and 4.15) which, as described in E, above, can be a highly subjective matter. Secondly, the pressure gauges are usually located in the flow string at quite different depths from the formation under test; either above or below but usually the former. This requires that pressures measured at the gauge depth must be extrapolated to the reservoir datum level using the pressure gradient of the fluids in the flow string between the gauges and perforations; but the nature of the fluids in this section of the string and their gradient are sometimes uncertain. Consequently, the use of RFT pressures, if available, is to be preferred.

(b) Development well testing

This is the routine form of test (usually pressure buildups) conducted in each development well throughout the lifetime of the project. For such tests, which have received the bulk of attention in the popular literature, the order of priority in data collection is normally:

(A) Measurement of the average pressure within the drainage area of the well ($\bar{p}$, psia)
(B) Calculation of the skin factor (S, dimensionless)
(C) Determination of formation properties (permeability, fractures)

A. Average pressure

Measurement of the pressure, which was assigned the lowest priority in appraisal well testing is now elevated to prime position. The RFT, which is basically an open-hole tool, can no longer be used; besides it would prove too expensive to run on a routine basis, even if it were physically possible. Instead, well pressures are measured by conducting buildup or two-rate flow tests in homogeneous acting reservoirs or flowmeter surveys in layered reservoirs in which there is a lack of pressure equilibrium across the formation.

Under reasonably stable production conditions in a reservoir, each well will tend to "carve-out" its own drainage area, as described in section 4.6a, and it is the main purpose in the test analysis to ascertain the average pressure, $\bar{p}$, within this area at the time of the survey. In this manner, the pressure decline record for each well and of the reservoir as a whole can be determined, the latter by applying the pressure averaging process described in section 4.6a. A reservoir model, either analytical or numerical, is then structured and applied in an attempt to history match the declining pressures and if this can be achieved the model may be considered sufficiently reliable to predict future performance — this being one of the most fundamental techniques in the subject of reservoir engineering. Yet the accurate determination of pressures in development wells has proven, over the years, to be one of the most difficult tasks confronting engineers in the whole subject of test interpretation. The reason for this is the same as described in section 4.4a, (E): boundary conditions and the requirement to specify the area drained by each well, its shape and the location of the well with respect to the boundaries, which can be a highly subjective assessment yet is necessary to calculate the average well pressure (section 4.19e). Since the advent of numerical simulation modelling, the emphasis in this subject has switched to matching measured test pressures to those calculated in each grid block containing a well at the time the individual surveys were conducted (section 4.19f). While this innovation would appear to offer an advantage over the more traditional methods its accuracy is downgraded in practice by the application of the technique of "history matching on pressures" (Chapter 3, section 3.8c). The evaluation of development well pressures for different flowing conditions is described in section 4.19.

B. Skin factor

Evaluation of the skin factor once again occupies second place in order of priority and therefore, on aggregate, is the single most important parameter obtained in either appraisal or development well testing. Wells can become damaged for a variety of reasons which may be categorised as both operational and pertaining to the formation itself and its fluid production. To quote Krueger [9]:

> "Laboratory and field studies indicate that almost every operation in the field-drilling, completion, workover, production and stimulation — is a potential source of damage to well productivity."

Production itself can cause damage due to the deposition of movable fines, chemical reactions, scaling and gas blockage around the well when producing below the bubble point pressure. Whatever the reason for the damage, the first step in affecting a cure must be the calculation of the magnitude of the skin factor, S, by detection and analysis of the initial, linear pressure trend on semi-log buildup plots (sections 4.12 and 13) or through the application of type-curve analysis (section 4.21). If the skin factor is large and the nature of the damage can be identified, then a remedial workover treatment, such as acidisation, can be performed to reduce or eliminate the skin and thus enhance the PI of the well-provided it is economically justified to do so.

C. Formation properties

The effective permeability of development wells can decrease during the producing lifetime of the reservoir and, as for the skin factor, its value can be determined using either semi-log buildup plots or type-curve analysis.

In a depletion-type field one of the most common causes of permeability reduction is reservoir compaction (Chapter 3, section 3.10). This can have such a severe effect on well productivities that additional development wells may require drilling to maintain the desired oil offtake rate. Since it is very difficult to anticipate this eventuality, based on static data collected at the appraisal stage, this phenomenon can cause particular problems in offshore projects where the scope for drilling additional development wells is often strictly limited. In waterdrive fields, reduction in effective permeability is usually associated with the breakthrough of injected water in production wells which diminishes the oil relative permeability.

In low-permeability reservoirs, where commercial production rates can only be sustained through hydraulic fracturing, it is necessary to check in routine well surveys whether the latest stimulation treatment has proved effective and the fractures remained open. The subject of test interpretation for fractured wells has received considerable attention in the literature [10–12] but is not concentrated upon in this chapter. Parameters that can be obtained from the mathematically complex analysis (which generally relies on type-curve matching) include the fracture half length and conductivity (fracture permeability × width) together with the skin factor. If the well is closed-in for long enough, the conventional semi-log buildup plots described in sections 4.12 and 4.13 can be applied to calculate the skin factor: a large negative value of -3 to -5 indicating the continued efficacy of the fractures.

Concerning the relevance of the other measurements which were important in appraisal well testing, the oil rate is regularly monitored for each producer throughout the lifetime of the project and although oil samples are collected routinely, they are seldom used for PVT analysis on account of uncertainty in the location of origin of the sample in the reservoir. Designing buildup tests to locate boundaries/fault positions is unnecessary because their positions can usually be inferred by direct observation of the production/pressure performance of individual wells or by performing simple interference tests (section 4.17). Therefore, the main objectives in routine development well testing are the determination of average well pressures and the investigation of effects that may diminish well productivity.

4.5. BASIC, RADIAL FLOW EQUATION

The steps involved in the derivation of the radial diffusivity equation (solutions of which form the basis of most test analysis techniques) were described in some detail in chapter 5 of reference 3 and the reader will therefore be spared any repetition of the mathematical derivations required in the formulation of the equation. Instead, this section reviews the physical assumptions implicit in deriving the equation and investigates the soundness of its assumed linearity.

(a) Radial diffusivity equation

The physical model considers the horizontal flow of a single-phase fluid inward to a wellbore located at the centre of a radial volume element. The assumptions implicit in the derivation of the radial flow equation are that:

- the formation is both homogeneous and isotropic
- the central well is perforated across the entire formation thickness
- the pore space is 100% saturated with any fluid.

The first assumption may appear restrictive but, as pointed out in sections 4.2b and c, provided there is pressure equilibrium across the formation during the test (no restriction to vertical fluid movement) then even if the reservoir is heterogeneous in terms of variation in permeability/porosity in the vertical section, the reservoir will "present itself" as homogeneous and formation properties determined in the analysis will be thickness averaged values. Fortunately, this is the most common reservoir condition encountered in testing and can be recognised by detailed inspection of the core/log data and observation of the degree of pressure equilibrium demonstrated by RFT surveys run under dynamic, producing conditions. Dual porosity systems (section 4.2b), in which the low-permeability intervals produce by cross-flow into the better quality sands and thence into the wellbore; and the even more complex dual permeability reservoirs, in which flow from the poorer sands can be both by cross-flow and directly into the wellbore, do not satisfy the first of the above physical requirements and the test analysis is more complex analytically [1].

The second assumption, that the well is fully perforated across the section, could perhaps be better stated as a necessary condition. The combination of the first two assumptions means that flow into the wellbore is purely radial and, in fact, reduces the mathematical description to one dimensional, radial flow. Sometimes, as in the case or reservoirs subject to basal waterdrive, there is no alternative to partial penetration but unless it is necessary the practice should be avoided since it leads to an indeterminacy in the calculated PI (Chapter 2, section 2.9) and reduces the prospect of attaining any meaningful test analysis results.

The requirement of 100% saturation of any fluid is simply a convention adopted in well testing: that all volumes used in calculations are the total pore volume, PV. The fact that the reservoir contains an irreducible water saturation that can change in volume, as can the pore space itself, is accommodated by using an effective compressibility

$$c = c_o S_o + c_w S_{wc} + c_f \tag{4.2}$$

which, when used in the definition of compressibility: $dV = c[PV]\Delta p$, is multiplied by the total pore volume. This differs from the effective compressibility defined by equation 3.26, which is multiplied by the hydrocarbon pore volume: $dV = c[HCPV]\Delta p$, but it will be noted that the values of dV calculated in both cases are equivalent. As mentioned above, the difference is simply a matter of convention: in the subject of material balance (Chapter 3) HPCVs are used whereas in test analysis it is PVs.

If the above assumptions/conditions are satisfied, then combining the basic physical principles of mass conservation, Darcy's law and isothermal compressibility [3], the basic radial flow equation may be derived as:

$$\frac{1}{r} \frac{\partial}{\partial r} \left(\frac{k\rho}{\mu} r \frac{\partial p}{\partial r} \right) = \frac{\phi c \rho}{0.000264} \frac{\partial p}{\partial t} \tag{4.3}$$

which is a second-order differential equation relating the dependent variable, the pressure, p, to the position, r, in the radial element and the time, t. Unfortunately, the equation is non-linear, meaning that it contains coefficients, $k\rho/\mu$, and $\phi c \rho$, which are themselves pressure dependent. Because of this complication, it is not possible to determine direct analytical solutions for use in well test analysis. It is first necessary to linearize the equation so that it can be formulated in such a manner that it contains no pressure dependent coefficients. The traditional method in which the linearization has been affected for liquid flow (undersaturated oil or water) is by the process of deletion of terms [3,13]. In this, the left-hand side of the equation is expanded by the chain rule for differentiation then, provided the following physical conditions are satisfied:

- the parameters μ, k, ϕ, c are largely independent of pressure

- the pressure gradient $\partial p/\partial r$ is small, so that the square of this term, which appears on the expanded left-hand side of the equation, is negligible
- the product $cp \lll 1$.

equation 4.3 is reduced to the form

$$\frac{1}{r}\frac{\partial}{\partial r}\left(r\frac{\partial p}{\partial r}\right) = \frac{\phi\mu c}{0.000264\,k}\frac{\partial p}{\partial t} \tag{4.4}$$

which is the radial diffusivity equation and, in form, is one of the most common in the entire subject of physics. The reciprocal of the coefficient on the right-hand side, $k/\phi\mu c$, is the hydraulic diffusivity constant and is a fundamental grouping of parameters that plays a major role in the whole subject of Reservoir Engineering, as already described in connection with the application of material balance (Chapter 3, section 3.3). In the context of well testing, the higher the value of the constant then usually the greater the depth of investigation into the reservoir so that, even in tests of moderate duration, boundary effects such as sealing faults exert their influence on the pressure response observed on a gauge located in the wellbore. It should be noted that the compressibility, c, in equation 4.4 and in all others throughout the chapter is the effective value defined by equation 4.2.

(b) Investigation of the validity of linearizing the basic radial flow equation by the method of deletion of terms

This approach to linearization of equation 4.3 has been traditionally accepted in the subject on account of its simplicity, the alternative: the application of integral transformations, being regarded as more complex and requiring the use of computers in test interpretation. Nowadays, however, computer packages of great mathematical sophistication are used universally and since the resolution of pressure gauges has vastly improved since the early 1970's, it is worthwhile investigating the validity of the method of linearization by deletion of terms and, in particular, check if its accuracy is commensurate with that aimed at in modern test analysis with such improved tools/techniques at our disposal.

It was Dranchuk and Quon [13] who stated the most stringent condition for the traditional linearization, that $cp \lll 1$, and to emphasise the point they used three "less-than" symbols; but precisely "how small is small" is left to the reader to decide, based on the level of accuracy required. They noted, however, that while a value of $cp = 0.10$ will introduce a 10% error in the differential equation, this is not necessarily reflected in its eventual solution. The extreme case, of course, is for the flow of a real gas for which, to a first approximation [3], the compressibility equals the reciprocal of the pressure so that the cp-product is practically equal to unity. Under these circumstances, Al-Hussainy et al. [14] introduced in 1966 the concept of the real gas pseudo-pressure:

$$m(p) = 2\int_{p_0}^{p} \frac{p\,\mathrm{d}p}{\mu Z} \tag{4.5}$$

in which the lower limit of integration, p_o, is some chosen base pressure. As described in chapter 8 of reference 3, when this integral transformation is substituted directly in the non-linear differential equation 4.3, mere cancellation of terms results directly in the evolution of the diffusivity equation 4.4, in which the pseudo-pressure, $m(p)$, replaces the pressure, p, as the dependent variable. In reaching this stage it is not necessary to invoke any of the three conditions stated above for linearization by deletion of terms. Solutions of the modified diffusivity equation, expressed in terms of $m(p)$ functions are then used directly in gas well test analysis.

A similar mathematical trick (for that is all it amounts to) can be also applied to undersaturated oil, which may prove necessary if the cp-product is not small. Consider defining an integral transformation (pseudo-pressure) of the form [13]:

$$m(p) = \int_{p_o}^{p} \frac{\rho}{\mu} \, dp \qquad (4.6)$$

incorporating the pressure dependent parameters, ρ and μ in the integrand (p_o is again a convenient base pressure). Then

$$\frac{\partial m(p)}{\partial r} = \frac{\partial m(p)}{\partial p} \cdot \frac{\partial p}{\partial r} = \frac{\rho}{\mu} \frac{\partial p}{\partial r}$$

and

$$\frac{\partial m(p)}{\partial t} = \frac{\partial m(p)}{\partial p} \cdot \frac{\partial p}{\partial t} = \frac{\rho}{\mu} \frac{\partial p}{\partial t}$$

which, on substitution into equation 4.3 and after cancelling terms, leads directly to the equation

$$\frac{1}{r} \frac{\partial}{\partial r} \left(r \frac{\partial m(p)}{\partial r} \right) = \frac{\phi \mu c}{0.000264 \, k} \frac{\partial m(p)}{\partial t} \qquad (4.7)$$

without having to satisfy the conditions necessary for linearization by deletion of terms other than that k and ϕ are constants. The equation is referred to as quasi-linear in that there may be a pressure dependence in the μc-product in the diffusivity constant. For undersaturated oil, μ increases with pressure while c decreases, so that their product is reasonably constant. If it does display significant variation, however, as can occur when testing low-permeability reservoirs, then it is necessary to generate and use a second integral transformation [15,16], referred to as the pseudo-time:

$$t_A = \int_{t_o}^{t} \frac{dt}{\mu c} \qquad (4.8)$$

and substitution of this into equation 4.7 results in the formulation of the diffusivity equation as

$$\frac{1}{r}\frac{\partial}{\partial r}\left(r\frac{\partial m(p)}{\partial r}\right) = \frac{\phi}{0.000264\,k}\frac{\partial m(p)}{\partial t_{A}} \tag{4.9}$$

which is strictly linear, provided ϕ and k are constant. Other authors [17,18] have even attempted to cater for pressure dependence of these two variables but the difficulty is in defining such a relationship based on laboratory compaction experiments which may bear little relationship to the *in situ* condition in the reservoir, close to the wellbore.

In an attempt to check on the necessity for using pseudo-pressure and time integral transformations, data from a deep (high-pressure) reservoir containing a highly volatile oil have been examined for which the maximum value of the cp-product is $cp = 29.91 \times 10^{-6}$ (psi^{-1}) $\times$ 7514 (psia) $= 0.225$, which according to the standards of Dranchuk and Quon [13], is a large number. The oil has a formation volume factor of $B_{ob} = 3.469$ rb/stb at the bubble point pressure of 5217 psia and a solution gas–oil ratio of $R_{si} = 3470$ scf/stb and is undersaturated by almost 2800 psi at initial conditions. The degree of volatility can be seen in Fig. 4.5a which is a plot of the oil formation volume factor. Immediately below the bubble point the function declines extremely rapidly: the shrinkage ($B_{ob} - B_o$) amounting to 32% for a 1000 psi pressure reduction. This is the type of reservoir in which the pressure must not be allowed to fall below the bubble point (Chapter 2, section 2.2) either in the reservoir itself or particularly near the wellbore, where the oil shrinkage could lead to a severe decline in well productivity. It is therefore the intention to develop the accumulation by pressure maintenance above the bubble point (5217 psia) through water injection.

The PVT properties (ρ, μ) in the undersaturated pressure range are listed in Table 4.1. Values of the integrand, ρ/μ, and its average value over each pressure step, $\overline{\rho/\mu}$, are then calculated from which the pseudo-pressure may be evaluated as (column 6):

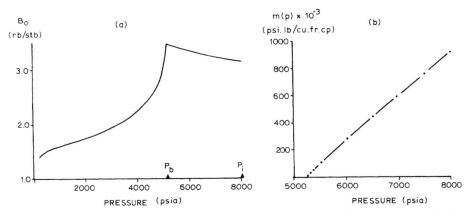

Fig. 4.5. (a) Formation volume factor. (b) Oil pseudo-pressure for an undersaturated, volatile oil.

$$m(p) = \int\limits_{5217}^{p} \frac{\rho}{\mu}\,\mathrm{d}p = \sum_{5217}^{p} \frac{\bar{\rho}}{\mu}\,\Delta p$$

The plot of $m(p)$ versus pressure is shown in Fig. 4.5b. As can be seen, apart from the region close to the bubble point (5217–5500 psia) the relationship between $m(p)$ and p is essentially linear, meaning that even for this highly volatile oil, it is unnecessary to resort to the use of an oil pseudo-pressure to linearize equation 4.3. Values of B_o are listed in column 7 of Table 4.1, from which the oil compressibility is evaluated as described in Chapter 3, section 3.7a as:

$$c_o = \frac{1}{B_{oi}} \frac{B_o - B_{oi}}{p_i - p} \tag{3.24}$$

and the effective compressibility, c, is calculated using equation 4.2 for $c_w = 3\times10^{-6}/$ psi, $c_f = 8 \times 10^{-6}/$psi, $S_{wc} = 0.10$ PV. Finally, the μc-product is listed in column 10 which displays a maximum variation of just over 1%, implying that the application of the pseudo-time integral transformation is also unnecessary for this undersaturated oil.

Therefore, by considering the case of a high-volatility, high-pressure oil with large cp-product, all this section has established is that even for this case the linearization of equation 4.3 using the traditional technique of deletion of terms is valid and the method is "robust" — more so than many would give it credit for. It must be appreciated, however, that this demonstration does not constitute a proof and when testing in a new reservoir the $m(p)$ function should be plotted and its linearity checked as a function of pressure and variation in the μc-product must also be inspected. If necessary, either or both of the integral transformations, equations 4.6 and 4.8, should be used in the linearization of the basic radial flow equation 4.3 and many of the computer packages on the market cater for this option. In the remainder of this chapter it is assumed that linearization by deletion of terms is appropriate and this has been checked in all the exercises/examples provided.

4.6. CONSTANT TERMINAL RATE SOLUTION OF THE RADIAL DIFFUSIVITY EQUATION

It is solutions of the radial diffusivity equation 4.4, or 4.9, if necessary, that are applied in the majority of well test analysis techniques. This, in itself, presents problems because for any second-order differential equation there is an infinite number of possible solutions, dependent on the choice of initial and boundary conditions. Therefore, we may eventually anticipate an infinite number of technical papers on the subject of Well Test Analysis and, at the time of writing, it is estimated that only about half of them have been written! Confronted with such a plethora of technical advice, the engineer must decide which technique is best suited to the test being planned. Fortunately, the choice is not as difficult as may be imagined. Invariably, the mathematical formulations presented in technical papers are perfectly correct but in an attempt to force the mathematical description on

TABLE 4.1

Generation of the oil pseudo-pressure and μc-product for an undersaturated, volatile oil

$\overline{p}$ (psia)	ρ (lb/cu.ft)	μ (cp)	ρ/μ (lb/cu.ft·cp)	$\overline{\rho/\mu}\Delta p$ (lb/cu.ft·cp)	$m(p)$ (psi·lb/cu.ft·cp)	B_o (rb/stb)	c_o (psi^{-1})	c (psi^{-1})	μc (cP·psi)
8014	33.46	0.1103	303.4	153750	924085	3.166			
7514	33.09	0.1062	311.6	174942	770335	3.204	24.01×10^{-6}	29.91×10^{-6}	3.18×10^{-6}
6961	32.59	0.1015	321.1	145062	595393	3.252	25.80×10^{-6}	31.52×10^{-6}	3.20×10^{-6}
6515	32.15	0.0976	329.4	167960	450331	3.296	27.39×10^{-6}	32.95×10^{-6}	3.22×10^{-6}
6014	31.65	0.0928	341.1	174450	282371	3.352	29.37×10^{-6}	34.73×10^{-6}	3.22×10^{-6}
5514	31.03	0.0870	356.7	54360	107921	3.417	31.71×10^{-6}	36.84×10^{-6}	3.20×10^{-6}
5363	30.84	0.0849	363.3	18597	53561	3.440	32.65×10^{-6}	37.69×10^{-6}	3.20×10^{-6}
5312	30.78	0.0841	366.0	17981	34964	3.448	32.96×10^{-6}	37.96×10^{-6}	3.19×10^{-6}
5263	30.65	0.0833	367.9	16983	16983	3.459	33.64×10^{-6}	38.58×10^{-6}	3.21×10^{-6}
5217	30.53	0.0824	370.5			3.469	34.22×10^{-6}	39.10×10^{-6}	3.22×10^{-6}

the physics of the reservoir and fluid system, far too many assumptions have to be invoked which may seem quite acceptable in the academic environment of a university or research institute but in the field fall far short of being practical.

The solution of the diffusivity equation which may be regarded as the "basic building block" in all test interpretation, upon which more complex analyses may be structured, is called the constant terminal rate (CTR) solution. This describes the pressure response observed on a gauge located in a wellbore resulting from producing a well at a constant rate, q, from time $t = 0$. This, it will be recognised, is an idealised solution; because those who have attended a well test, particularly at the appraisal stage, will appreciate how difficult it is to stabilize the flow rate from time $t = 0$. The ideal solution, however, can be modified to cater for variable rate history, as described later in this section. From its definition, the reader may also imagine that the CTR solution will only be appropriate for the description of a test conducted in a newly discovered reservoir in the exploration or subsequently drilled appraisal wells. This is not the case, however, and it does not matter whether the well is in a developed field and has been producing for twenty years at the time the survey is conducted, the CTR solution is still the basis of the mathematical description of the pressure response. This may not seem obvious and therefore, at present, it might prove helpful for the reader to consider the development of the following basic theory in terms of appraisal well testing. The extension of the application of the CTR solution to testing development wells in mature fields will then be described in section 4.19.

The pressure response in the wellbore due to producing a well at a constant rate can be described in terms of two extreme physical conditions which relate to the outer boundary condition of the system under test, these are:

(a) Bounded reservoir condition

For this, the well is surrounded by a no-flow boundary of arbitrary shape. The rate and separate phases of the wellbore pressure decline are as depicted in Fig. 4.6.

The point to note in this solution is that the pressure declines continuously as a function of time from the initial value of $p = p_i$ at $t = 0$. Since the system is bounded, this is perfectly logical and consistent with material balance considerations. It is this form of CTR solution that has been primarily featured in the literature and indeed, the following schematics appear in practically every text book on Well Testing, including this author's former work [3]. The pressure decline may be subdivided into three phases as follows:

A. Transient: 	During this initial pressure decline from $p = p_i$, a pressure recorder suspended in the wellbore is totally unaffected by the presence of any faults or boundaries out in the reservoir. In this respect, the system appears to be infinite in extent.
B. Late transient: 	If the boundary surrounding the well is, for instance, a $2:1$ rectangle in which the well is asymmetrically located, then during the period of late transience some but not all of the boundaries affect the pressure response in the wellbore.

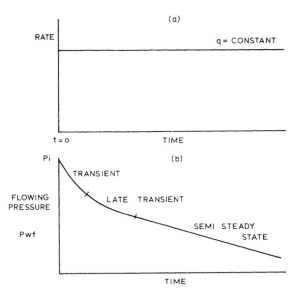

Fig. 4.6. Constant terminal rate performance. (a) Production rate. (b) Decline in bottom hole flowing pressure.

C. Semi-steady state: During this phase, all the outer boundaries influence the pressure response and if the well is producing at a constant rate, the rate of change of pressure with respect to time is also constant.

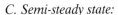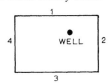

$$\frac{\mathrm{d}p_{\mathrm{wf}}}{\mathrm{d}t} = \text{constant}$$

A. Transience

The duration of the initial period of transience is dependent on the magnitude of the diffusivity constant, $k/\phi\mu c$, described in the previous section. Generally, the larger its value, the sooner discontinuities in the reservoir will influence the wellbore pressure and therefore the period of transience will be short. Conversely, in very-low-permeability reservoirs, the transient phase may extend for months rather than hours. Any pressure disturbance caused in the reservoir, such as opening a well to flow, closing it in or even changing its rate will induce a transient pressure response, identification and isolation of which permits the engineer to apply the simple transient solution of the diffusivity equation to the pressure–time record to calculate the permeability and skin factor of the formation under test. This is, therefore, the opening move in any test analysis because, quite apart from their intrinsic value, a knowledge of these parameters is necessary before indulging in any more sophisticated form of interpretation.

As mentioned in section 4.3, perhaps one of the more serious errors that has been traditionally made in the subject is the false *assumption* that in many forms of testing the wellbore pressure response is always in the transient condition. As mentioned previously, it is quite inadmissible that the engineer should make such assumptions and especially in the case of well testing where the whole purpose of the exercise is to investigate and learn about the system under test rather than impose conditions upon its behaviour. The historical reason for making this assumption is both obvious and understandable since the transient solution of the diffusivity equation is very simple and, as such, was essential to use in the era of the "slide rule" to obtain any form of result from the analysis. There is, however, no longer any reason for making this simplifying assumption although it still seems to be as widespread as ever. The danger is, of course, that during the period when the reservoir is assumed to be acting in a transient manner it may have, in fact, slipped into a more complex state such as late transience or even semi-steady state, in which case the mathematics describing transience would be inappropriate to describe the physical condition. The alternative approach advocated in this book is that the engineer must somehow *prove* that the condition of transience prevails before applying the simple mathematical description to this state.

Since the late 1970s, the word "transience" appears to have been superseded in the literature by the more convoluted expression "infinite acting radial flow" (IARF). Since the latter does little to clarify the physical state the term transience will be adhered to in this text.

B. Late transience

This wellbore pressure response is one that is difficult to describe mathematically. Even if the location of the boundaries were known, which is seldom the case, the complexity lies in the fact that the boundary conditions vary as a function of time during this period as their effect successively influences the pressures in the wellbore. Because of this difficulty, the late transient phase was largely neglected in test analysis in the past, if not in theory, then in practice. Even when describing theoretical behaviour, however, there was a proliferation of papers appearing in the literature on the subject of the anticipated pressure response of wells located at the centre of circles or squares. While there may be practical reasons for considering such in relation to pattern development of fields, locating the well at the centre of a regular geometrical configuration removes the problems associated with late transience altogether because, on account of the symmetry of the boundary with respect to the well, transience is followed directly by semi-steady state with no intervening late transient response. It was not until 1971 that a simple equation representing not just late transience but the whole of the CTR pressure decline (Fig. 4.6b) for a bounded reservoir system was derived [19]. It is based on the classical work of Matthews, Brons and Hazebroek [20] and description of the method and its application will be deferred until section 4.19a. In the meantime it will not be overlooked that late transience may occur, is complicated, and must somehow be distinguished from other pressure responses.

C. Semi-steady state

Periods of transient and late transient pressure response are to be anticipated when testing appraisal wells. In the latter case, the existence of faults close to the

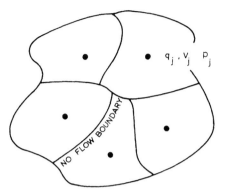

Fig. 4.7. Wells draining a reservoir under the semi-steady-state condition.

well may remove the condition of transience although they are not necessarily the outer boundaries of the system under test. The condition of semi-steady state (also referred to as pseudo- or quasi-steady state) should, hopefully, not be encountered in appraisal well testing for it implies that all the outer boundaries are influencing the pressure in the wellbore and this results in a stable rate of pressure decline throughout the system. Obviously, if this state were observed in an appraisal test, in which only a relatively small amount of fluid is withdrawn from the reservoir, it would also mean that the volume of STOIIP was necessarily small and most likely not worth developing, especially if located offshore. The condition of semi-steady state is most appropriately applied to describe reservoirs which have been under development for some time, as depicted in Fig. 4.7.

If individual well rates are maintained at reasonably stable levels, which is usually the aim, then each will carve out its own territory surrounded by a no flow boundary separating it from other wells. Then, by differentiating the depletion material balance, equation , in each cell with respect to time gives

$$q_j = \frac{\mathrm{d}V_j}{\mathrm{d}t} = cV_j\frac{\mathrm{d}p_j}{\mathrm{d}t} \tag{4.10}$$

in which q_j, V_j and p_j represent the oil rate, volume and average pressure in the area drained by the jth well. But, provided the condition of semi-steady state prevails, or approximately so, then by definition $\mathrm{d}p_j/\mathrm{d}t \sim$ constant and therefore:

$$q_i \propto V_j \tag{4.11}$$

This states the useful result that if a reservoir is being produced under reasonably stable conditions, the rate of each well is directly proportional to the volume it is draining and application of this principle has already been described in greater detail in Chapter 3, section 3.3.

The description of wells producing under semi-steady-state conditions has been adequately described in the literature [3–5], largely because the concentration in theme has been on the testing of development wells under depletion conditions (section 4.3). The subject is presented in this chapter in section 4.19.

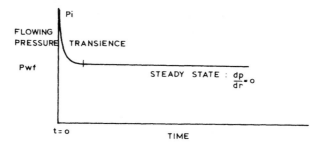

Fig. 4.8. Steady-state wellbore pressure response.

(b) Steady-state condition

This may be regarded as the opposite extreme CTR solution of the radial diffusivity equation to that described above for bounded systems. The wellbore pressure drop resulting form flow at a constant rate, q, from time $t = 0$ is depicted in Fig. 4.8.

Following what is usually a brief period of transience, stability of pressure is observed such that at the wellbore and throughout the volume under test $dp/dt = 0$. Naturally, this condition is observed in reservoirs in which pressure is maintained by the injection of water or gas, which makes it particularly relevant for development well testing in large offshore fields where such secondary recovery operations are commonly practised (section 4.19c).

In addition, and sometimes quite surprisingly, the steady-state response is also frequently observed while testing at the appraisal stage when Nature supplies the energy. Four such examples of natural pressure maintenance are illustrated below.

High flow capacity reservoirs

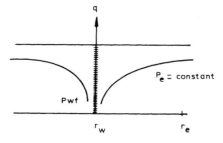

If the flow capacity of the formation is high ($kh > 50,000$ mD·ft) then, irrespective of the nature of the fluid, it is not uncommon to observe the steady-state condition. Flowing the well at a constant rate, q, causes a pressure drop at the wellbore, $r = r_w$, which is propagated into the reservoir until it is arrested at an outer radius, r_e, at which the pressure, p_e, is maintained, constant. The pressure recorded in the wellbore, p_{wf} is constant as are all pressures along the drawdown profile, $p_e - p_{wf}$. This author has witnessed this type of pressure response in appraisal well tests in areas such as the North Sea, the Middle East and Australia where flow capacities can be abnormally high. Therefore, even if the fluid produced is undersaturated oil, the cumulative volume removed during an appraisal test is usually so small relative to the STOIIP that the fluid outwith the constant pressure boundary, $r = r_e$, finds it easy to expand through the high flow capacity and fill the void, in spite of its low compressibility.

Gas reservoirs

The steady-state pressure response is most frequently observed in gas well tests, even though the formation flow capacity may be low. This is attributed to the high compressibility of the gas which is at least one order of magnitude greater than for undersaturated oil.

Gas cap reservoirs

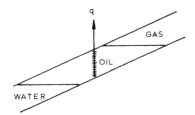

Pressure stability [4] during the drawdown can also result from the proximity of a gas cap to the test interval. The schematic depicts an appraisal well test over a thin, 10 m, oil column in which the formation permeability was only about 150 mD. Yet complete pressure maintenance was observed during the test presumably on account of the nearby, high-compressibility gas cap. There can also be a component of pressure support resulting from liberation of solution gas from the oil column — which is initially at its saturation pressure. In comparison, expansion and pressure support from the low-compressibility edge water is usually small and retarded compared to the support provided by the gas.

Basal waterdrive reservoirs

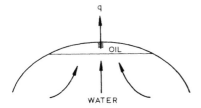

This situation usually arises during the appraisal of "marginal" oil accumulations in which the well has penetrated a massive porous sand section containing only a small volume of trapped oil. Pressure maintenance may be observed in this situation due to the large product of vertical permeability × area giving rise to a strong component of basal waterdrive.

The phenomenon of steady-state flowing pressure, when it occurs under natural conditions, can be attributed to high flow capacities, high fluid compressibilities or a combination of both. The situation illustrated in Fig. 4.8 is the classic case of complete stability of pressure, which is often observed, even within the high resolution of modern pressure gauges. To a lesser degree, partial pressure maintenance may occur in which the decline is intermediate between that of a fully bounded and a true steady-state system. Identification of such a flowing condition is described in section 4.10.

Sometimes the engineer may find it difficult to visualize exactly what is happening out there in the reservoir to give rise to the stability of flowing pressure or more precisely, to quantify the effect in terms of physical and mathematical boundary conditions. Rather than become too concerned with some of these finer points the engineer is advised to approach the analysis of such well tests in a purely pragmatic manner. That is, if stability of pressure has been observed during the flow period of well test, then the analysis must incorporate the steady-state solution of the diffusivity equation. Failure to do so can result in some rather strange conclusions being drawn from the test analysis, as described in section 4.18.

There has been a considerable amount of research on the general topic of pressure support and, in particular, engineered pressure maintenance but, as described in section 4.3, it has to date been confined largely to the "specialist" literature. For instance, on page 14 of reference 5, after stating the steady-state solutions of the diffusivity equation for different geometrical configurations, the author concludes that — "Linear and radial steady-state flow usually only occur in laboratory situations" — which tends to relegate the importance of the phenomenon. Perhaps one reason for its neglect arises from the suggestion made in section 4.3b that the majority of the literature is focused on development well testing. Usually in such routine tests a lubricator is rigged-up to the wellhead and the pressure gauge is run in against the flowing well stream. Once on station, the bottom hole flowing pressure is recorded over a short period, a few minutes, prior to well closure for a pressure buildup. Because of this only the final flow pressure is recorded rather than any extensive history of the pressure response. Consequently, there has been no tradition in the subject of inspecting and responding to the nature of the flowing pressures in the ensuing buildup analysis. This is evident from a cursory inspection of the literature. There are so many papers on the subject of pressure buildup analysis yet hardly any which, when illustrating the text with examples, ever show a plot of the flowing pressure history or even allude to it; invariably they merely state the final flowing pressure.

It must be stressed, however, that analysing a pressure buildup consists of implicitly solving two equations, one for the pressure drawdown during the flowing period and one for the pressure buildup. And, as pointed out in section 4.8, this in itself can cause sufficient ambiguity in analysis without adding to the problem by not even acknowledging the nature of the flowing pressures. When testing at the appraisal stage, there can be no excuse for such an oversight because usually drillstem tests are conducted in which the entire flowing pressure history is recorded.

At the time of writing, steady-state testing has become rather fashionable on account of engineered pressure maintenance gaining in popularity as a recovery mechanism and also because of the greater concentration of activity offshore where, in many areas, only high flow capacity reservoirs can be economically developed, which leads to the occurrence of the phenomenon when testing at the appraisal stage. Test analysis techniques appropriate for this situation are described in section 4.18 but in no sense do these add complication to the engineer's normal repertoire since the steady-state solution of the diffusivity equation is one of the very simplest to derive and apply.

4.7. THE TRANSIENT CONSTANT TERMINAL RATE SOLUTION OF THE RADIAL DIFFUSIVITY EQUATION

This is usually referred to as the line source CTR solution of equation 4.4 for which the initial and boundary conditions may be stated as follows:

$$p = p_i \quad \text{at } t = 0, \qquad \text{for all } r$$

$$r\frac{\partial p}{\partial r} = 141.2\frac{q\mu B_o}{kh}, \qquad \text{for } t > 0$$
$$\lim r \to 0$$

$$p = p_i \quad \text{at } r = \infty, \qquad \text{for all } t$$

The first is the initial condition while the second is the inner boundary condition and is a statement of Darcy's law at the wellbore, as r tends to zero. The third is the transient or infinite acting outer boundary condition. The expression "for all t" means for the period when transience prevails, during which pressure at the apparently infinite outer boundary equals the initial pressure, p_i. This imposes no condition on how long transience persists, it may be seconds or years, dependent on the magnitude of the diffusivity constant. But when transience finishes the third condition is inappropriate to describe the pressure response and the transient CTR solution is no longer valid.

The mathematical solution of the diffusivity equation for the above conditions, using the Boltzmann transform, is detailed both in reference 3 (chapter 7) and 4 (appendix A) and will not be repeated here. The solution, expressed in field units, is:

$$7.08 \times 10^{-3}\frac{kh}{q\mu B_o}(p_i - p_{r,t}) = \tfrac{1}{2}ei(x) \tag{4.12}$$

in which, the arrangement of parameters on the left-hand side is such that the group is dimensionless, as demonstrated on page 167 of reference 3; $p_{r,t}$ is the pressure at any radial position, r, at time, t, during the period of transience and the function $ei(x)$ is the exponential integral defined as

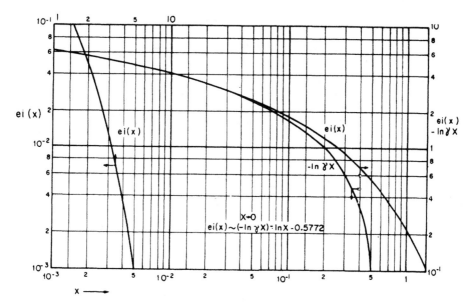

Fig. 4.9. Graph of the *ei*-function for $0.001 \leq x \leq 5.0$.

$$ei(x) = \int_{x}^{\infty} \frac{e^{-s}}{s} \, ds \qquad (4.13)$$

in which s is a dummy variable of integration and the lower limit of integration is defined as:

$$x = \frac{\phi \mu c r^2}{4 \times 0.000264 \, kt} \quad (t, \text{ hours}) \qquad (4.14)$$

which, as also demonstrated on page 167 of reference 3, is a dimensionless group of parameters.

The *ei*-function, which is a standard integral, may be evaluated for any value of its argument, x, using Table 4.2 or even more conveniently by use of pocket calculators, many of which can be directly programmed to evaluate equation 4.13. The numerical value of the function decreases as the lower limit of integration, x, which is its argument, increases, as shown in Fig. 4.9. The function also has the property that provided $x < 0.01$.

$$ei(x) \approx -\ln(\gamma x) \qquad (4.15)$$

in which the term γ is a constant which appears frequently in the solutions of the diffusivity equation and may be evaluated as

$$\gamma = e^{(\text{Euler's constant})} = e^{0.5772} = 1.781 \qquad (4.16)$$

Examination of equation 4.14 reveals that if the radius, r, at which the pressure is recorded is small then the condition that $x < 0.01$ will be satisfied rather quickly

TABLE 4.2

Values of the exponential integral function [21]

$-Ei(-x)$, $0.000 < 0.209$, interval $= 0.0001$

x	0	1	2	3	4	5	6	7	8	9
0.00	+∞	6.332	5.639	5.235	4.948	4.726	4.545	4.392	4.259	4.142
0.01	4.038	3.944	3.858	3.779	3.705	3.637	3.574	3.514	3.458	3.405
0.02	3.355	3.307	3.261	3.218	3.176	3.137	3.098	3.062	3.026	2.992
0.03	2.959	2.927	2.897	2.867	2.838	2.810	2.783	2.756	2.731	2.706
0.04	2.681	2.658	2.634	2.612	2.590	2.568	2.547	2.527	2.507	2.487
0.05	2.468	2.449	2.431	2.413	2.395	2.377	2.360	2.344	2.327	2.311
0.06	2.295	2.279	2.264	2.249	2.235	2.220	2.206	2.192	2.178	2.164
0.07	2.151	2.138	2.125	2.112	2.099	2.087	2.074	2.062	2.050	2.039
0.08	2.027	2.015	2.004	1.993	1.982	1.971	1.960	1.950	1.939	1.929
0.09	1.919	1.909	1.899	1.889	1.879	1.869	1.860	1.850	1.841	1.832
0.10	1.823	1.814	1.805	1.796	1.788	1.779	1.770	1.762	1.754	1.745
0.11	1.737	1.729	1.721	1.713	1.705	1.697	1.689	1.682	1.674	1.667
0.12	1.660	1.652	1.645	1.638	1.631	1.623	1.616	1.609	1.603	1.596
0.13	1.589	1.582	1.576	1.569	1.562	1.556	1.549	1.543	1.537	1.530
0.14	1.524	1.518	1.512	1.506	1.500	1.494	1.488	1.482	1.476	1.470
0.15	1.464	1.459	1.453	1.447	1.442	1.436	1.431	1.425	1.420	1.415
0.16	1.409	1.404	1.399	1.393	1.388	1.383	1.378	1.373	1.368	1.363
0.17	1.358	1.353	1.348	1.343	1.338	1.333	1.329	1.324	1.319	1.314
0.18	1.310	1.305	1.301	1.296	1.291	1.287	1.282	1.278	1.274	1.269
0.19	1.265	1.261	1.256	1.252	1.248	1.243	1.239	1.235	1.231	1.227
0.20	1.223	1.219	1.215	1.210	1.206	1.202	1.198	1.195	1.191	1.187

TABLE 4.2 (continued)

$-Ei(-x)$, $0.000 < x < 2.09$, interval $= 0.01$

x	0	1	2	3	4	5	6	7	8	9
0.0	$+\infty$	4.038	3.335	2.959	2.681	2.468	2.295	2.151	2.027	1.919
0.1	1.823	1.737	1.660	1.589	1.524	1.464	1.409	1.358	1.309	1.265
0.2	1.223	1.183	1.145	1.110	1.076	1.044	1.014	0.985	0.957	0.931
0.3	0.906	0.882	0.858	0.836	0.815	0.794	0.774	0.755	0.737	0.719
0.4	0.702	0.686	0.670	0.655	0.640	0.625	0.611	0.598	0.585	0.572
0.5	0.560	0.548	0.536	0.525	0.514	0.503	0.493	0.483	0.473	0.464
0.6	0.454	0.445	0.437	0.428	0.420	0.412	0.404	0.396	0.388	0.381
0.7	0.374	0.367	0.360	0.353	0.347	0.340	0.334	0.328	0.322	0.316
0.8	0.311	0.305	0.300	0.295	0.289	0.284	0.279	0.274	0.269	0.265
0.9	0.260	0.256	0.251	0.247	0.243	0.239	0.235	0.231	0.227	0.223
1.0	0.219	0.216	0.212	0.209	0.205	0.202	0.198	0.195	0.192	0.189
1.1	0.186	0.183	0.180	0.177	0.174	0.172	0.169	0.166	0.164	0.161
1.2	0.158	0.156	0.153	0.151	0.149	0.146	0.144	0.142	0.140	0.138
1.3	0.135	0.133	0.131	0.129	0.127	0.125	0.124	0.122	0.120	0.118
1.4	0.116	0.114	0.113	0.111	0.109	0.108	0.106	0.105	0.103	0.102
1.5	0.1000	0.0985	0.0971	0.0957	0.0943	0.0929	0.0915	0.0902	0.0889	0.0876
1.6	0.0863	0.0851	0.0838	0.0826	0.0814	0.0802	0.0791	0.0780	0.0768	0.0757
1.7	0.0747	0.0736	0.0725	0.0715	0.0705	0.0695	0.0685	0.0675	0.0666	0.0656
1.8	0.0647	0.0638	0.0629	0.0620	0.0612	0.0603	0.0595	0.0586	0.0578	0.0570
1.9	0.0562	0.0554	0.0546	0.0539	0.0531	0.0524	0.0517	0.0510	0.0503	0.0496
2.0	0.0489	0.0482	0.0476	0.0469	0.0463	0.0456	0.0450	0.0444	0.0438	0.0432

$-Ei(-x)$, $2.0 < x < 10.9$, interval $= 0.1$

x	0	1	2	3	4	5	6	7	8	9
2	4.89×10^{-2}	4.26×10^{-2}	3.72×10^{-2}	3.25×10^{-2}	2.84×10^{-2}	2.49×10^{-2}	2.19×10^{-2}	1.92×10^{-2}	1.69×10^{-2}	1.48×10^{-2}
3	1.30×10^{-2}	1.15×10^{-2}	1.01×10^{-2}	8.94×10^{-3}	7.89×10^{-3}	6.87×10^{-3}	6.16×10^{-3}	5.45×10^{-3}	4.82×10^{-3}	4.27×10^{-3}
4	3.78×10^{-3}	3.35×10^{-3}	2.97×10^{-3}	2.64×10^{-3}	2.34×10^{-3}	2.07×10^{-3}	1.84×10^{-3}	1.64×10^{-3}	1.45×10^{-3}	1.29×10^{-3}
5	1.15×10^{-3}	1.02×10^{-3}	9.08×10^{-4}	8.09×10^{-4}	7.19×10^{-4}	6.41×10^{-4}	5.71×10^{-4}	5.09×10^{-4}	4.53×10^{-4}	4.04×10^{-4}
6	3.60×10^{-4}	3.21×10^{-4}	2.86×10^{-4}	2.55×10^{-4}	2.28×10^{-4}	2.03×10^{-4}	1.82×10^{-4}	1.62×10^{-4}	1.45×10^{-4}	1.29×10^{-4}
7	1.15×10^{-4}	1.03×10^{-4}	9.22×10^{-5}	8.24×10^{-5}	7.36×10^{-5}	6.58×10^{-5}	5.89×10^{-5}	5.26×10^{-5}	4.71×10^{-5}	4.21×10^{-5}
8	3.77×10^{-5}	3.37×10^{-5}	3.02×10^{-5}	2.70×10^{-5}	2.42×10^{-5}	2.16×10^{-5}	1.94×10^{-5}	1.73×10^{-5}	1.55×10^{-5}	1.39×10^{-5}
9	1.24×10^{-5}	1.11×10^{-5}	9.99×10^{-6}	8.95×10^{-6}	8.02×10^{-6}	7.18×10^{-6}	6.44×10^{-6}	5.77×10^{-6}	5.17×10^{-6}	4.64×10^{-6}
10	4.15×10^{-6}	3.73×10^{-6}	3.34×10^{-6}	3.00×10^{-6}	2.68×10^{-6}	2.41×10^{-6}	2.16×10^{-6}	1.94×10^{-6}	1.74×10^{-6}	1.56×10^{-6}

so that the approximation stated in equation 4.15 can be applied. Specifically for pressures measured in the wellbore

$$p_{r,t} = p_{wf}$$

$$r = r_w$$

and, as demonstrated in reference 3 (chapter 7, exercise 7.1), using typical oilfield parameters ($k = 50$ mD : $r_w = 0.5$ ft) the value of x becomes less than 0.01 after a mere 15 seconds. In higher permeability environments, the time required for the condition to be satisfied will be proportionately less. Therefore, it is almost always safe to assume that for pressures measured in the wellbore the logarithmic approximation of the *ei*-function may be applied. The reservation is expressed because using high-resolution pressure gauges it has now become fashionable to attempt to analyse the very early pressure response at the start of a test using time increments of seconds. Prior to applying the logarithmic approximation under such circumstances, the engineer should first check its validity and if unsatisfied, use the *ei*-function instead.

Most of this chapter is devoted to the analysis of DSTs and production tests in which pressures are measured in the wellbore ($r = r_w$) and for these it will be assumed that the condition stated in equation 4.15 is satisfied.

There are, however, many applications of equation 4.12, such as interference testing between wells and, in a theoretical sense, the application of the method of images, as described in sections 4.17 and 4.16, when the *ei*-function must be fully evaluated rather than replaced by its logarithmic approximation. In an interference test, for instance, a pressure disturbance is caused in one well by changing its rate and the resulting pressure response is recorded in a second well which may be located hundreds or thousands of feet away. In this case, the r^2 term appearing in equation 4.14 is large and the condition necessary to apply the logarithmic approximation of the *ei*-function, that $x < 0.01$, may not be satisfied even at the end of the test. An example of the usefulness of applying the *ei*-function in interference testing is illustrated in section 4.17 and for determining the distance to a sealing fault in Exercise 4.2. When using the exponential integral in this manner, it must be remembered, above all else, that it only applies when transience prevails and the system under test appears to be infinite in extent. As soon as the effect of boundaries begins to be felt a more complex CTR of the diffusivity equation must be used and when all the boundaries affect the pressure response and stability is achieved the CTR solution must also include a material balance term.

For the present, attention will be focused on the most common form of testing in which pressures used in the analysis are recorded in the wellbore. In this case the logarithmic approximation of the *ei*-function, equation 4.15, may be legitimately applied to reduce equation 4.12 to the form:

$$7.08 \times 10^{-3} \frac{kh}{q\mu B_o}(p_i - p_{wf}) = \tfrac{1}{2}\ln\frac{4 \times 0.000264\,kt}{\gamma\phi\mu cr_w^2} + S \tag{4.17}$$

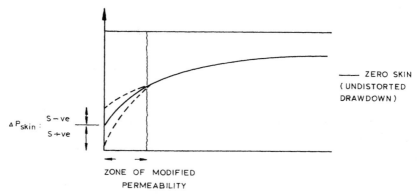

Fig. 4.10. Perturbed pressure drawdown across the zone of modified permeability close to the wellbore.

in which p_{wf} is the bottom hole flowing pressure. Since the first two terms in the equation are dimensionless then the skin factor, S, which is also dimensionless may be included intuitively on the right-hand side to represent the fact that part of the total pressure drawdown, $p_i - p_{wf}$, may be attributed to a component of pressure drop across the skin, Δp_{skin}. As indicated in Fig. 4.10 this will be either positive or negative dependent on whether the well has been damaged or stimulated. In either case, the pressure drop across the skin is defined through equation 4.18 as

$$\Delta p_{skin} = 141.2 \frac{q \mu B_o}{kh} S \qquad (4.18)$$

Equation 4.17 may be expressed in a more compact form by invoking the concept of dimensionless time, defined as

$$t_D = 0.000264 \frac{kt}{\phi \mu c r_w^2} \quad (t, \text{hours}) \qquad (4.19)$$

which, for other parameters being constant, is a direct relationship between the dimensionless and real time. The rather awkward constant, 0.000264, is commensurate with the real time, t, being measured in hours. Since, as described previously, equation 4.14 is dimensionless, then so too is equation 4.19. Inserting this in equation 4.17 yields

$$7.08 \times 10^{-3} \frac{kh}{q \mu B_o} (p_i - p_{wf}) = \tfrac{1}{2} \ln \tfrac{4t_D}{\gamma} + S \qquad (4.20)$$

which is the *transient*, CTR solution of the radial diffusivity equation and is referred to frequently throughout the remainder of the chapter.

In a more general sense, the CTR solution which is appropriate for any value of the flowing time and any type of flowing pressure response: transient, late transient, semi-steady-state, steady-state or any intermediate condition, is

$$7.08 \times 10^{-3} \frac{kh}{q \mu B_o} (p_i - p_{wf}) = p_D(t_D) + S \qquad (4.21)$$

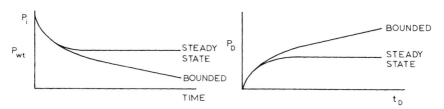

Fig. 4.11. Pressure drawdowns and equivalent p_D-functions.

which is the defining expression for the p_D-function, known as the dimensionless pressure, which is a function of the dimensionless time t_D. Equation 4.21 is by far the most important in the whole subject of well test analysis since it is the basic building block upon which all methods are based, as will be repeatedly demonstrated. The p_D-function itself is one that is characteristic of the formation and fluids under test and whose shape must be the inverse of the drawdown function since as p_{wf} decreases, p_D increases, as dictated by equation 4.21 and illustrated in Fig. 4.11.

For simple physical conditions, such as the period when pure transience prevails, the p_D-function may be readily evaluated since comparison of equations 4.20 and 4.21 reveals that

$$p_D(t_D) = \tfrac{1}{2} \ln \tfrac{4 t_D}{\gamma} \tag{4.22}$$

which contains no explicit dimensions or shape of the outer boundary since during the period when equation 4.20 is appropriate, the boundary appears to be infinite in extent. As soon as the flowing time exceeds that when equation 4.20 can be applied difficulties arise in the evaluation of the p_D-function.

Provided the engineer is quite satisfied that the nature of the formation under test: the well characteristics and the boundary conditions are known, then the p_D-function can be accurately determined for all values of the time argument. This is because the literature on well testing abounds with theoretical p_D-functions expressed in chart or tabular form for the following conditions, amongst others:

– wells located in the vicinity of fault configurations in an otherwise infinite system. These are referred to as semi-infinite geometries and influence tests conducted during field appraisal (section 4.16)

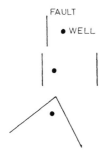

- bounded geometries: Matthews, Brons and Hazebroek studied the CTR response of wells located at different positions within a variety of bounded systems [20], from which p_D-functions were generated for 24 geometrical configurations (section 4.19a)

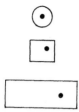

- steady state-and mixed (open-bounded) systems (section 4.19c)
- layered, fractured and dual porosity formations
- for wells with different wellbore storage and skin factor. There are various sets of type curves described in section 4.21.

And if these prove insufficient, the engineer may generate functions by, for instance, applying the method of images (sections 4.19a) to generate p_D-functions for different boundary conditions than catered for in the literature. For a defined system, therefore, the generation of p_D-functions is simply a matter of application of the appropriate mathematics. The problem is, however, how does the engineer know how to define the system under test in order to choose or generate the appropriate p_D-function? The short answer is that in the majority of cases he doesn't and therefore to select a p_D-function from the literature for application in the analysis involves making assumptions about the system under test. Ideally, identification of the nature of the formation and the boundary conditions affecting the pressure response should be the outcome of a successful interpretation — not the input. Thanks to modern computational techniques there are methods of attempting to gain such knowledge. These are iterative and proceed from a first guess at a p_D-function. An attempt is then made to match the rate–pressure–time record of the test using it and, if it fails, guidance will be provided on the choice of a more appropriate p_D-function, until hopefully one is eventually found to match the observations. This method is further described in sections 4.11 and 16d.

Throughout the remainder of the chapter, the basic theory of well test analysis is stated in terms of the dimensionless time and pressure functions defined by equations 4.19 and 4.21 respectively. This has been common practice in the Industry since the 1960's and serves not only to simplify but also to generalize the mathematics.

4.8. DIFFICULTIES IN APPLICATION OF THE CONSTANT TERMINAL RATE SOLUTION
OF THE RADIAL DIFFUSIVITY EQUATION

It is regrettable to commence any section of a textbook with the word "difficulties" but there are three technical problems associated with the application of general CTR solution

$$7.08 \times 10^{-3} \frac{kh}{q \mu B_o} (p_i - p_{wf}) = p_D(t_D) + S \tag{4.21}$$

which must be constantly borne in mind by the engineer to avoid making fundamental errors in test analysis. In much of the literature, these points are either referred to obliquely or else totally ignored. The three difficulties are:

(A) If all other parameters are known or may be independently determined, then equation 4.21 contains *three* unknowns k, S and the p_D-function.
(B) In applying equation 4.21 to test analysis, there can be a lack of mathematical uniqueness.
(C) The p_D-function is only physically defined through equation 4.21 during the period when a well is produced at constant rate, q, from time $t = 0$. Yet the application of the equation, in most practical forms of testing, requires extrapolation of the function beyond the time for which it is defined.

All three of the mathematical difficulties are inter-related which will be described, as appropriate, throughout the chapter. The three unknowns in equation 4.21 are

k, S which are generally regarded as constants
$p_D(t_D)$ which usually, although not necessarily, a variable, dependent upon the time and boundary conditions.

The lack of mathematical uniqueness simply relates to the number of unknowns (three) compared to the number of equations available to resolve their values. In the case of multi-rate testing, for instance (section 4.9), the analysis relies upon the manipulation of a single equation and one equation containing three unknowns is a hopeless situation — implying that there exists an infinite number of mathematically valid solutions — but which is the engineer to select as being physically correct.

Largely because of this mathematical uncertainty, most operators choose to test wells using the method of pressure buildup analysis, upon which the remainder of the chapter is largely focused. One of the many advantages in the analysis of such tests is that the determination of the formation parameters, k and S, implicitly relies upon the simultaneous solution of two equations: one for the drawdown, equation 4.21, and one for the buildup, equation 4.23. Furthermore, this type of test affords the best opportunity of identifying a transient pressure response which is manifest as a linear section of the buildup plot, shortly after well closure, when the data are plotted on a conventional semi-log plot (sections 4.12 and 13). Therefore, during this period, the p_D-function corresponding to closed-in time, Δt, assumes the simplest form

$$p_D(\Delta t_D) = \tfrac{1}{2}\ln\frac{4\Delta t_D}{\gamma} = \tfrac{1}{2}\ln\frac{4 \times 0.000264 k\,\Delta t}{\gamma\phi\mu c r_w^2} \tag{4.23}$$

in which, the unknown is the permeability, k. Consequently, the interpretation is reduced to solving two equations containing two unknowns, k and S, which should lead to a unique solution — provided the correct, linear, transient pressure response has been selected on the buildup plot so that application of equation 4.23 is valid.

The one thing the engineer must avoid at all cost is the unwarranted assumption that the transient condition prevails thus falsely reducing the number of unknowns to two: k and S. Traditionally, this has been, and still is, one of the most common errors in the subject which permeates test analysis procedures at all levels and invalidates many of the applications we tend to take for granted. As described in section 4.3, perhaps the assumption was necessary in the past when the application of complex mathematics to test analysis was a prohibitive prospect but the same excuse cannot be used nowadays and the engineer is obliged to *prove* that the condition of transience prevails before applying its simple mathematical formulation for drawdown, equation 4.22, or buildup, equation 4.23. In the remainder of the chapter there are numerous examples presented of the damaging effect of the false assumption of transience on commonly accepted test analysis procedures.

The mathematical difficulty referred to in C, above: the necessity to extrapolate the p_D-function to times for which it has not been defined is one that simply has to be tolerated in the subject. The only form of test which is free from this complexity is the single-rate drawdown, described in section 4.10, but this form of test is not normally relied upon for the reason that it proves difficult in practice to maintain a strictly constant flow rate, especially at the start of the test, which causes problems in defining an early, transient pressure response from which k and S can be determined. Any other type of test requires the extrapolation of the p_D-function. This includes the most common and reliable form of test, the pressure buildup for which the p_D-function is only defined until the end of the flow period, $p_D(t_D)$ (Fig. 4.1). Yet in the analysis, the function must somehow be extrapolated throughout the flow and buildup periods as $p_D(t_D + \Delta t_D)$. As will be demonstrated, the manner in which this extrapolation is accomplished applying the traditional semi-log plotting techniques of Miller, Dyes, Hutchinson and Horner (sections 4.12 and 4.13) is perfectly safe but in the more recently developed techniques of type-curve analysis (section 4.21), there is a vulnerability to the mode of extrapolation of which the engineer must be aware.

4.9. SUPERPOSITION OF CTR SOLUTIONS

Application of the principle of superposition permits the development of a complex CTR solution of the radial diffusivity equation in cases where, by accident or design, the oil rate has not been maintained constant throughout the flowing period of the test, as depicted in Fig. 4.12.

Applied to well testing, the mathematical principle of superposition states that,

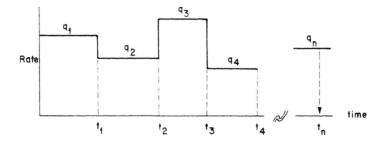

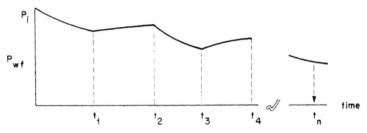

Fig. 4.12. Production history of a well showing both rate and bottom hole flowing pressure as functions of time.

if equation 4.21 is a solution of the linear, radial diffusivity equation, then so too is any linear combination of such solutions. This permits the generation of superposed CTR solution catering for the entire variable rate history. The recipe for determining an expression for the bottom hole flowing pressure at the end of the nth flow period following production at rate q_n is

$$7.08 \times 10^{-3} \frac{kh}{\mu B_o}(p_i - p_{wf_n}) = [\text{ rate change}] \times \begin{bmatrix} p_D\text{-function evaluated from} \\ t_n \text{ back to when the rate change} \\ \text{occurred + skin factor} \end{bmatrix}$$

that is

$$7.08 \times 10^{-3} \frac{kh}{\mu B_o}(p_i - p_{wf_n}) = (q_1 - 0)[p_D(t_{D_n} - 0) + S]$$
$$+ (q_2 - q_1)[p_D(t_{D_n} - t_{D_1}) + S]$$
$$\vdots$$
$$+ (q_j - q_{j-1})[p_D(t_{D_n} - t_{D_{j-1}}) + S]$$
$$\vdots$$
$$+ (q_n - q_{n-1})[p_D(t_{D_n} - t_{D_{n-1}}) + S]$$

which, if $\Delta q_j = q_j - q_{j-1}$, can be expressed as

$$7.08 \times 10^{-3} \frac{kh}{\mu B_o}(p_i - p_{wf_n}) = \sum_{j=1}^{n} \Delta q_j p_D(t_{D_n} - t_{D_{j-1}}) + q_n S \qquad (4.24)$$

In the summation, all the skin factor terms cancel, except for the last, $q_n S$. In principle, therefore, equation 4.24 may be regarded as the perfectly general CTR solution which can be applied to any complex rate–pressure–time record of a test, including periods of closure. In practice, however, its application is problematical. In attempting to use the equation, all three of the mathematical complications pointed to in the previous section affect matters: three unknowns (k, S, p_D) in the one equation, total lack of mathematical uniqueness and extrapolation of the p_D-function.

Concerning the last, the p_D-function evaluated for the maximum value of the time argument is $p_D(t_{D_n} - 0)$ occurring in the very first term in equation 4.24, yet the function has only been defined during the first drawdown period flowing at rate q for time t, as

$$7.08 \times 10^{-3} \frac{kh}{q_1 \mu B_o} = p_D(t_{D_1}) + S$$

How then is the extrapolation from $t = 0$ to t_n, required in the first term to be made? It is only possible if the p_D-function over the total time interval is known — which is seldom the case. Equation 4.24 is applied to multi-rate testing, particularly in gas wells, and to cater for variable rate history prior to well closure for a buildup in appraisal well testing. In both applications, however, the Industry has traditionally *assumed* that when equation 4.24 is applied to any well, anywhere in the world, for any test duration, then the p_D-function can always be evaluated under the *transient* flow condition so that, in analogy with equation 4.22

$$p_D(t_{D_n} - t_{D_{j-1}}) = \tfrac{1}{2} \ln \frac{4(t_{D_n} - t_{D_{j-1}})}{\gamma} \qquad (4.25)$$

One of the justifications invoked in making this assumption is that, provided the individual flow periods are of sufficiently short duration so that transience prevails in each, then the approach is valid. Unfortunately, this is not borne-out by equation 4.24 which states that transience must apply for the largest value of the time argument, $t_{D_n} - 0$, and therefore implies that the entire test duration must be sufficiently short to assure transience throughout. The assumption of transience may be appropriate in a very-low-permeability reservoir but certainly cannot be applied universally. In analysing a multi-rate test, both sides of the equation are divided by the rate, q_n, corresponding to the time t_n at which the calculation is performed. A plot is then made as depicted in Fig. 4.13a with the ordinate representing the variable group on the left-hand side of equation 4.24 and the abscissa the group on the right. The logarithmic term arises from the conventional assumption of transience (equation 4.25). The pressure–time points invariably lie on a straight line with slope and intercept on the ordinate being

$$m = 162.6 \frac{\mu B_o}{kh} \quad \text{and} \quad I = m \left[\log \frac{4}{\gamma} \frac{0.000264\,k}{\phi \mu c r_w^2} + 0.87 S \right]$$

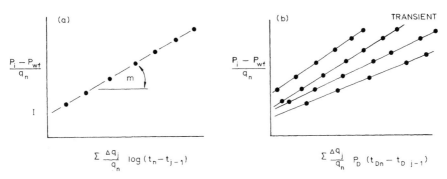

Fig. 4.13. Multi-rate drawdown analysis: (a) conventional transient assumption; (b) plotted for different, assumed p_D-function.

and if all other parameters are known, the kh-product is calculated from the slope and the skin factor from the intercept. The interpretation may appear convincing on account of the linearity of the plot but, as mentioned above, the assumption of transience is invariably unsound. Interpreting equation 4.24 in the correct manner, the full p_D-function, rather than its transient, logarithmic form, should be used in calculating abscissa values, Fig. 4.13b. Unfortunately, to do so requires a knowledge of the reservoir characteristics/boundary conditions which, of course, should be the outcome of the test — not the input. It happens that each time a new assumption is made concerning these physical conditions the same pressure–time points invariably lie on a straight line, their slopes and intercepts giving different values of kh and S. Such an interpretation is presented in exercise 7.8 in chapter 7 of reference 3, and illustrates the unfortunate ambiguity attached to multi-rate analysis. Should the engineer be obliged to conduct such tests, and in some countries/areas it is mandatory for gas well tests (being imposed by regulatory authorities) then as a precaution a pressure buildup should be conducted at the end of an oilwell test to reliably calculate kh and S. In the multi-rate testing of gas wells, two buildups will be required, following periods of different flow rate, to distinguish between the two components of skin factor resulting from formation damage and rate-dependent, turbulent (non-Darcy) flow (reference 3, Chapter 8). Although no valid interpretation of multi-rate flow tests is guaranteed under the assumption of transient flow, a more appropriate application is described in section 4.20c for Selective Inflow Performance (SIP) testing in development wells producing from layered reservoirs in which a reasonable degree of pressure maintenance, as opposed to transience, is the condition required to attain a meaningful quantitative analysis.

4.10. SINGLE-RATE DRAWDOWN TEST

This should provide the simplest means of testing a well: flowing it form time $t = 0$, at a constant rate q, for which the pressure response in the wellbore is described by equation 4.21. Such testing should only apply in exploration/appraisal wells. Unfortunately, it is a difficult form of testing to rely upon for the simple reason that it is not a straightforward matter to fulfil the basic requirement of flowing at a constant rate. To begin with, the test string is usually half filled with a water cushion at the start of a test so that on perforating there is an artificially low bottom hole pressure to encourage production and this distorts the initial flowing pressure response. Secondly, it can sometimes take hours for a sufficient degree of stability to be achieved to divert the flow through the test separator where the production rate and GOR can be monitored. Because of these practical difficulties, the initial data are often uncertain and this precludes meaningful interpretation of the most important early rate–pressure–time response when the condition of transience is to be expected. To overcome this difficulty, once stable flow through the separator has been achieved, the well should be closed-in until the wellhead pressure has stabilized when the production can be started again under controlled conditions. Since the flow period is the most important phase of any test (the subsequent pressure buildup (Fig. 4.1) serving merely as a reflection of events while flowing) it is worthwhile applying the above procedure to better define the early, transient test response. In the remainder of this section, it is assumed that an undistorted drawdown trend can be defined.

(a) Inspection of the flowing pressure

As mentioned in section 4.6b, there has been little attention focused in the literature on the importance of the nature of the pressure drawdown on the shape of the subsequent pressure buildup plot and, indeed, there are hardly any papers giving examples of pressure buildups that either show or even mention what happened during the drawdown. Yet, as demonstrated in section 4.13, the drawdown behaviour has a significant impact on the shape of any ensuing pressure buildup. It is therefore mandatory in test interpretation that the engineer should carefully examine the flowing pressures before ever embarking on the buildup analysis. Shown in Fig. 4.14a are three flowing pressure responses: for an infinite acting reservoir (A), for one which is bounded (B) and for one receiving a degree of pressure support (C); and their equivalent p_D-functions, which have the inverse shape, are shown in

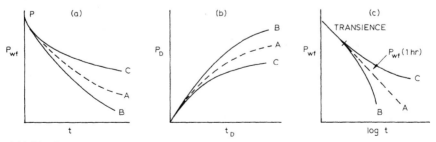

Fig. 4.14. Flowing pressure performance plots: (a) pure transience; (b) bounded system; (c) pressure support.

Fig. 4.14b. Sometimes it is unclear from inspection of the drawdown pressure alone precisely what the flowing condition is and it is therefore recommended that plot (c) be made of p_{wf} versus $\log t$. Since the general drawdown equation is

$$7.08 \times 10^{-3} \frac{kh}{q \mu B_o}(p_i - p_{wf}) = p_D(t_D) + S \qquad (4.21)$$

and

– for purely transient response (A): $p_D = \frac{1}{2} \ln \frac{4 t_D}{\gamma}$

– for a bounded system (B): $p_D > \frac{1}{2} \ln \frac{4 t_D}{\gamma}$

– for pressure support (C): $p_D < \frac{1}{2} \ln \frac{4 t_D}{\gamma}$

then the diagnostic shapes shown in Fig. 4.14c will result. Irrespective of the eventual condition, each drawdown must start with a linear, transient decline which, if the reservoir is infinite acting (A), will continue in this manner. If the system is bounded, however, the pressure points will deviate downwards from the initial linear trend (B), while for pressure support, the deviation will be upward (C). The initial pressure p_i, is determined at the start of the test, as described in section 4.14. Having identified the initial, linear drawdown trend representing the purely transient pressure response, the kh-product of the formation and its skin factor, S, may be calculated by measuring the slope of this trend, m — psi/log cycle, and solving the line-source solution of the radial diffusivity equation 4.17, which is appropriate for this initial phase of the drawdown, that is

$$kh = 162.6 \frac{q \mu B_o}{m} \quad \text{(mD·ft)} \qquad (4.26)$$

$$S = 1.151 \left[\frac{p_i - p_{wf(1\,hr)}}{m} - \log \frac{k}{\phi \mu c r_w^2} + 3.23 \right] \qquad (4.27)$$

In the latter, $p_{wf(1\,hr)}$ is the pressure after one hour read from the linear, transient trend or its extrapolation if the duration of transience is less than one hour (Fig. 4.14c). Analysis of the late-time drawdown data, when boundary effects influence the pressure response in the wellbore, is deferred until section 4.19. This type of test is illustrated in exercise 7.2 (p. 162) of reference 3.

(b) Time derivative of drawdown pressures

During the 1980's, the technique of analysing the time derivative of the pressure response of well tests became increasingly popular, being encouraged by the much improved pressure/time resolution of modern gauges. The main aim in calculating and plotting the pressure derivative data (superimposed on the p_{wf} versus $\log t$ plot, Fig. 4.14c) is to help in the identification of the linear transient phase of the drawdown when the derivative is constant. That is, differentiating the transient drawdown equation 4.17, with respect to $\log t$ gives

$$\frac{\mathrm{d}p_{wf}}{\mathrm{d}\log t} = m = \text{drawdown slope (psi/cycle)}$$

Alternative methods of plotting the derivative are

$$\frac{\mathrm{d}p_{wf}}{\mathrm{d}(\ln t)} = \frac{\mathrm{d}p_{wf}}{2.303\mathrm{d}(\log t)} = \frac{\mathrm{d}p_{wf}}{\mathrm{d}t}t = 0.434\,m \tag{4.28}$$

Expression in one of the latter forms is commensurate with the application of log–log type curves (section 4.21) to which form of analysis the description of derivative plotting has been largely confined in the literature. Application of the technique in semi-log plots, however, is equally if not more useful in defining linear trends in the pressure response since it does not suffer from the inevitable distortion in plotting the derivative data on a logarithmic scale.

There is another, popular method of drawdown testing, the two-rate test [4,5] but since this requires an understanding of the theory of pressure buildup analysis, its inclusion in the chapter is deferred until section 4.20a.

4.11. PRESSURE BUILDUP TESTING (GENERAL DESCRIPTION)

The pressure buildup is the most widely practised form of testing in the Industry because it is recognised as the most reliable in terms of acquiring meaningful results (value for money) for the following reasons

(A) It is the form of test that carries the highest probability of defining a period which contains the transient pressure response. This permits correct application of the transient CTR solution of the diffusivity equation 4.20, to describe this physical state in an unambiguous fashion and calculate kh and S.

(B) In routine tests conducted in development wells, it is one of the only methods that provides a means of determining the current average pressure in the drainage area of the well (another method of accomplishing this: two rate testing, is described in section 4.20a).

A schematic of the buildup test in an appraisal well is shown in Fig. 4.1. In essence, it is a two rate test in which the second rate is zero during the closed-in period when times are recorded as Δt. Such a rate ought to be controllable, provided measures are taken to reduce the effects of afterflow described in section 4.14b. Because of this, the transient period should be readily identifiable immediately after well closure and manifests itself as a linear section of the buildup when plotted in a semi-log form:

$$p_{ws} \text{ versus } \log[f(\Delta t)] \tag{1}$$

where p_{ws} is the increasing static pressure and $f(\Delta t)$ is some function of the closed-in time which varies dependent on the chosen method of interpretation.

The general form of the pressure buildup equation can be derived from the superimposed CTR solution of the diffusivity equation 4.24, using the nomenclature defined in Fig. 4.1, that is

$$7.08 \times 10^{-3} \frac{kh}{\mu B_o}(p_i - p_{ws}) =$$
$$(q - 0)[p_D(t_D + \Delta t_D) + S] + (0 - q)[p_D(\Delta t_D) + S]$$

in which t_D is the fixed dimensionless flowing time and Δt_D is the variable closed-in time at which the static pressures, p_{ws}, are recorded and read. If, for convenience, the constant σ (sigma) is introduced in field units as

$$\sigma = 7.08 \times 10^{-3} \frac{kh}{q\mu B_o} \tag{4.29}$$

which is used throughout the chapter, then the basic buildup equation may be reduced to the form

$$\sigma(p_i - p_{ws}) = p_D(t_D + \Delta t_D) - p_D(\Delta t_D) \tag{4.30}$$

in which it will be noted that the skin factor disappears by cancellation. In many computer programs designed to analyse well tests, this equation is used directly. That is, if the user specifies the nature of the system under test, for instance, the outer boundary conditions, then the program will generate the appropriate p_D-function. Taking differences in this for dimensionless time arguments $t_D + \Delta t_D$ and Δt_D, the program next evaluates p_{ws}, having first calculated the permeability as described in the following two sections.

The program then plots the calculated function, p_{ws} versus Δt and compares it with the observed buildup data. If there is a mis-match, then presumably the wrong boundary conditions have been chosen and the interpreter selects again generating a new p_D-function. The process may be repeated until an acceptable match between theoretical and observed data is obtained.

Prior to attempting such sophistication, however, it is a necessary opening move in any buildup analysis to isolate the period of transience from which the formation parameters k and S can be determined. This may be done by applying the semi-log plotting techniques of either Miller, Dyes, Hutchinson (MDH) or Horner, as described in sections 4.12 and 13, respectively. In studying these, the reader should consider the test as one conducted in a new reservoir at the appraisal stage of development. This simplifies the presentation of the basic theory without any significant loss of generality. When it comes to the matter of routine surveys conducted in development wells which have a lengthy production history, the necessary modifications will be described in section 4.19. It is also assumed that the complex afterflow effects which can predominate over other pressure responses at the start of the buildup have been largely eliminated. Since appraisal well testing is being considered, for which most operators run DST equipment, this can be readily achieved by closing in the well downhole in the tool assembly, as described in section 4.14b.

4.12. MILLER, DYES, HUTCHINSON (MDH) PRESSURE BUILDUP ANALYSIS

This provides the simplest means of identifying a period that includes transience following well closure. The original paper [22] was published in 1949 making it a pioneering treatise on the subject of pressure buildup analysis. If the basic equation

$$\sigma(p_i - p_{ws}) = p_D(t_D + \Delta t_D) - p_D(\Delta t_D) \tag{4.30}$$

is examined for small values of the closed-in time, Δt, when transience is to be expected, then applying equation 4.23,

$$p_D(\Delta t_D) = \tfrac{1}{2}\ln\frac{4\Delta t_D}{\gamma} \tag{4.31}$$

and provided

$$p_D(t_D + \Delta t_D) \approx p_D(t_D) = \text{constant} \tag{4.32}$$

then if these are substituted in the basic equation, it is reduced to

$$\sigma(p_i - p_{wsl}) = p_D(t_D) - \tfrac{1}{2}\ln\frac{4\Delta t_D}{\gamma} \tag{4.33}$$

which, after conversion from "ln" to "log" and collecting the constants together may be more conveniently expressed as

$$\sigma(p_i - p_{wsl}) = C - 1.151\log\Delta t \tag{4.34}$$

in which the constant C is

$$C = p_D(t_D) - 1.151\log\frac{4 \times 0.000264\,k}{\gamma\phi\mu c r_w^2} \tag{4.35}$$

Equation 4.34 indicates that provided conditions in equations 4.31 and 4.32 are satisfied, then the buildup equation may be reduced to a linear relationship between the static pressure and the logarithm of the closed-in time. The pressures in equation 4.34 are denoted by p_{wsl} to indicate that they are simply points anywhere on the straight line defined by the equation. There will be a period, for small values of Δt, when they match the observed pressures but there is nothing to prevent the extrapolation of the straight line outside the range of coincidence to either large or small values of Δt, when p_{wsl} has hypothetical values, as illustrated in Fig. 4.15. The usefulness of the extrapolation of equation 4.34 is demonstrated frequently throughout the chapter.

The two conditions in equations 4.31 and 4.32 that must be satisfied in order to reduce the basic equation 4.30 to its linear form, equation 4.34, require careful examination. The first is the purely transient condition and must apply immediately after well closure and whether it lasts for minutes or months, depends on the magnitude of the diffusivity constant, $k/\phi\mu c$, as described in section 4.5a. The second condition, 4.32, caters for the extrapolation of the p_D-function beyond the flowing time, t_D, for which it is defined and is not necessarily related to transience

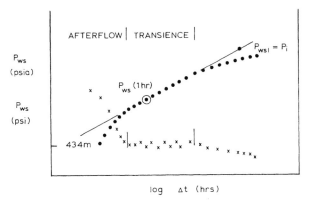

Fig. 4.15. Miller, Dyes, Hutchinson pressure buildup and time derivative plots. ······ = observed pressures (p_{ws}); —— = pressures on the straight line (p_{wsl}), equation 4.34; ××××× = derivative plot, equation 4.41.

at all. Since both conditions must be satisfied in order that the early buildup data lie on the straight line corresponding to equation 4.34, the duration of the linear section will be dependent upon which of the conditions fails first. One point that is evident from this consideration is that while the development of an early straight line on an MDH buildup implies the occurrence of transience, its duration is not necessarily related to the length of time for which transience prevails in the buildup. It may therefore be stated that the development of an early straight line on an MDH buildup plot is a necessary but not sufficient condition for transience [29]. The duration of the line depends to a large extent on the nature of the p_D-function characterising the formation. This matter is further highlighted in Exercises 4.2 and 4.3, which illustrate pressure buildup analyses for formations which have markedly different p_D-functions. Fortunately, the reader does not have to be too concerned about the time range for which conditions in equations 4.31 and 4.32 are valid since they are self-checking. When the conditions are simultaneously satisfied a straight line results which terminates when one or both conditions is invalid. Therefore, in connection with the concern expressed in section 4.8 over how the extrapolation of the p_D-function is handled for times beyond which it has been defined, the MDH method for developing the early, linear buildup trend is quite safe.

It will be noted from equation 4.34 that the slope of the straight line is $m = 1.151/\sigma$ and using equation 4.29 this may be evaluated as

$$m \ (\text{psi/cycle}) \ = \ \frac{1.151}{\sigma} \ = 162.6 \frac{q\mu B_o}{kh} \tag{4.36}$$

from which the kh-value can be calculated and if the reservoir has been perforated across its entire thickness, as suggested in section 4.5a, it should be, in order to satisfy the assumptions implicit in the derivation of the basic radial flow equation 4.3; then the kh-product may be split into its component parts. The value of k thus determined is the thickness averaged, effective permeability to oil in the presence of connate water.

The skin factor, S, can also be calculated by identification of the early straight line. Subtracting equation 4.34 from the drawdown equation 4.21 and solving explicitly for S at $\Delta t = 1$ hour, $p_{wsl} = p_{wsl(1\,hr)}$ yields, after some algebraic manipulation,

$$S = 1.151 \left[\frac{p_{wsl(1\,hr)} - p_{wf})}{m} - \log \frac{k}{\phi \mu c r_w^2} + 3.23 \right] \tag{4.37}$$

in which m is the slope of the linear buildup, psi/cycle, and p_{wf} is the final flowing pressure. This is the conventional expression for calculating the skin factor appearing in the literature in which p_{wsl} is evaluated after one hour of closure. It should be noted that this value is read from the straight line since equation 4.34 is used in the derivation of S. In Fig. 4.15 this point lies in the range of coincidence of the real pressures and the straight line but this is not necessarily always the case. If afterflow effects (section 4.14b) dominate the buildup pressure response for longer than one hour, the value of $p_{wsl(1\,hr)}$ will be read off the backward linear extrapolation of equation 4.34 at $\Delta t = 1$ hour. Alternatively, if there is pressure support (section 4.18), for which the buildup is usually very rapid, the transient response may have terminated in less than one hour meaning that the value of $p_{wsl(1\,hr)}$ will be read from the forward extrapolation of the straight line, as illustrated in Exercise 4.3. It should also be noted that in MDH analysis the evaluation of equation 4.37 is totally independent of the flowing time, t, unlike the similar formulation which is used in conjunction with Horner analysis, as described in the following section.

A useful technique in performing MDH analysis is the extrapolation of the early straight line to the closed-in time, Δt_s, at which the pressure p_{wsl} has risen to the value of the initial pressure p_i. Then, considering equation 4.33, it follows that

$$p_D(t_D) = \tfrac{1}{2} \ln \frac{4 \Delta t_{Ds}}{\gamma} \tag{4.38}$$

which provides a means of explicitly evaluating the p_D-function at the time of well closure. For instance, in the simple case of an appraisal well test in an infinite acting reservoir in which transience prevails throughout, then applying equation 4.22

$$p_D(t_D) = \tfrac{1}{2} \ln \frac{4 t_D}{\gamma} = \tfrac{1}{2} \ln \frac{4 \Delta t_{Ds}}{\gamma} \tag{4.39}$$

which implies that if the early linear section of the buildup is extrapolated to $\Delta t_s = t$, the fixed flowing time, then $p_{wsl} = p_i$ which determines the initial reservoir pressure. Care should be exercised in how the extrapolation of equation 4.33 is interpreted in practice. The equation states that if the early straight line has been defined correctly, which implies that k, S and $p_D(t_D)$ are also correctly determined (refer to Exercise 4.1), then the extrapolation of the line to the value of p_i (determined at the very start of the test with a short flow and buildup — section 4.14e) will occur at the closed-in time $\Delta)t_s$ defined by equation 4.38. Supposing, however, that a certain amount of depletion has occurred so that the initial pressure has fallen from p_i to a reduced averaged value $\bar{p}$, then, unfortunately, equation 4.43 by itself gives no guidance as to what time the extrapolation of equation 4.33 should be made to reach this pressure. What is required is inclusion of a material balance term, as described in sections 4.19a and b, to facilitate this extrapolation.

Furthermore, joining any two points on the buildup plot for either small or large values of Δt will provide a straight line from which values of k and S can be obtained that when inserted into the basic drawdown equation (4.21) will determine a value of $p_D(t_D)$. If this is used in equation 4.38, then again a value of Δt_s at which the straight line extrapolates to p_i. This is just an unfortunate reflection of the basic lack of uniqueness associated with the use of the CTR solution of the diffusion equation referred to in section 4.8. In spite of these difficulties, equation 4.38 proves to be the *key* to the successful application of the MDH analysis technique, as described throughout the remainder of the chapter.

Other, useful applications of this technique are described for appraisal well testing in sections 4.16 and 18c and illustrated in Exercises 4.2, and 4.3, and for development well testing in section 4.19b.

Another method of determining whether the correct linear part of the MDH buildup representing transience has been selected is to plot in conjunction with the conventional buildup the time derivative of the static pressures. That is, differentiating equation 4.34 with respect to log Δt gives:

$$\frac{\mathrm{d}p_{\mathrm{wsl}}}{\mathrm{d}(\log \Delta t)} = m \quad \text{(buildup slope)} \tag{4.40}$$

or, alternatively, to correspond with derivative plotting using log–log type curves (section 4.21)

$$p'_{\mathrm{ws}} = \frac{\mathrm{d}p_{\mathrm{ws}}}{2.303\,\mathrm{d}(\log \Delta t)} = \frac{\mathrm{d}p_{\mathrm{ws}}}{\mathrm{d}\Delta t} \cdot \Delta t = \frac{1}{2\sigma} = 0.434\,m \tag{4.41}$$

In practice when analysing a test, the derivative is calculated across the full time range resulting in the type of plot shown in Fig. 4.15. Then, whichever of the derivative expressions in equation 4.41 is used the function will have the property that during the early linear buildup the plot of the derivative versus log Δt or Δt will exhibit a near-plateau with numerical value $p_{\mathrm{ws}} = 0.434 \times$ buildup slope, the relationship between σ and m being established by equation 4.36. In fact, for the reasons stated in section 4.14a, it is preferable to use the description *near*-plateau rather than plateau.

It should be a matter of course for the engineer to plot both semi-log buildups and their time derivatives when analysing tests, in an attempt to confirm that the buildup and its derivative together, as illustrated in all the exercises in this chapter, in an attempt to confirm that the correct linear section of the buildup has been selected but unfortunately this does not guarantee that the transient response has been uniquely defined. That is, there may be more than one linear section in the buildup (as often proves to be the case) and equation 4.41 will apply to each since it merely states the obvious fact that the derivative of a straight line is a constant whose value is proportional to the magnitude of the slope. Nevertheless, derivative plotting often proves to be a useful method of establishing what is and what is not

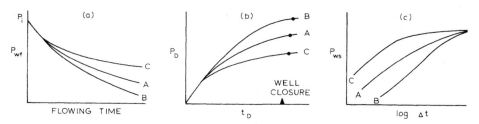

Fig. 4.16. Flowing pressure and MDH-buildup plots: A = pure transience, B = influence of faults, C = pressure support.

a straight line. Attempts have been made for many years to use time derivative techniques in test interpretation. Unfortunately, the application suffers from "scatter" in the data points arising from the fact that the derivative divides the differences in discretely measured pressures by the differences in discrete times, rather than using continuous data. Various methods have been presented in the literature [23,24] to smooth the derivative plots but these invariably assume that the scatter in the data points is random, which is not necessarily the case. Smoothing is not undertaken in this chapter since the time derivative plots are only used in a strictly qualitative manner to check trends in the conventional semi-log buildup plots, particularly the existence and duration of the early, linear trend containing transience.

As already mentioned in sections 4.6b and 10a, the shape of the pressure buildup plot is, to a large extent, dependent on the pressure response during the drawdown period. Figure 4.16 demonstrates the influence of three types of pressure drawdown on the subsequent MDH pressure buildups: A — a purely transient response; B — an increased drawdown due to the proximity of faults or other boundary effects; and C — an elevation of pressures resulting from partial pressure support. The corresponding p_D-functions, which have the inverse shape, are shown in Fig. 4.16b. If the well is closed-in for a buildup at the flowing dimensionless time t_D, the influence of the drawdown on the buildup can be determined by examining the effect of inclusion of the different p_D-functions in the MDH linear buildup equation 4.33. That is

$$A \text{ — transience,} \qquad p_D(t_D) = \tfrac{1}{2} \ln \frac{4 t_D}{\gamma}$$

$$B \text{ — boundary effects,} \quad p_D(t_D) > \tfrac{1}{2} \ln \frac{4 t_D}{\gamma}$$

$$C \text{ — pressure support,} \quad p_D(t_D) < \tfrac{1}{2} \ln \frac{4 t_D}{\gamma}$$

The three MDH pressure buildup plots are shown in Fig. 4.16c. For pure transience, the early, linear buildup is of very short duration, as illustrated in Exercise 4.1, and thereafter there is continued downward curvature for larger values of Δt. If the test is influenced by boundary effects, B, the right-hand side of equation 4.33 will be larger than in the infinite reservoir case, consequently values of p_{wsl} must be lower; nevertheless, the slope of the line is the same $(1.151/\sigma)$. For large values of the closed-in time, there may be some slight upward curvature of the plot before it eventually trends downwards. This type of buildup is illustrated in Exercise 4.2. Finally, in the case of partial pressure support, C, the right-hand side of equation 4.33 will be smaller than for a purely transient buildup. Therefore, p_{wsl} must be larger and the early, linear buildup response will lie above that for an infinite acting system but again must have the same slope. For large closed-in time the buildup displays strong downward curvature which, as pointed out in section 4.18c and illustrated in Exercise 4.3, is one of the advantages in applying the MDH buildup plotting technique to tests in which there is an element of pressure support.

It should be noted that the late time pressure response of any MDH plot must curve downwards, since the ultimate condition is

$$p_{ws} \rightarrow p_i \quad \text{as} \quad \Delta t \rightarrow \infty$$

Therefore, the final part of the plot is not used in any quantitative manner. Instead, the initial pressure in an appraisal well test or average pressure in a development well test (section 4.19b) is obtained by extrapolation of the early linear buildup trend, applying equation 4.38.

4.13. HORNER PRESSURE BUILDUP ANALYSIS

On reading the extensive literature on well testing, one cannot help but feel some sympathy for Horner. Invariably, when his analysis method is referred to, the reader is assiduously reminded that it was, in fact, the brain-child of Theis, a Groundwater Hydrologist, in 1935 [25]. Be that as it may, since first publication of his paper in 1951 [26], the Horner pressure buildup interpretation technique has been, and still remains, the most popular method applied in the Industry and has always carried his name. As in the approach adopted in the previous section, if the basic buildup equation

$$\sigma(p_i - p_{ws}) = p_D(t_D + \Delta t_D) - p_D(\Delta t_D) \tag{4.30}$$

is examined for small values of the closed-in time, Δt, when transience is to be expected, then the second term can be evaluated as:

$$p_D(\Delta t_D) = \tfrac{1}{2} \ln \frac{4\Delta t_D}{\gamma} \tag{4.31}$$

and substituting this in equation 4.30, while simultaneously adding and subtracting $\tfrac{1}{2}\ln(t_D + \Delta t_D)$, which leaves the equation unaltered, gives

$$\sigma(p_i - p_{ws}) = p_D(t_D + \Delta t_D) - \tfrac{1}{2} \ln \frac{4\Delta t_D}{\gamma} \pm \tfrac{1}{2} \ln(t_D + \Delta t_D) \tag{4.42}$$

which may be rearranged as

$$\sigma(p_i - p_{ws}) = \tfrac{1}{2} \ln \frac{t + \Delta t}{\Delta t} + p_D(t_D + \Delta t_D) - \tfrac{1}{2} \ln \frac{4(t_D + \Delta t_D)}{\gamma} \tag{4.43}$$

And again, for small values of Δt when the following conditions are satisfied

$$p_D(t_D + \Delta t_D) \sim p_D(t_D) = \text{constant} \qquad (4.32)$$

$$\ln(t_D + \Delta t_D) \sim \ln(t_D) = \text{constant} \qquad (4.44)$$

which follow, since the dimensionless flowing time, t_D, is a constant; then substituting these in equation 4.43 gives

$$\sigma(p_i - p_{wsl}) = \tfrac{1}{2}\ln\frac{t + \Delta t}{\Delta t} + p_D(t_D) - \tfrac{1}{2}\ln\frac{4\,t_D}{\gamma} \qquad (4.45)$$

which for plotting purposes is usually expressed as:

$$\sigma(p_i - p_{wsl}) = 1.151 \log\frac{t + \Delta t}{\Delta t} + p_D(t_D) - \tfrac{1}{2}\ln\frac{4\,t_D}{\gamma} \qquad (4.46)$$

This is the Horner equation which states that, provided conditions in equations 4.31, 4.32 and 4.44 are satisfied, then the static pressures will plot as a linear function of the logarithm of the Horner time ratio $t + \Delta t/\Delta t$, since the last two terms in equation 4.46 are constant. When the conditions are no longer satisfied the pressure points will deviate away from the straight line. As for the MDH linear buildup equation 4.33, the pressures in equation 4.46 are denoted as p_{wsl} to signify that they comprise a straight line. There will be a period when equation 4.46 matches the actual pressures but there are many interesting applications illustrated in the following sections and exercises in which equation 4.46 is extrapolated outside the range of coincidence with the pressures. Condition 4.31 is the statement of transience whereas equations 4.32 and 4.44 are not associated with transience. Therefore, as for the MDH plot, the development of an early linear section is a necessary but not a sufficient condition for transience. The early, linear section will persist until any one of the conditions breaks down but once again it is unnecessary to consider the time ranges for which the conditions are valid since they are self checking and departure of the actual pressures from the linear trend signifies their invalidity.

The two constants at the end of equation 4.46 influence the position and shape of the Horner buildup plot. Since both are dependent on the flowing time, it is instructive to consider the influence of the pressure behaviour during the flowing period on the buildup which follows. Three cases are illustrated in Fig. 4.17; A — for a purely transient drawdown, B — for a well whose drawdown is affected by boundary conditions such as sealing faults, and C — for a well receiving pressure support. In each case the rate is the same and remains constant during the production period. Also plotted are the equivalent p_D-functions for the three cases, which being proportional to $(p_i - p_{wf})$, equation 4.21, have the inverse shape. The influence of the drawdown behaviour on the shapes of the subsequent pressure buildups is as follows.

A. Transience

In this case, at the end of the flow period

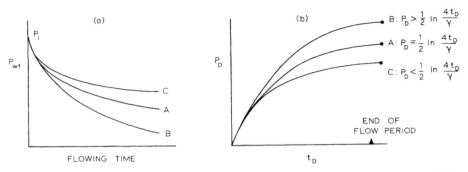

Fig. 4.17. Flowing well performance. (a) Bottom-hole flowing pressure. (b) Equivalent dimensionless pressure function.

$$p_D(t_D) = \tfrac{1}{2} \ln \frac{4\,t_D}{\gamma} \tag{4.22}$$

consequently, the last two items in equation 4.46 cancel and the linear buildup is reduced to

$$\sigma(p_i - p_{wsl}) = 1.151 \log \frac{t + \Delta t}{\Delta t} \tag{4.47}$$

which is Horner equation for an infinite acting reservoir. The slope of this straight line is

$$m \ (\text{psi/cycle}) = \frac{1.151}{\sigma} = 162.6 \frac{q\mu B_o}{kh} \tag{4.36}$$

which is precisely the same as for the MDH plot (section 4.12). Identification of the line and measurement of its slope permits the calculation of the kh-product. In this particular case, as can be seen from equation 4.47, extrapolation of the linear trend to infinite closed-in time gives:

$$\log \frac{t + \Delta t}{\Delta t} \rightarrow \log 1 = 0 \tag{4.48}$$

$$p_{wsl} \rightarrow p_i$$

that is, the buildup extrapolates to give the initial pressure, as shown in Fig. 4.18. This type of buildup is experienced during brief tests in large reservoirs of moderate to high flow capacity or in formations with low flow capacity.

B. Boundary effects

Appraisal well tests are often influenced by boundary effects such as the presence of sealing fault patterns in the reservoir. These may not be the outer boundaries of the system being tested but simply faults which remove the condition of pure transience, as described in greater detail in section 4.16. They have the effect of drawing down the flowing pressure to a greater extent than if the reservoir were infinite acting, Fig. 4.17a, so that the buildup starts at a lower level. In this case, at the end of the flow period

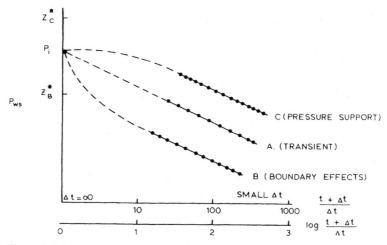

Fig. 4.18. Influence of drawdown behaviour on the shape of Horner buildup plots.

$$p_D(t_D) > \tfrac{1}{2} \ln \frac{4\,t_D}{\gamma}$$

as shown in Fig. 4.17b. When this condition is incorporated in equation 4.46, the right-hand side contains a positive constant so that for a given closed-in time, Δt, it will be larger than in the infinite reservoir case (equation 4.47), consequently p_{wsl} must be lower. During the period of initial linearity, data points on this early section of the buildup will lie below those of case A, Fig. 4.18, but will have the same slope, $1.151/\sigma$. For large values of the closed-in time, if the boundaries are not totally enclosing, so that depletion does not occur, the buildup will curve upwards to eventually reach the initial pressure p_i. Extrapolation of the transient, linear trend to infinite closed-in time gives a hypothetical value of pressure, $Z^{*\,\dagger}$, which is less than the initial reservoir pressure p_i. The pressure buildup can be analysed in an attempt to define the fault geometry as described in section 4.16 and illustrated in Exercises 4.2. This form of buildup analysis has been well documented in the popular literature.

C. Pressure support

This type of buildup response is observed in developed fields under pressure maintenance conditions, as described in section 4.19c. It also occurs for a variety of reasons while testing at the appraisal stage (section 4.18) when nature supplies the pressure support, for instance, due to the reservoir having a very high flow capacity. In this case, at the end of the flowing period

$$p_D(t_D) < \tfrac{1}{2} \ln \frac{4\,t_D}{\gamma}$$

[†] In the literature, this hypothetical pressure has always been referred to as p^* but there are very sound reasons, presented in section 4.15b, for changing the nomenclature and calling it Z^*

and therefore, the right-hand side of equation 4.46 now contains a negative constant. Consequently, for a given closed-in time it will be less than in the infinite reservoir case (equation 4.47), meaning that p_{wsl} will be larger. The initial, linear part of the buildup will therefore lie above that for case A but be parallel to it, again having slope $1.151/\sigma$, as shown in Fig. 4.18.

For large values of the closed-in time the buildup must naturally curve downwards towards the initial pressure p_i which would be reached at infinite closed-in time. In this case, extrapolation of the initial, linear trend gives a value of the hypothetical pressure, Z^*, which is in excess of the initial reservoir pressure p_i.

As mentioned in sections 4.3 and 4.6b, the topic of the analysis of buildup tests following a flow period in which there has been full or partial pressure support has received little attention in the popular literature. Because of this, errors are frequently made in the interpretation of such buildup responses, particularly in field areas where the flow capacities are high and in gas well testing since under both circumstances some degree of natural pressure support is frequently observed. The main source of error is that of selecting the initial linear response, the interpreter finds it difficult to accept that its linear extrapolation yields a value of the extrapolated pressure, Z^*, which is in excess of the initial reservoir pressure, p_i. A linear part of the buildup is therefore sought, and often found, for larger values of the closed-in time, Δt, which extrapolates towards p_i. As will be clear from Fig. 4.18, however, such a line will have a smaller slope than for transience thus giving a higher value of the permeability (equation 4.36) and, to balance the CTR solution (equation 4.21), a higher value of the skin factor also. There are two popular misconceptions involved in this error, namely

- that the correct straight line on a Horner plot should necessarily extrapolate to give p_i at infinite closed-in time; whereas, as described in A above, this only occurs in the case of a purely transient test (and a few other exceptional cases described later).
- that Z^* is a real pressure, whereas it is only a number with the dimensions of pressure.

The safest method of analysing buildups following a flow period in which there has been some degree of pressure support is described in section 4.18 and illustrated in Exercise 4.3, while the meaning of Z^* is described below and also in section 4.15b.

The conventional approach for calculating the skin factor in Horner analysis is to subtract the linear equation 4.46, from that describing the drawdown, equation 4.21, to give

$$S = \sigma(p_{wsl} - p_{wf}) + \tfrac{1}{2} \ln \frac{t + \Delta t}{\Delta t} - \tfrac{1}{2} \ln \frac{4\, t_D}{\gamma}$$

which, when evaluated for $\Delta t = 1$ hour, $p_{wsl} = p_{wsl(1\,hr)}$, and provided $t \gg 1$ hour may be reduced to the form

$$S = 1.151 \left[\frac{p_{wsl(1\,hr)} - p_{wf}}{m} - \log \frac{k}{\phi \mu c r_w^2} + 3.23 \right] \tag{4.37}$$

This, it will be recognised, is precisely the same expression as used with the MDH plotting technique (section 4.12). When applying equation 4.37 in Horner analysis, however, there will be a loss of accuracy if the flowing time is not considerably greater than one hour which does not act as a limitation when applying the same formula in MDH analysis, since the flowing time does not enter the equations used in the latter method. To overcome this slight weakness in applying equation 4.37 to Horner analysis, it is suggested that instead, for appraisal well testing, the transient linear part of the buildup be extrapolated to infinite closed-in time giving $p_{wsl} = Z^*$ which reduces equation 4.46 to the form

$$\sigma(p_i - Z^*) = p_D(t_D) - \tfrac{1}{2}\ln\frac{4\,t_D}{\gamma} \tag{4.49}$$

which when subtracted from equation 4.21 and solved for S gives

$$S = 1.151\left[\frac{Z^* - p_{wf}}{m} - \log\frac{kt}{\phi\mu cr_w^2} + 3.23\right] \tag{4.50}$$

While this is similar in form to the more commonly used equation 4.37, $p_{wsl\,(1\,hr)}$ is replaced by Z^* and the flowing time, t, is included in the logarithmic term. For the reasons stated in section 4.15a, the latter should be evaluated as

$$t = \frac{N_p\ (\text{stb})}{q\ (\text{stb/d})} \times 24 \quad (\text{hours}) \tag{4.51}$$

where N_p is the cumulative oil production and q the final rate. This expression preserves the material balance associated with production. As in MDH analysis, the time derivative of the Horner plot can be diagnostic in selecting the early, linear section of the buildup. That is, taking the derivative of equation 4.46 with respect to the log of the Horner time ratio gives

$$\frac{dp_{ws}}{d\left(\log\dfrac{t + \Delta t}{\Delta t}\right)} = \frac{1.151}{\sigma} = m \quad (\text{buildup slope})$$

or to make the derivative compatible with log–log type-curve analysis (section 4.21)

$$p'_{ws} = \frac{dp_{ws}}{2.303\,d\left(\log\dfrac{t + \Delta t}{\Delta t}\right)} = \frac{1}{2\sigma} = 0.434\,m \tag{4.52}$$

Therefore a plot of the derivative superimposed on the conventional Horner plot of p_{ws} versus $\log(t + \Delta t)/\Delta t$ will exhibit a near-plateau of magnitude $0.434\,m$ (psi/cycle) during the early, linear section of the buildup, helping to define this response but unfortunately, as for the MDH plot, the existence of such a plateau does not guarantee that the transient part of the buildup has been uniquely defined, since equation 4.52 will be satisfied for any linear section of the plot. The engineer must be guided by the fact that if there is more than one linear section on the buildup, it is the first, after well closure, that contains the transient response. Application of the Horner derivative plot is illustrated in Exercises, 4.1, 4.2 and 4.3.

4.14. SOME PRACTICAL ASPECTS OF APPRAISAL WELL TESTING

Before studying various exercises on appraisal well testing, it is first necessary to consider two practical aspects of pressure buildup testing: the determination of the initial pressure at the very start of the test and the control of well afterflow during buildup periods.

(a) Determination of the initial pressure

At the start of an appraisal well test, it is customary for most operators to flow the well for a few minutes and then to conduct a short buildup, solely to establish the initial reservoir pressure. The flow and buildup periods are typically 5 and 60 minutes respectively which result in a maximum value of the abscissa of a Horner plot of

$$\log \frac{t + \Delta t_{max}}{\Delta t_{max}} = \log \frac{5 + 60}{60} = 0.035$$

meaning that the extrapolation to infinite closed-in time is very short, providing a reliable determination of the initial pressure. To check for such effects as pressure depletion, through analysis of the main flow and buildup that follows, it is essential to have an accurate value of the initial pressure with the gauge selected for the interpretation of the main part of the test. That is, although the pressure may have been measured in advance during an RFT survey, it is pointless to attempt to compare this with the DST pressure since they are recorded with different gauges. And, while modern gauges are noted for their high resolution, it is quite possible that they could differ by, say, 10 psi in measurement of the absolute pressure. If this disparity were interpreted in terms of depletion, it could lead to a pessimistic conclusion concerning the size of the hydrocarbon accumulation.

The duration of the initial flow period should be kept as short as possible to reduce the length of the subsequent buildup necessary to obtain a short but accurate extrapolation to the initial pressure. Five minutes is usually sufficient to relieve any supercharged effects resulting from the test completion practices and ensure that sufficient hydrocarbons have been produced to rise above the pressure gauges in the flow string. The latter facilitates the calculation of the pressure gradient between the gauges and the formation and hence the extrapolation of the recorded pressure to any chosen datum depth.

(b) Afterflow

If, when conducting an appraisal test, the well is closed-in at the surface, which may be several kilometres away from the producing formation, the flow of fluids from the reservoir into the wellbore does not stop instantaneously, as depicted in Fig. 4.19. The reason is because as the hydrocarbons rise to the surface during the flow period they experience a reduction in pressure and temperature meaning that at any point in the flow string the fluid compressibility, c_T, is greater than the value in

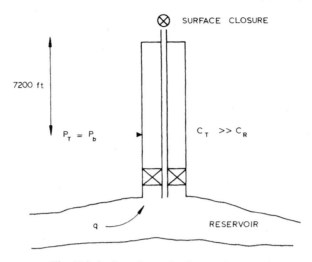

Fig. 4.19. Surface closure leading to afterflow.

the reservoir, c_R. Therefore, production from the formation will continue at an ever diminishing rate until the fluids in the drillpipe or tubing have been compressed into a state of equilibrium. This phenomenon, known as afterflow, may last for minutes or many hours dependent on the nature of the fluid properties and the capacity of the flow string. For instance, when testing a gas saturated oil from a deep reservoir, the combination of highly variable compressibility and large storage volume to the surface will provide conditions conducive to a lengthy period of afterflow.

The effect of afterflow is to distort the early part of the buildup as shown in Figs. 4.20a and b. In the former the duration of the phenomenon is slight and the early linear section of the Horner buildup plot is still readily discernible. The situation depicted in Fig. 4.20b, however, illustrates the extreme case in which the transient, linear part of the buildup is obscured by the afterflow response. The type of situation which can give rise to this complex form of buildup was described in a paper published in 1983 [27] — "Complexities of the Analysis of Surface Shut-In Drillstem Tests in an Offshore Volatile Oil Reservoir", a title which

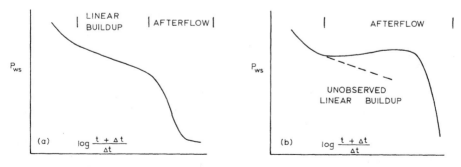

Fig. 4.20. Distortions in the early pressure buildup due to afterflow.

in itself reveals the problem. The reservoir contained a moderately volatile oil ($B_{oi} = 1.73$ rb/stb, $R_{si} = 1120$ scf/stb) but the main difficulty was that during the vertical flow, the pressure in the flow string, p_T, fell below the bubble point, p_b, some 7200 ft from the surface, above which there was a variable free gas saturation (Fig. 4.19). Following closure at the surface, the reservoir continued to produce for a considerable period while re-compressing the liberated gas into an equilibrium state. Matters were further complicated in that between the sea bed and the derrick floor the gas experienced a cooling effect on account of the proximity of the cold sea water. Free gas in the riser was therefore subjected to a slight but continuous decrease in compressibility which merely served to prolong the period of afterflow. Altogether, the authors isolated five phenomena in the sequence of tests which contributed to the complex afterflow observed and suggested lines of research to elucidate the many problems leading to a means of test interpretation which, it was anticipated, would be increasingly required especially when testing offshore. Instead, it is suggested that the more obvious solution (as stated in both the conclusions and recommendations in the paper [27]) is to *close in downhole*. In one simple stroke, this overcomes all the difficulties associated with complex afterflow effects since the re-compression of fluids in the tubing or drillpipe is avoided and the firm closure results in the cessation of flow almost immediately, or within a matter of minutes.

It is appreciated that the majority of operators are fully aware of this simple point and conduct tests at the appraisal stage using DST equipment which permits downhole closure to be affected with ease; compared to the less common practice of testing by making a temporary completion with a packer and production tubing which, unless the equipment is specially adapted, precludes downhole closure. At the appraisal stage, the elimination of afterflow is considered essential because the whole process is a matter of testing the unknown and since the duration of afterflow resulting from surface closure is uncertain, to encourage the phenomenon is merely adding to the potential complications. Also, as illustrated in Exercise 4.3, which applies to a rapid buildup in a fairly high flow capacity reservoir, the transient, linear part of the buildup lasts for only ten minutes and could easily have been missed had the well not been closed-in downhole.

Alternatively, it could be argued that afterflow itself can provide useful information on the formation being tested and also that analytical techniques have been devised to account for such complications as gas break-out in the flow string. Both statements are justified. In fissured reservoirs the afterflow includes storage in the fractures intersecting the well as described by Gringarten [1]. The study of the early-time behaviour in testing such a formation can prove diagnostic in evaluating flow performance close to the well while, concerning the occurrence of free gas in the tubing or drillpipe, Fair [28] has presented a means of accounting for this in idealised systems. Nevertheless, Operators still prefer to affect downhole closure when testing at the appraisal stage, if possible. This is because these are usually the most important tests conducted during the lifetime of a reservoir and provide a base-line against which all subsequent development tests are compared. Furthermore, on account of the high cost of such tests, especially when conducted offshore, the uncertainty in interpretation must be minimised so that the results lead to useful development decisions (section 4.4a). Therefore, the deliberate encouragement of afterflow, which can only be described under ideal conditions and not when there are complications such as described in reference 27, is to be avoided at all cost.

The analysis methods of MDH and Horner cannot be applied to interpret the pressure behaviour during the period of afterflow, they merely indicate when the response terminates. The accepted method is through the use of Type Curves and Derivative Type Curves, as described in section 4.21. Accounting for afterflow in development well testing, where it is much more prevalent due to the common practice of surface closure, is described in section 4.19g.

In the event that the start of a pressure buildup is affected by afterflow the start of the semi-log straight line on a Horner or MDH plot can be located by applying "Ramey's one and a half cycle rule". If a plot is made of $\log(p_{ws} - p_{wf})$ versus $\log \Delta t$, during the period of complete dominance of afterflow, the points will lie on a 45° degree straight line. As soon as the points start to deviate below this initial trend, then moving one and a half log cycles along the $\log \Delta t$ scale will identify the approximate start of the semi-log straight line. The method, which is surprisingly accurate, is also incorporated in log–log type-curve interpretation (section 4.21a).

Exercise 4.1: Pressure buildup test: infinite acting reservoir

Introduction

This is a DST conducted in an appraisal well in which a transient pressure response prevailed throughout. The purpose in analysing this simple test is to contrast the two semi-log plotting techniques of Miller, Dyes, Hutchinson (MDH) and Horner for this condition.

Question

An appraisal well was tested by producing a cumulative 810 stb of oil, the final flow rate being 1870 stb/d. During the production period, the pressure declined continuously as a function of time and flow was terminated by firm down-hole closure thus largely eliminating afterflow. The buildup data are listed in Table 4.3 and the reservoir and fluid data required to analyse the test are as follows:

p_i = 7245 psia (5 minutes flow/1 hour buildup) c_w = 3.0 × 10^{-6}/psi
p_{wf} = 5500 psia (final pressure) c_f = 6.0 × 10^{-6}/psi
h = 25 ft (fully perforated well) S_{wc} = 0.28 PV
r_w = 0.354 ft (8½ inch hole) μ_o = 0.226 cp
ϕ = 0.22 B_{oi} = 1.740 rb/stb
c_o = 17.24 × 10^{-6}/psi

- Analyse the test using both the MDH and Horner buildup plotting techniques together with their time derivatives.

Solution

The effective flowing time required in the Horner analysis can be calculated using equation 4.51 as:

TABLE 4.3

Pressure buildup data, Exercise 4.1 ($t = 10.4$ hours)

Δt (hours)	p_{ws} (psia)	MDH		Horner	
		$\log \Delta t$	$p'_{ws\,(MDH)}$ [a]	$\log(t + \Delta t)/\Delta t$	$p'_{ws\,(Horner)}$ [b]
0.10	6640	−1.000		2.021	
0.13	6749	−0.886	417.8	1.908	418.8
0.17	6796	−0.770	176.3	1.794	179.0
0.20	6815	−0.699	117.2	1.724	117.9
0.25	6839	−0.602	108.0	1.629	109.7
0.30	6859	−0.523	110.0	1.552	112.8
0.35	6875	−0.456	104.0	1.487	106.9
0.40	6890	−0.398	112.5	1.431	116.3
0.50	6913	−0.301	103.5	1.338	107.4
0.70	6947	−0.155	102.0	1.200	107.0
0.90	6972	−0.046	100.0	1.099	107.5
1.00	6982	0	95.0	1.057	103.4
1.30	7008	0.114	99.7	0.954	109.6
1.60	7028	0.204	96.7	0.875	109.9
1.90	7044	0.279	93.3	0.811	108.6
2.20	7057	0.342	88.8	0.758	106.5
2.50	7068	0.398	86.2	0.713	106.1
3.00	7084	0.477	88.0	0.650	110.3
3.50	7096	0.544	78.0	0.599	102.2
4.00	7107	0.602	82.5	0.556	111.1
4.50	7116	0.653	76.5	0.520	108.6
5.00	7123	0.699	66.5	0.489	98.0
6.00	7136	0.778	71.5	0.437	108.6
7.00	7147	0.845	71.5	0.395	113.7
8.00	7155	0.903	60.0	0.362	105.3
9.00	7162	0.954	59.5	0.334	108.6
10.00	7168	1.000	57.0	0.310	108.6

[a] $p'_{ws\,(MDH)}$: equation 4.41; [b] $p'_{ws\,(Horner)}$: equation 4.52.

$$t = \frac{N_p}{q} \times 24 = \frac{810}{1870} \times 24 = 10.4 \text{ hours} \tag{4.51}$$

which is used in Table 4.1 to calculate the abscissa of the Horner plot. Also listed are the time derivative data which are the average values during each time step. That is, the value listed at $\Delta t = 0.70$ hours is the average between $\Delta t = 0.50$ and 0.70 hours. For this purely transient test, the Horner analysis method is more straightforward than that of MDH and is therefore described first.

Horner. The pressure buildup and time derivative plots are presented in Fig. 4.21. On account of the down-hole closure, the period of afterflow is short-lived and after 12 minutes ($\Delta t = 0.20$ hours) the buildup becomes linear and remains so throughout the test. Extrapolation of this trend to infinite closed-in time gives a pressure of 7245 psia which is identical with the initial reservoir pressure established during the brief five minutes flow and buildup at the start of the test. This

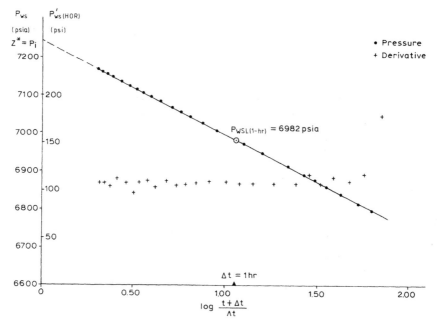

Fig. 4.21. Horner plot (Exercise 4.1).

proves that the entire test, both flow and buildup, occurred under purely transient conditions for which equation 4.47 is appropriate. From the linear section it can be determined that

slope (m) = 250.5 psi/log cycle

$p_{ws(1\,hr)}$ = 6982 psia (Table 4.3, Fig. 4.21)

In spite of the inevitable scatter in the data points, the derivative plot for $\Delta t > 0.20$ hours maintains a plateau, the average value being 108.1 psi, which is approximately $0.434\,m$, as required by equation 4.52. The kh-product can be evaluated using equation 4.36, as

$$kh = 162.6\frac{q\mu B_o}{m} = \frac{162.6 \times 1870 \times 0.226 \times 1.740}{250.5} = 477 \text{ mD·ft}$$

and since the well is fully perforated across the 25 ft thick reservoir then $k = 19$ mD.

The skin factor can be evaluated using the value of $p_{ws(1\,hr)}$ = 6982 psia in the conventional expression used by the Industry (equation 4.37), that is

$$S = 1.151\left[\frac{p_{ws(1\,hr)} - p_{wf})}{m} - \log\frac{k}{\phi\mu c r_w^2} + 3.23\right] \qquad (4.37)$$

in which the effective compressibility is evaluated using equation 4.2 as

$$c = c_o S_o + c_w S_{wc} + c_f$$

$$= (17.24 \times (1 - 0.28) + 3.0 \times 0.28 + 6.0) \times 10^{-6} = 19.25 \times 10^{-6}/\text{psi}$$

giving

$$S = 1.151 \left[\frac{6982 - 5500}{250.5} - \log \frac{19}{0.22 \times 0.226 \times 19.25 \times 10^{-6} \times 0.354^2} + 3.23 \right]$$

$$= 1.09$$

Alternatively, the skin factor may be determined using equation 4.50, which does not rely on the condition being satisfied that $t \gg 1$ hour, which is a requirement in the above formulation. That is

$$S = \left[\frac{Z^* - p_{wf}}{m} - \log \frac{kt}{\phi \mu c r_w^2} + 3.23 \right] \tag{4.50}$$

in which $Z^* = p_i = 7245$ psia and $t = 10.4$ hours, giving

$$S = 1.151 \left[\frac{7245 - 5500}{250.5} - \log \frac{19 \times 10.4}{0.22 \times 0.226 \times 19.25 \times 10^{-6} \times 0.354^2} + 3.23 \right]$$

$$= 1.13$$

This result is only some 4% higher than the value obtained using equation 4.37, which in this example is a negligible difference. Nevertheless, considering the intrinsic importance of accurate determination of the skin factor (section 4.4a) and the overall accuracy in results aimed at using high-resolution pressure recorders and computer analysis methods, the use of equation 4.50 is to be preferred over equation 4.37. The pressure drop across the skin can be calculated using equation 4.18, that is

$$\Delta p_{skin} = 141.2 \frac{q \mu B_o}{kh} S \tag{4.18}$$

but since

$$m = 162.6 \frac{q \mu B_o}{kh} \tag{4.36}$$

then it follows that

$$\Delta p_{skin} = 0.87 \, mS = 0.87 \times 250.5 \times 1.13 = 246 \text{ psi} \tag{4.53}$$

which comprises some 14% of the maximum drawdown of $p_i - p_w = 7245 - 5500 = 1745$ psi.

The observed PI at the end of the flow period is

$$PI = \frac{q}{p_i - p_{wf}} = \frac{1870}{7245 - 5500} = 1.07 \text{ stb/d/psi} \tag{4.1}$$

and the ideal

$$PI_{ideal} = \frac{q}{p_i - p_{wf} - \Delta p_{skin}} = \frac{1870}{7245 - 5500 - 246} = 1.25 \text{ stb/d} \tag{4.1}$$

which is 17% higher. Since these are both evaluated under transient conditions, before any degree of stability was achieved, they represent maximum values of the PI which will decline further with time.

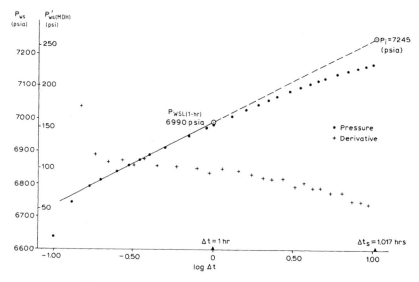

Fig. 4.22. Miller, Dyes, Hutchinson plot (Exercise 4.1).

Miller, Dyes, Hutchinson (MDH). The pressure buildup and time derivative plots are presented in Fig. 4.22. As can be seen, the nature of the plot is quite different from that of Horner. The latter clearly demonstrated that transience prevailed throughout the test since $Z^* = p_i$ yet in the MDH plot, the early, linear part of the buildup lasts for less than one hour. This anomaly is explained below but it is clear that in this instance the engineer would be at a severe disadvantage in selecting the correct, transient straight line using the MDH plotting technique alone. Since it is known from the Horner analysis, however, that the test is purely transient then it also follows from equation 4.39 that the extrapolation of the MDH transient section will yield the initial pressure of $p_i = 7245$ psia when $\Delta t_s = t = 10.4$ hours, which occurs at a value of $\log \Delta t_s = 1.017$ (Fig. 4.22). Knowledge of this helps to identify the straight line which otherwise would be difficult. Assistance can also be gained from inspection of the derivative plot which, in spite of the scatter in data points, demonstrates a plateau level of 107.6 psia between $0.20 < \Delta t < 0.70$ hours followed by a steady decline implying continued downward curvature of the buildup for larger values of the closed-in time.

The early straight line, once identified, has precisely the same slope of $m = 250.5$ psi/log cycle as the Horner plot, consequently the kh-product will also be the same. In calculating the skin factor using equation 4.37, the value of $p_{wsl(1\,hr)}$ read from the extrapolated straight line is 6990 psia. It should be noted, as shown in Fig. 4.22, that this point lies outside the range of coincidence of the straight line (from which it is read) and the buildup data points which have already started to curve downwards. Using this value of $p_{wsl(1\,hr)} = 6990$ psia in equation 4.37 gives a value of $S = 1.13$ which is identical to the correct value determined in the Horner analysis using equation 4.50 containing Z^* rather than $p_{wsl(1\,hr)}$. This merely confirms the point made in section 4.12, that the use of equation 4.37 is quite accurate because it does not depend on the requirement that $t \gg 1$ hour, since the flowing time does not feature in the MDH analysis. All other results are the same as determined in the Horner interpretation.

Conclusions

In the first place, the author is obliged to admit that, whereas the pressure buildups illustrated in Exercises 4.2 and 4.3 are genuine field examples, the present one is contrived. That is, the reservoir and its fluid properties have been "invented", following which the transient equation 4.22 has been applied for the total flow and buildup periods to generate the required p_D-function which, in turn, has been used in equation 4.30 to predict the buildup performance of this hypothetical, infinite acting reservoir system. The reason for adopting this approach was to ensure that the pressure response studied was for strictly transient conditions to permit valid comparison of the Horner and MDH plotting techniques for this well defined but simple physical condition.

On viewing the buildup plots for the two different approaches the reader may wonder why bother considering the MDH technique any further: applying it in isolation to a transient test such as this one could lead to serious error in mistaking the early linear buildup trend and overestimating kh and S by choosing a later, apparent linear section with smaller slope [29]. But at no stage of this chapter is it ever suggested that either MDH or Horner plots should be used exclusively — the engineer must try both techniques and, if applied correctly, the results obtained should be identical. There are circumstances, such as in the present exercise, in which Horner clearly offers the simpler, less ambiguous interpretation but there are others, such as when there is an element of pressure support (Exercise 4.3), when the roles are reversed and the MDH plot appears much safer than Horner. The difference in duration of the early linear section of both plots lies simply in the mathematical conditions governing linearity, as described below. What is important in a diagnostic sense is that in the present test, for instance, the marked difference in the duration of early linearity of the buildup clearly defines the test as one performed on an infinite acting system, which is not always clear using just one of the plots in isolation.

In generating the early, linear buildup equations for MDH, equation (4.33), and Horner (4.46), the mathematical condition implicit in each is that the second term in the basic buildup equation 4.30 could be evaluated under transient conditions so that

$$p_D(\Delta t_D) = \tfrac{1}{2} \ln \frac{4\Delta t_D}{\gamma} \tag{4.31}$$

Then, in deriving the Horner equation there was the additional requirement (equation 4.43) that

$$p_D(t_D + \Delta t_D) - \tfrac{1}{2} \ln \frac{4(t_D + \Delta t_D)}{\gamma} = \text{constant} \tag{4.54}$$

But the latter must always be satisfied in a purely transient test: the constant being zero, which reduces the linear equation to the form expressed by equation 4.47. Therefore, for a test of an infinite acting system, the Horner straight line is governed purely by the transient condition expressed by equation 4.31.

In the derivation of the MDH straight line, however, the condition other than transience that must be satisfied is that expressed by equation 4.32, that is,

$$p_D(t_D + \Delta t_D) \approx p_D(t_D) = \text{constant} \tag{4.32}$$

and clearly in the case of a transient test, this condition breaks down rather rapidly causing a deviation of the pressure points below the linear trend. For the transient test in the system under study

$$p_D(t_D') = \tfrac{1}{2} \ln \frac{4\,t_D'}{\gamma} = \tfrac{1}{2} \ln \frac{4 \times 0.000264 kt'}{\gamma \phi \mu c r_w^2}$$

$$= \tfrac{1}{2} \ln \frac{4 \times 0.000264 \times 19t'}{1.781 \times 0.22 \times 0.226 \times 19.25 \times 10^{-6} \times 0.354^2}$$

$$p_D(t_D') = 5.7251 + \tfrac{1}{2} \ln t'$$

where t' is an arbitrary time argument. If $p_D(t_D)$ in equation 4.33 were to be evaluated as a constant at the end of the flowing period ($t = 10.4$ hours), then applying the above expression: $p_D(t_D) = 6.8960$. Alternatively, considering the variation of the function after one hour of well closure (by which time the pressures have begun to deviate below the straight line — Fig. 4.22) then $t + \Delta t = 11.4$ hours and $p_D(t_D + \Delta t_D) = 6.9419$. The increase compared to $p_D(t_D)$ is only +0.7% yet this is sufficient to increase the right-hand side of equation 4.33 and so reduce the pressures below the linear trend.

What this demonstrates is that the duration of the MDH early linear trend is not governed by the condition of transience alone as it is, in this particular example, for the Horner plot. But for the latter, it must be stressed that this only applies for a purely transient test. In all other cases, the condition stated by equation 4.54 must also be satisfied. Because of this, the statement is made frequently in this chapter — that the early straight line "contains" transience — but is not manifest of this condition alone.

4.15. PRACTICAL DIFFICULTIES ASSOCIATED WITH HORNER ANALYSIS

There are two features in Horner analysis which, if not fully understood, can and do lead to serious errors in the test interpretation itself and in the application of results of further reservoir engineering calculations. These are the inclusion of the flowing time, t, and the meaning and use of the extrapolated pressure, p^* (redefined in this section, for safety reasons, as Z^*). Many mistakes are made arising from the misunderstanding of these two facets of Horner analysis. Both t and Z^* are interrelated and this section describes their influence in appraisal testing, while section 4.19e accounts for the same in development well testing in which their correct handling can prove even more troublesome.

(a) Flowing time/superposition

In deriving the equation of the Horner straight line in section 4.13, the simultaneous addition and subtraction of $0.5\ln(t_D + \Delta t_D)$ to the right-hand side of equation 4.30 is the step that introduces the (largely unnecessary) complication of accounting for the flowing time, t, in Horner buildup analysis; the inclusion of which, it will be noted, is not required in developing the corresponding MDH linear equation 4.33. If an appraisal well could be produced at a constant rate prior to closure, the flowing time would be defined unambiguously as simply the length of the production period. Regrettably, for a variety of practical reasons, it is difficult to maintain a constant rate in a new well (section 4.10) and the actual production profile is often as depicted in Fig. 4.23.

Looking back on events from any time, Δt, measured from the instant of closure, t_N, the equation to determine the static pressure during the buildup may be expressed as

$$\sigma(p_i - p_{ws}) = \sum_{j=1}^{N} \frac{\Delta q_j}{q} p_D(t_{D_N} + \Delta t_D - t_{D_{j-1}}) - p_D(\Delta t_D) \tag{4.55}$$

which it will be recognised is a re-statement of the superposed CTR solution of the radial diffusivity equation 4.24, in which both sides have been divided by the final rate, q, prior to closure. Since the last two terms on the right-hand side are

$$\frac{(q - q_{N-1})}{q} \left[p_D(t_{D_N} + \Delta t_D - t_{D_{N-1}}) + S \right] + \frac{(0 - q)}{q} \left[p_D(\Delta t_D) + S \right]$$

the skin factor disappears completely in the summation. Re-tracing the steps taken in section 4.13, for the derivation of the Horner straight line (equation 4.46) requires the addition and subtraction of $0.5\ln(t_{D_N} + \Delta t_D)$ to the right-hand side of equation 4.55 and the evaluation of $p_D(\Delta t_D)$ for transient conditions, giving

$$\sigma(p_i - p_{ws}) = 1.151 \log \frac{t_N + \Delta t}{\Delta t}$$

$$+ \sum_{j=1}^{N} \frac{\Delta q_j}{q} p_D(t_{D_N} + \Delta t_D - t_{D_{j-1}}) - \frac{1}{2} \ln \frac{4(t_{D_N} + \Delta t_D)}{\gamma} \tag{4.56}$$

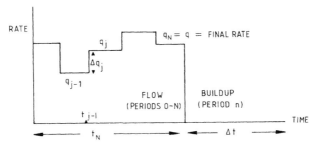

Fig. 4.23. Variable rate history during an appraisal well test divided into N discrete periods.

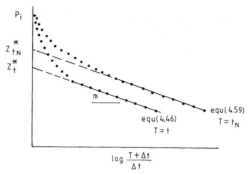

Fig. 4.24. Pressure buildup using two values of the flowing time: t_N = actual time, and t = effective flowing time.

which, provided

$$\sum_{j=1}^{N} \frac{\Delta q_j}{q} p_D(t_{D_N} + \Delta t_D - t_{D_{j-1}}) \approx \sum_{j=1}^{N} \frac{\Delta q_j}{q} p_D(t_{D_N} - t_{D_{j-1}}) = \text{constant} \qquad (4.57)$$

and

$$\ln(t_{D_N} + \Delta t_D) \approx \ln t_{D_N} = \text{constant} \qquad (4.58)$$

reduces equation 4.56 to the linear form, for small Δt

$$\sigma(p_i - p_{wsl}) = 1.151 \log \frac{t_N + \Delta t}{\Delta t} + \sum_{j=1}^{N} \frac{\Delta q_j}{q} p_D(t_{D_N} - t_{D_{j-1}}) - \tfrac{1}{2} \ln \frac{4\, t_{D_N}}{\gamma} \qquad (4.59)$$

in which the last two terms, when conditions in equations 4.57 and 4.58 prevail, are constants that, as described in section 4.13, merely alter the position of the straight line on the Horner plot, dependent on the nature of the p_D-function (refer Fig. 4.18).

On the other hand, suppose the variable rate profile shown in Fig. 4.23 were averaged as:

$$N_p = \frac{q \times t}{24} \text{ (stb)}$$

in which q is the final rate (stb/d) and, t, the "effective" flowing time in hours, which can be greater or less than the actual total flowing time, t_N, dependent on the magnitude of q with respect to the other rates. Then the complex form of equation 4.59 can be replaced by

$$\sigma(p_i - p_{wsl}) = 1.151 \log \frac{t + \Delta t}{\Delta t} + p_D(t_D) - \tfrac{1}{2} \ln \frac{4\, t_D}{\gamma} \qquad (4.46)$$

which is precisely as stated in section 4.13 for constant production. Two pressure buildups for the same observed data points (p_{ws}, Δt) but interpreted using t_N and t respectively are shown in Fig. 4.24.

Also shown are the two linear sections, for small closed-in time, represented by equations 4.59 and 4.46 (in this case the final rate, q, must be greater than the average during the variable rate history since the latter buildup is displaced towards the left). If these are both extrapolated to infinite closed-in time they become

$$\sigma(p_i - Z^*_{t_N}) = \sum_{j=1}^{N} \frac{\Delta q_j}{q} p_D(t_{D_N} - t_{D_{j-1}}) - \tfrac{1}{2} \ln \frac{4 t_{D_N}}{\gamma} \tag{4.60}$$

and

$$\sigma(p_i - Z^*_t) = p_D(t_D) - \tfrac{1}{2} \ln \frac{4 t_D}{\gamma} \tag{4.61}$$

But since both lines have precisely the same slope, $m = 1.151/\sigma$, it will be appreciated from Fig. 4.24 that

$$Z^*_{t_N} - Z^*_t \approx m \log \frac{t_N}{t} = \frac{1.151}{\sigma} \log \frac{t_N}{t} = \frac{1}{2\sigma} \ln \frac{t_N}{t} \tag{4.62}$$

Consequently, subtracting equation 4.60 from 4.61 and inserting this condition implies that

$$\sum_{j=1}^{N} \frac{\Delta q_j}{q} p_D(t_{D_N} - t_{D_{j-1}}) = p_D(t_D) \tag{4.63}$$

It follows from the above argument that it is quite unnecessary to include the complexity of all the variable rate history when analysing a pressure buildup using the Horner method. That is, whether one attempts to use the full equation 4.55, with linear section defined by 4.59, or the simpler, constant rate equation 4.30 which reduces to the linear form 4.46 is immaterial because

- Both have exactly the same slope, $m = 1.151/\sigma$, which contains the final rate, q. Consequently, the kh-values determined will be equal
- Whether equation 4.50, using Z^*, or equation 4.37, using $p_{wsl(1\,hr)}$, is applied to calculate the skin factor, the same result will be achieved whichever of the two buildups is employed. This is guaranteed through condition 4.63.
- If the well is closed-in for a long period, then both buildups must eventually converge towards the initial pressure, p_i: the only physically real pressure in the reservoir.

Therefore, to simplify matters, with no loss of accuracy at all, the pressure buildups illustrated in the exercises in this chapter are performed using the combination of effective flowing time, t, and the final rate, q.

Notwithstanding the simple argument above, it has become fashionable in recent years to replace the normal abscissa of the Horner plot by what is referred to as the "superposed" or "superposition" time defined as

$$\text{Superposed time} = \sum_{j=1}^{N} \frac{q_j}{q} \log \frac{(t_N + \Delta t - t_{j-1})}{(t_N + \Delta t - t_j)} \tag{4.64}$$

which is derived from the general buildup equation 4.55, catering for the variable rate history, using the nomenclature defined in Fig. 4.23. In order to obtain this expression, however, it is necessary that the condition of pure *transience* prevails throughout the *entire* flow and buildup periods so that the p_D-function in equation 4.55 can be expressed as

$$p_D(t_{D_N} + \Delta t_D - t_{D_{j-1}}) = \tfrac{1}{2} \ln \frac{4}{\gamma}(t_{D_N} + \Delta t_D - t_{D_{j-1}}) \tag{4.65}$$

Then, with some algebraic manipulation, the reader can verify that the buildup equation, equation 4.55, is reduced to

$$\sigma(p_i - p_{ws}) = 1.151 \sum_{j=1}^{N} \frac{q_j}{q} \log \frac{(t_N + \Delta t - t_{j-1})}{(t_N + \Delta t - t_j)} \tag{4.66}$$

It must be remembered, however, that as clearly stated in italics in Earlougher's SPE Monograph [5] (p. 55), this equation is only relevant for "*infinite acting systems*".

Yet, in spite of Earlougher's sound advice, one frequently sees tests analysed incorrectly using this method, as depicted in Fig. 4.25. The application of superposed time is correct in schematic (a), since the reservoir must be infinite acting and equation 4.65 applies for all values of the time argument. In case (b), however, this is clearly not so because the buildup is not totally linear, meaning that the p_D-function is more complex than stated by equation 4.65. Something, perhaps a nearby fault, as described in section 4.16, has drawn down the pressure by an extra amount so that the buildup starts at a lower level and after the initial linear period demonstrates upward curvature.

Before the introduction of computer systems to analyse tests, engineers were protected from making such errors as illustrated in Fig. 4.25b for the simple reason that to calculate the superposed time manually was far too tedious an exercise. Nowadays, the calculation is simple and has become so routine that one hardly ever sees a normal Horner buildup plot any more — irrespective of the nature of the p_D-function. But the reader should bear in mind that producing plots such as Fig. 4.25b, is implicitly applying the wrong mathematics to describe the physical condition in the reservoir and, while it may appear sophisticated, such practice does little to establish one's credibility as an engineer.

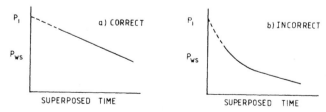

Fig. 4.25. Pressure buildup plots using the superposed time abscissa.

(b) The meaning of p*

The extrapolation of the early linear section of a Horner buildup plot to infinite closed-in time has always been referred to in the literature as p^* — which implies that it is a pressure and it has been described as the "false" or "hypothetical" pressure — which still imputes it with the properties of a pressure. But the main point to be stressed in connection with p^* is that — *it is not a pressure at all.* Instead, it is simply a number that has the dimensions and units of pressure but whose physical meaning is quite different and whose magnitude is completely arbitrary. In fact, if anybody stops you in the street demanding a definition of p^* then the closest you can get would be through the quotation of equation 4.49

$$\sigma(p_i - p^*) = p_D(t_D) - \tfrac{1}{2} \ln \frac{4\,t_D}{\gamma} \tag{4.49}$$

which, as described in section 4.13, results from the extrapolation of Horner's linear equation 4.46, to $\Delta t = \infty$ and serves as the defining expression for p^*.

Perhaps, when first introduced, if Horner [26] had not referred to this number as p^*, which symbolically implies pressure, but instead had used Z^*, for instance, many of the practical difficulties associated with its use would never have arisen. Therefore, throughout this chapter, it is considered preferable and safer to perform the simple "re-christening" that

$$p^* = Z^*$$

This is not some pedantic whim on the part of the author but a genuine attempt to make engineers pause and consider carefully before ever using p^* in reservoir engineering calculations. The problem is that the quoting of p^* is so widespread in the Industry in test interpretation reports that those unfamiliar with its true meaning (or lack of it) tend to use it directly in field development calculations. It is not uncommon to find people using p^* in material balance or numerical simulation studies as somehow representing a real pressure. But if, instead, it were referred to as Z^* then perhaps engineers might be forced to think twice before indulging in a twenty year reservoir performance history match using this parameter as a pressure — it remains to be seen.

One popular misconception in the interpretation of Z^* is illustrated in Fig. 4.26. According to the original definition, diagram (a) is correct and the extrapolation of

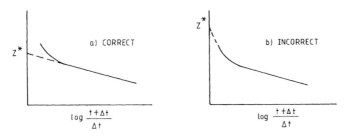

Fig. 4.26. Interpretation of Z^* in Horner buildup plotting.

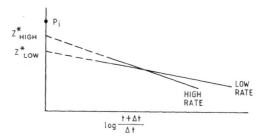

Fig. 4.27. Influence of the final rate on the value of Z^*.

this early, linear part of the buildup is described by equation 4.49. The interpretation shown in (b) is, however, incorrect on two counts. In the first place it deviates from the definition of Z^* in that it is the linear extrapolation of the late-time part of the buildup. Secondly, as described in sections 4.13 and with a few notable exceptions, there is nothing in the theory of well testing which guarantees that the late-time buildup, and particularly it's extrapolation, should be linear. Quite the contrary, in most cases this part of the buildup is distinctly non-linear. Therefore, in the majority of cases, the incorrect extrapolation shown in Fig. 4.26b, leads to a value of Z^* which is without definition in the context of the basic physics and mathematics of well testing.

The numerical value of Z^* has no particular significance in well test analysis and in application it is merely an intermediate step in obtaining various test results. This point has already been illustrated in connection with the evaluation of the flowing time to use in the analysis. As illustrated in Fig. 4.24, for a given set of buildup data (p_{ws} and Δt), varying the flowing time will produce different values of Z^* but eventually the same values of k and S. The value of Z^* is similarly affected by the final rate prior to closure, q. For a well test in which a volume of oil, N_p (stb), is produced, the buildup plot will be dependent on whether the final rate is high or low, as dictated by equation 4.36 and demonstrated in Fig. 4.27. As can be seen, the buildup following the higher-rate production has a steeper slope and a smaller value of t, resulting in an elevation of Z^* compared to the low-rate buildup. Nevertheless, since the slope is proportioned to the rate, the kh-product evaluated in both buildups will be the same and using a similar argument to that presented in (a), above, it can be shown that the skin factor determined in both cases will also be the same.

The value of Z^* is therefore to be treated — not as an end result — but as a means to an end. If understood and used correctly, the extrapolation of the early straight line on a Horner buildup to give the number Z^* at infinite closed-in time has many useful applications and four of these described in this chapter are

- calculation of the skin factor in an appraisal well using Z^* and t (already described in section 4.13)
- determining the distance of a single fault from a well (section 4.16b).
- derivation of a p_D-function for all values of the flowing time in a bounded system (section 4.19a).

– calculating the average pressure within the drainage area of a development well using the Matthews, Brons, Hazebroek technique (section 4.19a).

It will be noted that in the derivation of the MDH early, linear buildup trend (section 4.12), the flowing time is not included in the derivation of equations 4.33 and 34 by the simultaneous addition and subtraction of $0.5 \ln(t_D + \Delta t_D)$ to the right-hand side of equation 4.30 — as was the case in generating equation 4.46 for Horner buildup analysis. As a result, neither t nor Z^* are involved in the MDH analysis which automatically eliminates the complications associated with their interpretation described above. MDH analysis implies that no matter how long a well has been produced, or at what rate, the mere act of closure will cause a transient pressure response which is manifest as an early straight line on a semi-log buildup plot of p_{ws} versus $\log \Delta t$. Even following a variable rate history, equation 4.57 will apply immediately after closure, which not only guarantees early linearity of the buildup but again negates the influence of the rate history: the MDH linear equation being expressed as

$$\sigma(p_i - p_{wsl}) = \sum_{j=1}^{N} \frac{\Delta q_j}{q} p_D(t_{D_N} - t_{D_{j-1}}) \underbrace{- \tfrac{1}{2} \ln \frac{4\Delta t_D}{\gamma}}_{[\text{constant}]} \qquad (4.67)$$

in which the second term on the right is the expression of $p_D(\Delta t_D)$ in transient form.

Furthermore, in performing MDH analysis, there is no temptation to extrapolate the early straight line, or any other section of the plot, to infinite closed-in time to evaluate Z^* or p_i. Instead, the initial linear trend is extrapolated to a value of Δt_s, as dictated by equation 4.38, to determine directly the initial pressure in appraisal wells and the current average pressure in development wells (section 4.19b). In the remainder of this chapter, it is demonstrated that anything that can be accomplished in terms of formation evaluation using the Horner plot can equally well be achieved using the MDH analysis technique — only in a simpler manner and with the avoidance of errors associated with the evaluation and use of t and Z^*, as required in Horner analysis.

4.16. THE INFLUENCE OF FAULT GEOMETRIES ON PRESSURE BUILDUPS IN APPRAISAL WELL TESTING

(a) General description

If a fault, or a system of faults, is located in the vicinity of an appraisal well it will influence the flowing pressure during the drawdown period which in turn affects the subsequent pressure buildup. Typical fault configurations encountered are depicted in Fig. 4.28.

Appraisal well testing is usually of relatively short duration so the faults which influence the test are those located close to the well and are not necessarily the outer boundary faults of the reservoir. Because of this, the sealing faults are in what

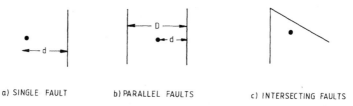

a) SINGLE FAULT   b) PARALLEL FAULTS   c) INTERSECTING FAULTS

Fig. 4.28. Fault geometries, semi-infinite space.

is described as "semi-infinite" space, that is, remove them and the reservoir would appear infinite in extent.

(b) Single fault

To examine the influence of faults on the pressure response during a test, consider the simplest yet most commonly observed effect: that due to a single sealing fault, as first described by Horner [26]. The most straightforward way of accounting for the presence of the fault is by applying the method of images, the principle being illustrated in Fig. 4.29.

The presence of the single sealing fault can be simulated by placing an image well producing at the same constant rate, q, an equal distance, d, on the other side of the fault. Then, the potential distribution between the wells is such as to create a no-flow boundary at the position of the fault which may then be effectively removed, leaving the real and image wells in an otherwise infinite space. Consequently the pressure drawdown at the real well can be expressed as the sum of two transient solutions of the diffusivity equation (section 4.7), that is

$$\sigma(p_i - p_{wf}) = \tfrac{1}{2} \ln \frac{4 t_D}{\gamma} + \tfrac{1}{2} ei \left(\frac{\phi \mu c d^2}{0.000264 \, kt} \right) + S \tag{4.68}$$

in which the first term and skin factor represent the drawdown component due to the production of the real well and the second results from the production of image well located a distance $2d$ away. The exponential integral is required in the second term since, if d is large, its argument, x (equation 4.14), may not satisfy the requirement that $x < 0.01$ necessary for application of the logarithmic approximation as described in section 4.7. In direct analogy with equation 4.21, it can be see that for this real-image well system, the p_D-function is

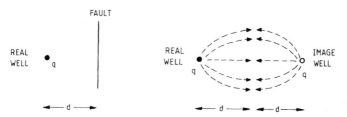

Fig. 4.29. Application of the method of images to single fault geometry.

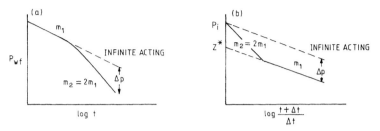

Fig. 4.30. Influence of a single, sealing fault: (a) on ideal drawdown performance; (b) on a Horner buildup plot.

$$p_{\mathrm{D}}(t_{\mathrm{D}}) = \tfrac{1}{2}\ln\frac{4\,t_{\mathrm{D}}}{\gamma} + \tfrac{1}{2}ei\left(\frac{\phi\mu c d^2}{0.000264\,kt}\right) \tag{4.69}$$

For small flowing times the argument of the *ei*-function is large so that the second term in equation 4.69 is negligible; whereas if the flowing time is large it can be represented by its logarithmic equivalent expression (equation 4.15). Consequently, if the drawdown could be conducted under ideal conditions ($q = $ constant from $t = 0$), the presence of a nearby fault would be manifest as a doubling of the slope in a drawdown plot of p_{wf} versus $\log t$ (equation 4.68) from $1.151/\sigma$ to $2.303/\sigma$. Furthermore, compared to a purely infinite acting system, the influence of the fault can be evaluated by subtracting equation 4.20 from 4.68 to give

$$\sigma\Delta p = \tfrac{1}{2}ei\left(\frac{\phi\mu c d^2}{0.000264\,kt}\right) \tag{4.70}$$

Therefore, at any stage of the drawdown, if Δp is measured (Fig. 4.30a), then equation 4.70 may be solved directly to calculate d or, as described below for buildup testing, the logarithmic approximation of the *ei*-function is usually quite acceptable (equation 4.15) in which case the distance to the fault may be calculated as

$$d = \left[\frac{0.000264\,kt}{\gamma\phi\mu c\,e^{0.0142kh\Delta p/q\mu B_{\mathrm{o}}}}\right]^{\frac{1}{2}} \tag{4.71}$$

The foregoing argument is to demonstrate a fact not emphasised sufficiently in the literature that the most important phase of a flow-buildup test is the drawdown period. In fact, the drawdown is the test, when all the action takes place, the ensuing buildup is merely its reflection. Therefore, in answer to the question frequently posed in the technical literature — how long is it necessary to close-in a well to detect a fault? The t time is until $\Delta t = 0$: it is unnecessary to perform a buildup at all. This does not just apply to a single fault analysis but to any system tested. Since the equation for the actual flowing pressure

$$\sigma(p_{\mathrm{i}} - p_{\mathrm{wf}}) = p_{\mathrm{D}}(t_{\mathrm{D}}) + S \tag{4.21}$$

and for the initial, linear transient drawdown flowing response is:

$$\sigma(p_{\mathrm{i}} - P_{\mathrm{wfl}}) = \frac{1}{2}\ln\frac{4t_{\mathrm{D}}}{\gamma} + S \tag{4.20}$$

in which P_{wfl} represents hypothetical pressures on the extrapolated early trend. Then subtracting equation 4.21 from 4.20 and solving explicitly for $p_D(t_D)$ gives:

$$p_D(t_D) = \sigma \Delta p + \frac{1}{2} \ln \frac{4t_D}{\gamma}$$

in which $\Delta p = p_i - P_{wfl}$, which can be readily measured from any flowing time, as shown in Fig. 4.30a for single fault analysis. Since k can be determined from the initial, transient drawdown, as described ion section 4.10, then the $p_D(t_D)$ can be evaluated at any time during the drawdown by solving this equation.

If the p_D-function for a single fault (equation 4.69) is substituted in the basic buildup equation 4.30, the result is

$$\sigma(p_i - p_{ws}) = 1.151 \log \frac{t + \Delta t}{\Delta t}$$

$$+ \tfrac{1}{2}ei\left(\frac{\phi \mu c d^2}{0.000264\,k(t + \Delta t)}\right) - \tfrac{1}{2}ei\left(\frac{\phi \mu c d^2}{0.000264\,k\,\Delta t}\right) \tag{4.72}$$

The procedures for analysing a pressure buildup using the methods of Horner and MDH are then as follows:

Horner. Examining equation 4.72 for small Δt, the final term becomes vanishingly small so that the equation is reduced to

$$\sigma(p_i - p_{ws}) = 1.151 \log \frac{t + \Delta t}{\Delta t} + \tfrac{1}{2}ei\left(\frac{\phi \mu c d^2}{0.000264\,kt}\right) \tag{4.73}$$

in which the ei-function is evaluated for $t + \Delta t \sim t$ and is a constant. The equation demonstrates that the buildup starts with a linear section of slope $m_1 = 1.151/\sigma$, just as described previously in section 4.13, for the infinite reservoir case. As Δt becomes large, the arguments of both ei-functions in equation 4.72 become very small so that they may be replaced by their logarithmic equivalent forms (refer section 4.7) which reduces the equation to the form

$$\sigma(p_i - p_{ws}) = 2.303 \log \frac{t + \Delta t}{\Delta t} \tag{4.74}$$

which again is linear with slope $m_2 = 2.303/\sigma$. Thus the influence of a single fault on a Horner buildup plot is as shown in Fig. 4.30b. During the drawdown the flowing pressure is reduced with respect to an infinite acting system by the amount stated in equation 4.70 and consequently the buildup starts at a pressure which is depressed to the same extent with respect to that in an infinite acting system, as illustrated in Fig. 4.30b. The first part of the buildup is then linear with slope m_1 and this contains the transient pressure response. Thereafter, there is a non-linear section that eventually leads to a second straight line with slope $m_2 = 2m_1$, which in an appraisal well test should extrapolate to the initial pressure, p_i, at infinite closed-in time.

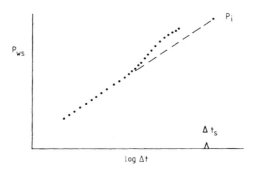

Fig. 4.31. MDH buildup plot, illustrating the influence of a single, sealing fault close to the well.

Determination of the initial slope m_1 and the extrapolated value of the early straight line, Z^*, permits the calculation of the permeability and skin factor, as described in section 4.13, and also the value of the additional component of drawdown due to the fault (Fig. 4.30b) as

$$\Delta p = p_i - Z^* \tag{4.75}$$

Inserting this value in equation 4.70 enables the distance to the fault, d, to be directly calculated using plots or table values of the ei-function (Table 4.2). Although one technical paper on the subject presents fault distances evaluated to two places of decimal (in feet); it should be remembered that this is not an exact science and therefore rounding-off the figure to the nearest 20–30 ft is more appropriate. Bearing this in mind, it is usually quite acceptable to express the ei-function using its logarithmic approximation in which case equation 4.71 can be used to calculate the fault distance.

Miller, Dyes, Hutchinson. Application of the MDH analysis technique to determine the distance to a single, sealing fault has received little attention in the literature although the method is quite straightforward as illustrated in Fig. 4.31 The buildup commences with a linear section, containing the transient response, the equation of which is

$$\sigma(p_i - p_{wsl}) = p_D(t_D) - \tfrac{1}{2} \ln \frac{4\Delta t_D}{\gamma} \tag{4.33}$$

in which the p_D-function is represented by equation 4.69 and, being evaluated at the end of the flow period, is a constant. For larger values of the closed-in time there is upward curvature of the buildup but unlike the Horner plot there is no characteristic doubling of the slope. If the early straight line is extrapolated to $p_{wsl} = p_i$, determined as described in section 4.14a, then equation 4.38 may be solved as

$$p_D(t_D) = \tfrac{1}{2} \ln \frac{4\Delta t_{Ds}}{\gamma} = \tfrac{1}{2} \ln \frac{4t_D}{\gamma} + \tfrac{1}{2} ei \left(\frac{\phi \mu c d^2}{0.000264\, kt} \right)$$

in which Δt_s is the closed-in time at which the linear extrapolation has risen to the initial pressure, as illustrated in Fig. 4.31. This equation can be reduced to

$$\ln \frac{\Delta t_s}{t} = ei \left(\frac{\phi \mu c d^2}{0.000264 kt} \right) \tag{4.76}$$

which can be solved directly to determine the distance to the fault, d. Alternatively, using the logarithmic approximation of the ei-function.

$$d = \sqrt{\frac{0.000264\, k}{\gamma \phi \mu c \Delta t_s} \times t} \tag{4.77}$$

Both the Horner and MDH analysis techniques are illustrated in the following exercise.

Exercise 4.2: Pressure buildup test: single fault analysis

Introduction
     This appraisal well test is influenced by the presence of a single sealing fault. The analysis illustrates and compares the separate methods of Horner and Miller, Dyes, Hutchinson in calculating the distance to the barrier. The drawdown data were not reliable enough for detailed analysis.

Question
     A DST was performed in an appraisal well during which it produced a cumulative of 5320 stb of oil, the final rate being 3500 stb/d. Throughout the flow period, the pressure declined continuously as a function of time. The well was then closed-in downhole for a 29 hour pressure buildup and the resulting static pressure and time data during this period are listed in Table 4.4. The remainder of the data required to analyse the buildup are as follows

p_i = 3460 psia (5 minutes flow/1 hour buildup): c = 17×10^{-6}/psi
                                                        μ = 1.0 cp
p_{wf} = 2970 psia (final flow): B_{oi} = 1.30 rb/stb
h = 25 ft (fully perforated well): r_w = 0.510 ft ($12\frac{1}{4}$ inch hole)
ϕ = 0.25

- Analyse the test using both the Horner and MDH plotting techniques and their time derivatives, p'_{ws}.

Solution
     The effective flowing time required to calculate the Horner time ratio is

$$t = \frac{N_p}{q} \times 24 = \frac{5320}{3500} \times 24 = 36.5 \text{ hrs} \tag{4.51}$$

TABLE 4.4

Pressure and time-derivative data, Exercise 4.2

Δt (hours)	p_{ws} (psia)	$\log \Delta t$	$p'_{ws\,(MDH)}$ [a] (psi)	$\log(t + \Delta t)/\Delta t$	$p'_{ws\,(Horner)}$ [b] (psi)
0.050	3284.1	−1.699		2.864	
0.117	3310.6	−0.932	33.03	2.495	31.18
0.183	3321.9	−0.738	25.68	2.302	25.42
0.250	3329.1	−0.602	23.27	2.167	23.16
0.317	3333.6	−0.499	19.04	2.065	19.16
0.383	3337.1	−0.417	18.56	1.984	18.76
0.450	3340.3	−0.347	19.89	1.914	19.85
0.650	3347.5	−0.187	19.80	1.757	19.91
0.850	3352.8	−0.071	19.88	1.643	20.19
1.050	3357.1	0.021	20.43	1.553	20.75
1.250	3360.6	0.097	20.12	1.480	20.82
1.517	3364.2	0.181	18.65	1.399	19.30
2.050	3370.0	0.312	19.41	1.274	20.15
2.517	3374.1	0.401	20.05	1.190	21.19
3.050	3377.9	0.484	19.84	1.113	21.43
4.050	3383.8	0.607	20.95	1.001	22.87
5.050	3388.8	0.703	22.75	0.915	25.25
6.050	3392.4	0.782	19.98	0.847	22.99
7.050	3395.4	0.848	19.65	0.791	23.26
8.050	3398.5	0.906	23.41	0.743	28.04
9.050	3401.2	0.957	23.09	0.702	28.59
10.050	3403.6	1.002	22.92	0.666	28.95
11.050	3405.7	1.043	22.16	0.634	28.50
12.050	3408.0	1.081	26.57	0.605	34.44
13.117	3410.0	1.118	23.59	0.578	32.16
14.050	3411.6	1.148	23.29	0.556	31.58
15.050	3413.3	1.178	24.74	0.535	35.15
16.050	3414.9	1.205	24.88	0.515	34.74
17.050	3416.3	1.232	23.17	0.497	33.77
18.050	3417.6	1.256	22.82	0.480	33.20
20.050	3420.1	1.302	23.81	0.450	36.18
21.050	3421.2	1.323	22.61	0.437	36.74
22.050	3422.3	1.343	23.71	0.424	36.74
23.050	3423.3	1.363	22.55	0.412	36.18
24.050	3424.2	1.381	21.20	0.401	35.53
25.050	3425.1	1.399	22.10	0.390	35.53
26.050	3426.0	1.416	23.00	0.380	39.08
27.050	3426.8	1.432	21.24	0.371	38.60
28.050	3427.6	1.448	22.04	0.362	38.60
29.000	3428.3	1.462	21.02	0.354	37.99

[a] $p'_{ws\,(MDH)}$: equation 4.41; [b] $p'_{ws\,(Horner)}$: equation 4.52.

The data required for both Horner and MDH plots are listed in Table 4.4 together with their time derivatives calculated using equations 4.52 and 4.41 respectively. The latter are evaluated at the mid-point of each time interval and are plotted as such in Figs. 4.32 and 4.33.

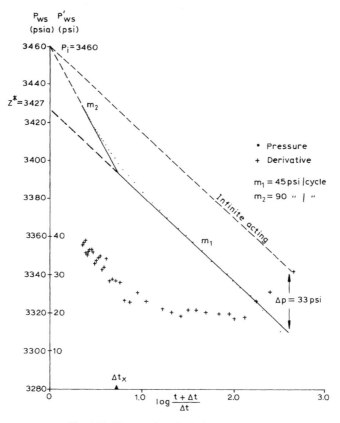

Fig. 4.32. Horner plot, single fault analysis.

Horner. The buildup plot and its derivative are presented in Fig. 4.32. Afterflow ceases after about a quarter of an hour resulting from the practice of downhole closure. A well developed linear section containing the transient response can then be identified on both the buildup and derivative plots. This lasts until $\log(t + \Delta t)/\Delta t \sim 1.4(\Delta t \sim 1.5$ hours), has a slope of $m_1 = 45$ psi/log cycle and extrapolates to give a value of $Z^* = 3427$ psia at infinite closed-in time. This is followed by a lengthy non-linear transition period lasting until $\log(t + \Delta t)/\Delta t \sim 0.44(\Delta t \sim 21$ hours) when a second linear section develops which lasts for the remainder of the 29 hour buildup. The slope of this line is $m_2 = 90$ psi/log cycle, exactly double the value of the initial linear section, and its extrapolation to infinite closed-in time gives a pressure which coincides with the initial value of $p_i = 3460$ psi. The development of the second plateau level is not clear by inspection of the derivative plot (Fig. 4.32) on account of the logarithmic nature of the time scale, but if the same data are plotted with a linear time scale (Fig. 4.34), the occurrence of a late time plateau is evident.

From the initial, transient pressure response, the basic parameters k and S can be calculated using equations 4.36 and 4.50 as:

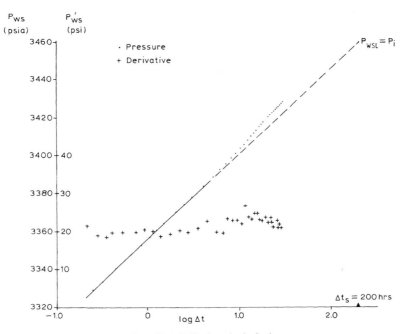

Fig. 4.33. MDH plot, single fault.

$$kh = 162.6\frac{q\mu_o B_o}{m_1} = \frac{162.6 \times 3500 \times 1.0 \times 1.30}{45} = 16440 \text{ mD·ft}$$

$$k = 16440/25 = 658 \text{ mD}$$

$$S = 1.151\left[\frac{3427 - 2970}{45} - \log\frac{658 \times 36.5}{0.25 \times 1.0 \times 17 \times 10^{-6} \times 0.51^2} + 3.23\right] = 3.5$$

Furthermore

$$\Delta p_{skin} = 0.87 \text{ mS} = 0.87 \times 45 \times 3.5 = 137 \text{ psi} \tag{4.53}$$

$$\text{PI} = \frac{q}{p_i - p_{wf}} = \frac{3500}{3460 - 2970} = 7.1 \text{ stb/d/psi} \tag{4.1}$$

and the ideal but non-stabilized PI is:

$$\text{PI} = \frac{q}{p_i - p_{wf} - \Delta p_{skin}} = \frac{3500}{3460 - 2970 - 137} = 9.9 \text{ stb/d/psi} \tag{4.1}$$

The complete doubling of slope in the Horner buildup plot manifests the existence of a single, sealing fault close to the well, as also suggested by the current geological model. The distance to this fault can be calculated using the approxima tion

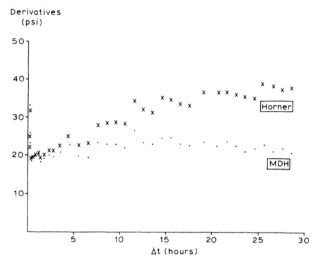

Fig. 4.34. Comparison of Horner and MDH derivative plots single fault analysis.

$$d = \sqrt{\frac{0.000264\,kt}{\gamma\phi\mu c\, e^{0.0142kh\Delta p/q\mu B_o}}} \qquad (4.71)$$

$$d = \sqrt{\frac{0.000264 \times 658 \times 36.5}{1.781 \times 0.25 \times 1.0 \times 17 \times 10^{-6}\, e^{0.0142 \times 16440 \times 33/3500 \times 1.0 \times 1.3}}} = 390 \quad \text{(ft.)}$$

in which $\Delta p = p_i - Z^* = 3460 - 3427 = 33$ psi

Alternatively, using the more rigorous exponential integral method, equation 4.70

$$7.08 \times 10^{-3}\frac{kh}{q\mu B_o}\Delta p = \tfrac{1}{2}ei\left(\frac{\phi\mu cd^2}{0.000264\,kt}\right) \qquad (4.70)$$

$$\frac{7.08 \times 10^{-3} \times 16440 \times 33}{3500 \times 1.0 \times 1.3} = \tfrac{1}{2}ei\left(\frac{0.25 \times 1.0 \times 17 \times 10^{-6}d^2}{0.000264 \times 658 \times 36.5}\right)$$

$$1.688 = ei\,(0.670 \times 10^{-6}d^2)$$

which can be solved to give $d = (0.1162 \times 10^6/0.670)^{1/2} = 415$ ft.

Miller, Dyes, Hutchinson. The buildup and its derivative are plotted in Fig. 4.33 and the early, linear section of the plot is readily identifiable from both. This has a slope of 45 psi/log cycle, precisely the same as for the Horner interpretation and an average value of the derivative plateau of 19.6 psi/log cycle which satisfies the condition stated in equation 4.41. The duration of the linear section of the buildup is until $\log \Delta t \sim 0.7(\Delta t \sim 5.0$ hours) which in this case is 3.5 hours longer than in the equivalent Horner buildup. Then the slope of the buildup increases slightly until

$\Delta t \sim 15$ hours and then tends to decrease. This can be seen by careful inspection of the MDH buildup itself but is better demonstrated in the linear time scale version of the derivative (Fig. 4.34). Having identified the same early straight line as in the Horner plot the values of k and S determined by the separate methods will be identified.

The distance to the fault can be calculated, as described in section 4.16b, by first extrapolating the early straight line to $p_{wsl} = p_i = 3460$ psia, which occurs at a value of $\Delta t_s = 200$ hours (Fig. 4.33) and then applying the approximate formula, equation 4.77.

$$d = \sqrt{\frac{0.000264\,k}{\gamma\phi\mu c\,\Delta t_s}} \times t \tag{4.77}$$

$$= \sqrt{\frac{0.000264 \times 658}{1.781 \times 0.25 \times 1 \times 17 \times 10^{-6} \times 200}} \times 36.5 = 390\ \text{ft}$$

or, using the more exact expression, equation 4.76

$$\ln\frac{\Delta t_s}{t} = ei\left(\frac{\phi\mu c d^2}{0.000264 kt}\right) \tag{4.76}$$

$$\ln\frac{200}{36.5} = 1.701 = ei(0.670 \times 10^{-6} d^2)$$

giving $d = 415$ ft. Therefore, both the Horner and MDH methods confirm the existence of a sealing fault at a distance of 415 ft from the appraisal well location.

Conclusions

As in Exercise 4.1, it is interesting to compare the Horner and MDH buildups and, in particular, investigate the reason why the latter should, in this case, produce a longer initial straight line. Considering the Horner equation for this initial trend (equation 4.73), a condition for linearity is that

$$\frac{1}{2}ei\left(\frac{\phi\mu c d^2}{0.000264\,k(t+\Delta t)}\right) \approx \frac{1}{2}ei\left(\frac{\phi\mu c d^2}{0.000264\,kt}\right) = \text{constant} \tag{4.78}$$

In contrast, the early linear trend of the MDH plot is expressed by equation 4.33, in which the p_D-function is evaluated using equation 4.69 for single fault geometry. The condition for linearity is therefore

$$\frac{1}{2}\ln\frac{4(t_D+\Delta t_D)}{\gamma} + \frac{1}{2}ei\left(\frac{\phi\mu c d^2}{0.000264\,k(t+\Delta t)}\right)$$

$$\approx \frac{1}{2}\ln\frac{4\,t_D}{\gamma} + \frac{1}{2}ei\left(\frac{\phi\mu c d^2}{0.000264\,kt}\right) = \text{constant} \tag{4.79}$$

Using the data provided in the exercise, it can be determined that after 1.5 hours, when the Horner early linear section of the buildup terminates, the left-hand side of equation 4.78 exceeds the right by about 1.4%. But this same percentage increase of the left over the right-hand side of equation 4.79 does not occur until the closed-in time approaches 5 hours in the MDH interpretation, which explains why the early, linear buildup is 3.5 hours longer than that observed in the Horner plot. In neither plot, however, does the initial linear portion of the buildup represent the condition of "pure transience".

The flow rate during this test was slightly unsteady and consequently in their analysis the operator plotted the Horner buildup using the superposed time, equation 4.64, as the abscissa. But the exercise demonstrates the inappropriateness in attempting such apparent sophistication. Equation 4.64 only applies for pure transience throughout the flow and buildup periods but this condition is rapidly removed, after a few hours of flow, by the presence of the sealing fault.

The method presented in this section, and demonstrated in the exercise, for determining the distance to a fault using the Horner plot differs from the conventional methods described in the literature. These usually require the location of the closed-in time Δt_x (Fig. 4.32) at which the linear extrapolations of the early and late straight line trends intersect which, in the present case, would be for $\log(t + \Delta t_x)/\Delta t_x = 0.74$: $\Delta t_x = 8.2$ hours ($t = 36.5$ hours). Having defined this point, the popular method of Davis and Hawkins [31] can be used to calculate the distance to the fault as

$$d = \sqrt{\frac{1.48 \times 10^{-4} k \Delta t_x}{\phi \mu c}} = \sqrt{\frac{1.48 \times 10^{-4} \times 658 \times 8.2}{0.25 \times 1 \times 17 \times 10^{-6}}}$$

giving a value of $d = 430$ ft, which is slightly greater than the value determined in the present exercise.

(c) Some general considerations in defining fault positions

Exercise 4.2 demonstrates some general features in conducting and analysing tests to locate distances to faults, whether single or multiple, the latter being described in section 4.16d. Perhaps the main point to stress is that faults are "seen" during the drawdown period — not during the ensuing pressure buildup. Well production causes a continuous pressure drop at the fault which is reflected back and reduces the wellbore pressure by the amount expressed by equation 4.70. The pressure drop at the fault itself can be determined by evaluating the p_D-function at the position of the fault, d, when both terms in equation 4.69 are evaluated using the ei-function.

$$p_D(t_D) = \sigma(p_i - p_d) = \sigma \Delta p_d = ei\left(\frac{\phi \mu c d^2}{0.000264 \, kt}\right) \tag{4.80}$$

The pressure drop at the fault, Δp_d, can therefore be seen to be double the amount caused by the reflection in the wellbore (equation 4.70). As soon as the well is closed-in the pressure decline at the fault ceases as pressures increase throughout the reservoir. In the wellbore, the buildup starts at a depressed level with respect to an infinite acting reservoir: $\Delta p = p_i - Z^*$, which can be established by extrapolating the initial, linear buildup to infinite closed-in time giving Z^*. The distance to the fault can then be calculated using equation 4.70 for Horner analysis or 4.76 for MDH analysis.

Considering the conventional methods presented in the literature [26,31] for calculating distances to faults, which rely on the detection of the closed-in time, Δt_x, at which the initial and final linear sections of the Horner buildup intersect (Fig. 4.32); the impression in conveyed that it is necessary to close the well in for a sufficiently long period so that an exact doubling of the buildup slope is observed — otherwise "wrong results" will be obtained [30]. But even using these methods this condition is unnecessary. Provided the geological model indicates the likelihood of a single fault nearby and the initial pressure is determined at the start of the test (section 4.14a); then the depression of the early buildup trend will be detectable as will be the slope of the first linear section. Since, eventually the slope must double and extrapolate in a linear fashion to p_i, then the above observations are sufficient to define the closed-in time, Δt_x.

Applying the alternative methods suggested in this section, however, it can be seen that there is no reliance on the length of the buildup in the Horner or MDH analyses. In fact, provided there is confidence in the geological model and the initial pressure has been determined, then considering the data in Exercise 4.2, it would only be necessary to close-in the well for, say, one hour during the early period of transience to calculate the fault distance, whereas to confirm the exact doubling of slope would require closure for the full 29 hours. Besides, as illustrated in the MDH analysis-doubling of the buildup slope is not a prerequisite in the analysis. In planning such tests, surface read-out of pressures is of assistance in deciding when a buildup can be safely terminated.

Because faults are detected during the flowing period, it doesn't matter how long a well is closed-in for a buildup, the fault will not be "seen" unless the well has been produced for a sufficient period to cause a clearly detectable incremental drawdown, $\Delta p = p_i - Z^*$, in the wellbore. This point should be noted because some engineers seem to believe that the fault directly influences the closed-in pressures and analyse lengthy buildups following a short flow period — which is impossible unless the fault is extremely close to the well. The buildup shape is merely a reflection of what happened during the drawdown. In order to produce a significant incremental pressure drop in the wellbore, it is recommended that the well be flowed for long enough so that the radius of investigation is at least four times the distance to the fault [30]. This radius may be calculated as

$$r_{inv} = 0.03\sqrt{\frac{kt}{\phi\mu c}} \quad (t, \text{hours}) \tag{4.81}$$

which derives from the work of Matthews et al. [20], as explained in section 4.19i. Essentially what it determines is the distance that would be "seen" out into the reservoir by flowing a well at a constant rate for a time, t, provided purely *transient* conditions prevail. That is, the reservoir must be infinite acting which, in fact, makes it a hypothetical calculation if there is a fault present but nevertheless it is performed to establish the distance seen, supposing the fault were not present. For instance using the rock properties and PVT data for Exercise 4.2, then after the 36.5 hour effective flowing time the radius of investigation would be 2255 ft, which is 5.4 times the distance to the fault of 415 ft Because of this, the incremental pressure drop of 33 psi at the wellbore is clearly detectable and the distance to the fault can be unambiguously determined. In this respect it is worthwhile examining the dependence of the pressure drop at the well, caused by the fault, on the flowing time and also the radius of investigation, the data used are from Exercise 4.2. The pressure drop may be evaluated using equation 4.70 which reduces to

$$\sigma \Delta p = \tfrac{1}{2} ei(x): \quad 0.02558 \Delta p = \tfrac{1}{2} ei \left(\frac{4.2136}{t} \right)$$

and the radius of investigation using equation 4.81 which becomes

$$r_{inv} = 373.3 \sqrt{t}$$

As can be seen in Fig. 4.35, which plots the pressure drop at the wellbore, Δp, and the ratio of the radius of investigation to the distance to the fault, r_{inv}/d, both as functions of the flowing time, t; for the recommended value of $r_{inv}/d = 4$, the pressure drop at the well would be 23 psi which is sufficient to enable a valid analysis to be performed to locate the fault and would require flowing the well for 20 hours. To double the incremental drawdown to 46 psi would require extending the flow period to 73 hours for which $r_{inv}/d = 7.8$. The plots would therefore suggest that a minimum value of $r_{inv}/d = 4$ should be catered for but it is suggested that in planning a test to detect a fault calculations should be made, as in Table 4.5, to decide upon a reasonable value of the ratio. Formation and PVT data can be taken from neighbouring wells, if available, or better still the formation characteristics

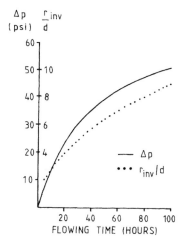

Fig. 4.35. Δp and r_{inv}/d vs. flowing time.

TABLE 4.5

Calculation of the pressure drop at the wellbore due to a fault and radius of investigation (data: Exercise 4.2)

t (hours)	$ei(x)$	Δp (psi)	r_{inv} (ft)	r_{inv}/d
5	0.288	6	835	2.0
10	0.670	13	1180	2.8
20	1.180	23	1670	4.0
30	1.521	30	2045	4.9
36.5	1.696	33	2255	5.4
50	1.979	39	2640	6.4
100	2.633	51	3730	9.0

can be determined from core analysis, which can often be performed in advance of the test and PVT parameters can be assessed from provisional wellsite analysis. It should also be planned to test the well by flowing at the highest rate practicable through the surface test facilities, this has the effect of reducing the coefficient σ (equation 4.70) which, in turn, increases the component of pressure drop, Δp, at the wellbore.

There is a popular misconception that fault patterns can be best delineated by conducting tests in high-permeability reservoirs. The belief is that, since the radius of investigation, equation 4.81, is proportional to the square root of the permeability, such tests will reveal the fault quickly and clearly. To demonstrate that this is not necessarily the case, however, the test data presented in Exercise 4.2 have been re-worked for permeabilities ranging between 100 and 4000 mD with everything else remaining unchanged. Results, expressed in terms of the incremental pressure drop in the wellbore and the ratio r_{inv}/d, are presented in Table 4.6 and plotted in Fig. 4.36 as Δp versus $\log k$. The calculations for the 36.5 hours of flow and a final rate of 3500 stb/d, have been conducted applying equation 4.70 expressed

TABLE 4.6

Calculation of Δp and r_{inv}/d (data: Exercise 4.2 for different permeabilities)

k (mD)	$ei(x)$	Δp (psi)	r_{inv} (ft)	r_{inv}/d
100	0.335	43	879	2.1
200	0.741	48	1243	3.0
300	1.044	45	1523	3.7
500	1.454	37	1966	4.7
658	1.696	33	2255	5.4
1000	2.074	27	2780	6.7
2000	2.731	18	3932	9.5
3000	3.128	13	4815	11.6
4000	3.411	11	5560	13.4

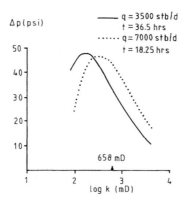

Fig. 4.36. Δp vs. $\log k$.

as

$$\Delta p = \frac{1}{2\sigma} ei(x) = \frac{12853}{k} ei\left(\frac{75.961}{k}\right)$$

while the radius of investigation is $r_{inv} = 87.9\sqrt{k}$, and $d = 415$ ft. The plot displays a definite peak, for which $\Delta p = 48$ psi, that occurs for a permeability of about 200 mD. For higher values, Δp declines significantly, basically on account of the much larger area affected when the permeability is high and even the value of $k = 658$ mD determined in Exercise 4.2 is above the optimum level. For higher permeabilities, however, the flow rate could be increased thus reducing the length of the flow period. To demonstrate this effect, the rate has been doubled to 7000 stb/d and the flow time halved to 18.25 hours. As can be seen, this shifts the plot in Fig. 4.36 towards the right, raising the optimum permeability to induce the maximum pressure drop to 350 mD. Therefore, in testing wells to detect the presence of faults, it usually proves beneficial to select reservoirs of moderate rather than high permeability.

The engineer must also check carefully whether there is any component of natural pressure support during the flow period — as described at the commencement of section 4.18. If there is, then none of the techniques described in this section is appropriate for the test analysis. As presented, the application of the method of images relies on the assumption that placement of the image well and removal of the fault leaves the two wells in an infinite acting system in which transience prevails. Clearly, if there is a degree of pressure support then the analysis becomes much more complex, the difficulty being in defining the p_D-function under these circumstances. The interpretation could be attempted using the technique described in section 4.19c which implicitly applies the method of images for mixed (open and closed) boundary conditions. Furthermore, since natural pressure support is associated with high-permeability reservoirs (sections 4.6b), this provides a further argument for avoiding such reservoirs as candidates for tests intended to detect faults

It must be appreciated that any attempt in reservoir engineering to "see" away from the wellbore contains a healthy element of "black-art" and the engineer is obliged to ask the question at the planning stage — is the expenditure involved in attempting to locate the fault justified in terms of leading to useful field development decisions? Such tests are necessarily of considerable duration and therefore expensive, especially when conducted offshore. Amongst the uncertainties to be faced in deciding whether to test are the fact that although a fault may be detected at a certain distance from the well, its bearing is not determined. Furthermore, as pointed out in the following section, for complex fault systems there is often a lack of mathematical uniqueness in deciding between one fault pattern and another. It is also sometimes difficult to distinguish between the influence of faults and other factors, such as dual porosity, which can also cause upward curvature in the late time pressure buildup behaviour. In this respect, Gringarten advises [1] that if faults in a reservoir are located extremely close to the wells tested then it may very well be dual porosity that is influencing the tests instead. The obvious way to check this is, of course, by careful examination of the core permeabilities. Finally, it must be remembered that during most tests, the pressure drop imposed at the fault is slight. As noted earlier (equation 4.80) the pressure drop at the fault is double the amount imposed by the fault in the wellbore; in Exercise 4.2, for instance, it would amount to 66 psi which is only 13% of the total test drawdown. For such a relatively small pressure differential the fault may very well prove to be a sealing, no-flow barrier, as it is implicitly assumed to be in the analysis method described in this section. But, when the field is developed much larger pressure differentials, 500–1000 psi, may be deliberately or accidentally imposed across the fault under which circumstance it may be ruptured and lo longer act as a seal. Because of this some of the barriers detected during the appraisal stage tend to be of the "academic" variety and cause little concern when the field is producing.

(d) Definition of more complex fault geometries

It is a relatively simple matter to calculate the distance to a single sealing fault but for more complex geometrical configurations the analytical methods become so mathematically involved (usually requiring iterative solution) that the engineer is obliged to resort to the use of computer analysis. The general method of approach adopted is as outlined in section 4.11: after examination of the latest geological model, a fault pattern is selected for which a p_D-function is generated. Having determined k and S from Horner or MDH analysis, the function is used to calculate the increasing static pressure in the basic buildup equation 4.30. Failure to attain a match between theoretical and observed pressures means that the choice of fault pattern was incorrect and the p_D-function is altered, often in an automatic fashion, and the process repeated until a satisfactory match is achieved.

There have been numerous papers written on the subject of defining fault patterns in reservoirs from which the following are notable for semi-infinite systems.

(A) Delineation of parallel faults [32] and elongate linear flow systems [33] (Fig. 4.28b).
(B) Positioning of a well between intersecting faults [34] (Fig. 4.28c).
(C) Quantifying the effect of a partially sealing fault [35].

It is not the intention in the present text to duplicate the mathematical descriptions provided in such papers but instead to alert the reader, again, of some of the practicalities associated with the application of such theoretical models to define fault systems.

– The approach amounts to solving differential equations to determine the boundary conditions (section 4.4a).
– Any such testing demands a long flowing period.

The first of these again raises the question of the validity of applying the technique of mathematical curve-fitting to define the physical state of a complex underground system and in this respect it is worthwhile quoting from an excellent paper on this subject — "Pitfalls in Well Test Analysis", (Ershaghi and Woodbury [36]) concerning the validation of mathematical models and the inherent lack of uniqueness in their application.

"Real-life examples of pressure data to fit a given idealised model are often non-existent. Consequently, many authors use synthetic data to point out the use of their proposed technique or model. In fact, practising engineers are now reading about many techniques and models for which there may never be examples of actual data to fit. Some people may even criticise the enormous effort toward prediction of pressure response in certain idealised models. One must note, however, that all idealised cases published to date and yet to be published are opening our eyes to response similarities that may exist among the performance of completely different systems".

Concerning the first part of this lucid statement, this author has been frequently requested by developers of software to provide field examples of tests influenced by fault patterns to permit them to calibrate their models and, while this can be done for such a straightforward case as the single fault example illustrated in Exercise 4.2, to go beyond that in providing more complex examples is treading on thin ice on account of the lack of mathematical uniqueness. It is not just that different systems will produce similar p_D-functions but also that quite distinct p_D-functions can produce similar differences:

$$p_D(t_D + \Delta t_D) - p_D(\Delta t_D)$$

which are used in buildup analysis (equation 4.30) and as a result different fault configurations (or other reservoir complexities: dual porosity, free gas breakout) can produce similar theoretical test responses.

The second point referred to above, the requirement of having a long drawdown period to see any significant distance away from the wellbore, was quantified in the previous section and is obviously equally relevant for complex fault geometries. It is during the drawdown that boundaries influence pressures recorded in the wellbore — not during the buildup.

On account of the lack of uniqueness and the expense of running long tests this author would advise against planning tests in appraisal wells to detect boundaries unless it is definitely believed that the results will have an impact on development decision making for the field. Even then the order of priorities must be: geological advice first (plus all the other observations listed in section 4.2), mathematical modelling second. In this respect the geologists should also be warned not to unreservedly change fault positions on their maps based solely on test interpretation.

4.17. APPLICATION OF THE EXPONENTIAL INTEGRAL

While still on the subject of describing pressure behaviour in infinite and semi-infinite acting systems, it is worth reconsidering the application of the *ei*-function. It was introduced in section 4.7 as the basic CTR solution of the radial diffusivity equation during the initial phase of flow when transience prevails and the *ei*-function (a standard mathematical integral) is evaluated as

$$ei(x) = ei \left(\frac{\phi \mu c r^2}{4 \times 0.000264\, kt} \right)$$

Provided the argument, x, is less than 0.01, which is satisfied very quickly when recording pressures in the wellbore ($r = r_w$, which is small), then the *ei*-function can be replaced by its logarithmic equivalent, equation 4.15, but if applied to determine pressure responses remote from a wellbore in which a rate perturbation is induced, then it is usually necessary to apply the *ei*-function directly. Although, as demonstrated in Exercise 4.2, sometimes the difference between use of the *ei*-function or its logarithmic approximation is outwith the level of accuracy being sought in the calculations.

In this section, use of the *ei*-function is illustrated in what might be broadly described as interference calculations between individual wells in a reservoir and between reservoirs in separate hydrocarbon accumulations. In such applications what is usually required is the superposition of individual *ei*-functions, as illustrated by the example shown in Fig. 4.37. Three development wells have been drilled in

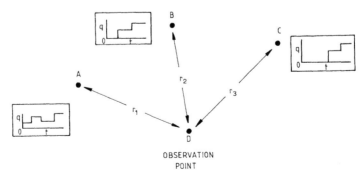

Fig. 4.37. Superposition in rate and time of three wells in an infinite acting reservoir.

a new reservoir in the sequence A, B, C and the requirement is to predict the pressure drop that should be recorded at the time of drilling the fourth well at location D. This will be influenced by the production histories of the three earlier well completions which requires the application of the principle of superposition. If the basic CTR equation 4.12 is expressed in the form

$$\Delta p = (p_i - p_{r,t}) = 70.6 \frac{q \mu B_o}{kh} ei \left(\frac{\phi \mu c r^2}{4 \times 0.000264 \, kt} \right) \tag{4.82}$$

Then applying the "recipe" for superposition described in section 4.9 to the well configuration shown in Fig. 4.37 gives

$$\Delta p_D = \sum_{i=1}^{3} \sum_{j=1}^{n} 70.6 \frac{\Delta q_j \mu B_o}{kh} ei \left(\frac{39.46 \phi \mu c r_i^2}{k(T - t_{j-1})} \right) \tag{4.83}$$

in which:

Δq_j = the rate change, $q_j - q_{j-1}$, (stb/d) in the ith well
T = the fixed time (days) at which the pressure will be measured in well D
t_{j-1} = the time (days) at which the rate change, Δq_j, occurred in the ith well
r_i = the distance between the ith well and location D
Δp_D = the overall pressure drop (psi) at location D at time T.

If the eventual pressure drop recorded at the new location does not match that predicted using equation 4.83, then the numerous parameters in the equation can be fine-tuned until correspondence is achieved. The parameter which exerts the greatest influence is the average permeability, k, since it appears both in the coefficient and argument of the ei-function, yet its effect is difficult to predict. If it is decreased, for instance, the coefficient increases and so to does the argument, x, which decreases the value of $ei(x)$ (Fig. 4.9). Consequently, the coefficient and ei-function exert opposite influences on the calculated pressure drop Δp.

Equation 4.83 has many useful applications in gaining an understanding of pressure communication at a distance although the engineer must be aware that results fall in the "rough and ready" category. One reason is because of the number of potential unknowns in the equation another is that application is strictly only valid for *transient* conditions. If, in the above example, for instance, there had been a substantial withdrawal of fluids through wells A, B and C prior to the pressure measurement in D then there would also be a material balance pressure drop in the system to be catered for. Notwithstanding these obvious limitations, if applied in a sensible fashion, equation 4.83 can provide useful results as illustrated in the example below

(a) Example: interference between oilfields

Use of the ei-function is not restricted to quantifying interference effects between wells within a single reservoir but, as in the present example, can also be applied to investigate interference between reservoirs in adjacent fields. The situation is

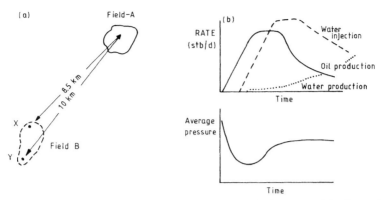

Fig. 4.38. (a) Relative locations of fields A and B. (b) Oil production, water injection and pressure profiles of a typical, offshore waterdrive field.

depicted in Fig. 4.38a. An offshore oil accumulation (field B) was discovered with the drilling of well X, eight years after the start of continuous production of field A, located some 8.5 km to the north east. An RFT survey run in well X indicated pressure depletion of $\Delta p_X = 220$ psi with respect to the initial pressure regime in the area, as established in field A. One year later, an RFT survey in appraisal well Y to the south of field B and 10 km from field A revealed depletion of $\Delta p_Y = 185$ psi. These observed pressure drops in wells X and Y can be intuitively attributed to the production history of field A and can be quantified through application of equation 4.83, catering for superposition in both rate and time.

Field A has been developed by engineered waterdrive for which typical production, injection and pressure profiles are shown in Fig. 4.38b. If the accumulation was discovered at undisturbed initial pressure for the area, then at the start of the development oil production is deliberately encouraged without supportive water injection to propagate a significant pressure drop throughout the reservoir which will be detected by running RFT surveys in each new development well. The technique, which is described in greater detail in Chapter 5, section 2, aims at establishing the degree of areal and vertical pressure communication in the reservoir (or lack of it) prior to finalizing the planning of water injection and, in particular, the locations of the injection wells and their completion intervals. The overall effect of this strategy is to cause a pressure drop in the reservoir and aquifer system which perturbs pressures at distant locations. In this respect, field A acts as an underground "beacon" influencing the initial pressures measured in wells X and Y.

The production and injection profiles for field A are listed in Table 4.7 in which q_o and q_{wp} are the oil and water production rates and q_{wi} the water injection rate: all in Mstb/d. The total underground withdrawal rate, which is the equivalent of the qB_o term in equation 4.82, may be expressed as:

$$[UW] = q_o B_o + q_{wp} - q_{wi} \quad (\text{Mrb/d})$$

Also listed are the changes in withdrawal $\Delta[UW]_j = [UW]_j - [UW]_{j-1}$ required

TABLE 4.7

Production/injection profiles, Field A (B_w = 1.0 rb/stb)

Time (years)	q_o (Mstb/d)	B_o (rb/stb)	q_{wp} (Mstb/d)	q_{wi} (Mstb/d)	[UW] (Mrb/d)	Δ[UW] (Mrb/d)
0						91
1	62	1.467			91	37
2	154	1.478		100	128	52
3	255	1.490		200	180	–
4	253	1.503		200	180	−60
5	251	1.512		260	120	−30
6	250	1.520	10	300	90	−96
7	220	1.520	20	360	−6	10
8	220	1.520	30	360	4	–
9	200	1.520	40	340	4	

in equation 4.83 which, with times expressed in months, becomes

$$\Delta p = \sum_{j=1}^{n} 70.6 \frac{\Delta[UW]_j \mu_w}{kh} ei \left(\frac{1.294 \phi \mu_w c r^2}{k(T - t_{j-1})} \right) \tag{4.84}$$

in which for either well, X or Y, the summation is taken over all the historic rate changes in field A. Since the communication between fields is mainly through water, and the oil and water properties (mobilities) are similar the values of the remaining parameters in the equation are aquifer properties between the fields assessed as:

$\mu_w = 0.4cP$ $\qquad\qquad \phi = 0.24$

$c = c_w + c_f = 7.0 \times 10^{-6}/\text{psi}$ $\quad h = 200$ ft

and inserting these values in equation 4.84 reduces it to

$$\Delta p = \sum_{j=1}^{n} 0.1412 \frac{\Delta[UW]_j}{k} ei \left(\frac{0.8696 \times 10^{-6} r^2}{k(T - t_{j-1})} \right) \tag{4.85}$$

which may be applied to wells X and Y as follows

Well X: $\quad r \quad$ = 8.5 km (27880 ft)

$\qquad\qquad T \quad$ = 8 years (96 months)

$$\Delta p_X = \sum_{j=1}^{8} 141.2 \frac{\Delta[UW]_j}{k} ei \left(\frac{675.91}{k(96 - t_{j-1})} \right)$$

Well Y: $\quad r \quad$ = 10 km (32800 ft)

$\qquad\qquad T \quad$ = 9 years (108 months)

$$\Delta p_Y = \sum_{j=1}^{9} 141.2 \frac{\Delta[UW]_j}{k} ei \left(\frac{935.52}{k(108 - t_{j-1})} \right)$$

TABLE 4.8

Calculation of pressure drops at locations X and Y, Field B, for different assumed values of the average aquifer permeability

Time (months)	$\Delta[UW]$ (Mrb/d)	$T - t_{j-1}$ (months)	$k = 50$ mD			$k = 100$ mD			$k = 150$ mD		
			x	$ei(x)$	Δp (psi)	x	$ei(x)$	Δp (psi)	x	$ei(x)$	Δp (psi)
Well X											
0	91	96	0.141	1.52	391	0.070	2.16	278	0.047	2.53	217
12	37	84	0.161	1.40	146	0.080	2.03	106	0.054	2.40	84
24	52	72	0.188	1.27	186	0.094	1.88	138	0.063	2.26	111
36	0	60									
48	−60	48	0.282	0.96	−163	0.141	1.52	−129	0.094	1.88	−106
60	−30	36	0.376	0.74	−63	0.188	1.27	−54	0.125	1.62	−46
72	−96	24	0.563	0.49	−133	0.282	0.96	−130	0.188	1.27	−115
84	10	12	1.127	0.18	5	0.563	0.49	7	0.376	0.74	7
96											
Total pressure drop (psi)					369			216			152
Well Y											
0	91	108	0.173	1.34	344	0.087	1.95	251	0.058	2.34	200
12	37	96	0.195	1.24	130	0.097	1.85	97	0.065	2.23	78
24	52	84	0.223	1.14	167	0.111	1.73	127	0.074	2.11	103
36	0	72									
48	−60	60	0.312	0.88	−149	0.156	1.43	−121	0.104	1.79	−101
60	−30	48	0.390	0.72	−61	0.195	1.24	−53	0.130	1.59	−45
72	−96	36	0.520	0.54	−146	0.260	1.02	−138	0.173	1.34	−121
84	10	24	0.780	0.32	9	0.390	0.72	10	0.260	1.02	10
96	0	12									
108											
Total pressure drop (psi)					294			173			124

(N.B.: $\Delta[UW]$ is expressed in Mrb/d in these two equations, Table 4.7; $ei(x)$-values are taken from Table 4.2.

The results of application of equation 4.85 to the two wells are listed in Table 4.8. The aquifer permeability, which is the major unknown, has been assigned values of 50, 100 and 150 mD, which are considered to bracket the correct value, and pressure drops at X and Y have been calculated for each, the trends being plotted in Fig. 4.39. Comparison with the observed pressure drops of 220 and 185 psi at locations X and Y respectively reveals that the average permeability required to match both is slightly less than 100 mD.

Although consistent, the results of such calculations are to be regarded as a reasonable approximation, particularly in treating field A as a line-source (section 4.7) but the high degree of symmetry of production wells across the accumulation tends to justify considering the fluid withdrawal at a"centroid" point. The results should be of encouragement to the operator of field B who is in the privileged position of appraising the accumulation under dynamic conditions on account of the established pressure communication with field A. In comparison to appraisal under

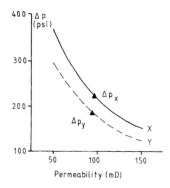

Fig. 4.39. Calculated pressure drops at well locations X and Y

purely static conditions, which reveals nothing concerning the degree of communication throughout the accumulation, the results of these simple calculations indicate an openness in the system: not only between fields A and B but also within the latter. This, in turn, implies that waterdrive in field B should prove successful and that injection wells can be safely located in the peripheral aquifer, which is a distinct advantage in early development planning.

Strictly speaking, the CTR-line source solution of the diffusivity equation is only valid for production of a single phase fluid, not for a produced oil volume with pressure transmission through water. Nevertheless, since the oil and water mobilities are similar in the above example, the line source solution has been applied using aquifer properties. If the mobilities are significantly different, however, such as in a gas accumulation with surrounding aquifer then equation 4.82 cannot be applied in such a meaningful fashion.

4.18. PRESSURE SUPPORT DURING APPRAISAL WELL TESTING

Partial or complete maintenance of bottom hole flowing pressure is to be expected in fields being developed by secondary recovery flooding, either waterdrive (Chapter 5) or gas drive (Chapter 6), in which one of the basic aims is to energise the reservoir by supporting pressure. But also, as described in section 4.6b, the condition is frequently observed when testing new reservoirs at the appraisal stage when nature supplies the energy. The circumstances described when this can occur are in testing high flow capacity reservoirs (even though they contain low-compressibility, undersaturated oil), gas reservoirs or oil reservoirs with gas cap and oil accumulations underlain by massive basal aquifers. Tests of this type are not difficult to analyse but since their interpretation has received little attention in the popular literature (section 4.3) serious and costly errors can be and are made through the application of techniques appropriate for infinite acting (transient) or bounded systems (depletion) to tests which demonstrate an element of pressure support during their flow periods. Some of the more popular misconceptions concerning this type of testing and the errors to which they lead are described in this section.

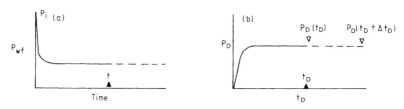

Fig. 4.40. Steady-state flow. (a) Drawdown pressures. (b) Corresponding p_D-function.

(a) Pressure buildup performance

If the general pressure buildup equation

$$\sigma(p_i - p_{ws}) = p_D(t_D + \Delta t_D) - p_D(\Delta t_D) \tag{4.30}$$

is subtracted from the drawdown equation

$$\sigma(p_i - p_{wf}) = p_D(t_D) + S \tag{4.21}$$

the result is

$$\sigma(p_{ws} - p_{wf}) = p_D(\Delta t_D) + S + [p_D(t_D) - p_D(t_D + \Delta t_D)] \tag{4.86}$$

But, as illustrated in Fig. 4.40, for the case of complete steady-state flow — which is frequently observed in testing; if the well is closed in at time t (dimensionless time, t_D), then since the p_D-function is "flat" its extrapolation to $t_D + \Delta t_D$ will be such that

$$p_D(t_D + \Delta t_D) \approx p_D(t_D)$$

and the term in parenthesis in equation 4.86 vanishes giving a pressure buildup of the form

$$\sigma(p_{ws} - p_{wf}) = p_D(\Delta t_D) + S \tag{4.87}$$

which, it will be recognised, is the mirror image of the drawdown equation 4.21 as shown in Fig. 4.41. This observation implies the following:

– the duration of the early, linear sections on semi-log plots of the drawdown and buildup should be the same

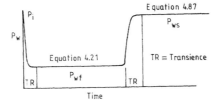

Fig. 4.41. Pressure drawdown and buildup responses during a steady-state test.

- there is a lack of mathematical uniqueness in the test interpretation
- the plotting technique of Miller, Dyes and Hutchinson should prove the most appropriate for buildup analysis.

And, in the case that the pressure support is partial rather than complete, all the above arguments fall in the category of being "approximately correct".

Equations 4.21 and 4.87 imply that while transience lasts for either the drawdown or buildup phases of a steady-state test

$$p_D(t_D) = p_D(\Delta t_D) = \tfrac{1}{2} \ln \frac{4\,t'_D}{\gamma} \qquad (4.88)$$

where t'_D represents the dimensionless time argument during either period. Therefore, there will be early linear sections of equal length at the start of the drawdown and buildup if plots are made of the wellbore pressure versus $\log t$ or $\log \Delta t$ respectively. Furthermore, the duration of the linear sections will be solely dependent on pure transient condition prevailing — nothing else.

Since equation 4.87, during the period of transience, represents a direct relationship between p_{ws} and $\log \Delta t$, it explains why the MDH plotting technique proves to be the more appropriate for buildup analysis. In comparison, the Horner plot is more complex and error prone, as illustrated in Exercise 4.3.

The lack of mathematical uniqueness has been referred to previously (section 4.8) as being a persistent difficulty in well testing in general but, in the case of steady-state testing, is at its very worst and any degree of pressure support tends to lead towards this condition. The reason for this is quite simple. The test analysis requires implicitly the simultaneous solutions of the drawdown and buildup equations 4.21 and 4.87 but since there are identical in form and contain *three unknown constants* (k, S and p_D) they are intractable to unique mathematical solution. For different physical conditions in which the p_D-function varies with time this difficulty is lessened but the closer the flow approaches the steady-state condition the more pronounced becomes the lack of mathematical uniqueness. The reason for this is something we are taught in school at a very early stage of our mathematical development: two equations, three unknown constants — hopeless situation. Yet somehow the realisation of this simple fact seems to have eluded the Industry. In a series of technical papers describing test analyses of Prudhoe Bay wells in Alaska [37–39], for instance, the complexity is referred to frequently. The oil accumulations are overlain by gas caps and although none of the papers illustrates the drawdown behaviour of any of the wells tested, it is quite obvious from the concave downward shapes of the pressure buildups (Fig. 4.42) that the high-compressibility gas provided a significant degree of support of the flowing pressures. While all the papers note and complain about the lack of uniqueness in the analyses (on one occasion referred to as bewildering) the obvious reason for this condition is not explained, although the consequences are [37]:

"If the well is damaged, the combined effects of the gas cap, afterflow and damage make the buildup analysis non unique. Unlimited combinations of wellbore damage [S] and formation permeability [k] give nearly the same buildup behaviour".

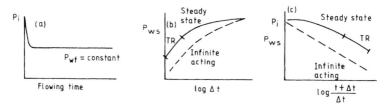

Fig. 4.42. Steady-state appraisal test. (a) Flowing pressure. (b) MDH buildup. (c) Horner buildup. *TR* = transient response.

The second sentence in the above quotation is perfectly correct and is demon-strated in Exercise 4.3 for a steady-state test. Apart from analysing the test using the correct, transient straight line, two other lines (tangents) to the MDH buildup are arbitrarily selected. While each has markedly different values of k and S, it is demonstrated that they not only can be used to identify identical values of the initial pressure but also satisfy both the drawdown and buildup equations 4.21 and 4.87, as they must do; in fact there is an infinite number of solutions that will do the same. Faced with this mathematical anomaly, the engineer must select the correct straight line representing transience for sound physical reasons: it must be the *first* linear section that develops after well closure — none other. The means of selecting the correct, early linear section on either MDH or Horner buildups is by applying the double time derivative plot as descripbed in section 4.14a and illustrated in Excer-cises 4.1 and 2. Since this initial pressure response is usually very rapid, sometimes lasting only a few minutes, it is imperative in appraisal well testing to run a DST or similar assembly to affect downhole closure, otherwise the transient response can be missed completely. Then, as shown in Fig. 4.42 the brief, transient part of the buildup is parallel to but elevated above that for a purely infinite acting reservoir for the sound mathematical reasons argued in sections 4.12 and 4.13. Then, both the MDH and Horner buildups are concave downwards for the remainder of closed-in period although this trend is much more pronounced for the former, which is one of its salving features.

(b) Dimensionless pressure-radius of investigation

The solution of the radial diffusivity equation 4.4 for steady-state flow ($\partial P/\partial t = 0$) can be accomplished by integration by parts, as described in chapter 6 of reference 3. But a much simpler approach is to apply Darcy's law directly to the radial geometry and drawdown shown in Fig. 4.43. That is, the flow at any radius r may be described, in absolute units, as

$$q = \frac{kA}{\mu}\frac{\partial p}{\partial r} = \frac{2\pi kh}{\mu}r\frac{\partial p}{\partial r}$$

which, upon separating the variables and integrating becomes

$$p_e - p_{wf} = \frac{q\mu}{2\pi kh}\ln\frac{r_e}{r_w}$$

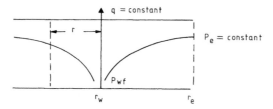

Fig. 4.43. Pressure drawdown for radial, steady-state flow.

This may be expressed in dimensionless form and field units as

$$7.08 \times 10^{-3} \frac{kh}{q\mu B_o}(p_e - p_{wf}) = \ln \frac{r_e}{r_w} + S \tag{4.89}$$

in which the skin factor has been incorporated using its defining expression, equation 4.18. If applied to an initial well test, the steady-state pressure at the constant pressure outer boundary, p_e, is replaced by the initial pressure p_i. Once p_i, kh and S have been determined by conventional buildup analysis (sections 4.12 and 4.13) then equation 4.89 can be solved for r_e — the radius of the constant pressure boundary, which is equivalent to the radius of investigation for steady-state flow.

By direct analogy with equation 4.21, it can be seen that the p_D-function for steady-state, radial flow expressed in equation 4.89 (for $p_e = p_i$) is simply

$$p_D(t_D) = \ln \frac{r_e}{r_w} = \text{constant} \tag{4.90}$$

and Earlougher [5] has presented p_D-functions for other geometries including linear flow and regular five spot patterns.

One of the more common mistakes in analysing tests, in which there has been an element of pressure support, is to apply the "conventional" expression for determining the radius of investigation as

$$r_e = 0.03 \sqrt{\frac{kt}{\phi\mu c}} \tag{4.81}$$

using the entire flowing time, t, in the calculation. But, as pointed out in section 4.16c, equation 4.81 is only valid provided the pressure is continuously declining in a purely *transient* fashion. If equation 4.81 is to be applied at all to steady-state tests, then the time used in the calculation must be that for which transience is observed during the flowing period. Yet, at the time of writing, every commercial computer system for test analysis examined by the author only applies equation 4.81 when the "radius of investigation" button is pushed — and this usually leads to a significant overestimation of this parameter in comparison to that calculated using equation 4.90. The error is one of applying incorrect mathematics to describe the physical situation and, if stability of pressures is observed during the flow period of a test then, no matter what cause it may be attributed to, it is obvious that the steady-state solution of the diffusivity equation 4.89 must be applied rather than the conventional expression for transient flow, equation 4.81.

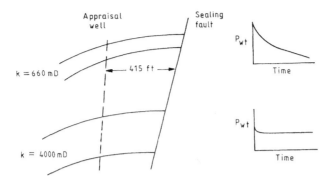

Fig. 4.44. North Sea, appraisal oilwell test sequence.

An example of this error is illustrated in a test sequence conducted in an appraisal well in a North Sea oilfield as illustrated in Fig. 4.44. The well was drilled in the vicinity of a sealing fault whose existence was known with certainty, because it acted as the structural trap for the hydrocarbon accumulation, yet whose exact position was unknown. The test in the upper reservoir ($k = 660$ mD) clearly established the distance of the fault from the well (this is the test described in detail in Exercise 4.2) as being 415 ft. The lower of the two reservoirs was an excellent Jurassic sand which was twice the thickness of the upper and had an average permeability of 4000 mD. The test on the lower consisted of a twelve hour flowing period at an average rate of 15,000 stb/d, which is exceptionally high for an offshore rig in which the test separator and flare capacities are limited. Yet the flowing pressure remained constant throughout the entire production period. The engineer, analysing the data, defined correctly the position of the fault in the test on the upper reservoir but for the lower, which characteristically demonstrated a continuous downward curvature during the entire buildup period, he mistakenly applied equation 4.81 to calculate the "radius of investigation", which turned out to be $\geq$4000 ft. This mis-match of the mathematics to the physics led to an interesting debate between the test interpreter and the geologists! The error is also illustrated in Exercise 4.3, in which the misapplication of equation 4.81 to calculate the radius of investigation in a steady-state test provides an overestimation of its true value, obtained using equation 4.89 by a factor of 2.26.

The paradoxical conclusion to be reached is that the better the reservoir, in terms of high flow capacity and therefore natural pressure support, the more limited is the depth of investigation out into the formation.

(c) Miller, Dyes, Hutchinson interpretation

In deriving the early, straight-line equation for the MDH plot in section 4.12, it was noted that the conditions which must be satisfied to assure linearity are that

$$p_D(\Delta t_D) = \tfrac{1}{2} \ln \frac{4 \Delta t_D}{\gamma} \tag{4.31}$$

$$p_D(t_D + \Delta t_D) \approx p_D(t_D) = \text{constant} \tag{4.32}$$

the first of these being the transient condition. But, for a test in which there is complete pressure support during the flow period (Fig. 4.40b), the latter expression, equation 4.32, should, in principle, be valid for any value of the closed-in time. In fact, referring to the concern expressed in section 4.8, regarding the handling of the extrapolation of the p_D-function for times beyond which it is physically measured — that required for steady-state flow should be the soundest extrapolation in the business. Consequently, the duration of the early straight line on an MDH buildup plot (equation 4.33) following complete steady-state flow is governed entirely by transience. As soon as equation 4.31 is no longer satisfied, the pressure points will deviate away from and below the straight line, as illustrated in Fig. 4.42b. Furthermore, the downward curvature of the plot is usually maintained throughout the remainder of the buildup thus removing the confusion in selecting the purely transient response that frequently occurs in Horner plotting, as described in the following section. Because of this dependence on transience alone for definition of the early straight line, it is considered that the MDH plotting technique is both the soundest and safest for analysing steady-state tests and the same is true even if there is only partial pressure support. One of its endearing features is that there is no temptation to extrapolate to give the reservoir pressure at infinite closed-in time which causes so many errors in Horner plotting. For a complete steady-state test in an appraisal well, the initial pressure can be determined by applying equation 4.38, which for radial geometry, becomes

$$p_D(t_D) = \tfrac{1}{2} \ln \frac{4\Delta t_{Ds}}{\gamma} = \ln \frac{r_e}{r_w} \tag{4.38}$$

and implies that extrapolation of the early, linear buildup trend to a time

$$\Delta t_s = \frac{1686.6\phi\mu c r_e^2}{k} \quad \text{(hours)} \tag{4.91}$$

will yield the initial pressure. The technique is illustrated in Exercise 4.3.

Visual identification of the period of initial transience in an MDH buildup is eased by the observation made in section 4.18a (Fig. 4.41) that the duration of transience should be the same as for the pressure drawdown. Due to instability in the initial production rate it is not usually possible to analyse the transient drawdown data quantitatively, especially if using a water cushion at the start of the test to promote a high initial drawdown and flow rate. Under these circumstances, the test should be stopped when oil first reaches the surface and resumed shortly afterwards with the flow string full of oil to remove the initial distortion in flowing pressures.

(d) Horner interpretation

For linearity of the early Horner buildup plot three conditions must be satisfied, as described in section 4.13, namely

$$p_D(\Delta t_D) = \tfrac{1}{2}\ln\frac{4\Delta t_D}{\gamma} \tag{4.31}$$

$$p_D(t_D + \Delta t_D) \approx p_D(t_D) = \text{constant} \tag{4.32}$$

$$\ln(t_D + \Delta t_D) \approx \ln(t_D) \quad = \text{constant} \tag{4.44}$$

Consequently, even for a buildup following a period of complete steady-state flow, for which condition 4.32 is always satisfied, the duration of the straight line will be dependent on the breakdown of one or both conditions in equations 4.31 and 4.44. Therefore, early linearity of the Horner buildup in a steady-state test is not solely dependent on transience prevailing, as for the MDH plot, but is more complex and in many cases — confusing. While having the concave downward shape which is so characteristic following a flow period with pressure support (refer section 4.13), the Horner plot does have the tendency to produce an apparent although rather convincing linear section for large values of the closed-in time. The latter is clearly demonstrated in the test presented in Exercise 4.3 (Fig. 4.48b) in which the initial linear section lasts for only 10 minutes whereas the later "apparent" straight line persists for 4.2 hours. Not only that, the early straight line quite naturally extrapolates to a value of Z^* in excess of the initial pressure while the later linear trend extrapolates to a value which is very close to p_i and seems to convince many engineers (section 4.13) that the second straight line is the correct one. Selection of the later straight line with its smaller slope leads to a serious overestimation of both the permeability (equation 4.36) and skin factor (equation 4.50) both of which must balance on opposite sides of the basic drawdown equation 4.21. In areas such as the North Sea, where high flow capacity reservoirs are commonplace, the error has reached epidemic proportions and in the worst example noted by this author the mistaken and actual test results are as detailed in Table 4.9. Considering the abnormally high permeability determined in the mistaken analysis, which in terms of sedimentary rock classification must place the formation in the "see through" variety; the calculated skin factor of 110 casts some doubt on the nature of the job performed: was it a well test or a cement squeeze? The worst error, however, is the one order of magnitude overestimation of the ideal productivity index which, as suggested in section 4.4, is the most important result determined in an appraisal well test. This arises from the calculation of far too large a pressure drop across

TABLE 4.9

Severe example of the error in selecting the incorrect straight line on a Horner buildup plot for a steady-state appraisal well test

	Mistaken interpretation	Correct interpretation
Duration of linear section (hours)	0.5–6.0	0–0.3
Permeability (mD)	18000	1270
Skin factor	≥ 110	0
Productiviey index (stb/d/psi)	13	13
Ideal productivity index (stb/d/psi)	125	13

the fictitious damaged zone close to the well (equation 4.18), and its inclusion in the denominator of equation 4.1. For all well tests, and this type in particular, a check should be made with the core data to see if the results are realistic. In the above test, however, only the mathematical interpretation was relied upon. That is, the engineer was satisfied that the values of k and S determined matched both the buildup and drawdown pressure responses but, on account of the total lack of mathematical uniqueness associated with steady-state testing, described in section 4.18a and illustrated in Exercise 4.3, there is an infinite number of values of k and S that will satisfy both phases of such a test.

There is an apparent anomaly in the basic CTR-solution of the radial diffusivity equation

$$7.08 \times 10^{-3} \frac{kh}{q\mu B_o}(p_i - p_{wf}) = p_D(t_D) + S \tag{4.21}$$

(an equation that must implicitly be satisfied in all test interpretations) that if the kh-product on the left-hand side is inadvertently calculated as being too large, on account of choosing the wrong linear section of the buildup with too small a slope; then to balance the equation the skin factor on the right-hand side must also be overestimated. In other words, high permeabilities go hand in hand with high skin factors which may be mathematically sound but is not very convincing from the physical point of view. The engineer should therefore be cautious in this respect and if an excellent reservoir is also diagnosed as being severely damaged check carefully whether the correct linear section of the buildup has been selected for the calculation of k and S. The author has noticed several tests in which k and S have been overestimated by choosing the wrong straight line and the results prompted the operator to perform expensive acid stimulation to remove the hypothetical skin. The inevitable result is that the formation flow capacity is enhanced so that in a post-workover pressure buildup, if the incorrect straight line is again chosen, it has an even smaller slope which leads to the calculation of a higher skin factor than before the stimulation!

Yet another error associated with the Horner buildup plot and its tendency to produce a convincing looking straight line for large values of the closed in time, following a flow period in which there was an element of pressure support, is that people determine values of k and S from the early and late linear trends and impute them to different parts of the reservoir. That is, the values from the initial, transient line are attributed to formation characteristics in the immediate vicinity of the well, whereas k and S determined from the second straight line are assumed to be somehow representative of conditions remote from the wellbore. But there are two sound physical arguments against such a hypothesis:

- If there is pressure stability during the flow period then the radius of investigation is strictly limited (section 4.18b), therefore formation parameters "remote" from the well cannot be evaluated in such a test.
- The skin factor determined from the second straight line is a concept that is not defined in the subject of well testing.

The first of these has been described already: no matter what the flow pattern, radial or spherical, the radius of investigation is usually small and the greater the degree of pressure support the more restricted it becomes. There are several physical causes that can give rise to a measured skin factor: formation damage, partial perforation, non-Darcy flow, well inclination but whichever of these is active it must be appreciated that the skin factor is ONLY defined in the immediate vicinity of the wellbore — nowhere else. As such it is considered that the component of pressure drop across the skin will adjust instantaneously following a rate change and therefore the skin factor can be included in the basic drawdown equation 4.21 as a rate independent perturbation. But nowhere in the extensive literature on well testing is a skin factor defined remote from the well (mistakenly referred to as deep formation damage) nor is it possible to do so. Away from the wellbore the skin factor is zero and even if the formation was for some reason damaged this would appear in equations as a modified permeability across which pressure disturbances would have a time dependence.

This false concept in attributing the second linear trend of a Horner plot as somehow characterising the formation away from the well is simply a postulate which cannot be mathematically proven [38]. In section 4.13 the steps are followed that demonstrate that provided conditions in equations 4.31, 4.32 and 4.44 are satisfied an early, transient straight line will develop from which k and S can be determined; but how this analysis can be extended to a second straight line on a Horner plot is not evident. The shape of the Horner plot following a flow period in which there is any degree of pressure support is obvious and has been explained already in section 4.13 and illustrated in Figs. 4.18 and 4.42. The early, transient line must be located above that for an infinite acting reservoir, consequently as the buildup progresses the pressure points will inevitably curve downwards (referred to as the "roll-over" effect) as the initial reservoir pressure is approached. If a section of this late time response happens to be linear, as unfortunately is repeatedly the case with Horner plots, then it is advised that no quantitative analysis of this trend should be attempted.

There has always been a potentially dangerous tradition in reservoir engineering to seek out straight-line trends and impute them with some physical meaning. Perhaps the worst example of this is the notorious P/Z-plot in gas material balance interpretation, described in Chapter 6, section 3, in which it is demonstrated that: just because an apparent straight-line relationship between P/Z and cumulative gas production can be defined, it does not necessarily mean that the reservoir is of the volumetric depletion type. Similarly, in Horner buildup plotting: just because there happens to be a late-time linear trend it does not mean that it represents properties remote from the wellbore and caution should be attached to any such assumption. In the case of the complete doubling of slope caused by the presence of a nearby sealing fault, for instance (Exercise 4.2), nobody would attribute this effect to a halving of permeability and reduction of skin away from the well, the effect is merely due to the depression of the flowing pressures during the drawdown period. Therefore, in the case of a roll-over buildup, why should it be necessary to consider any late straight line on a Horner plot as being other than the natural consequence of pressure support during the drawdown (as described in section 4.13) and the nature of the plot itself which, in this case, proves more confusing than the MDH interpretation technique. The best way to avoid this error is to carefully examine the drawdown behaviour, as described in section 4.10a, to establish whether there has been any element of pressure support. Unfortunately, such advice cannot be regarded as covering all eventualities. In the present section the pressure support is considered in an infinite acting system but, as illustrated in the theoretical example (section 4.19d) for an extended test in a well confined between faults yet receiving pressure support, it is the faults that dominate the early pressure drawdown response and the effect of the pressure maintenance is not evident until much later during the flow period.

(e) Variable skin factor (well clean-up)

Considering the basic drawdown equation

$$\sigma(p_i - p_{wf}) = p_D(t_D) + S \tag{4.21}$$

it is conceivable that what is assumed to be an element of partial pressure support, complete pressure maintenance or, sometimes, an increase of flowing pressure with time could be due to well clean-up. That is, the p_D-function increases steadily with time but S decreases in such a manner as to compensate for its variation giving the impression of pressure support. This immediately involves an element of subjectivity in the test analysis: the engineer believes that the skin factor is constant and there is some degree of pressure maintenance or, alternatively, that there is no support and the skin is decreasing and unfortunately no amount of theorising will assist in deciding which assumption is correct. Instead, the matter must be settled by observation of system under test. If the formation is of low flow capacity and contains a highly undersaturated oil (no free gas) then it would seem fairly safe to assume well clean-up because there would be no apparent reason for pressure maintenance, whereas in a high flow capacity formation containing high-compressibility fluids, the decision may very well be the opposite.

A variable skin factor will further complicate application of the variable-rate CTR solution of the diffusivity equation (equation 4.24) since the skin factor in each of the superposed terms will not systematically cancel as they do in deriving the equation if the skin is constant. For a pressure buildup, however, the basic equation 4.30 is still valid since, even if the skin is a variable, its value in the two superposed terms which make up the equation is that at the end of the flow period and consequently cancellation occurs in adding the terms. The buildup can then be analysed in the usual manner to calculate the kh-value of the formation and its skin factor at the time of well closure. In principle, the rate of increase of the skin factor during the drawdown period can then be determined by direct solution of equation 4.21 as

$$S(t) = \sigma(p_i - p_{wf}) - p_D(t_D) \tag{4.92}$$

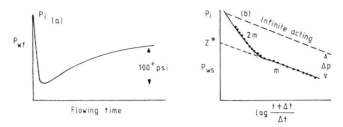

Fig. 4.45. (a) Drawdown pressure. (b) Horner pressure buildup.

If the double time derivative technique has been correctly applied to define the correct early straight line on the buildup (section 4.14a), then kh and S will be correctly determined, the latter being the value at the end of the flow period. This will lead to the evaluation of $p_D(t_D)$ also immediately prior to well closure using equation 4.21. Comparison of this with the transient value of $p_D(t_D)$, evaluated using equation 4.22, will at least alert the engineer to whether there has been an element of pressure support or not. If there has been then the transient function will be in excess of the test value (refer to Exercise 4.1) meaning not all of the stable or rising flowing pressure is due to well clean-up.

Larsen [40] has described the analysis of such tests in which the skin variation could be empirically fitted as a simple hyperbolic function of time. In the analysis, however, the usual assumption of transient flow appears to have been made for which the p_D-function is evaluated using equation 4.22 but more complex situations can be catered for. In one test analysed by the author the rate of well clean-up was so great that the bottom hole flowing pressure increased by over 100 psi during an eight hour flow period as shown in Fig. 4.45a yet a lengthy pressure buildup displayed a doubling of slope which was interpreted as due to the presence of a sealing fault (Fig. 4.45b). The buildup could be interpreted, as illustrated in Exercise 4.2 to determine kh, Z^*, $\Delta p = p_i - Z^*$ and the skin factor at the end of the flowing period. The distance to the fault, d, was then calculated using equation 4.70 which, in turn, permitted the p_D-function to be evaluated throughout the entire flowing period using equation 4.69. The final step was to insert this function in equation 4.92 to evaluate the rate of decrease of the skin factor with time during the flow period. This particular test could be interpreted uniquely simply because it proved possible to define independently the p_D-function for inclusion in equation 4.92 but in cases for which this cannot be done there will be ambiguity in distinguishing between the two variables on the right-hand side of equation 4.21.

Exercise 4.3: Pressure buildup test: steady-state flow condition

Introduction

This is an appraisal well test in which complete stability of pressure was observed during the flowing period. The subsequent pressure buildup is analysed using the MDH and Horner plotting techniques and illustrates the common error in applying the latter in isolation to this type of test.

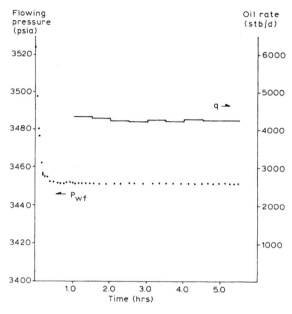

Fig. 4.46. Test production history (Exercise 4.3).

Question

An appraisal well was tested by producing a cumulative 930 stb of oil, the final rate being 4250 stb/d. The rate and bottom hole flowing pressures during the production period are plotted in Fig. 4.46. Although there was a slight instability in the flowing pressures at the start of the test, it is clear that the transient flow period lasted about 15 minutes and certainly less than half an hour. During the remainder of the production period, stability of the bottom hole flowing pressure was observed in which slight fluctuations of less than 1 psi occurred while flowing at a controlled constant rate. The well was closed-in down hole and the pressure data recorded during a 5.5 hour buildup are listed in Table 4.10. Other data required to interpret the test are as follows:

p_i = 3524.60 psia (5 minutes flow/1 hour buildup) $c = 15.0 \times 10^{-6}$/psi
p_{wf} = 3451.50 psia (average value) $\mu = 0.70$ cp
h = 60 ft (fully perforated well) $B_{oi} = 1.22$ rb/stb
ϕ = 0.20 $r_w = 0.510$ ft ($12\frac{1}{4}$ inch hole)

- Analyse the test using both the MDH and Horner plotting techniques together with their time derivatives.

Solution

The effective flowing time required in the Horner analysis can be calculated as

$$t = \frac{N_p}{q} \times 24 = \frac{930}{4250} \times 24 = 5.25 \quad \text{(hours)} \tag{4.51}$$

TABLE 4.10

Pressure buildup data, Exercise 4.3 ($t = 5.25$ hours)

Δt (hours)	p_{ws} (psia)	$\log \Delta t$	$p'_{ws\,(MDH)}$ [a] (psi)	$(\log t + \Delta t)/\Delta t$	$p'_{ws\,(Horner)}$ [b] (psi)
0.008	3509.22	−2.097		2.818	
0.017	3512.12	−1.770		2.491	
0.025	3513.15	−1.602	2.70	2.324	2.68
0.031	3514.09	−1.509	4.39	2.231	4.38
0.036	3514.56	−1.444	3.15	2.167	3.19
0.040	3514.84	−1.398	2.66	2.121	2.64
0.044	3515.10	−1.357	2.73	2.080	2.75
0.050	3515.42	−1.301	2.51	2.025	2.53
0.056	3515.74	−1.252	2.83	1.977	2.89
0.062	3516.02	−1.208	2.75	1.933	2.76
0.067	3516.23	−1.174	2.71	1.900	2.76
0.074	3516.48	−1.131	2.52	1.857	2.52
0.081	3516.71	−1.092	2.55	1.818	2.54
0.087	3516.89	−1.060	2.52	1.788	2.61
0.094	3517.10	−1.027	2.71	1.755	2.89
0.103	3517.33	−0.987	2.52	1.716	2.56
0.111	3517.52	−0.955	2.54	1.684	2.58
0.127	3517.88	−0.896	2.68	1.627	2.74
0.143	3518.15	−0.845	2.28	1.576	2.30
0.161	3518.41	−0.793	2.20	1.526	2.26
0.194	3518.83	−0.712	2.26	1.448	2.34
0.244	3519.31	−0.613	2.10	1.352	2.17
0.278	3519.59	−0.556	2.15	1.299	2.29
0.328	3519.91	−0.484	1.94	1.231	2.04
0.361	3520.10	−0.442	1.98	1.192	2.12
0.428	3520.42	−0.369	1.88	1.123	2.01
0.494	3520.69	−0.306	1.89	1.065	2.02
0.528	3520.80	−0.277	1.65	1.039	1.84
0.628	3521.08	−0.202	1.62	0.971	1.79
0.711	3521.27	−0.148	1.53	0.923	1.72
0.778	3521.41	−0.109	1.56	0.889	1.79
0.861	3521.56	−0.065	1.48	0.851	1.71
0.944	3521.68	−0.025	1.30	0.817	1.53
1.094	3521.89	0.039	1.43	0.763	1.69
1.294	3522.08	0.112	1.13	0.704	1.40
1.361	3522.14	0.134	1.19	0.686	1.45
1.428	3522.20	0.155	1.25	0.670	1.63
1.528	3522.29	0.184	1.33	0.647	1.70
1.644	3522.38	0.216	1.23	0.623	1.63
1.761	3522.45	0.246	1.02	0.600	1.32
1.878	3522.52	0.274	1.09	0.579	1.45
1.978	3522.57	0.296	0.96	0.563	1.36
2.061	3522.61	0.314	0.97	0.550	1.34
2.128	3522.64	0.328	0.94	0.540	1.30
2.288	3522.71	0.359	0.97	0.518	1.38
2.361	3522.74	0.373	0.96	0.508	1.30
2.478	3522.79	0.394	1.03	0.494	1.55

TABLE 4.10 (continued)

Δt (hours)	p_{ws} (psia)	$\log \Delta t$	$p'_{ws\,(MDH)}$ [a] (psi)	$(\log t + \Delta t)/\Delta t$	$p'_{ws\,(Horner)}$ [b] (psi)
2.644	3522.86	0.422	1.07	0.475	1.60
2.811	3522.92	0.449	0.98	0.458	1.53
3.044	3522.98	0.483	0.75	0.435	1.13
3.394	3523.06	0.531	0.74	0.406	1.20
3.664	3523.12	0.564	0.78	0.386	1.30
4.144	3523.22	0.617	0.81	0.355	1.40
4.644	3523.31	0.667	0.79	0.328	1.45
4.987	3523.36	0.698	0.70	0.312	1.36
5.494	3523.43	0.740	0.72	0.291	1.45

[a] $p'_{ws\,(MDH)}$: equation 4.41; [b] $p'_{ws\,(Horner)}$: equation 4.52

which is used to calculate the Horner time ratio in Table 4.10. It will be noted in this table that the static pressures are quoted in psia to two places of decimal. The use of a high-resolution pressure gauge justifies this accuracy which is required in such a fast buildup over a limited pressure range. As can be seen, for instance, there are 16 points included over a one psi pressure interval from 3522 to 3523 psia. Also listed in Table 4.10 are the time derivative data, equations 4.41 and 4.52. These are the average values for each time interval. As described in section 4.18c, the safest way to interpret such a test is by applying the MDH technique, as follows.

Miller, Dyes, Hutchinson. The MDH and its derivative plot are presented in Fig. 4.47. Since, as noted in section 4.18a, the buildup should be the mirror image of the drawdown, it follows from inspection of the flowing pressure history, Fig. 4.46 that the linear, transient part of the buildup be sought certainly within the first half hour following well closure. Such a transient response can be readily identified as starting after $\Delta t = 0.031$ hours (2 minutes), when afterflow ceases, extending to $\Delta t = 0.161$ hours (10 minutes). If the well had not been closed-in downhole, it may have been difficult to detect this brief linear section of the buildup. Although the derivative data display considerable scatter, they do confirm that the only plateau period occurs for $\Delta t < 10$ minutes ($\log \Delta t < -0.8$), thus confirming the linearity. Thereafter, there is a steady downward drift of the points indicating a roll-over of the buildup such that the rate of change of pressure with respect to time is continuously decreasing. From the linear section of the plot $(A-B)$ it can be determined that

slope $= m = 6.05$ psi/log cycle $p_{ws\,(1\,hr)} = 3523.30$ psia

The kh-product can then be calculated using equation 4.36 as

$$kh = \frac{162.6q\mu_o B_o}{m} = \frac{162.6 \times 4250 \times 0.7 \times 1.22}{6.05} = 97550 \text{ mD·ft} \qquad (4.36)$$

and therefore $k = 1625$ mD. The skin factor is determined using equation 4.37 as

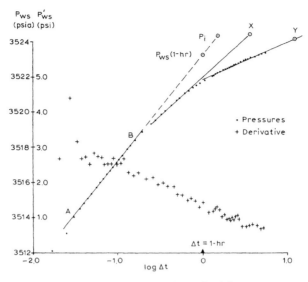

Fig. 4.47. MDH plot (Exercise 4.3).

$$S = 1.151 \left[\frac{3523.30 - 3451.5}{6.05} - \log \frac{1625}{0.2 \times 0.7 \times 15 \times 10^{-6} \times 0.51^2} + 3.23 \right]$$
$$= 6.5 \qquad (4.37)$$

Assuming radial geometry, the radius of investigation (constant pressure outer boundary) can be calculated using equation 4.89, the steady-state solution of the diffusivity equation during the flow period, as:

$$\frac{7.08 \times 10^{-3} \times 97550}{4250 \times 0.7 \times 1.22}(3524.60 - 3451.50) = \ln \frac{r_e}{0.51} + 6.5 \qquad (4.89)$$

which gives $r_e = 840$ ft. It is interesting to note that if the conventional formula for calculating the radius of investigation were used inadvertently, for the full flowing time of 5.25 hours, the result would be

$$r_e = 0.03 \sqrt{\frac{kt}{\phi \mu c}} = 0.03 \sqrt{\frac{1625 \times 5.25}{0.2 \times 0.7 \times 15 \times 10^{-6}}} = 1900 \text{ ft} \qquad (4.81)$$

which is more than double the correct value. The stabilized PI is:

$$\text{PI} = \frac{q}{p_i - p_{wf}} = \frac{4250}{3524.60 - 3451.50} = 58.1 \text{ stb/d/psi} \qquad (4.1)$$

and since the pressure drop across the skin is:

$$\Delta p_{skin} = 0.87mS = 0.87 \times 6.05 \times 6.5 = 34.2 \text{ psi} \qquad (4.53)$$

then the ideal PI is:

$$\text{PI}_{\text{ideal}} = \frac{q}{p_i - p_{\text{wf}} - \Delta p_{\text{skin}}} = \frac{4250}{3524.6 - 3451.5 - 34.2} = 109.3 \text{ stb/d/psi} \qquad (4.1)$$

The closed-in time to which the early linear trend of the MDH plot must be extrapolated to reach the initial pressure can be evaluated applying equation 4.91 as

$$\Delta t_s = \frac{1686.6 \phi \mu c r_e^2}{k} \qquad (4.91)$$

$$= \frac{1686.6 \times 0.2 \times 0.7 \times 15 \times 10^{-6} \times 840^2}{1625} = 1.54 \text{ hours}$$

and, as can be seen in Fig. 4.47, extrapolation of the line A–B to the equivalent, $\log \Delta t_s = 0.188$, gives a value of $p_{\text{wsl}} = p_i = 3524.4$ psia, practically identical with the measured value. This coincidence does not imply, however, that A–B must necessarily be the correct MDH straight line, on account of the lack of mathematical uniqueness referred to in section 4.18a. To illustrate this point, suppose the interpreter had inadvertently chosen either lines X or Y as representing the linear pressure responses — although, as can be seen from the time derivative plot, such linear sections are "apparent". The comparative buildup results, applying the same formula as for line A–B, are as listed in Table 4.11. As can be seen, although the values of k and S from the three lines are markedly different, they eventually provide values of p_i by extrapolation which are within 0.4 psi of the correct value.

Furthermore, if the three different values of r_e are used to calculate p_D-functions for steady-state flow, using equation 4.90, and these are inserted in the general drawdown equation 4.21, for the different values of kh and S; the reader can verify that precisely the same stable flowing pressure ($p_{\text{wf}} = 3451.50$ psia) will be calculated in each case. Consequently, all three lines selected will mathematically satisfy both drawdown and buildup equations and, indeed, there is an infinite number of compatible values of kh and S that will do so. The correct straight line must be therefore selected for sound physical rather than mathematical reasons: it is the first definable linear section and has the same duration as the transient response of the drawdown period.

Horner. The buildup plot and its derivative are shown in Fig. 4.48. The correct linear section of the buildup is between points A and B. This corresponds to the MDH transient response terminating at $\Delta t = 0.161$ hours (10 minutes) and has the same slope of $m = 6.05$ psi/log cycle. As explained in section 4.13, for a buildup following

TABLE 4.11

Demonstrating the lack of uniqueness in a steady-state buildup

Line	Slope (psi/cycle)	kh (mD·ft)	k (mD)	S	r_e (ft)	Δt_s (hours)	p_i (psia)
A–B	6.05	97550	1625	6.5	840	1.54	3524.4
X	4.45	132620	2210	10.9	1537	3.78	3524.5
Y	2.16	273220	4554	29.8	4848	11.89	3524.2

a flow period in which there is any degree of pressure support, the linear part of the buildup extrapolates to give a value of Z^* at infinite closed-in time which is in excess of the initial pressure, p_i. In this case, $Z^* = 3527.65$ psia which is 3 psi higher than the initial pressure. If the early, transient part of the buildup $(A–B)$ is identified correctly using both the pressure and derivative data, the results of the interpretation will be precisely the same as obtained from the MDH analysis. In particular, using the combination of $Z^* = 3527.65$ psia and the effective flowing time of $t = 5.25$ hours in equation 4.50 leads to the calculation of the skin factor as:

$$S = 1.151 \left[\frac{3527.65 - 3451.5}{6.05} - \log \frac{1625 \times 5.25}{0.2 \times 0.7 \times 15 \times 10^{-6} \times 0.51^2} + 3.23 \right]$$
$$= 6.5 \tag{4.50}$$

which is just the same as calculated from the MDH interpretation.

Unfortunately, in the analysis, the engineer mistakenly chose the section $C–D$ as being the transient part of the buildup, as is so often the case in this type of test. The reasons for this selection were that the element $C–D$ appeared to be a much "better quality" straight line than $A–B$ lasting for 4.2 hours ($\Delta t = 1.294 - 5.494$ hours) rather than a meagre 10 minutes. Furthermore, section $C–D$ appeared to extrapolate to give a value of $Z^* \sim p_i$ which seemed to be a comforting feature — although quite mistaken, as described in section 4.13, since this only occurs in a purely transient test. The derivative plot, although demonstrating considerable scatter does indicate a "reasonable" plateau corresponding to section $C–D$ of the

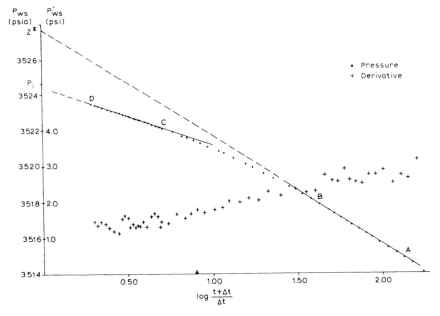

Fig. 4.48. Horner plot (Exercise 4.3).

buildup.

It is worthwhile investigating the magnitude of the errors in calculating the formation properties using the incorrect choice of straight line. The slope of section $C-D$ (Fig. 4.48) is $m = 3.25$ psi/log cycle giving by proportionality

$$kh = 181590 \text{ mD ft:} \quad k = 3025 \text{ mD}$$

which is nearly double the correct value of $k = 1625$ mD. Using the extrapolated value of $Z^* = 3524.40$ psia for line $C-D$, the skin factor resulting from the application of equation 4.50, is

$$S = 1.151 \left[\frac{3524.4 - 3451.5}{3.25} - \log \frac{3025 \times 5.25}{0.2 \times 0.7 \times 15 \times 10^{-6} \times 0.51^2} + 3.23 \right] = 17.5$$

which is almost three times the correct value. The pressure drop across this skin is

$$\Delta p_{skin} = 0.87 \times 3.25 \times 17.5 = 49.5 \text{ psi} \tag{4.53}$$

and hence PI_{ideal} (equation 4.1) is calculated as 180.0 stb/d/psi which is 65% too large. The final error made by the operator was to use the conventional formula, equation 4.81, to calculate the radius of investigation, using the full flowing time, as

$$r_e = 0.03 \sqrt{\frac{kt}{\phi \mu c}} = 0.03 \sqrt{\frac{3025 \times 5.25}{0.2 \times 0.7 \times 15 \times 10^{-6}}} = 2600 \text{ ft} \tag{4.81}$$

instead of 840 ft.

Conclusion

If the initial straight line is correctly defined on either the MDH or Horner interpretation plots then, obviously, the exact same results will be obtained in terms of kh-product and skin factor. The danger is, with the Horner analysis, that the initial straight line is overlooked in favour of a more convincing later, linear development which extrapolates towards the initial pressure and so can understandably deceive engineers. In this respect, the MDH plotting technique is considered the safer method of unambiguously defining the early linear buildup trend which, in this case, represents the condition of pure transience.

4.19. WELL TESTING IN DEVELOPED FIELDS

This section describes interpretation techniques for well tests conducted on a routine basis throughout the lifetime of a field. These are invariably pressure buildup tests the main aims of which have already been outlined in section 4.4a as: the determination of the average pressure $\bar{p}$, within the drainage area of each well and the reservoir as a whole and also to monitor any change in the well's PI which measures its flow efficiency. There are two methods of analysing pressure buildups in mature production wells: that of Horner–Matthews, Brons, Hazebroek

(MBH) [20] and the equivalent approach of Miller, Dyes, Hutchinson (MDH)–
Dietz [41]. These are both described below, in the first instance for wells draining
from completely bounded systems and later this constraint is relaxed to cater for
mixed boundary conditions in which there can be a combination of closed and open
boundaries, the latter giving rise to pressure support.

(a) Pressure buildup analysis method of Horner-MBH for bounded reservoir systems

This subject has already been extensively covered in reference 3, but will be
described again briefly here for completeness. In applying the method, it is assumed
that the well being tested is completely surrounded by a no-flow boundary such that
there is no influx of fluids and production leads to a continuous pressure decline.
The physical reality of this model has already been described in section 4.6a, that if
wells in a depletion type field are produced at reasonably steady rates then each will
carve-out its own drainage area surrounded by a no-flow boundary such that when
the semi-steady-state condition prevails ($dp_{wf}/dt=$ constant) the oil volume drained
by each well is directly proportional to its production rate.

This type of system was comprehensively described by Matthews, Brons and
Hazebroek in 1954 in one of the classic papers [20] on the subject of well testing.
They applied the method of images (section 4.16a) to wells located at different
positions within nine geometrical bounded shapes (Fig. 4.50).

What MBH were aiming at was devising a method of calculating the average
well pressure, $\bar{p}$, in cases when the well had not been closed-in for long enough
to observe the flattening of the pressure response that eventually occurs, as shown
in Fig. 4.49a. They expressed their results in terms of dimensionless pressure plots
for the different geometrical shapes and degrees of well asymmetry. A schematic of
such plots for a well at the centre of a square and one located in one quadrant is
shown in Fig. 4.49b. Since these functions have been published in such a widespread
manner in the literature, both in graphical [3–5] and tabular [5,42] presentations, it
is not the intention to reproduce them again in this text but merely to describe their
physical significance and application. The MBH dimensionless pressure is defined as

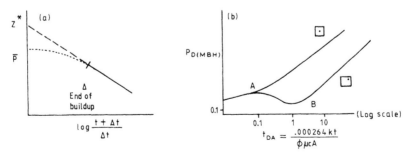

Fig. 4.49. (a) Typical Horner buildup plot for a bounded system; (b) MBH dimensionless pressure
functions for a well draining from within a bounded square.

$$p_{D\,(MBH)} = 0.0142\frac{kh}{q\mu B_o}(Z^* - \overline{p}) = 2\sigma(Z^* - \overline{p}) \tag{4.93}$$

and to appreciate how they were generated in the first place it is necessary to combine the material balance for a well in a bounded drainage volume with the defining expression for Z^*, equation 4.49. The former may be stated as

$$\frac{5.615}{24}q B_o t = cAh\phi(p_i - \overline{p})$$

and multiplying both sides of this by σ, equation 4.29, and rearranging the terms gives

$$7.08 \times 10^{-3}\frac{kh}{q\mu B_o}(p_i - \overline{p}) = 0.00166\frac{kt}{\phi\mu cA} = 2\pi \times \frac{0.000264\,kt}{\phi\mu cA}$$

$$\sigma(p_i - \overline{p}) = 2\pi t_{DA} \tag{4.94}$$

in which t_{DA} is the dimensionless time used in conjunction with MBH analysis and is related to the wellbore dimensionless time, t_D (equation 4.19) as follows

$$t_{DA} = \frac{0.000264\,kt}{\phi\mu cA} = t_D\frac{r_w^2}{A} \tag{4.95}$$

Equation 4.94 is the statement of material balance for depletion type reservoirs and demonstrates the important fact that buildups in such reservoirs must eventually flatten to give a plateau whose value is dependent on the flowing time, as shown in Fig. 4.49b and illustrated in the example well test (section 4.19d, Fig. 4.57). This implies that the extrapolation of any early linear trend on a Horner plot to infinite closed-in time, in the hope of determining a meaningful reservoir pressure, which is a common practice, is likely to overestimate the average pressure compared to the true plateau value.

Subtracting the defining expression for Z^*

$$\sigma(p_i - Z^*) = p_D(t_D) - \tfrac{1}{2}\ln\frac{4\,t_D}{\gamma} \tag{4.49}$$

from equation 4.94 yields

$$\sigma(Z^* - \overline{p}) = 2\pi t_{DA} - p_D(t_D) + \tfrac{1}{2}\ln\frac{4\,t_D}{\gamma} \tag{4.96}$$

But, the left-hand side of this equation is $\tfrac{1}{2}p_{D\,(MBH)}(t_{DA})$, consequently, equation 4.96 may be solved for $p_D(t_D)$ to give

$$p_D(t_D) = 2\pi t_{DA} + \tfrac{1}{2}\ln\frac{4\,t_D}{\gamma} - \tfrac{1}{2}p_{D\,(MBH)}(t_{DA}) \tag{4.97}$$

This is a most important expression [43] for it provides a means of generating a p_D-function for any defined, bounded system for all values of the flowing time

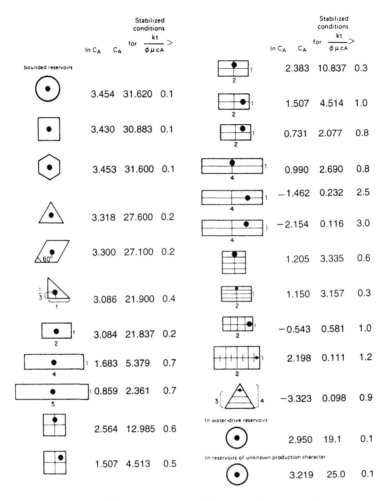

Fig. 4.50. MBH bounded geometrical shapes, Dietz shape factors.

covering: transience, late transience and semi-steady state, all in one simple equation which relies for evaluation on the MBH dimensionless pressure functions. In evaluating these functions in the first place, Matthews, Brons and Hazebroek separately determined p_D-functions by applying the method of images, as mentioned above, to the geometries shown in Fig. 4.50.

This requires the superposition of exponential integral solutions of the radial diffusivity equation for an infinite network of image wells which simulates the presence of a total no-flow boundary surrounding the real well. Therefore, removal of the boundary leaves the real well and network of image wells in an otherwise infinite acting system, just as described in the case of single fault analysis in section 4.16a. The general p_D-function for any of the MBH geometries may then be expressed as

$$p_D(t_D) = \tfrac{1}{2} \ln \frac{4\, t_D}{\gamma} + \tfrac{1}{2} \sum_{j=2}^{\infty} ei \left(\frac{\phi \mu c d_j^2}{4 \times 0.000264\, kt} \right) \tag{4.98}$$

in which the first term on the right-hand side represents the component due to the production of the real well (section 4.7) and the second results from the influence of the network of image wells for which the j-th is separated from the real well by distance d_j. The summation is evaluated until addition of further wells has no effect on the calculated value of the p_D-function. In the MBH paper it was values of p_D so determined that were inserted in equation 4.96 to generate the $p_{D\,(MBH)}$-functions which was their means of presenting results. To use these charts or tables to determine $\bar{p}$, the steps are as follows:

- Plot the Horner pressure buildup (Fig. 4.49a) and identify the early linear trend containing the transient pressure response. This permits the calculation of kh, k, and the skin factor and extrapolation to infinite closed-in time determines the value of Z^*.
- Calculate the dimensionless flowing time, t_{DA}, using equation 4.95. Then enter the appropriate MBH chart, selected on the grounds of both geometry and well asymmetry, and for the value of t_{DA} read the ordinate value of $p_{D\,(MBH)}$. Substitute this in the defining equation 4.93 to explicitly calculate $\bar{p}$.

There are difficulties associated with the method, particularly with regard to geometrical definition, that will be deferred until section 4.19e for description. The technique is illustrated in section 4.19d.

To consider further the shapes of the MBH plots for different geometries; in the first place, the time abscissa (t_{DA}, dimensionless flowing time — Fig. 4.49b) is plotted on a logarithmic scale which distorts the more usual concave downward shape of p_D-functions when plotted on a linear time scale. Initially, all the plots have upward curvature and this represents the purely transient pressure response. Eventually, all the plots become linear on the semi-log presentation and when the linearity first occurs it indicates the onset of the semi-steady-state flow condition. In between transience and the later stabilized flow, there is a non-linear section of the plots which represents the highly complex late transient pressure response, first described in section 4.6a, which characterises the period when the boundary conditions are varying with time. For a well located at the centre of one of the regular geometries depicted in Fig. 4.50, the period of late transience does not exist because the boundary, which is uniformly spaced from the well, exerts its influence instantaneously at the wellbore. There is therefore a direct change from transience to semi-steady state. On the other hand, if the well is asymmetrically located within the no-flow boundary there can be an extensive period of late transience ($A–B$, Fig. 4.49b) as the boundaries sequentially exert their influence on pressures in the wellbore. All the complexity contained in the late transient response is expressed in the relatively simple equation 4.97, whose application, in conjunction with the general equation for pressure buildup (equation 4.30), is illustrated in section 4.19d.

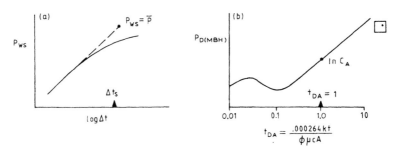

Fig. 4.51. (a) MDH pressure buildup plot for a bounded system. (b) Determination of Dietz shape factors from MBH functions.

(b) Pressure buildup analysis method of MDH-Dietz for bounded reservoir systems

In 1965, Dietz presented a simple but effective method of circumventing the use of the MBH-functions by introducing the concept of shape factors [41]. The method is only appropriate once the semi-steady-state condition prevails but in practice this is usually no severe limitation because wells are generally produced until such stability is achieved between successive pressure buildup surveys. The Dietz shape factors, for different geometrical configurations and well asymmetries are defined in such a way that entering the appropriate MBH chart for the value of $t_{DA} = 1$ (Fig. 4.51b) gives

$$p_{D\,(MBH)}(t_{DA} = 1) = \ln C_A \tag{4.99}$$

In obtaining these values of C_A, for the majority of the charts, the value of $t_{DA} = 1$ occurs after the onset of the linear part of the functions representing semi-steady-state flow but if this is not the case, as for highly asymmetric well locations, then the value of C_A is obtained from the backward extrapolation of the eventual linear trend to the value of $t_{DA} = 1$. The shape factors and the dimensionless times when semi-steady state commences are listed in Fig. 4.50 in which Dietz's original values [41] have been updated by those presented by Earlougher [42]. As noted in the previous section, the MBH plots are linear when semi-steady-state conditions prevail but not only that, they all have a unit slope in the semi-log presentation such that

$$\frac{dp_{D\,(MBH)}}{d\ln t_{DA}} = 1$$

Consequently, once stability has been achieved, it follows from equation 4.99 that the MBH-functions may be simply evaluated as

$$p_{D\,(MBH)}(t_{DA}) = \ln(C_A t_{DA}) = 2.303 \log(C_A t_{DA}) \tag{4.100}$$

If this relationship for stabilized flow is inserted into the general expression for generating p_D-functions for bounded systems, equation 4.97, then some slight algebraic manipulation will reveal that

$$p_{\mathrm{D}} = 2\pi t_{\mathrm{DA}} + \tfrac{1}{2}\ln\frac{4A}{\gamma C_{\mathrm{A}} r_{\mathrm{w}}^2} \tag{4.101}$$

which applies for semi-steady-state flow: the first term representing the material balance component while the second is purely geometrical.

In analysing pressure buildup tests using the combination of MDH and Dietz, the first step is to plot the buildup (p_{ws} versus $\log \Delta t$) and identify the early straight line from which kh, k and S can be determined in the usual manner (section 4.12). Following this, it is necessary to calculate the value of Δt_s (Fig. 4.51a) which represents the closed-in time to which the initial straight line must be extrapolated to reach the average pressure, $\bar{p}$, which is determined by setting $p_{\mathrm{wsl}} = \bar{p}$ and $\Delta t_{\mathrm{D}} = \Delta t_{\mathrm{Ds}}$ in the MDH linear equation 4.33, giving

$$\sigma(p_{\mathrm{i}} - \bar{p}) = p_{\mathrm{D}}(t_{\mathrm{D}}) - \tfrac{1}{2}\ln\frac{4\Delta t_{\mathrm{Ds}}}{\gamma} \tag{4.102}$$

But the left-hand side of this equation is simply the material balance, equation 4.94, and evaluating the p_{D}-function for semi-steady-state flow conditions using equation 4.101, gives

$$2\pi t_{\mathrm{DA}} = 2\pi t_{\mathrm{DA}} + \tfrac{1}{2}\ln\frac{4A}{\gamma C_{\mathrm{A}} r_{\mathrm{w}}^2} - \tfrac{1}{2}\ln\frac{4\Delta t_{\mathrm{Ds}}}{\gamma}$$

which, after cancellation of terms may be solved to give

$$\Delta t_s = 3788\frac{\phi\mu c A}{k C_{\mathrm{A}}} \text{ (hours)} \tag{4.103}$$

This is an interesting result for it states that — provided a well has been produced at a stable rate for a sufficient period so that the semi-steady condition prevails prior to closure — the buildup can be analysed to evaluate the average pressure, $\bar{p}$, without involvement of the flowing time, t. Similarly, using the Horner plot, there is an equivalent method for determining the time, Δt_s, to which the early, linear buildup trend must be extrapolated to reach $\bar{p}$. That is, extrapolating Horner's linear equation 4.46 to $p_{\mathrm{wsl}} = \bar{p}$ gives

$$\sigma(p_{\mathrm{i}} - \bar{p}) = 2\pi t_{\mathrm{DA}} = 1.151\log\frac{t + \Delta t_s}{\Delta t_s} + p_{\mathrm{D}}(t_{\mathrm{D}}) - \tfrac{1}{2}\ln\frac{4\,t_{\mathrm{D}}}{\gamma} \tag{4.104}$$

and evaluating the p_{D}-function for semi-steady-state conditions, using equation 4.101, gives

$$\log\frac{t + \Delta t_s}{\Delta t_s} = \log(C_{\mathrm{A}} t_{\mathrm{DA}}) \tag{4.105}$$

Alternatively, if the semi-steady-state condition has not been reached at the time of the survey, the times to which the MDH or Horner plots must be extrapolated to reach $\bar{p}$ can be determined by direct substitution of the general p_{D}-function for bounded systems, equation 4.97, into equations 4.102 and 4.104, respectively, giving

$$\text{MDH:} \qquad \Delta t_s = t\,\mathrm{e}^{-p_{\mathrm{D(MBH)}}} \tag{4.106}$$

Horner: $2.303 \log \dfrac{t + \Delta t_s}{\Delta t_s} = p_{D\,(MBH)}(t_{DA})$ (4.107)

Generally, the time between successive buildup surveys is sufficiently long so that the semi-state condition prevails (or, at least, that is the conventional assumption). In this case the MDH-Dietz technique offers the simpler approach to calculating $\bar{p}$, since it avoids the inclusion of the flowing time which can cause unnecessary confusion in interpreting tests in mature producing fields, as described in section 4.19e.

(c) Buildup analysis for systems with constant pressure or mixed boundary conditions

The description of test analysis methods in sections 4.19a and b was restricted to wells that were assumed to be individually surrounded by complete no-flow boundaries, thus being appropriate for depletion type fields operating under primary recovery conditions. If there is an element of natural pressure support or if secondary recovery (water or gas drive) is being practised, which is increasingly common, then it is necessary to modify the methods of MBH and Dietz to accommodate this condition. In this section, test analysis is described for wells which have outer boundaries that are completely open to influx or have a mixture of no-flow and open boundaries. With the promotion of pressure maintenance schemes in the Industry, there has been an ever increasing interest in such analysis in recent years and perhaps one of the clearest and simplest approaches to the subject is that presented by Larsen [8] in a paper written in 1984. Since the technique relies heavily on the theory relating to bounded systems, described in sections 4.19a and b, the reader would be advised to be quite familiar with the contents of these sections before proceeding.

The method applies the principle of superposition to generate a p_D-function for a constant pressure/mixed boundary condition system by superposing on the geometry an equivalent totally bounded system with the same shape. The result is the generation of a new shape which is completely bounded. A simple example for determining the p_D-function for a well at the centre of a constant pressure square is shown in Fig. 4.52 and demonstrates the technique and nomenclature.

In the schematic, superposition of the required (R) constant pressure square with the equivalent old (O) fully bounded system means that injection and production wells will eliminate one another resulting in a new (N) totally bounded system which has double the area and whose real and image wells have double the rate. In terms of p_D-functions, the superposition may be expressed as

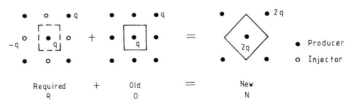

Fig. 4.52. Generation of a p_D-function for a well at the centre of a constant pressure square.

$$p_D^R(t_D, A) = 2p_D^N(t_D, 2A) - P_D^O(t_D, A) \tag{4.108}$$

The factor "2" multiplying the new p_D-function is required because the rate for this configuration on the left-hand side of the defining expression for the function, equation 4.21, has been doubled. The equation is perfectly general in form for all geometries that are amenable to the application of the principle of superposition and therefore requires full evaluation.

Since the two p_D-functions on the right-hand side of equation 4.108 are for totally bounded systems, they may be evaluated for all values of the flowing time by applying equation 4.97: in particular, since $t_{D(2A)} = \frac{1}{2}t_{DA}$, then

$$p_D^N(t_D, 2A) = 2\pi \frac{t_{DA}}{2} + \frac{1}{2}\ln\left(\frac{4\,t_D}{\gamma}\frac{1}{2} \times 2\right) - \frac{1}{2}p_{D\,(MBH)}^N\left(\frac{t_{DA}}{2}, A\right)$$

or

$$p_D^N(t_D, 2A) = \frac{1}{2}\ln 2 + p_D^N\left(\frac{t_D}{2}, A\right) \tag{4.109}$$

and substituting this in equation 4.108 while again applying equation 4.97 for bounded systems gives

$$p_D^R(t_D, A) = \ln 2 + 2\pi t_{DA} + \ln\frac{4\,t_D}{\gamma 2} - p_{D\,(MBH)}^N\left(\frac{t_{DA}}{2}, A\right)$$

$$- 2\pi t_{DA} - \frac{1}{2}\ln\frac{4\,t_D}{\gamma} + \frac{1}{2}p_{D\,(MBH)}^O(t_{DA}, A)$$

which, on gathering terms yields

$$p_D^R(t_D, A) = \frac{1}{2}\ln\frac{4\,t_D}{\gamma} - \frac{1}{2}\left[2p_{D\,(MBH)}^N\left(\frac{t_{DA}}{2}, A\right) - p_{D\,(MBH)}^O(t_{DA}, A)\right] \tag{4.110}$$

This is the p_D-function of the required constant pressure/mixed boundary condition system for any flowing time. In the event that the time is sufficiently long that the $p_{D\,(MBH)}$-functions can be evaluated under semi-steady-state conditions, which is often the case in practice, then substituting equation 4.100 into 4.110 reduces the latter to

$$p_D^R(t_D, A) = \frac{1}{2}\ln\frac{4\,t_D}{\gamma} - \frac{1}{2}\left[2\ln C_A^N\frac{t_{DA}}{2} - \ln C_A^O t_{DA}\right]$$

$$= \frac{1}{2}\ln\frac{4\,t_D}{\gamma} - \frac{1}{2}\ln\left[\frac{(C_A^N)^2}{C_A^O}\frac{t_{DA}}{4}\right]$$

and defining a new shape factor appropriate for constant pressure/mixed boundary conditions [8] as

$$C_A' = (C_A^N)^2/C_A^O \tag{4.111}$$

in which C_A^N and C_A^O relate to the new and old bounded geometrical shapes, will finally reduce the equation to

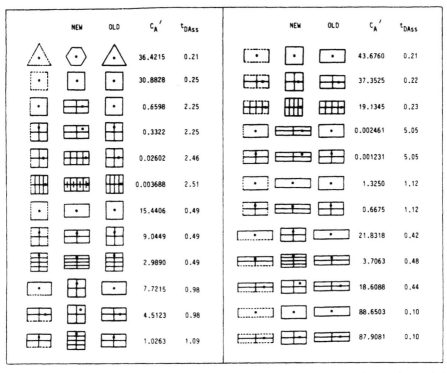

Fig. 4.53. New and old bounded geometries required to generate p_D-functions for steady-state/mixed boundary condition systems (after Larsen).

$$p_D^R(t_D, A) = \tfrac{1}{2}\ln\frac{16\,A}{\gamma\,C_A'\,r_w^2} \tag{4.112}$$

which, on account of the lack of time dependence, will be recognised as the p_D-function appropriate when steady-state conditions eventually prevail in the system. In his paper [8] Larsen acknowledges that the most difficult aspect in applying the technique is in trying to establish the shape of the "new" bounded system and position of the well with respect to the boundary. He has therefore supplied a useful chart (Fig. 4.53) indicating these geometries and well asymmetries, together with shape factors, C_A', and dimensionless times, $t_{DA_{ss}}$, when application of equation 4.112 for steady-state conditions first becomes appropriate. In viewing this chart, which caters for most of the MBH-Dietz shapes (Fig. 4.50), it should be recalled that the area of the "new" geometrical shape is always double that of the "old". The value of Larsen's work is that whereas equation 4.112 had previously been defined only for simple systems, such as a well at the centre of a square [7], his paper extends the application of the equation to all the geometrical configurations shown in Fig. 4.53 for which the C_A' values are provided. It will also be recognised that for the simple case of an appraisal well test in an infinite acting reservoir under steady-state conditions (section 4.18b), the p_D-function for such a system, $\ln r_e/r_w$,

is practically the same as evaluated from the general expression, equation 4.112
$(C'_A = C_A = 31.62, A = \pi r_e^2)$.

The pressure buildup may be analysed using either the plotting technique of
Miller, Dyes, Hutchison or Horner: identification of the early, linear buildup trend
permitting the calculation of the kh-product and skin factor. Using the MDH plot,
the reservoir pressure can be determined by extrapolating the early, linear trend
(equation 4.33) to $p_{wsl} = p_i$ which will be reached at a closed-in time of Δt_s.
Extrapolation of the equation gives

$$p_D(t_D) = \tfrac{1}{2} \ln \frac{4 \Delta t_{Ds}}{\gamma} \qquad (4.38)$$

and provided the flow period has been sufficiently long so that steady-state conditions prevail, then the p_D-function can be evaluated using equation 4.112 giving

$$\tfrac{1}{2} \ln \frac{16A}{\gamma C'_A r_w^2} = \tfrac{1}{2} \ln \frac{4 \Delta t_{Ds}}{\gamma}$$

which can be reduced to

$$\Delta t_{DA_s} = \frac{4}{C'_A} \text{ or } \Delta t_s = 15152 \frac{\phi \mu c A}{k C'_A} \qquad (4.113)$$

Similarly, when using the Horner plot an equivalent closed-in time can be
determined when the early straight line defined by equation 4.45 extrapolates to the
level $p_{wsl} = p_i$ giving

$$\tfrac{1}{2} \ln \frac{t + \Delta t_s}{\Delta t_s} = \tfrac{1}{2} \ln \frac{4 t_D}{\gamma} - p_D(t_D) \qquad (4.114)$$

and again, evaluating the p_D-function for steady-state flow, using equation 4.112
yields

$$\log \frac{t + \Delta t_s}{\Delta t_s} = \log \frac{C'_A t_{DA}}{4} \qquad (4.115)$$

Again it will be noted that provided the well is flowing under stabilized conditions
prior to closure, the extrapolation of the MDH straight line to give the initial static
pressure, equation 4.113, is quite independent of the flowing time, whereas the
extrapolation of the Horner linear trend, equation 4.115, contains this additional
complication. This matter is referred to further in section 4.19e. In the event
that stable conditions have not been attained during the flow period, then the
p_D-functions in equations 4.38 and 4.114 would require evaluation using the more
complex expression: equation 4.110.

In secondary recovery schemes, such as engineered waterdrive, the pressure
is usually dropped by an amount $\Delta p = p_i - \bar{p}$ and maintained at the lower
level throughout the flood. In this event, the above equations relating to pressure
drawdown/buildup analysis are equally relevant with the initial pressure p_i replaced
by the reduced average pressure $\bar{p}$.

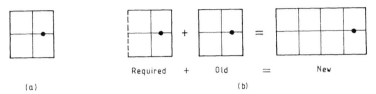

Fig. 4.54. (a) Bounded square geometry (test A). (b) Square with one open boundary illustrating the generation of a "new" bounded geometry by superposition (Test B).

(d) Example well test

In this section theoretical well tests are analysed using both completely bounded and mixed boundary conditions. The purpose is to illustrate the method of generation of p_D-functions and to demonstrate the techniques described in sections 4.19a–c for analysing such tests using the buildup plots of MDH and Horner. The first geometry considered (test A) is that of a well located in one half of a completely bounded square, Fig. 4.54a, while the second (test B) is for the same geometry and well asymmetry but with the boundary remote from the well open to pressure support, Fig. 4.54b. Also shown is the superposition of a well in a completely bounded square to produce the "new" geometrical bounded system which, as shown in Figs. 4.53 and 4.54b is a 2:1 rectangle of twice the area of the square with the well located one eighth of the distance from one end.

The reservoir and fluid properties are as follows

k $= 75$ mD $B_o = 1.35$ rb/stb
h $= 100$ ft $\mu = 1.5$ cp
$p_i = 6000$ psia $c = 19 \times 10^{-6}$/psi
$A = 120$ acres $= 5.227 \times 10^6$ ft² $r_w = 0.51$ ft
$\phi = 0.23$

The skin factor is zero and there are no afterflow effects. The well's history consists of six months (4380 hours) of production at an average rate of 2500 stb/d, prior to closure for a pressure buildup. The steps taken in analysing the well's performance are as follows:

(A) Generate p_D-functions for both boundary conditions and inspect the flowing pressures
(B) Use the p_D-functions to predict theoretical buildup performance using the general buildup equation 4.30.
(C) Analyse the buildups using the MDH and Horner plotting techniques

Bounded reservoir (test A)

The analysis data for this boundary condition are presented in Table 4.12. The first two columns are the dimensionless time, t_{DA}, and $p_{D\,(MBH)}$-function taken from Earlougher's paper [42]. It is customary to work with at least four places of decimal to minimise round off errors in the calculations. The two dimensionless times

TABLE 4.12

Generation of p_D-function, flowing and buildup pressures for the bounded square (the time in column 3 is used for both flowing and buildup periods)

1	2	3	4	5	6	7	8	9	10
t_{DA}	$p_{D\,(MBH)}$	Time (hours)	p_D	p_{wf} (psia)	$p_D(t_D + \Delta t_D)$	Δp_D	p_{ws} (psia)	$\log \Delta t$	$\dfrac{\log(t + \Delta t)}{\Delta t}$
0.0010	0.0126	1.730	5.3580	5489.7	23.4403	18.0823	4277.8	0.238	3.404
0.0015	0.0188	2.596	5.5610	5470.4	23.4435	17.8825	4296.9	0.414	3.227
0.0200	0.0251	3.461	5.7048	5456.7	23.4466	17.7418	4310.3	0.539	3.103
0.0030	0.0314	5.191	5.9106	5437.1	23.4529	17.5423	4329.3	0.715	3.927
0.0040	0.0503	6.922	6.0513	5423.7	23.4592	17.4079	4342.1	0.840	2.802
0.0050	0.0628	8.652	6.1629	5413.1	23.4655	17.3026	4352.1	0.937	2.705
0.0060	0.0754	10.38	6.2539	5404.4	23.4717	17.2178	4360.2	1.016	2.626
0.0070	0.0880	12.11	6.3310	5397.1	23.4780	17.1470	4367.0	1.083	2.560
0.0080	0.1005	13.84	6.3978	5390.7	23.4843	17.0865	4372.7	1.141	2.502
0.0090	0.1130	15.57	6.4567	5385.1	23.4906	17.0339	4377.7	1.192	2.451
0.0100	0.1254	17.30	6.5095	5380.1	23.4969	16.9874	4382.2	1.238	2.405
0.0200	0.2402	34.61	6.8616	5346.5	23.5597	16.6981	4409.7	1.539	2.106
0.0300	0.3333	51.91	7.0805	5325.7	23.6225	16.5420	4424.6	1.715	1.931
0.0400	0.4108	69.22	7.2485	5309.7	23.6854	16.4369	4434.6	1.840	1.808
0.0500	0.4791	86.52	7.3887	5296.3	23.7482	16.3595	4442.0	1.936	1.714
0.0600	0.5413	103.8	7.5114	5284.6	23.8109	16.2995	4447.7	2.016	1.635
0.0700	0.5991	121.1	7.6224	5274.1	23.8737	16.2513	4452.3	2.083	1.570
0.0800	0.6531	138.4	7.7249	5264.3	23.9366	16.2117	4456.0	2.141	1.514
0.0900	0.7038	155.7	7.8213	5255.1	23.9994	16.1781	4459.2	2.192	1.464
0.1000	0.7516	173.0	7.9128	5246.4	24.0622	16.1494	4462.0	2.238	1.420
0.1500	0.9583	259.6	8.3268	5207.0	24.3766	16.0498	4471.4	2.414	1.252
0.2000	1.1314	346.1	8.6980	5171.6	24.6907	15.9927	4476.9	2.539	1.135
0.2500	1.2854	432.6	9.0465	5138.4	25.0048	15.9583	4480.2	2.636	1.046
0.3000	1.4257	519.1	9.3815	5106.5	26.3188	15.9373	4482.2	2.715	0.975
0.4000	1.6720	692.2	10.0306	5044.7	25.9474	15.9168	4484.1	2.840	0.865
0.5000	1.8797	865.2	10.6663	4984.2	26.5755	15.9092	4484.8	2.937	0.783
0.6000	2.0563	1038	11.2963	4924.2	27.2029	15.9066	4485.1	3.016	0.718
0.7000	2.2083	1211	11.9254	4864.2	27.3311	15.9057	4485.2	3.083	0.664
0.8000	2.3411	1384	12.5537	4804.4	28.4592	15.9055	4485.2	3.141	0.620
0.9000	2.4586	1557	13.1819	4744.6	29.0874	15.9055	4485.2	3.192	0.581
1.0000	2.5638	1730	13.8099	4684.8	29.7155	15.9056	4485.2	3.238	0.548
2.0000	3.2569	3461	20.0936	4086.3	36.0006	15.9070	4485.1	3.539	0.355

required are

$$t_D = \frac{0.000264\,kt}{\phi\mu c r_w^2} \quad \frac{0.000264 \times 75 \times t}{0.23 \times 1.5 \times 19 \times 10^{-6} \times 0.51^2} = 11613t \tag{4.19}$$

$$t_{DA} = \frac{t_D r_w^2}{A} = \frac{11613 \times 0.51^2 \times t}{5.227 \times 10^6} = 5.7787 \times 10^{-4} t \tag{4.95}$$

From the latter relationship, the real time has been calculated (column 3) and then the p_D-function, using equation 4.97 for the totally bounded system (column 4), as

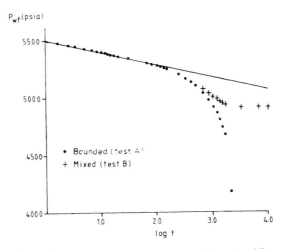

Fig. 4.55. Flowing pressure performance, Tests A and B.

$$p_D = 3.6309 \times 10^{-3}t + 5.084 + \tfrac{1}{2}\ln t - \tfrac{1}{2}p_{D\,(MBH)}(t_{DA})$$

The flowing pressure is next calculated using the defining expression for the p_D-function, equation 4.21 as

$$7.08 \times 10^{-3}\frac{kh}{q\mu B_o}(p_i - p_{wf}) = p_D(t_D) + S \tag{4.21}$$

$$\frac{7.08 \times 10^{-3} \times 75 \times 100}{2500 \times 1.5 \times 1.35}(6000 - p_{wf}) = p_D(t_D) + 0$$

$$p_{wf} = 6000 - \frac{p_D(t_D)}{0.0105} \tag{4.116}$$

The flowing pressures (column 5) are shown in Fig. 4.55, plotted in terms of p_{wf} versus $\log t$. The initial response during the period of transience (equation 4.20) lasts for about 40 hours ($\log t = 1.6$) after which the pressures fall below the linear trend, representing the depletion condition (section 4.19a). The slope of the early straight line is

$$m = 162.6\frac{q\mu B_o}{kh} = \frac{162.6 \times 2500 \times 1.5 \times 1.35}{75 \times 100} = 109.8 \text{ psi/cycle}$$

The pressure buildup is evaluated by applying equation 4.30

$$\sigma(p_i - p_{ws}) = p_D(t_D + \Delta t_D) - p_D(\Delta t_D) = \Delta p_D \tag{4.30}$$

$$0.0105\,(6000 - p_{ws}) = \Delta p_D \tag{4.117}$$

Semi-steady-state conditions prevail after a dimensionless time $t_{DA} = 0.6$ (Fig. 4.50): $t = 1038$ hours and therefore the dimensionless pressure $p_D(t_D + \Delta t_D)$ can be

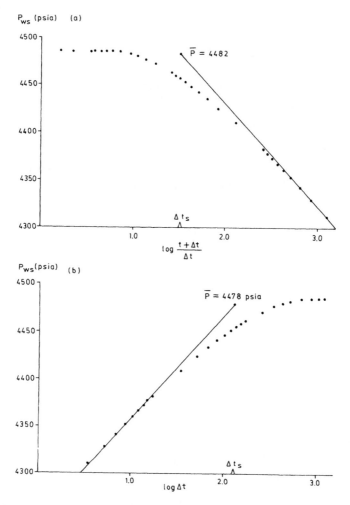

Fig. 4.56. (a) Horner and (b) MDH buildups — bounded reservoir (Test A).

evaluated using equation 4.101 ($C_A = 12.985$, Fig. 4.50)

$$p_D(t_D + \Delta t_D) = 2\pi(t_{DA} + \Delta t_{DA}) + \tfrac{1}{2}\ln\frac{4A}{\gamma C_A r_w^2}$$
$$= 3.6309 \times 10^{-3}(t + \Delta t) + 7.5307$$

and these figures are listed in column 6 of Table 4.12 and values of Δp_D (column 6–column 4) in column 7, leading to the determination of p_{ws} as listed in column 8. Both the Horner and MDH plots are presented in Figs. 4.56a and b; both have slopes of approximately 109 psi/log cycle. The flowing time of 4380 hours represents a dimensionless time of $t_{DA} = 2.5311$ (equation 4.95). Therefore, the average pressure may be determined by extrapolation of the early, linear trend of the Horner plot to

$$\log \frac{t + \Delta t_s}{\Delta t_s} = \log(C_A t_{DA}) = \log(12.985 \times 2.5311) = 1.517 \tag{4.105}$$

giving a value of $\overline{p} = 4482$ psia (Fig. 4.56a) which is 3 psia below the actual value of 4485 psia (Table 4.12). Similarly, the MDH plot must be extrapolated to

$$\Delta t_S = 3788 \frac{\phi \mu c A}{k C_A} = \frac{3788 \times 0.23 \times 1.5 \times 19 \times 10^{-6} \times 5.227 \times 10^6}{75 \times 12.985} \tag{4.103}$$

that is $\Delta t_s = 133.27$ hours, $\log \Delta t_s = 2.125$ for which $\overline{p} = 4478$ psia (Fig. 4.56b). As can be seen from the buildup plots, the well would need to be closed-in for about 500 hours to observe the flattening of the pressures towards the average value $\overline{p}$, whereas the early linear trend is defined for only the first 17 hours. Consequently, a buildup of relatively short duration would be sufficient to define $\overline{p}$ for this system.

Mixed boundary condition (test B)

Values of t_{DA} and $p_{D(MBH)}$ have been taken from reference 42 and are listed in columns 1 and 2 of Table 4.13. The $p_{D(MBH)}$-function relates to the "new", 2:1 bounded rectangular geometry shown in Fig. 4.54b. (N.B.: In table 2 of reference 42, it appears that the $p_{D(MBH)}$ data for ☐• and ☐• have been accidentally listed in the wrong columns and need reversing). The p_D-function is column 4 is the "required" value for the mixed boundary condition and is evaluated using equation 4.110 expressed as

$$p_D(t_D) = 5.084 + \tfrac{1}{2} \ln t - \tfrac{1}{2} \left[2 p_{D(MBH)}^N \left(\frac{t_{DA}}{2} \right) - p_{D(MBH)}^O(t_{DA}) \right] \tag{4.110}$$

in which the first term in parenthesis is evaluated using column 2 of Table 4.13 and the second from column 2 of Table 4.12. The "required" p_D-function has then been used in equation 4.116 to calculate the flowing bottom hole pressure, p_{wf}, during the production period. The drawdown history is plotted in Fig. 4.55 and demonstrates that the (water) influx through the single open boundary does eventually provide stability of pressure. From Fig. 4.53, it can be seen that the onset of the steady-state pressure response should occur for a value of $t_{DA_{ss}} = 2.46$ and, using equation 4.95, this corresponds to a flowing time of $t = 4257$ hours, which is just short of the total flowing time of 4380 hours. On account of this condition, the value of the function $p_D(t_D + \Delta t_D)$ in the buildup equation 4.30 can be evaluated for the steady-state condition applying equation 4.112 in which (Fig. 4.50)

$$C_A' = (C_A^N)^2 / C_A^O = (0.581)^2 / 12.985 = 0.026 \tag{4.111}$$

Therefore,

$$p_D(t_D + \Delta t_D) = \tfrac{1}{2} \ln \frac{16 A}{\gamma C_A' r_w^2} = \tfrac{1}{2} \ln \frac{16 \times 5.227 \times 10^6}{1.781 \times 0.026 \times 0.51^2} = 11.3306 \tag{4.112}$$

The buildup equation may then be evaluated as

TABLE 4.13

Generation of p_D-function, flowing and buildup pressures for the mixed boundary condition (the time in column 3 is used for both flowing and buildup periods

1	2	3	4	5	6	7	8	9
t_{DA}	$p_{D\,(MBH)}$	Time (hours)	p_D (time)	p_{wf} (psia)	Δp_D	p_{ws} (psia)	$\log \Delta t$	$\dfrac{\log(t + \Delta t)}{\Delta t}$
0.0010	0.0126	1.730					0.238	3.404
0.0015	0.0188	2.596	5.5547	5471.0	5.7759	5449.9	0.414	3.227
0.0020	0.0251	3.461	5.7047	5456.7	5.6259	5464.2	0.539	3.103
0.0030	0.0314	5.191	5.9044	5437.7	5.4262	5483.2	0.715	3.927
0.0040	0.0502	6.922	6.0514	5423.7	5.2792	5497.2	0.840	2.802
0.0050	0.0626	8.652	6.1660	5412.8	5.1646	5508.1	0.937	2.705
0.0060	0.0745	10.38	6.2602	5403.8	5.0704	5517.1	1.016	2.626
0.0070	0.0858	12.11	6.3342	5396.7	4.9964	5524.2	1.083	2.560
0.0080	0.0962	13.84	6.3978	5390.7	4.9328	5530.2	1.141	2.502
0.0090	0.1050	15.57	6.4568	5385.1	4.8738	5535.8	1.192	2.451
0.0100	0.1144	17.30	6.5095	5380.1	4.8211	5540.9	1.238	2.405
0.0200	0.1589	34.61	6.8618	5346.5	4.4688	5574.4	1.539	2.106
0.0300	0.1633	51.91	7.0888	5324.9	4.2418	5596.0	1.715	1.931
0.0400	0.1492	69.22	7.2491	5309.6	4.0815	5611.3	1.840	1.808
0.0500	0.1224	86.52	7.3926	5295.9	3.9380	5625.0	1.936	1.714
0.0600	0.0862	103.8	7.5126	5284.5	3.8180	5636.4	2.016	1.635
0.0700	0.0437	121.1	7.6256	5273.8	3.7050	5647.1	2.083	1.570
0.0800	−0.0028	138.4	7.7264	5264.2	3.6042	5656.7	2.141	1.514
0.0900	−0.0512	155.7	7.8241	5254.8	3.5065	5666.0	2.192	1.464
0.1000	−0.1004	173.0	7.9140	5246.3	3.4166	5674.6	2.238	1.420
0.1500	−0.3322	259.6	8.3223	5207.4	3.0833	5713.5	2.414	1.252
0.2000	−0.5189	346.1	8.6735	5174.0	2.6571	5746.9	2.539	1.135
0.2500	−0.6580	432.6	8.9779	5145.0	2.3527	5775.9	2.636	1.046
0.3000	−0.7555	519.1	9.2551	5118.6	2.0755	5802.3	2.715	0.975
0.4000	−0.8547	692.2	9.7088	5076.1	1.6218	5845.5	2.840	0.865
0.5000	−0.8671	865.2	10.0633	5041.3	1.2673	5879.3	2.937	0.783
0.6000	−0.8284	1038	10.3402	5015.2	0.9904	5905.7	3.016	0.718
0.7000	−0.7620	1211	10.5429	4995.9	0.7877	5925.0	3.083	0.664
0.8000	−0.6820	1384	10.7256	4978.5	0.6050	5942.4	3.141	0.620
0.9000	−0.5969	1557	10.8496	4966.7	0.4810	5954.2	3.192	0.581
1.0000	−0.5115	1730	10.9609	4956.1	0.3697	5964.8	3.238	0.548
2.0000	0.1507	3461	11.2986	4923.9	0.0320	5997.0	3.539	0.355
4.0000	0.8436	6922	11.3296	4921.0	0.0010	6000.0	3.840	0.213
8.0000	1.5370	13844	11.3300	4921.0	0.0006	6000.0	4.141	0.119

$$0.0105(6000 - p_{ws}) = 11.3306 - p_D(\Delta t_D) = \Delta p_D \qquad (4.118)$$

In which $p_D(\Delta t_D)$ is taken from column 4 of Table 4.13 and the values of Δp_D and p_{ws} are listed in columns 6 and 7 respectively.

The Horner and MDH buildup plots are shown in Fig. 4.57a and b, the initial, linear trend of each has the required slope of approximately 109 psi/log cycle. Thereafter, there is sharp upward curvature in each plot until the initial plateau pressure of $p_i = 6000$ psia is reached. In spite of the stability of pressure towards

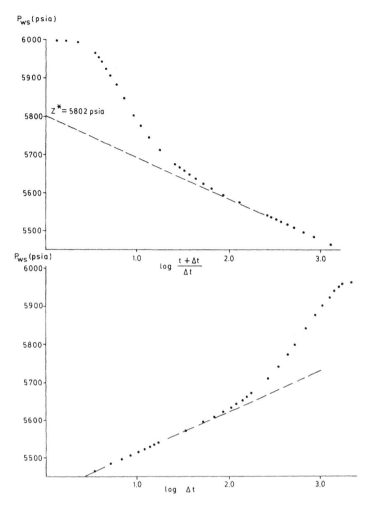

Fig. 4.57. (a) Horner and (b) MDH buildups — mixed boundaries (Test B).

the end on the flow period (Fig. 4.55) the initial buildup points are depressed with respect to the infinite reservoir case and therefore the initial linear buildup trends fall below those for an infinite acting system, as described in sections 4.12 and 13. Both early linear trends can be extrapolated to determine the initial pressure; in the case of Horner, this is to the value

$$\log \frac{t + \Delta t_s}{\Delta t_s} = \log \frac{C'_A t_{DA}}{4} = \log \frac{0.026 \times 2.5311}{4} = -1.784 \qquad (4.115)$$

and since $Z^* = 5802$ psia (Fig. 4.57a) and the buildup slope is $m = 109.3$ psi/log cycle, then the extrapolation can be calculated to give a value of $p_i = 5997$ psia: 3 psi below the true value. In the case of the MDH plot, the extrapolation is to

$$\Delta t_s = 15152\frac{\phi\mu cA}{kC_A'} = \frac{15152 \times 0.23 \times 1.5 \times 19 \times 10^{-6} \times 5.227 \times 10^6}{75 \times 0.026} \quad (4.113)$$

$\log \Delta t_s = 5.425$, and since the extrapolated pressure is $p_{wsl} = 5733$ psia at $\log \Delta t = 3.0$, and the buildup slope is 110 psi/log cycle, the initial pressure can be calculated as $p_i = 6000$ psia.

(e) Practical difficulties in testing development wells

Sections 4.19a–d described and illustrated the application of techniques for analysing well tests in mature, developed fields requiring the incorporation of complex boundary conditions. There are, however several practical difficulties associated with this type of interpretation which are described in this section.

Flowing time/Z*

These related parameters, as described in section 4.15, cause considerable confusion in initial well testing which, if anything, is exacerbated in development well testing. Difficulties only arise in conjunction with Horner analysis for, provided there is reasonable stability in flowing pressure prior to the buildup, then the more direct MDH-plotting method does not depend on the flowing time, t, nor is there any requirement to extrapolate to infinite closed-in time to determine Z^*, as there is in Horner plotting. As noted in section 4.15a, it is the arbitrary inclusion of the term $\pm 0.5 \ln(t_D + \Delta t_D)$ to the right-hand side of the basic buildup equation 4.30 in generating the Horner linear equation 4.46, that drags both t and Z^* into the analysis.

One difficulty confronting engineers is — suppose a well has been producing for ten years at different flow rates, including periods of closure, what should be used as the flowing time? Is it the total time, the effective flowing time (equation 4.51), the superposed rate-time (equation 4.64) or the time to reach stabilised flow (Fig. 4.50, semi-steady state; Fig. 4.53, steady-state flow). The effect of choosing different values of the flowing time for a well draining from a bounded reservoir element is shown in Fig. 4.58a. There is one real pressure, $\overline{p}$ — the average in the drainage

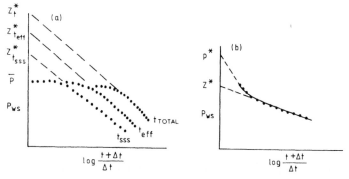

Fig. 4.58. Horner pressure buildups in development wells (a) illustrating the influence of the flowing time on the buildup. (b) Common error in extrapolation of the Horner plot.

area of the well at the time of the survey and provided $t > t_{sss}$ then its value will be correctly determined, applying the MBH technique, irrespective of the flowing time used. This argument has been presented in chapter 7 of reference 3 (p. 198) and is not dissimilar from that in section 4.15a of this book. Applying the defining equation for $p_{D\,(MBH)}$

$$p_{D(MBH)} = 0.0142 \frac{kh}{q\mu B_o}(Z^* - \overline{p})$$

(4.93)

will produce the correct value of $\overline{p}$, irrespective of the magnitude of Z^*, since $p_{D\,(MBH)}$ is a linear function of $\log t_{DA}$ (equation 4.100) once stability of the flowing pressure has been reached.

Figure 4.58b illustrates one of the most persistent and damaging mistakes in the whole subject of test interpretation: the extrapolation of the late time buildup pressures in a linear fashion to give a value of p^* — which is used in reservoir engineering calculations as a real pressure. This mater has already been described in section 4.15b but the warning is worth repeating. In the first place, with the exception of a few special cases (infinite acting system (Exercise 4.1), single fault buildup response (Exercise 4.2)) there is nothing in the mathematics of buildup analysis to suggest that the late time Horner buildup should be linear. In particular, for the closed and mixed boundary conditions described in this section, the pressures must necessarily flatten to give the average pressure, $\overline{p}$, as illustrated by the example in section 4.19d, and any linear extrapolation of the late time pressures prior to the flattening would be completely wrong under these circumstances. The only valid, linear extrapolation of pressures is of the early, linear trend containing transience to the value of Z^*, as defined by equation 4.49. Z^* is not a pressure in itself, although it has the same dimensions and, as illustrated in this chapter, has many useful applications in test analysis. Intrinsically, however, Z^* has no clear physical meaning, other than through its defining equation, and therefore should *never* be quoted in test analysis reports.

Boundary conditions

The mathematics describing pressure buildup behaviour for closed/mixed boundary conditions is, as usual, perfectly correct and, in this authors opinion, has a certain elegance in its structure. Yet interpretation is, and always has been, dominated by the complication of fitting boundary conditions. As stated in sections 4.16c–d, as soon as the engineer strays from the wellbore, the problem becomes one of trying to solve second-order differential equations to determine the boundary condition, which is a back-to-front procedure in comparison to the normal methods applied in the solution of such equations in science and engineering. The difficulty is that to apply either MBH-Horner or MDH-Dietz, the engineer must select not just the shape and areal extent drained by the well but also its position with respect to the boundaries. In such analyses the advance of computer software has done much to alleviate the difficulties. As described in section 4.16d an initial estimate is made of the shape, area and well asymmetry, for which a p_D-function is generated. This is used in the basic buildup equation 4.30 to calculate the static pressure, p_{ws}, which is compared with the actual buildup. If there is no correspondence then the boundary conditions are changed, often automatically by the program, until a match is found. But even so, there is an inevitable lack of uniqueness associated with the technique.

What should be avoided, at all costs, is taking short-cuts in the analysis. That is, operators often choose some arbitrary value of the closed-in time, Δt_s, to which either the Horner or MDH-plot should be extrapolated so that $p_{wsl} = \bar{p}$, to be applied to all wells in a field or even, in some cases, a producing area. But since Δt_s (equations 4.103, 4.113) contains the parameters: ϕ, μ, c, A, k, C_A, it is simply impossible to prescribe a universal value of Δt_s that will apply for all wells — let alone reservoirs or fields. In one area where the author worked, the "magic" number was $\Delta t_s = 360$ hours, while in another it was $\Delta t_s = 4400$ hours and it is suggested that if such a primitive approach is to be applied then the latter figure is preferable since it represents six months and nobody is going to close a well in for such a period to check on the veracity of results — play it safe! Unfortunately, the difficulty of describing boundary conditions is not one that is going to diminish and it can only be hoped that advances in computer technology will, at least, provide the engineer with the flexibility for exploring different options.

(f) Relationship between wellbore and numerical simulation grid block pressures

Since the early 1970's there have bene numerous papers written on the subject of relating flowing/buildup pressures measured in the wellbore to the average pressure in a numerical simulation grid block at the time the measurement is made — it being assumed that the dimensions of the grid block are considerably less than the overall reservoir volume being drained. (Fig. 4.59a).

The authoritative papers on this subject appear to be those of Peaceman [44,45] who determined from detailed simulation modelling of a square grid around a wellbore, that the radius, r_o, at which the flowing pressure equalled the average block pressure could be defined for such a cell with sides of length Δx as

$$r_o = 0.208\Delta x \tag{4.119}$$

which applies for the steady-state flow condition at the block boundary.

It is common practice when simulating to compare wellbore pressures from buildup surveys with the current grid block pressures; in which case it must be determined at what closed-in time, Δt_s, does the extrapolation of the early, linear

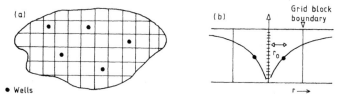

Fig. 4.59. (a) Numerical simulation model with wells in square grid blocks. (b) Illustrating the radius, r_o, at which the pressure equals the average grid block pressure.

MDH plot equal the average grid block pressure at radius r_0. This can be evaluated for steady-state conditions by applying equation 4.113 in which $A \approx \pi r_0^2 = \pi(0.208\Delta x)^2$ and $C_A' = 30.88$ (Fig. 4.53), giving

$$\Delta t_s = \frac{15152\phi\mu c\pi(0.208\Delta x)^2}{k \times 30.88} = \frac{66.7\phi\mu c\Delta x^2}{k} \tag{4.120}$$

The result is practically the same as achieved by Peaceman [44] who used a somewhat different approach to determine the constant as 67.5. Previous authors [3,46], failing to appreciate the significance of the relationship stated by equation 4.119, had overestimated the constant by a factor of three. Peaceman has also demonstrated that the method can be applied for unsteady-state flow conditions and for the use of rectangular grid blocks of sides Δx, Δy, in which case, equation 4.119 is replaced by

$$r_0 = 0.14(\Delta x^2 \times \Delta y^2)^{\frac{1}{2}} \tag{4.121}$$

The fact that there is a 1% discrepancy between the constant in equation 4.120 and that of Peaceman is neither here nor there because, as described in Chapter 3, section 3.8c, the whole process of "history matching on pressure" in any reservoir engineering calculations is intrinsically inaccurate. It is necessitated by the structure of simulation models in which pressures are "solved for" at the end of each time step rather than treated as "known data" to be used as input to calculations. Consequently, application of the technique tends to denigrate all attempts (such as in this chapter) to present methods of accurately determining pressures and therefore, any minor differences in the application of equations such as 4.120 are usually of little significance in comparison to those associated with application of the technique of matching pressures itself.

The prevalent attitude in the business is that because of the common use of numerical simulation modelling for history matching reservoir performance, the determination of average pressures in the drainage area of wells, $\overline{p}$, has become redundant. This is a completely false impression, however, because apart from the inherently clumsy technique of history matching on grid block pressures, numerical simulation models, as normally operated, can hardly be regarded as an investigative tool for defining reservoir drive mechanisms or the volume of hydrocarbons in place: they merely reflect the consequences of the assumed input volumetrics (maps/logs) and drive mechanisms (aquifer/gas cap). The best means of defining the volume and dynamics of the system is through the application of material balance, described for oil reservoirs in Chapter 3 and for gas in Chapter 6. The advantage in the technique is that it does not require a geological model and simply through manipulation of the production, pressure and PVT data, the hydrocarbons in place and drive mechanism can be defined. Of particular importance is that pressures are treated as *known* input data and, therefore, whether there is uniformity of pressure decline in the reservoir or not (so that rate averaged well pressures must be used to define the pressure decline) it is essential that average well pressures, $\overline{p}$, be determined frequently and accurately for use in material balance calculations, which is an essential step in advance of constructing credible numerical simulation models.

(g) Afterflow

There is no excuse, other than tool failure, for tolerating afterflow in appraisal well testing (section 4.14b) since generally such tests are conducted with DST tools which permit down hole closure. In development wells, however, completion is with a permanent packer and tubing and is not usually designed for down hole closure — although the means of doing so in production wells has been available to the industry for decades. Whether it is necessary to affect down hole closure and so eliminate afterflow is dependent upon the severity of its effect. If the Horner/ MDH early straight line is entirely obscured by afterflow, then some measures must be taken to reduce its influence. Severe and worsening afterflow will occur, for instance, in reservoirs in which depletion occurs below the bubble point pressure, which is particularly relevant in reservoirs with a gas cap. In this case, the amount of free gas in the tubing will increase as depletion continues, to the extent that the uninterpretable afterflow effects will dominate the buildup performance. Under these circumstances, it will be necessary to complete "selected wells "in the field with the mechanical facility for down hole closure and restrict well testing to those wells.

(h) Extended well testing

One feature that the author has noted as common to all extended tests it that no matter for how long they are conducted — it is never quite long enough! Such testing has become increasingly popular, especially offshore during the appraisal stage, the main aims being to establish the hydrocarbons in place and the drive mechanism in advance of committing to a costly, full field development. For an oil test, the STOIIP is invariably calculated by application of the depletion material balance (Chapter 3, section 3.7) as

$$N_p B_o = N B_{oi} c \Delta p \tag{3.25}$$

in which N_p is the cumulative production (stb), N the STOIIP (stb) and c the effective compressibility (1/psi) (equation 3.26). The pressure drop is $\Delta p = p_i - \overline{p}$ (psi), in which the initial and final pressures are determined from buildups at the beginning and end of the test. Unfortunately, the conventional assumption of straightforward depletion quite overlooks the fact that there might also be an element of pressure support so that the correct material balance equation ought to be expressed as

$$N_p B_o = N B_{oi} c \Delta p + \text{(water/gas influx)} \tag{4.122}$$

And, since there is often a commercial need to test high flow capacity reservoirs to break-even in the cash-flow; then there is an accompanying probability of some finite influx during any extended flow period. If there is, then since there are two unknowns on the right-hand side of equation 4.122 (N, influx), there is no unique mathematical solution and the best means of attempting to distinguish between the unknowns is by the application of the technique of Havlena and Odeh, described in Chapter 3, section 3.8b. If equation 3.25 is inadvertently applied to the test when there has been an element of pressure support then the STOIIP can be significantly overestimated and the drive mechanism misunderstood. To illustrate this point, suppose a reservoir with a STOIIP of $N = 120$ MMstb is tested in such a way that

$N_p = 0.75$ MMstb $B_o = 1.36$ rb/stb $c = 16 \times 10^{-6}$/psi
$W_e = 0.20$ MMrb $B_{oi} = 1.34$ rb/stb $\Delta p = 320$ psi

Then if equation 3.25 were applied to these data, assuming depletion alone, the STOIIP would be calculated as almost 150 MMstb: an overestimation of 25% and the greater the influx the more significant the error.

A similar approach to determining the STOIIP, which also depends on material balance, requires the identification of any semi-steady-state pressure decline for which p_{wf} plots as a linear function of the flowing time (section 4.6a). If the p_D-function for this condition, equation 4.101, is substituted in the general drawdown equation 4.21, the result is

$$\sigma(p_i - p_{wf}) = 2\pi t_{DA} + \tfrac{1}{2} \ln \frac{4A}{\gamma C_A r_w^2} + S$$

and evaluating σ and t_{DA}, using equations 4.29 and 4.95 respectively, and differentiating with respect to the flowing time, t, gives

$$\frac{dp_{wf}}{dt} = -0.234 \frac{q B_o}{c A h \phi} = -0.0417 \frac{q B_o (1 - S_{wc})}{c B_{oi} [\text{STOIIP}]} \quad \left(\frac{\text{psi}}{\text{hour}}\right) \qquad (4.123)$$

Consequently, if a linear pressure decline can be defined on a plot of p_{wf} versus t, then measurement of its slope will permit an estimation of the STOIIP (stb) using equation 4.123. This type of analysis is illustrated in Exercise 7.2 of reference 3 (p. 162). Application of the technique, however, is again only valid if the system is completely bounded, which is implicit in the derivation of the material balance term, $2\pi t_{DA}$ (section 4.19a), used in the generation of equation 4.123. A perfectly reasonable linear pressure decline, in appearance, may also result if there is an element of pressure support, the duration of such a trend being dependent on the strength of the influx, duration of the flow period, boundary effects and general accuracy of the test measurements. The slope of the apparent linear pressure decline would be too small, however, on account of the pressure support, leading once more to an overestimation of the STOIIP through the misapplication of equation 4.123.

The problem is a difficult one that must be anticipated in advance of the test by the engineer. If, as suggested above, the best means of attempting to distinguish between the STOIIP and drive mechanism is through application of the method of Havlena and Odeh then the test design must incorporate the following:

- accurate measurement of the initial pressure (section 4.14a)
- determination of reliable PVT functions
- frequent rate gauges and continuous recording of the wellbore pressure

In particular, the test should be designed so that the well is flowed for as long as practicably possible at high rate to maximise the underground withdrawal (and profitability if the oil can be marketed). Interrupting the test with lengthy pressure buildups is counter productive (a waste of time and money) yet there is a need to monitor the average reservoir pressure during the extended test to acquire several values for use in the Havlena-Odeh plot (Fig. 3.14a). It is suggested that this can be accomplished by conducting brief pressure buildups throughout this test or, better still, by varying the production rate and performing the two-rate analysis as described in section 4.20 (a). As illustrated in the example in section 4.19d buildups need only last for a few hours: sufficient time to define the initial, linear trend for extrapolation to $\overline{p}$. The determination of average well pressures from two rate flow tests does not appear to have been described in the literature (at least, not to the knowledge of this author) but is obviously possible applying buildup analysis to a test in which the second rate is not zero. Earlier descriptions of two rate testing [4,5] have made the restrictive assumption of transience throughout but if this is relaxed then analysis to determine $\overline{p}$ is possible for bounded, open or mixed boundary conditions. Before relying entirely on two rate analysis, however, a rate change fairly early in the extended test should be followed by well closure for a buildup to compare the values of $\overline{p}$ determined with the different methods. Both, however, are dependent on (subjective) estimates of the area drained, its shape and the position of the well with respect to the boundaries, as described in the section 4.19e. Nevertheless, use of the average pressures in conjunction with the Havlena-Odeh material balance interpretation technique should define the drive mechanism, depletion or pressure support, and provide a reasonable estimate of the STOIIP which must be compared with the volumetric figure. Further, practical aspects of extended well testing are described in sections 2.10 and 4.22.

(i) Radius of investigation

This is, to a large extent, a hypothetical concept that the author would prefer to omit form this text altogether were it not for the fact that its misuse in reservoir engineering calculations, particularly in connection with estimation of the STOIIP drained, can lead to errors affecting development planning. It is therefore worthwhile investigating its origin and meaning. The radius may be defined in the following terms: suppose a well is produced at a constant rate for a flowing time, t, the radius of investigation is the distance "seen" into the reservoir provided "infinite acting" conditions prevail. It is its restriction to purely transient systems that limits the practical application of the concept.

For a well/reservoir defined by parameters, k, ϕ, μ, c, r_w; if the radius $r = r_{inv}$ is considered as being the distance to a physical, no flow radial boundary, then for a flowing time, t, its minimum value may be determined from the MBH-chart for a well at the centre of a circle (Fig. 4.60a) at a value of $t_{DA} = 0.1$, which is the dimensionless time at which the change from transience directly to semi-steady-state flow occurs. Consequently, the radius of investigation can be calculated as the

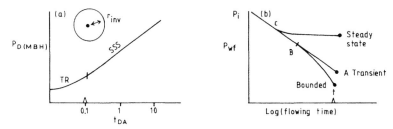

Fig. 4.60. (a) MBH dimensionless pressure function for a well at the centre of a circle. (b) Drawdown behaviour for different flowing conditions.

solution of the equation

$$t_{DA} = 0.1 = \frac{0.000264\, kt}{\phi \mu c \pi r_{inv}^2}$$

giving

$$r_{inv} = 0.03 \sqrt{\frac{kt}{\phi \mu c}} \quad \text{(ft)} \tag{4.81}$$

in which t is in hours. This is identical with the minimum value determined by van Poollen [47], using a different means of derivation. In an updated paper on the subject [48], Johnson has reviewed former attempts at interpreting and calculating the radius of investigation (drainage) and settled on minimum and maximum values of the constant in equation 4.81 varying between 0.03 and 0.04 but the Industry seems to have selected van Poollen's minimum value for common application.

The misconceptions/misapplications of the concept are:

– equation 4.81 is frequently applied using the total flowing time, irrespective of the drawdown condition, whereas it is only appropriate for a purely transient pressure decline.
– the radius of investigation is often used in calculating the STOIIP drained.

Considering the former, three flowing conditions for a transient, bounded and steady-state system are shown in Fig. 4.60b. While it is appropriate to calculate r_{inv} for the full flowing time (A) for the purely transient response, in a bounded reservoir the calculation must be made using the time at the end of transience (B) and for a steady-state drawdown at point (C). In the case of the bounded reservoir, the radius calculated represents the distance to the nearest boundary while for steady-state flow it is the radius of the constant pressure boundary. The latter may be alternatively calculated using the steady-state solution of the diffusivity equation, as described in section 4.18b. But to inadvertently use the total flowing time in calculating the radius for a bounded or steady-state test would lead to a significant overestimation in its value, as demonstrated in Exercise 4.3.

There is no validity in attempting to relate the STOIIP being drained to the radius of investigation. In the case of testing a bounded or steady-state system this statement is obvious but even in the case of an infinite acting reservoir, the same applies. the STOIIP being tested could be much larger than calculated from the radius of investigation if the flow period is short and in the case of a long drawdown, the oil volume could be less than calculated if the diffusivity constants $(k/\phi\mu c)$ of the oil column and aquifer are similar. Any attempt to calculate the STOIIP must be based on the application of material balance to an observed pressure drop in the reservoir, as described in section 4.19h (which is difficult enough); it cannot possibly be based on a concept such as the radius of investigation which is derived for infinite acting systems.

The only application of the concept of radius of investigation in this chapter is in the planning of a test to locate the position of a single, sealing fault, as described in section 4.16c, in which it was noted that in order that the fault should cause a significant component of drawdown in the wellbore, it is necessary that the radius of investigation be about four times the distance to the fault. Therefore, if the fault distance is known approximately from seismic/geological modelling as d (ft), then the well should be produced for a time commensurate with a radius $4d$, calculated using equation 4.81. This, it will be appreciated, is a somewhat tenuous calculation, nevertheless it is about the most useful application of the concept. As with the parameter $Z^*(p^*)$ (section 4.15b), under no circumstances should the radius of investigation be quoted in test analysis reports (in case someone tries to use it) and it should never be applied in any quantitative calculations upon which development decisions might be made.

4.20. MULTI-RATE FLOW TESTING

The inherent difficulty in attempting to apply multi-rate testing techniques which rely upon the use of equation 4.24 (the superposed CTR solution of the radial diffusivity equation) have already been described in section 4.9 and also in reference 3. The equation contains three unknowns, k, S and the p_D-function meaning there is a complete lack of uniqueness in its general application: choice of different boundary conditions affects the p_D-function which, in turn, dictates the values of k and S determined in the analysis. Since there is an infinite number of ways of selecting a p_D-function then there is also an infinite number of values of k and S that can result from a single set of rate–pressure–time data recovered from a test. The Industry has always managed to ignore this difficulty through the blithe assumption that multi-rate tests, no matter under what circumstances they are conducted, always occur under *transient* flow conditions throughout so that the p_D-function can be evaluated using equation 4.22. This has the effect of reducing the number of unknowns to two permitting an apparent unique solution to be obtained using the plotting technique described in section 4.9 but the assumption of transience removes confidence in the use of equation 4.24, in a general sense, for the analysis of multi-rate tests.

Yet there are ways of designing tests so that application of equation 4.24 becomes tractable and leads to unique results in terms of k and S through the correct identification of a transient pressure response. The most celebrated of these is the pressure buildup test which is, in fact, a two-rate flow test in which the second rate is zero (section 4.11). There are also other forms of multi-rate test which are amenable to interpretation using equation 4.24 and two of these are described in this section: the two-rate flow test and general multi-rate tests in which there is a degree of pressure support during each separate flow period. While both can be conducted in either appraisal or development wells, normal practice is to restrict their use to the latter.

(a) Two-rate flow testing

This form of test has always been popular for the obvious reason that the amount of deferred production is reduced compared to a pressure buildup survey. It is also appropriate for wells which, for one reason or another, demonstrate a reluctance to flow following periods of closure. Since the test consists of a change from one finite rate to another, instead of complete closure, there will be inevitably an element of afterflow which is mechanically uncontrollable but, as advised in reference 4, the duration of afterflow is usually less than for a pressure buildup (following surface closure) and is diminished further if the rate change is a reduction rather than an increase. The rate–pressure–time records for such a test are shown in Fig. 4.61. The basic principle is the same as for buildup testing: irrespective of the nature of the rate history of a well or its duration, the act of changing its production rate will cause a transient pressure response and the analysis technique seeks to isolate this through appropriate semi-log plotting and so determine kh and S. It is also possible to calculate the initial pressure in an appraisal well test or the average pressure within the drainage area of the well in a development well test. Applying equation 4.24 to the test depicted in Fig. 4.60 gives

$$7.08 \times 10^{-3} \frac{kh}{\mu B_o}(p_i - p_{wf}) = q_1[p_D(t_D + \Delta t'_D) + S]$$
$$+ (q_2 - q_1)[p_D(\Delta t'_D) + S] \tag{4.124}$$

The conventional approach in dealing with this equation presented in the literature [4,5] is (again) to assume transience throughout the test so that all the p_D-functions can be evaluated using equation 4.22 which reduces the equation to the form

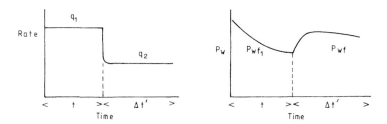

Fig. 4.61. Rate–pressure–time records for a two-rate flow test.

$$\sigma(p_i - p_{wf}) = \tfrac{1}{2} \ln \frac{t + \Delta t'}{\Delta t'} + \tfrac{1}{2}\frac{q_2}{q_1} \ln \frac{4\Delta t'_D}{\gamma} + \frac{q_2}{q_1} S \tag{4.125}$$

in which

$$\sigma = 7.08 \times 10^{-3} \frac{kh}{q_1 \mu B_o}$$

and p_{wf} is the pressure recorded during the second flow period at time $\Delta t'$, measured since the rate change. The equation may be expressed in a more practical form as

$$\sigma(p_i - p_{wf}) = 1.151 \left[\log \frac{t + \Delta t'}{\Delta t'} + \frac{q_2}{q_1} \log \Delta t' \right]$$

$$+ 1.151 \frac{q_2}{q_1} \left[\log \frac{k}{\phi \mu c r_w^2} - 3.23 + 0.87S \right] \tag{4.126}$$

which implies that a plot of

$$p_{wf} \text{ versus } \log \frac{t + \Delta t}{\Delta t} + \frac{q_2}{q_1} \log \Delta t' \tag{4.127}$$

should be linear with slope

$$m = \frac{1.151}{\sigma} = 162.6 \frac{q_1 \mu B_o}{kh} \tag{4.128}$$

which is exactly the same as for pressure buildup analysis (equation 4.36). Finally, subtracting equation 4.126 from the drawdown equation 4.17 and solving explicitly for S at $\Delta t' = 1$ hour, $p_{wf} = p_{wfl(1\,hr)}$ (read from the straight line or its linear extrapolation) gives the skin factor equation as

$$S = 1.151 \left[\frac{(p_{wfl(1\,hr)} - p_{wf_i})}{m} \left(\frac{q_1}{q_1 - q_2} \right) - \log \frac{k}{\phi \mu c r_w^2} + 3.23 \right] \tag{4.129}$$

The above, it will be recognised, is the equivalent of the Horner plot but the analysis is incomplete in its *ad hoc* assumption that $p_D(t_D + \Delta t_D)$ can be evaluated under transient conditions and therefore equation 4.126 is not necessarily the correct equation of the early straight line. A more rigorous approach is to follow the argument presented in section 4.13 for the derivation of the Horner straight line for pressure buildup analysis: requiring the addition of $\pm[(q_1 - q_2)/q_1]\tfrac{1}{2} \ln(t_D + \Delta t'_D)$ to the right-hand side of equation 4.124 and its inspection for small values of $\Delta t'$, when transience is to be expected. This would result in the reformulation of equation 4.125 as

$$\sigma(p_i - p_{wfl}) = \left(\frac{q_1 - q_2}{q_1} \right) \tfrac{1}{2} \ln \frac{t + \Delta t'}{\Delta t'} + \frac{q_2}{q_1} S + p_D(t_D) - \left(\frac{q_1 - q_2}{q_1} \right) \tfrac{1}{2} \ln \frac{4\,t_D}{\gamma}$$

and since the last two terms are evaluated for the fixed flowing time at rate q_1 both terms are constants which influence the position of the initial straight line on the plot. This argument, which would eventually lead to the determination of

$\bar{p}$, the average pressure in the drainage area of a development well will not be pursued further because a more direct method is presented below. Suffice to say that provided the plot of equation 4.127 is linear soon after the rate change then the slop of the line gives the kh-product using equation 4.128 and the skin factor from equation 4.129 but from the conventional description of the analysis in the literature [4,5] no method is suggested for estimating the average pressure $\bar{p}$.

A much simpler approach to the analysis is the equivalent of the MDH method for buildup interpretation. Following the steps described in section 4.12 for the derivation of the early straight line equation on an MDH buildup plot requires the inspection of equation 4.124 for small values of $\Delta t'$ when transience is to be expected. Then

$$p_D(t_D + \Delta t'_D) \approx p_D(t_D) = \text{constant} \tag{4.32}$$

and

$$p_D(\Delta t'_D) = \tfrac{1}{2} \ln \frac{4\Delta t'_D}{\gamma} \tag{4.31}$$

reducing equation 4.124 to

$$\sigma(p_i - p_{wfl}) = p_D(t_D) - \left(\frac{q_1 - q_2}{q_1}\right) \tfrac{1}{2} \ln \frac{4\Delta t'_D}{\gamma} + \frac{q_2}{q_1} S \tag{4.130}$$

which can be presented in a more practical form as

$$\sigma(p_i - p_{wfl}) = \text{constant} - 1.151 \left(\frac{q_1 - q_2}{q_1}\right) \log \Delta t' \tag{4.131}$$

in which

$$\text{constant} = p_D(t_D) - 1.151 \left(\frac{q_1 - q_2}{q_1}\right) \left[\log \frac{k}{\phi \mu c r_w^2} - 3.23\right] + \frac{q_2}{q_1} S \tag{4.132}$$

Equation 4.131 suggests that the more complex Horner-type plot (equation 4.127) can be replaced by the much simpler MDH equivalent plot of

$$p_{wf} \text{ versus } \log \Delta t' \tag{4.133}$$

It should be noted that p_{wfl} in equation 4.131 denotes pressures on the straight line defined by the equation and is the analogy of p_{wsl} (equation 4.34) for MDH buildup analysis. At the start of the second flow period (once afterflow ceases) there will be a period when equation 4.133 coincides with the measured pressures but the straight line can be extrapolated beyond the maximum time of coincidence defining the hypothetical pressures p_{wfl}. As with buildup analysis, the time for which the actual pressures define an early linear trend is not solely dictated by the condition of transience but also depends on the rate of change of $p_D(t_D + \Delta t_D)$ and when it can no longer be regarded as a constant.

The slope of the early linear trend is

$$m' = \frac{1.151}{\sigma}\left(\frac{q_1 - q_2}{q_1}\right) = 162.6\frac{q_1\mu B_o}{kh}\left(\frac{q_1 - q_2}{q_1}\right) \tag{4.134}$$

and subtracting equation 4.130 from the drawdown equation

$$\sigma(p_i - p_{wf_1}) = p_D(t_D) + S \tag{4.21}$$

and reading the pressure $p_{wfl\,(1\,hr)}$ from the straight line at $\Delta t' = 1$ hour enables the skin to be determined as

$$S = 1.151\left[\frac{(p_{wfl\,(1\,hr)} - p_{wf_1})}{m'} - \log\frac{k}{\phi\mu c r_w^2} + 3.23\right] \tag{4.135}$$

in which p_{wf_1} is measured at the end of the first flow period. But since, according to equation 4.134,

$$m' = m\left(\frac{q_1 - q_2}{q_1}\right) \tag{4.136}$$

Equations 4.135 and 4.129 can be seen to be equivalent.

In the case of conducting a two-rate test in an exploration or appraisal well, the initial reservoir pressure, p_i, can be calculated directly using equation 4.21 during the period of transience of the first flow period [4]. Since the two-rate test is largely confined to development wells, however, there is a much greater need to find a means of calculating the average pressure, $\bar{p}$, in the drainage area of the well at the time of the survey — a topic which appears to have been overlooked in the popular literature.

Considering the case of a well producing from within a bounded reservoir element under volumetric depletion conditions, as described in section 4.19b, the technique for calculating the average pressure in a flow test is the same as practised for the MDH buildup: the determination of the time, $\Delta t'_s$, at which the early linear trend during the second flow period must be extrapolated so that $p_{wfl} = \bar{p}$. If the material balance equation

$$\sigma(p_i - \bar{p}) = 2\pi t_{DA} \tag{4.94}$$

is subtracted from the straight-line equation 4.130, the result is

$$\sigma(\bar{p} - p_{wfl}) = p_D(t_D) - \left(\frac{q_1 - q_2}{q_1}\right)\frac{1}{2}\ln\frac{4\Delta t'_D}{\gamma} + \frac{q_2}{q_1}S - 2\pi t_{DA}$$

and extrapolating this so that $p_{wfl} = \bar{p}$, $\Delta t'_D = \Delta t'_{Ds}$ and evaluating the $p_D(t_D)$-function using the general expression for bounded geometries (equation 4.97) yields

$$0 = \frac{1}{2}\ln\frac{4t_D}{\gamma} - \frac{1}{2}p_{D\,(MBH)}(t_{DA}) - \left(\frac{q_1 - q_2}{q_1}\right)\frac{1}{2}\ln\frac{4\Delta t'_D}{\gamma} + \frac{q_2}{q_1}S \tag{4.137}$$

Finally, provided the well is flowing under semi-steady-state conditions at the end of first flow period, then

$$p_{D(MBH)}(t_{DA}) = \ln(C_A t_{DA}) \tag{4.100}$$

and with some algebraic manipulation equation 4.137 can be reduced to

$$\Delta t'_{Ds} = \frac{\gamma}{4} \left(\frac{4A}{\gamma C_A r_w^2} \right)^{\frac{q_1}{q_1-q_2}} \exp\left[(2\frac{q_2}{q_1} s \right] \tag{4.138}$$

It will be appreciated that for a pressure buildup test ($q_2 = 0$) this equation is reduced to the form expressed in equation 4.103 for the extrapolation of the MDH straight line. Similarly, for a well producing within mixed/open boundaries, so that the steady-state condition prevails at the end of the first flow period equation 4.138 is replaced by

$$\Delta t'_{Ds} = \frac{\gamma}{4} \left(\frac{16A}{\gamma C'_A r_w^2} \right)^{\frac{q_1}{q_1-q_2}} \exp\left[2\frac{q_2}{q_1} s \right] \tag{4.139}$$

in which C'_A is the modified Dietz shape factor (equation 4.111, Fig. 4.53). For a pressure buildup ($q_2 = 0$) equation 4.139 is reduced to the form of equation 4.113.

Although it was avoided earlier in this section, a similar argument could be presented for determining the average pressure using the equivalent Horner method. The analysis and final result are, however, somewhat more complex than using the MDH approach on account of the unnecessary inclusion of the flowing time and the derivation will therefore be left as an exercise for the reader.

Equation 4.138 has been derived under the assumption that the flowing condition at the time of the rate change is that of semi-steady state which preferably it should be. This simplifies matters since equation 4.100 can be applied to express the MBH-function. If a more complex condition prevailed, however, such as late transience, then the method of determining $\bar{p}$ is still valid except that the MBH-function has to be expressed in non-linear form and the same applies for steady-state flow.

(b) Example well test

To illustrate the techniques described above and validate them theoretically, a two-rate test pressure response will first be generated using the data provided for the Example test described in section 4.19d for a well draining from an asymmetric position within a bounded square

$C_A = 12.985$ (Fig. 4.50)

The steps in the exercise are as follows

(A) Generate a theoretical pressure response for the second flow period of a two-rate test (p_{wf} versus $\Delta t'$) using the p_D-function defined in column 4 of Table 4.12 for the above geometrical configuration. The flow rates are $q_1 = 2500$ stb/d, $q_2 = 2000$ stb/d and the rate change occurs after $t = 1038$ hours, by which time the semi-steady-state condition prevails.

(B) Analyse the pressure–time response for the first 20 hours after the rate change using the equivalent Horner and MDH plots: equations 4.127 and 4.131, respectively to determine kh, S and the average pressure $\bar{p}$.

The bottom hole flowing pressure during the second flow period may be calculated by direct solution of equation 4.124 expressed in the more convenient form

$$\sigma(p_i - p_{wf}) = p_D(t_D + \Delta t'_D) - p_D(\Delta t'_D) + \frac{q_2}{q_1}(p_D(\Delta t'_D) + S) \tag{4.140}$$

in which: p_i = 6000 psia, S = 0. The remainder of formation/fluid properties (section 4.19d) are: k = 75 mD, h = 100 ft, A = 120 acres (5.227×10^6 ft^2), B_o = 1.35 rb/stb, μ = 1.5 cp, c = 19×10^{-6}/psi, ϕ = 0.23, r_w = 51 ft. The function $p_D(t_D + \Delta t'_D)$ is obtained by linear interpolation of its values presented in Table 4.12 between t = 1038 hours (time of the rate change) and t = 1211 hours giving

$$p_D(t_D + \Delta t'_D) = 3.6364 \times 10^{-3} t + 7.5217$$

The period of transience (section 4.19d) is considerably longer than the 20 hours for which the pressure response is studied following the rate change. Consequently

$$p_D(\Delta t'_D) = \tfrac{1}{2} \ln \frac{4\Delta t'_D}{\gamma} = \tfrac{1}{2} \ln \frac{4 \times 0.000264 \times 75 t}{1.781 \times 0.23 \times 1.5 \times 19 \times 10^{-6} \times 0.51^2}$$

$$= \tfrac{1}{2} \ln(26082\, t)$$

and the coefficient σ has the value

$$\sigma = 7.08 \times 10^{-3} \frac{kh}{q_1 \mu B_o} = \frac{7.08 \times 10^{-3} \times 75 \times 100}{2500 \times 1.5 \times 1.35} = 0.0105$$

Equation 4.140 may then be reduced to ($S = 0$)

$$0.0105\,(6000 - p_{wf}) = p_D(t_D + \Delta t'_D) - \left(\frac{q_1 - q_2}{q_1}\right) p_D(\Delta t'_D)$$

$$= 3.6364 \times 10^{-3} t + 7.5217 - 0.1 \ln(26082 \Delta t')$$

which can be solved directly for p_{wf}, the values being listed in column 5 of Table 4.14. The abscissae required for MDH (equation 4.131) and Horner (equation 4.127) interpretations are listed in columns 6 and 7 (α) respectively. The corresponding pressure plots are shown in Fig. 4.62. Both demonstrate an early linear trend which lasts for about 2.5 hours on each plot. Thereafter, the pressure points curve downwards, below the early linear trend which simply represents the breakdown of the conditions necessary for linearity, which in the case of the MDH plot means that equation 4.32 is not valid since transience prevails for longer than the 20 hour period studied. The results are

	MDH	Horner
Slope (psi/cycle)	21.4	107.0
$p_{wf\,(1\,hr)}$	5020.7	5020.7

TABLE 4.14

Example well test: calculation of the pressure during the second flow period and the times required for
equivalent MDH (equation 4.131) and Horner (α) (equation 4.127) plots ($t = 1038$ hours)

$\Delta t'$ (hours)	$p_D(\Delta t'_D)$	$t + \Delta t'$ (hours)	$p_D(t_D + \Delta t'_D)$	p_{wf} (psia)	$\log \Delta t'$	α (equation 4.127)
0		1038.000	11.2963			
0.125	4.0448	1038.125	11.2967	5001.2	−0.903	3.197
0.25	4.3914	1038.250	11.2972	5007.7	−0.602	3.137
0.375	4.5941	1038.375	11.2976	5011.5	−0.426	3.102
0.50	4.7359	1038.500	11.2981	5014.2	−0.301	3.077
0.75	4.9407	1038.750	11.2990	5018.0	−0.125	3.041
1.00	5.0845	1039.000	11.2999	5020.7	0	3.017
1.25	5.1961	1039.250	11.3008	5022.7	0.097	2.997
1.50	5.2872	1039.500	11.3017	5024.4	0.176	2.982
1.75	5.3643	1039.750	11.3026	5025.7	0.243	2.968
2.00	5.4311	1040.000	11.3036	5026.9	0.301	2.957
2.50	5.5426	1040.500	11.3054	5028.9	0.398	2.938
3.00	5.6338	1041.000	11.3072	5030.4	0.477	2.922
3.50	5.7109	1041.500	11.3090	5031.7	0.544	2.909
4.00	5.7776	1042.000	11.3108	5032.8	0.602	2.897
4.50	5.8365	1042.500	11.3126	5033.8	0.653	2.887
5.00	5.8892	1043.000	11.3145	5034.6	0.699	2.878
6.00	5.9804	1044.000	11.3181	5036.0	0.778	2.863
7.00	6.0575	1045.000	11.3217	5037.1	0.845	2.850
8.00	6.1242	1046.000	11.3254	5038.0	0.903	2.839
10.00	6.2358	1048.000	11.3326	5039.5	1.000	2.820
15.00	6.4385	1053.000	11.3508	5041.6	1.176	2.787
20.00	6.5824	1058.000	11.3690	5042.6	1.301	2.764

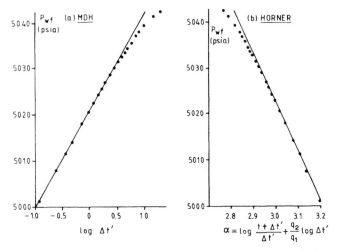

Fig. 4.62. Equivalent (a) MDH and (b) Horner plots of the pressure response during the second flow
period.

the difference in slopes being consistent with equation 4.136. From either plot the value of k can be calculated as approximately 75 mD (equation 4.128 or 4.134 for Horner or MDH, respectively) and using a final flowing pressure before the rate change of $p_{wf_1} = 4924.2$ psia after 1038 hours (Table 4.12) the skin factor can be calculated as zero using equation 4.135 for MDH and equation 4.129 for Horner analysis. These results validate the use of either interpretation technique.

The time to which the early linear trend of the MDH plot must be extrapolated so that $p_{wfl} = \overline{p}$ can be determined by application of equation 4.138 which, for $S = 0$, is reduced to

$$\Delta t_{Ds} = \frac{\gamma}{4}\left(\frac{4A}{\gamma C_A r_w^2}\right)^{\frac{q_1}{q_1-q_2}} = \frac{1.781}{4}\left(\frac{4 \times 5.227 \times 10^6}{1.781 \times 12.985 \times 0.51^2}\right)^{\frac{2500}{2500-2000}}$$
$$= 2.259 \times 10^{32}$$

This extraordinarily large number results from the small slope of the MDH plot but is quite realistic. Inserting it directly in the linear equation 4.130 ($\ln \Delta t_{Ds} = 74.4977$, $p_D(t_D) = 11.2963$) gives

$$0.0105(6000 - \overline{p}) = 11.2963 - 0.1\left(\ln\frac{4}{1.781} + 74.4977\right)$$

which may be solved to give $\overline{p} = 5641$ psia. The check whether this figure is correct, consider a pressure buildup conducted after 1038 hours of flow instead of a rate change from 2500 to 2000 stb/d. Since, at this time, semi-steady-state conditions prevail then, as demonstrated in the example in section 4.19d, the time to which the linear buildup must be extrapolated so that $p_{wsl} = \overline{p}$ is $\Delta t_s = 133.27$ hours: $\Delta t_{Ds} = 1.5477 \times 10^6$ and inserting this value in the linear buildup equation 4.33, gives

$$0.0105(6000 - \overline{p}) = 11.2963 - \frac{1}{2}\left(\ln\frac{4}{1.781} + 14.2523\right)$$

which may be solved to give $\overline{p} = 5641$ psia which is identical to the value determined from the two-rate flow test.

The main advantage that pressure buildup testing offers over two-rate testing is that for the former the second rate is zero which should be operationally easier to control than adjusting to a second, finite rate, provided downhole closure is affected. Buildup testing is preferable for exploration/appraisal wells, especially offshore, in which the produced oil has to be flared and therefore closure is more convenient than a rate change. Two-rate tests are best suited to development wells in which the avoidance of well closure is beneficial to field operation. In this respect, Matthews and Russell [4] offer sound advice in that engineers and field personnel should be familiar with the flow characteristics of wells before undertaking two-rate tests and, in particular, the ease and amount by which their rates can be changed without inducing significant afterflow.

Nevertheless, this author has noted a certain reluctance amongst operators to attempt two-rate testing because of a widespread belief that this form of test cannot be used to determine the average pressure within the drainage area of a well, $\bar{p}$, which as illustrated above is not the case, the mathematical formulations being only slightly more cumbersome than in buildup analysis. Operators in high cost areas, such as the North Sea, where generally speaking the oil remains in a highly undersaturated state on account of water injection (thus minimising afterflow) could profit from applying two-rate tests rather than indulging in lengthy pressure buildups. Unfortunately, the main difficulty associated with buildup tests in development wells: the definition of the drainage area, A, and shape factor, C_A, described in section 4.19e, are precisely the same for two-rate testing.

(c) Selective inflow performance (SIP) testing

This form of test is conducted in delta top reservoir environments, as described in Chapter 5, section 7, in which the total section consists of discrete productive layers separated by impermeable barriers (shales). There is invariably a lack of areal correlation of the layers between wells and of pressure equilibrium across the section, the latter being established from RFT surveys run under dynamic conditions in each new development well. Conventional tests of development wells in such an environment, by pressure buildup or two-rate flow testing, is an unacceptable means of attempting to evaluate formation characteristics since the basic assumption of formation homogeneity implicit in the derivation of the radial diffusivity equation (section 4.5a) is violated by the lack of pressure equilibrium across the section. Closure for a buildup, for instance, leads to severe cross-flow in the wellbore between the high-pressure, low-permeability layers and the more productive, high-permeability layers which have lower pressure. This usually renders conventional tests of the whole section uninterpretable.

The sensible alternative is to run production logging (PLT) surveys to establish the relative contribution of flow from the individual layers and to attempt to determine the different average pressures in each layer at the time of the survey. The logging tool is passed across the section with the well flowing at different rates. The bottom hole flowing pressure will be common to all layers (in hydrostatic equilibrium) at each different rate but the survey will determine the individual rates from each layer. Then, considering the performance of any one of the layers in isolation, a plot is made of its rate as a function of the flowing wellbore pressure during the survey, as shown in Fig. 4.63 for three separate flow rates. The technique is then to draw the best linear trend through the points and extrapolate back to $q = 0$ at which point the pressure, p_0, is assumed to represent the average static pressure in the particular layer at the time of the survey, $\bar{p}$. The question arises, however, of — under what physical circumstance is it valid to regard the points as comprising a straight line such that backward extrapolation gives a meaningful value of $\bar{p}$? The trend need not necessarily be linear (although some operators use only two different flow rates to ensure that it is!) and a least squares fit is often applied incorrectly under the assumption that the trend should form a straight line.

To investigate this matter, it is necessary to examine the appropriate multi-rate flow equation 4.24, expanded in such a manner as to isolate the latest rate q_n,

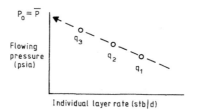

Fig. 4.63. PLT survey to determine individual layer pressures.

evaluated at time t_n when the wellbore flowing pressure is p_{wf_n}, that is

$$\sigma'(\bar{p} - p_{wf_n}) = \left[\sum_{j=1}^{n-1} \Delta q_j p_D(t_{D_n} - t_{D_{j-1}}) - q_{n-1} p_D(t_{D_n} - t_{D_{n-1}})\right]$$
$$+ q_n(p_D(t_{D_n} - t_{D_{n-1}}) + S) \qquad (4.141)$$

in which $\bar{p}$ is the average stable pressure in the layer prior to the survey and, in this case

$$\sigma' = 7.08 \times 10^{-3} \frac{kh}{\mu B_o}$$

In order that the equation represent a linear relationship between p_{wf_n} and q_n such that its backward extrapolation to $q = 0$ gives $p_{wf_0} = p_0 = \bar{p}$ two conditions must be satisfied, namely

(A) the slope $(p_D(t_{D_n} - t_{D_{n-1}}) + S)/\sigma'$ must be constant

(B) the expression $\left[\sum_{j=1}^{n-1} \Delta q_j p_D(t_{D_n} - t_{D_{j-1}}) - q_{n-1} p_D(t_{D_n} - t_{D_{n-1}})\right] = 0$

but the problem is defining a flow condition that will comply with both. Any time dependent response such as transience (even if the state could be proven) would satisfy neither. It would appear that the only appropriate condition is that of steady-state flow described in sections 4.18 and 4.19c, for which the p_D-function is constant: having the form

$$p_D(t_D) = \ln \frac{r_e}{r_w} = C = \text{constant}$$

for any value of the dimensionless time argument, provided the flow is radial in each layer: r_e being the radius of the constant pressure boundary. Applying this in the above two conditions gives

(A) slope $= (C + S)/(\sigma') = \text{constant}$

(B) $C \left[\sum_{j=1}^{n-1} \Delta q_j - q_{n-1}\right] = 0$

To illustrate the validity of the latter, in a three-rate flow test, B would be evaluated as

$$C[(q_1 - 0) + (q_2 - q_1) - q_2] = 0$$

It appears, therefore, that the conventional SIP interpretation is strictly only valid for completely stabilized, steady-state flow conditions.

In attempting to apply any form of reservoir engineering technique (including well testing and analysis) to a complex delta top environment, there is inevitably a certain loss of control on account of the normal policy of perforating large sections, which eventually removes pressure equilibrium between individual layers. Yet it would be unwise to restrict perforations to a limited number of layers with compatible permeabilities in any given well, since this would amount to a pre-judgement on the part of the engineer on the correlation of layers from well to well across the field. But since there is usually a strong element of randomness in correlation, it is perhaps best to perforate the entire section, especially in water injection projects, to maximise the probability of flooding the greatest number of sands — even though success may never be directly observed.

The loss of reservoir engineering control implies the acceptance of lower standards in general and, under the circumstances, the application of the SIP testing technique must be regarded as acceptable in — making the best of a bad job. Nevertheless, the above analysis suggests that, to improve the reliability of results, each flow period should be conducted for a sufficiently long period before pressure measurement to aim at attaining a "reasonable" degree of pressure equilibrium within each layer. SIP testing will therefore provide the best results in water injection projects (prior to water breakthrough, after which PLT surveys become increasingly more difficult to interpret) in which the aim is to achieve steady-state conditions and in gas wells in which, on account of the high compressibility, complete pressure stability is frequently observed (section 4.6b). The purpose in measuring individual layer pressures is to calibrate numerical simulation models to facilitate field performance predictions.

4.21. LOG–LOG TYPE CURVES

(a) Conventional type-curve interpretation

This form of test interpretation was introduced during the 1970's; those involved with the development of the method being Ramey [49], Earlougher [50], McKinley [51], Agarwal et al. [52] and Gringarten et al. [53]. But it is the presentation technique of Gringarten that has achieved the most widespread use because of its simplicity and the coincidence that at the time Gringarten et al. wrote their paper in 1979, well test analysis computer programs incorporating such techniques were becoming used increasingly throughout the Industry.

Gringarten solved the radial diffusivity equation for wells with a given skin factor and wellbore storage, the latter being defined as

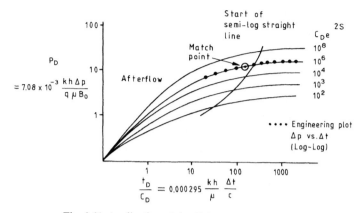

Fig. 4.64. Application of the Gringarten type curves.

$$C = \frac{dV}{dp} \quad \text{(bbl/psi)}$$

which is a measure of the well's capacity to store fluid during the period of afterflow. Related to this parameter is the dimensionless wellbore storage constant

$$C_D = \frac{5.615\,C}{2\pi\,\phi c_{\text{eff}} h r_w^2} = \frac{0.8937\,C}{\phi c_{\text{eff}} h r_w^2} \tag{4.142}$$

in which c_{eff} is the effective compressibility (equation 4.2). Gringarten's type curves are plots of p_D versus t_D/C_D (Fig. 4.64) and cater for both idealised wellbore storage, not allowed for in semi-log plotting techniques, and the period of transience (referred to in the appropriate literature as IARF — "infinite acting radial flow").

The demarcation between afterflow and transience is indicated by the solid line labelled "approximate start of the semi-log straight line" which corresponds with Ramey's one and a half cycle rule (section 4.14b). The reason for making the p_D-plots as a function of t_D/C_D, on log–log scales, is that during the period completely dominated by afterflow

$$p_D = \frac{t_D}{C_D} \tag{4.143}$$

and consequently all the functions converge on to a 45° straight line (Fig. 4.64) for small values of t_D/C_D. The p_D-function is defined as

$$p_D = \sigma \Delta p = 7.08 \times 10^{-3} \frac{kh}{q\mu B_o} \Delta p \tag{4.144}$$

which excludes the skin factor from the normal defining expression (equation 4.21) but the skin is catered for in that each of the theoretical curves is for a fixed value of the parametric group $C_D e^{2S}$. The abscissa of the type curves is determined from equations 4.19 and 4.142 as

$$\frac{t_D}{C_D} = 0.000295 \frac{kh}{\mu} \frac{\Delta t}{C} \tag{4.145}$$

The analysis is usually performed using computer programs but if attempting manual interpretation the engineer must plot either drawdown or buildup pressure differences $\Delta p = p_i - p_{wf}$ (drawdown) or $\Delta p = p_{ws} - p_{wf}$ (buildup) as a function of the flowing or buildup time (both denoted as Δt) on a piece of transparent log–log paper with the same log scales as the dimensionless type curves. The engineering plot is then moved vertically and laterally until coincidence is achieved with one of the curves. This amounts, in log–log space, to evaluating the constants in the following two equations

$$\log p_D = \log 7.08 \times 10^{-3} \frac{kh}{q\mu B_o} + \log \Delta p \tag{4.146}$$

and

$$\log \frac{t_D}{C_D} = \log 0.000295 \frac{kh}{\mu C} + \log \Delta t \tag{4.147}$$

By choosing a "match point" on the engineering curve for any value of Δp and Δt, equations 4.146 and 4.147 may be evaluated to determine kh and C, respectively as

$$kh = \frac{q\mu B_o}{7.08 \times 10^{-3}} \frac{p_D}{\Delta p}$$

$$C = 0.000295 \frac{kh}{\mu} \frac{\Delta t}{t_D/C_D}$$

for which p_D and t_D/C_D are the ordinate and abscissa values of the type curves corresponding to the selected match point. The dimensionless wellbore storage, C_D, is then evaluated using equation 4.142 and finally by noting upon which of the theoretical type curves the engineering plot best fits the value of the parametric group $C_D e^{2S}$ is established permitting the calculation of the skin factor.

There are, however, certain disadvantages in applying type curve analysis which should be appreciated. In the first place, the plotting of pressures on a logarithmic scale removes the advantage in using semi-log plots, that for either drawdown or buildup analysis there is a linear section, following any afterflow, which includes the transient pressure response. In employing a logarithmic pressure scale, not only is the resolution in pressure diminished but also the transient response is non-linear. Secondly, there is a difficulty in that the type curves are solutions of the radial diffusivity equation, for constant-rate drawdown, whereas they are primarily applied in the interpretation of pressure buildup tests. That is, the type curves have been generated as

$$\sigma(p_i - p_{wf}) = p_D(t_D) \tag{4.148}$$

for wells with different storage and skin factor but subtraction of the basic buildup equation 4.30 from this results in

$$\sigma(p_{\text{ws}} - p_{\text{wf}}) = [p_{\text{D}}(t_{\text{D}}) - p_{\text{D}}(t_{\text{D}} + \Delta t_{\text{D}})] + p_{\text{D}}(\Delta t_{\text{D}}) \tag{4.149}$$

Therefore, if a buildup plot of $\log(p_{\text{ws}} - p_{\text{wf}})$ versus $\log \Delta t$ is to match the type curves, then the difference in the p_{D}-functions in parenthesis must be negligible. There are obvious cases when this will be realised, such as in steady-state tests (sections 4.18 and 19c) but in general the difference may not be negligible. This exemplifies the difficulty referred to in section 4.8: that of extrapolating the p_{D}-function to times for which it has not been determined. In attempting to solve the problem, Gringarten et al. [53] applied correction factors to their theoretical type curves, which are seldom used in practice. An alternative solution was forwarded by Slider [54] who proposed the extrapolation of the bottom hole flowing pressure, $p_{\text{wf(ext)}}$, and hence the p_{D}-function so that equation 4.148 is replaced by

$$\sigma(p_{\text{i}} - p_{\text{wf(ext)}}) = p_{\text{D}}(t_{\text{D}} + \Delta t_{\text{D}})$$

and subtraction of the basic buildup equation 4.30 from this gives

$$\sigma(p_{\text{ws}} - p_{\text{wf(ext)}}) = p_{\text{D}}(\Delta t_{\text{D}}) \tag{4.150}$$

which means that if $\log \Delta p$ is expressed using the pressure difference on the left-hand side of this equation then direct use of the Gringarten curves for buildup analysis is appropriate. But such an extrapolation of the pressure drawdown, and implicitly the p_{D}-function, must be regarded as somewhat arbitrary. Perhaps the most commonly used method of attempting to overcome the difficulty in applying drawdown type curves directly to buildup analysis is that presented by Agarwal [55]. He *assumed* that equation 4.149 could always be evaluated for transient conditions as

$$\sigma(p_{\text{ws}} - p_{\text{wf}}) = \tfrac{1}{2} \ln \frac{4 \, t_{\text{D}}}{\gamma} - \tfrac{1}{2} \ln \frac{4(t_{\text{D}} + \Delta t_{\text{D}})}{\gamma} + \tfrac{1}{2} \ln \frac{4 \Delta t_{\text{D}}}{\gamma} \tag{4.151}$$

that is

$$\sigma(p_{\text{ws}} - p_{\text{wf}}) = \tfrac{1}{2} \ln \frac{4}{\gamma} \left(\frac{t_{\text{D}} \times \Delta t_{\text{D}}}{t_{\text{D}} + \Delta t_{\text{D}}} \right) \tag{4.152}$$

and consequently it was suggested by Agarwal that if an engineering plot is made of

$$\log \Delta p = \log(p_{\text{ws}} - p_{\text{wf}}) \text{ versus } \log \frac{t \times \Delta t}{t + \Delta t} = \log \Delta t_{\text{e}}$$

then the Gringarten plots could be used directly for buildup analysis using the so-called "equivalent time", Δt_{e}. Unfortunately, on account of the basic assumption, the method is restricted in application to tests in which the total flow and buildup periods occur under purely transient conditions so that $p_{\text{D}}(t_{\text{D}} + \Delta t_{\text{D}})$ in equation 4.149 can be evaluated in its transient form as in equation 4.151. At the time of writing this text, it appears that no sound method has been presented for applying drawdown type curves to buildup analysis and the engineer must therefore exercise caution in their use. The reader will appreciate that the difficulty amounts to defining when the condition

$$p_D(t_D + \Delta t_D) \approx p_D(t_D) \tag{4.32}$$

breaks down to such an extent that the drawdown type curves cannot be matched with buildup data. This same condition occurs in semi-log plotting techniques (Horner and MDH) but causes no difficulty because when equation 4.32 is no longer satisfied, the pressure points will deviate away from the initial, linear buildup trend: either above or below. In this respect the condition is self checking. In type curve analysis, however, the is no such insurance because on the log–log plot there is no well defined shape characterising the period containing transience when equation 4.32 is also satisfied. Consequently, it is easy to mismatch an engineering buildup plot with a drawdown type curve without being aware of the error.

(b) Time derivative type curves

In 1983, Bourdet et al. [56] presented a paper in which he took the (dimensionless) time derivative of the Gringarten drawdown type curves, which were plotted in terms of

$$p_D' \frac{t_D}{C_D} \text{ versus } \frac{t_D}{C_D}$$

in which $p_D' = dp_D/d(t_D/C_D)$. The reason for this choice of presentation was because if, during the period of pure afterflow, equation 4.143 is differentiated with respect to t_D/C_D, the result is simply $p_D' = 1$ and multiplying both sides of this by t_D/C_D gives

$$p_D' \frac{t_D}{C_D} = \frac{t_D}{C_D} \tag{4.153}$$

meaning, as with the Gringarten type curves, that for the period dominated by afterflow, the left-hand side of the equation is a linear function of t_D/C_D (45º straight line — Fig. 4.65). Consequently, the early part of the derivative plots

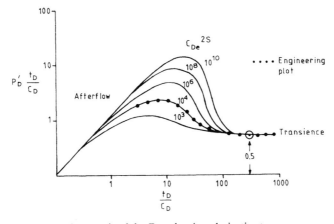

Fig. 4.65. Schematic of the Bourdet time derivative type curves.

superimpose on the normal type curves. More significantly, during the period of transience, the derivative of equation 4.22 with respect to t_D/C_D is

$$p'_D \frac{t_D}{C_D} = 0.5 \tag{4.154}$$

Therefore, for this condition all the derivative plots, which are for different values of t_D/C_D, converge on to a horizontal straight line (Fig. 4.65) with ordinate value of 0.5. It follows that if a well test contains the two elements of pure afterflow and transience, then matching the engineering plot against the two asymptotic extremes of the derivative type curves should assure a unique match on one of the distinctive curves, each of which is characterised by a given value of the parametric group $C_D e^{2S}$.

If the derivative curves are used for drawdown analysis (for which the convention associated with type curve analysis is again used: that the flowing time is denoted as Δt) then differentiating the basic drawdown equation 4.20, for transient flow leads to

$$\sigma \frac{dp_{wf}}{d\Delta t} \Delta t = \frac{\sigma}{2.303} \frac{dp_{wf}}{d(\log \Delta t)} = |0.5| = p'_D \frac{t_D}{C_D} \tag{4.154}$$

which demonstrates the equivalence of the ordinate of the derivative curves and real pressure derivatives. Performing the analysis manually an engineering plot is made of either

$$\frac{dp_{wf}}{d\Delta t} \Delta t \quad \text{or} \quad \frac{dp_{wf}}{2.303 \, d(\log \Delta t)} \quad \text{versus} \quad \Delta t \tag{4.155}$$

on transparent paper with log–log scales exactly the same as for the dimensionless derivative curves. The engineering plot is then superimposed on the type curves and moved laterally and vertically until coincidence is attained between the horizontal, transient sections. Then for the know value of the value of the derivative of the actual flowing pressure (match point), equation 4.154 can be solved to evaluate σ and hence the kh-product. Using the corresponding time match point, the wellbore storage can be determined from the abscissa using equation 4.147, as for normal type curve analysis, and C_D using equation 4.142. Finally, noting which of the curves best matches the engineering values between the afterflow and transient asymptotes identifies the value of $C_D e^{2S}$ and therefore S.

Just as for the Gringarten type curves, the Bourdet derivative plots are more commonly applied to pressure buildup analysis. Then, by direct analogy with equations 4.41 and 4.52, the relationship between the real and dimensionless pressure derivatives during transience is

$$\frac{\sigma}{2.303} \frac{dp_{ws}}{d(\log \Delta t)} = \frac{\sigma}{2.303} \frac{dp_{ws}}{d\left(\log \dfrac{t + \Delta t}{\Delta t}\right)} = |0.5| = p'_D \frac{t_D}{C_D} \tag{4.156}$$

The first of these corresponds to the Miller, Dyes, Hutchinson form of pressure buildup, while the second to that of Horner. The engineering plots should therefore be constructed as

$$\frac{\mathrm{d}p_{ws}}{2.303\,\mathrm{d}(\log \Delta t)} \quad \text{or} \quad \frac{\mathrm{d}p_{ws}}{2.303\,\mathrm{d}\left(\log \dfrac{t+\Delta t}{\Delta t}\right)} \text{ versus } \Delta t \tag{4.157}$$

Either plot is made on transparent log–log paper for superimposition on the derivative type curves. Defining a match point on the horizontal, transient asymptote, the analysis technique proceeds exactly as described above for drawdown analysis to determine, kh, C and S. In his original paper [56], Bourdet presented an argument that the engineering plot for buildup analysis should be made in terms of

$$\left(\frac{\mathrm{d}p_{ws}}{\mathrm{d}\Delta t}\right)\Delta t\left(\frac{t+\Delta t}{t}\right) \text{ versus } \Delta t \tag{4.158}$$

both on log–log scales (t retains its usual meaning as the final flowing time in this expression). But some slight mathematical manipulation will demonstrate that this version of the engineering plot is precisely the same as the Horner expression in equation 4.157. The engineering plots suffer from an inevitable scatter in the data points and although smoothing algorithms have been devised [23,24] the difficulty is diminishing rapidly with the improved resolution of pressures using modern gauges. Besides, as described in section 4.12, derivatives are mainly used in a qualitative sense to define the period of transience.

(c) Practical aspects

In practice, the Gringarten and Bourdet type curves are usually presented in superimposed form in a single plot, as shown in Fig. 4.66. This is possible because

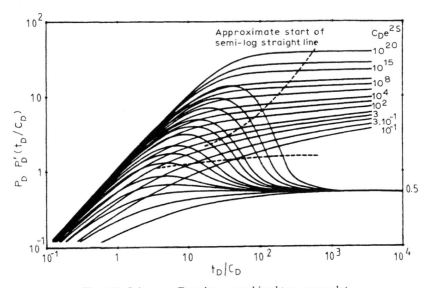

Fig. 4.66. Gringarten/Bourdet — combined type curve plot.

of their coincidence in form during the initial period of afterflow and permits the matching the pressure and pressure derivative engineering plots simultaneously on the dimensionless type curves which improves the uniqueness of the interpretation. This is the equivalent in log–log space of plotting the MDH or Horner buildups together with their time derivatives, in semi-log space, described in sections 4.12 and 4.13 and illustrated in Exercises 4.1, 4.2 and 4.3.

One of the advantages claimed for log–log type-curve analysis over the earlier semi-log interpretation methods is that it caters for afterflow associated with surface closure at the start of the buildup. It must be remembered, however, that this is "idealised" afterflow for a well with constant wellbore storage, C, for which each of the theoretical type curves has been generated. If there is any complication such as gas–oil segregation in the flow string, described in section 4.14b, then afterflow analysis using the type curves is quite unsuitable. The trend in exploration/appraisal well testing, using DST equipment, and to a lesser extent in routine development well testing (especially in gassy wells) is to close-in wells downhole thus reducing and in some cases almost eliminating the influence of afterflow. Under these circumstances the initial sections of the dimensionless pressure and derivative curves — which have the greatest character (resolution) and should facilitate the most accurate matches with the engineering plots, have greatly diminished use. In some cases, for instance, when there is firm downhole closure in a "hard" undersaturated oil reservoir and transience begins almost immediately after well closure, it becomes extremely difficult to distinguish with which of the type curves the test data coincide and therefore precludes the definition of the $C_D e^{2S}$ group and hence the skin factor itself. The situation is at its worst using the derivative plots, each of which converges on to the same horizontal straight line at the start of transience thus tending to remove uniqueness in the separate type curve matching. (The two tests described in Exercises 4.2 and 4.3 both have downhole closure leading to a lack of uniqueness by type curve analysis.) On account of this difficulty, proponents of log–log plotting techniques have suggested that wells should be deliberately closed-in at the surface to promote a significant degree of afterflow that would facilitate the use of the high-resolution early part of the type curves. Few operators would agree with this suggestion, especially when running expensive offshore tests at the exploration/appraisal phase. The reason is because at this stage, the degree and severity of afterflow is unpredictable and as an insurance against failure in obtaining a reliable and meaningful test analysis, downhole closure is favoured.

As noted in both the introduction to this chapter and in section 4.2a, there has been an increasing tendency in recent years to regard test analysis as a matter of curve fitting — encouraged by computer programs enabling engineers to manipulate complex mathematics in a manner undreamed of in the past. The most common way of doing this is by trying to match the Gringarten-Bourdet dimensionless log–log plots — particularly the latter, which is often referred to as the "global approach" to interpretation. The aim is to first isolate the transient pressure response and subsequently attempt to identify the nature of the system under test: boundary conditions, dual porosity behaviour etc. In doing so, the engineer specifies what he

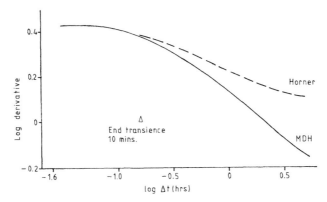

Fig. 4.67. Smoothed log–log derivative plots (Horner-MDH) of the pressure buildup data presented in Exercise 4.3 (steady-state test).

believes to be the system, based on the inspection of the derivative engineering plot. The computer program will then generate a compatible p_D plot and its derivative and attempt to match the actual test data. (This, it will be recognised, is the reverse of the manual curve matching described earlier in the section.) There are dangers implicit in applying this technique, however, not least of which is a dependence on the manner in which the engineering plot is constructed: using either the methods of Horner or MDH. the difference is illustrated in Fig. 4.67, using the buildup data for the steady-state test described in Exercise 4.3. As can be seen, the period containing transience lasts for about 10 minutes in both engineering plots but thereafter the shapes of the derivative trends differ markedly. Both buildup plots (Figs. 4.47 and 4.48) are concave downwards for large values of the closed-in time, meaning that their time derivatives decrease continuously; but the degree of downward curvature from the plateau is largely dependent on the manner in which the mathematical conditions (equations 4.31, 4.32 and 4.44) which govern the duration of the plateau, "break-down". The difference in the engineering plots has nothing to do with the nature of the system under test since it is the same in both cases — it is simply a manifestation of the slight difference in mathematical approach in generating the engineering plot: Horner or MDH. In this respect, the engineer must be particularly careful in not trying to "read" too much into the late-time derivative data. For instance, this author has frequently seen the derivative plots of a test including an element of pressure support confused with that for a dual porosity system [1]. The drawdown performance for the latter is illustrated in Fig. 4.68a. Section $A–B$ is the early, transient response as the better quality (fissured) rock produces directly into the wellbore (section 4.2b). $B–C$ represents the transition period as oil is produced from the tight matrix into the main flow channels while finally $D–E$, which is parallel to $A–B$, represents a further infinite acting response as the total system, matrix plus flow channels, contributes to production. The corresponding time derivative engineering plot is shown in Fig. 4.68b and naturally contains two transient plateaus separated by a dip representing the transition phase. The buildup

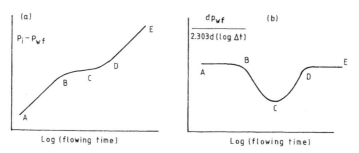

Fig. 4.68. Dual porosity flow: (a) pressure drawdown; (b) equivalent derivative plot.

performance of a dual porosity system displays exactly the same form of derivative plot illustrated in Fig. 4.68b for drawdown and the danger is that on viewing the buildup performance for a test with pressure support (Fig. 4.67), the engineer might confuse the pattern with that for a dual porosity test in which the drawdown was not continued beyond point C (Fig. 4.68b) so that full equilibrium of the matrix flow had not been achieved and the element *C–D* was present neither in the drawdown nor buildup responses.

4.22. CONCLUSIONS

This final section attempts to unify what the author regards as some of the more important technical arguments presented in the chapter. In particular, the means of avoiding many of the traditional errors in test design and interpretation are explained and suggestions made for generally *smartening-up* procedures that are intended to make testing more cost effective. Perhaps it is not surprising that the concentration of theme is on the meaning and identification of the early straight line on semi-log buildup plots. Failure to identify it correctly is regarded as the root cause of most of the difficulties and misunderstandings affecting test design and interpretation (section 4.15).

(a) The elusive straight line

Any rate perturbation in an oilwell will give rise to a transient pressure response during which the reservoir appears infinite in extent. It will be manifest as a linear relationship between the changing wellbore pressure and the logarithm of some time function, as follows:

Drawdown: p_{wf} vs. $\log t$ *Buildup (MDH):* p_{ws} vs. $\log \Delta t$

Buildup (HOR): p_{ws} vs. $\log \dfrac{t + \Delta t}{\Delta t}$

It is the essential opening move in any test interpretation to identify correctly the early straight line, the slope of which provides values of kh and S which, in turn, define the p_D-function characteristic of the formation under test (equation 4.21). Unless these can be quantified accurately, any attempt at more sophisticated interpretation is impossible.

The conditions for early linearity (sections 4.10, 12 and 13) are:

MDH drawdown	*MDH buildup*	*Horner buildup*	
$p_D(t_D) = \frac{1}{2} \ln \dfrac{4t_D}{\gamma}$	$p_D(\Delta t_D) = \frac{1}{2} \ln \dfrac{4\Delta t_D}{\gamma}$	$p_D(\Delta t_D) = \frac{1}{2} \ln \dfrac{4\Delta t_D}{\gamma}$	(4.22/23)
	$p_D(t_D + \Delta t_D) = p_D(t_D)$	$p_D(t_D + \Delta t_D) = p_D(t_D)$	(4.32)
		$\ln(t_D + \Delta t_D) = \ln(t_D)$	(4.44)

For a drawdown test the duration of the early straight line is dictated exclusively by the transient condition but for buildup testing matters are somewhat more complex in that additional conditions must be satisfied for linearity. Consideration of these leads to the following conclusions concerning the straight line:

– Horner potting is liable to be more complex than that of MDH
– early linearity of a buildup does not depend on transience alone
– there is, in fact, no such thing as a *pure*, initial, semi-log straight line.

The reason why the Horner buildup can be more complex results from the fact that there are three conditions that must be satisfied for linearity compared to only two for MDH analysis. This sometimes leads to a tendency for multiple straight lines to emerge on Horner plots, as illustrated by the steady-state buildup in Exercise 4.3, Fig. 4.48.

The popular literature gives the impression that the early straight line on a buildup is governed by transience alone but, as has frequently been referred to in this chapter, the straight line *contains* the element of transience but its duration is also dictated by other factors: conditions 4.32 and 4.44.

The third conclusion means that we engineers have been deceived since the early 1950's: the straight line on a buildup does not exist in an absolute sense. That is, suppose a well has been produced for 100 hours, then even after 10 minutes of closure the necessary conditions for linearity stated in equations 4.32 and 4.44 can never be exactly satisfied. For a considerable period after well closure they may be correct to the second or third place of decimal but they are never precise. In viewing semi-log buildup plots directly the resulting minor deviations are hardly evident but they do tend to be revealed in the more sensitive time derivative plotting, especially for "fast" buildups such as occur when there is an element of pressure support. Quite often, there is a slight tilt in the early derivative plateau.

(b) Saving money in well testing

In a conventional pressure buildup test, the test is effectively over at the end of the drawdown period when the dynamic *message* from the formation is already recorded on the pressure gauge in the wellbore, its expression being contained in the p_D-function which is characteristic of the reservoir under test (equation 4.21). Therefore, it should not require a lengthy buildup to unravel the message, which indeed is the case, as clearly emphasised with the MDH plotting technique. This asserts that the only requirement is the identification of the early straight line (equation 4.33 — which contains the p_D-function) and its extrapolation to either the initial or average pressure, as appropriate (equation 4.38), which is all that is necessary to define the system under test. The late-time MDH pressure data are discarded. The same is true for Horner analysis, if considered correctly. Equation 4.45, defining the early straight line is the only one used in the analysis and again, it will be noted, contains the *message* of the test: the p_D-function at the time of well closure. This is emphasised by equations 4.105 and 4.115 which apply the extrapolation of the early Horner straight line to calculate average well pressures in development well testing.

Therefore, why has the Industry always indulged in the practice of lengthy pressure buildups and where did the idea come from in the first place? It seems that the tradition stems from the original paper of Horner from 1951. This is an excellent treatise in which the author states his technical arguments in a clear, modest and almost apologetic manner, which was the practice at that time. Unfortunately, in selecting examples to illustrate his buildup plotting method, Horner chose two for which it is theoretically valid to extrapolate the late time trend in pressures in a linear fashion to infinite closed-in time to determine some meaningful reservoir pressure, in this case — p_i. These were for an infinite acting reservoir and for single fault analysis, as illustrated in Exercises 4.1 and 4.2 of this chapter (Figs. 4.21 and 4.32). Furthermore, the only buildup equations stated in Horner's paper were equations 4.47 and 4.72, for the two specialised cases studied. The basic equation of the early straight line:

$$\sigma(p_i - p_{wsl}) = \tfrac{1}{2} \ln \frac{t + \Delta t}{\Delta t} + p_D(t_D) - \tfrac{1}{2} \ln \frac{4t_D}{\gamma} \qquad (4.45)$$

(section 4.13) was not stated and did not appear in the literature until it was presented in Professor Ramey's classic paper on buildup analysis written in 1968 [29] "*A General Pressure Buildup Theory for a Well in a Closed Drainage Area*". As a consequence of regarding the examples in Horner's paper as stating the general rule rather than illustrating exceptional cases, it seems that the common belief is that it is correct to perform a linear extrapolation of the late time pressures on a Horner plot to determine some meaningful reservoir pressure, which is usually not the case. In the derivation of equation 4.45, described in section 4.13, the basic buildup equation, 4.30, is examined for *small* Δt such that the three conditions stated above are satisfied. Therefore, there is simply no reason for looking for or using linear trends in pressures on Horner plots for large values of the closed-in time. Nobody would attempt to use the late time buildup pressures on an MDH plot. Therefore, what is it that sanctifies the Horner plot? That somehow the unnecessary inclusion of the flowing time, t, permits the meaningful interpretation of the late time pressure data.

One of the strongest arguments against long pressure buildups is illustrated in Fig. 4.69a and b, which should be viewed in conjunction with the basic buildup

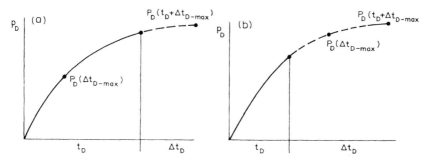

Fig. 4.69. p_D-functions: (a) long drawdown, short buildup, (b) short drawdown, long buildup.

equation:

$$\sigma(p_i - p_{ws}) = p_D(t_D + \Delta t_D) - p_D(\Delta t_D) \tag{4.30}$$

For any length of drawdown and buildup, the first p_D-function must be necessarily evaluated for a time which is greater than the total flowing time, t, for which it is physically defined while flowing at a constant rate (equation 4.21). As described in section 4.8, and 4.21a it is the extrapolation of the function from t to $t + \Delta t$ that can naturally cause difficulties in buildup interpretation. For a short buildup (Fig. 4.69a), while the first p_D-function in equation 4.30 is in this category, the second, even when evaluated for the maximum closed-in time, $p_D(\Delta t_{D\text{-max}})$, is still in the time range for which it has been measured. Therefore, the interpreter has one foot in reality, at least. For a buildup that is longer than the drawdown (Fig. 4.69b), however, the situation is quite different because now, at the end of the buildup, both $p_D(t_D + \Delta t_{D\text{-max}})$ and $p_D(\Delta t_{D\text{-max}})$ lie outwith the time range for which they have been defined. Under these circumstances the engineer has entered mathematical *fantasy land*. Using test analysis computer software packages, it is usually possible to empirically define some function that will curve-fit the difference in p_D-functions in equation 4.30, but how they can ever be validated against physical reality is always in doubt (refer to section 4.16d). Fortunately, development decisions are seldom based on the interpretation of minor "wiggles" in late-time Horner pressure buildups and therefore the indulgence can usually be regarded as simply an academic pursuit, albeit a very expensive one.

The above description is appropriate for exploration/appraisal wells but when it comes to development well testing in mature fields, the attempt to rely on the interpretation of late-time buildup pressures can be even more damaging. Consider the completely bounded reservoir shown in Fig. 4.7. Under stable producing conditions, each well carves-out its own no-flow boundary between itself and its neighbours such that, for semi-steady state flow, it drains a volume that to a first approximation is proportional to its rate. When any one of the wells is closed in for a routine pressure buildup it must be appreciated that what is being attempted is to determine the average pressure, $\bar{p}$, within the drainage area of the well *at the time of closure* and the methods for evaluating this pressure, described in sections 4.19a and b, rely on mathematics that is geared to do just that. As soon as the particular well is closed-in for a survey, its boundary condition starts to change, as will the boundary conditions for all the other wells, the rate at which the change occurs being proportional to the diffusivity constant $k/\phi\mu c$. Consequently, to attempt to interpret the late-time buildup pressures is to deal also with an unquantifiable variable boundary condition problem, which amounts to well interference effects. The worst mistake is to extrapolate any increasing late-time pressure trend in a linear fashion to determine something referred to as p^* which, as described in section 4.15b, has no evident physical meaning whatsoever.

The same applies in tests in wells that have open or mixed boundary conditions (section 4.19c) so that there is an element of pressure support. The author has noted this type of misinterpretation in extended well tests (EWTs). The well may be produced for several months and then closed-in for a similar period to determine the final pressure from the extrapolation of the late-time Horner buildup plot. But naturally during the lengthy buildup the continual influx of aquifer water into the reservoir may return the pressure to the initial value, or very close to it. The conclusion is therefore reached that there is strong natural water influx, which precludes the calculation of the STOIIP. This may be a correct observation but the relevant pressure for use in material balance calculations is the average pressure determined at the time of well closure, as described in section 4.19c not the pressure after a long restorable buildup.

What is the root of the problem is the unnecessary inclusion of the flowing time in Horner buildup analysis. With it comes the false impression of the need to extrapolate to infinite closed-in time which gives rise to the practice of lengthy buildups to minimise the length of the extrapolation. As illustrated in the chapter, however, provided a test could be run under ideal conditions, then the minimum time for which a well need be closed-in for a buildup is *zero hours*. That is because all the information from the test is contained in the p_D-function in the drawdown equation:

$$\sigma(p_i - p_{wf}) = p_D(t_D) + S \tag{4.21}$$

and as illustrated in exercise 4.2, drawdown analysis alone can be used to theoretically characterise the formation, in this case define the position of a single sealing fault, without the necessity of closing-in the well at all. Furthermore, equation 4.21 is much easier to handle than the basic buildup equation, 4.30, the latter expressing a difference in the p_D-function. The same applies to test design in mature fields. As illustrated in section 4.20a, simply changing the rate of a producing well will cause a pressure response, containing the element of transience, which can be interpreted to determine the average pressure within the drainage area of the well at the time of the rate change, which is the main purpose in development well testing: well closure is not necessary.

As soon as the subject of drawdown or two rate testing is mentioned, however, out comes the general complaint that flow tests cannot be controlled as easily as buildups. This is quite correct but perhaps the one of the reasons for this is because historically the same effort has not been applied in even attempting to control drawdown tests. Instead, most of the technological improvements have been directed at buildup testing in the development of sophisticated DST equipment with such features as the facility for downhole closure. It is suggested that if a similar effort were directed towards drawdown tests the whole subject of test design and interpretation could be improved.

Above all, the practice of indulging in lengthy pressure buildups should be given serious consideration by the engineer. Not only is it a waste of money but for the reasons stated throughout the chapter, and reinforced in this section, long buildups are not theoretically justified. If they could be eliminated then so too would the most common errors associated with the almost universal malpractice of extrapolating the late-time pressure trends of Horner plots. To get the best value for money, tests should be conducted with lengthy drawdown periods at as high a rate as practicably possible, followed by a brief buildup sufficient to unambiguously define the *early* straight line and extrapolate it, as substantiated by the basic mathematics of pressure buildup behaviour, to learn something useful about the formation under test.

(c) Identification of the correct early straight line

Considering the importance attached to the early straight line of a pressure buildup in this chapter, it would seem appropriate if some sound practical advice could be given on how best to identify it. Obviously, it must be the first linear section on an MDH or Horner plot that emerges following any period of afterflow. Furthermore, the time derivative of either buildup plot should have a near plateau value of *0.434 m* for the duration of the straight line, but these requirements together still do not appear to be sufficient to avoid errors occurring in its identification. There is a method, however, that has already been alluded to in the chapter, that should permit the identification of the linear section in an unambiguous fashion. This relies on the fact that the conditions for early linearity are different for MDH and Horner plotting. The latter requires that three conditions be satisfied, as defined by equations 4.23, 4.32 and 4.44, which are detailed above, while the MDH straight line is only dependent on the first two of these being valid. Therefore, the linear sections on both plots must be of different duration and this can be detected by comparing the time derivatives of the buildups. These are derived in sections 4.12 and 4.13 as:

$$p'_{ws\,(MDH)} = \frac{dp_{ws}}{2.303d(\log \Delta t)} \tag{4.41}$$

and:

$$p'_{ws\,(HOR)} = \frac{dp_{ws}}{2.303d(\log(t + \Delta t)/\Delta t)} \tag{4.52}$$

If a plot is made of the two derivatives together versus $\log \Delta t$, simply for demonstration purposes, then during the period dominated by afterflow and for the early period when all the conditions for linearity are satisfied, the derivative points on the MDH and Horner plots will be coincidental and this defines the early straight line. Eventually, the time derivative points will separate from one another which signifies the difference in the conditions governing linearity. There is always a tendency for the MDH derivative plot to curve downwards to a greater extent than that of Horner, which merely reflects the limiting condition for the former that as $p_{ws} \longrightarrow p_i$, $\Delta t \longrightarrow \infty$, which causes a continual reduction in slope of the basic MDH plot for large values of Δt. To illustrate the effectiveness of the technique, MDH and Horner derivative plots using the data from Exercises 4.1, 4.2 and 4.3 are presented in Fig. 4.70a–c, and may be described as follows.

The test in Exercise 4.1 is for a purely infinite acting reservoir for which the MDH and Horner time derivatives are plotted in conjunction in Fig. 4.70a. For this test, the Horner plot is linear for all values of Δt because for a purely transient test:

$$p_D(t_D + \Delta t_D) = \tfrac{1}{2} \ln \frac{4(t_D + \Delta t_D)}{\gamma}$$

for all values of the time argument, which reduces the buildup equation, 4.43, to:

$$\sigma(p_i - p_{wsl}) = 1.151 \log \frac{t + \Delta t}{\Delta t}$$

This continual plateau for the Horner derivative is evident in Fig. 4.70a. It is in this test that the MDH plotting technique proved to be so inadequate in providing a straight line of any duration. In fact, relying on the MDH plot alone, the engineer would have the greatest of difficulty in selecting the correct straight line. The problem is that its length is dictated entirely by the breakdown of the condition that $p_D(t_D + \Delta t_D) = p_D(\Delta t_D)$, which happens very early in the buildup. The derivative plots in figure 4.70a illustrate this point, with separation occurring after 40 minutes. The usefulness of this plot is that the continual Horner plateau and rapidly diminishing values of the MDH derivative points are diagnostic of a purely infinite acting system.

Exercise 4.2 is for a well located 415 ft from a single sealing fault. The derivative plots (Fig. 4.70b) indicate that separation occurs after about 90 minutes but in this case the situation is completely reversed compared to Exercise 4.1, in that the MDH plot retains a plateau for much longer than that of Horner. As described in Section 4.16b, the effect of a single fault is to double the Horner slope and therefore its derivative, whereas its influence on the MDH plots is more subtle. The natural tendency for the MDH plot to curve downwards for large Δt is offset by the increase of $p_D(t_D + \Delta t_D)$ due to the presence of the fault which leads to the extension of the apparent straight line.

In the third example, for the steady-state test described in Exercise 4.3, the difference between the Horner and the MDH derivative plots (Fig. 4.70c) is less pronounced than in the other two examples and both demonstrate downward curvature for large Δt which is the inevitable "roll over" effect following a flow period in which there is an element of pressure support. The two derivative plots are coincidental for only 10 minutes and in this case the duration of the early straight line is dictated entirely by the breakdown of the transient condition. This can be inferred from the MDH linear equation, 4.33, in which the condition that $p_D(t_D + \Delta t_D) = p_D(t_D)$ should, for a steady-state test, be satisfied for all values of

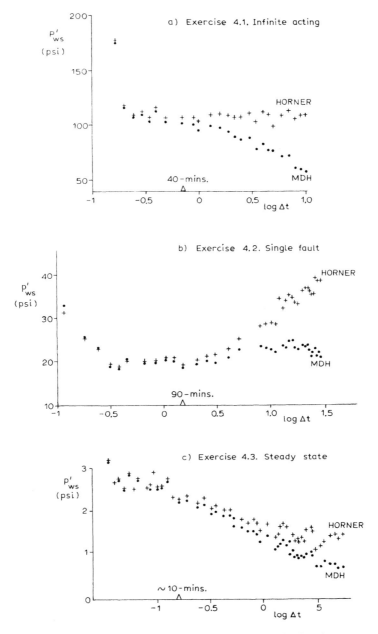

Fig. 4.70. Comparison of Horner and MDH-time derivative plots.

Δt, meaning that it can only be the departure from the transient condition that influences the duration of linearity. Since the Horner straight line lasts also for only about 10 minutes, the same conclusion should apply.

The MDH plotting technique has never been developed by the Industry to the same extent as Horner, largely because its poor performance in infinite acting tests gave people the impression that it always gave a shorter straight line than the Horner plot. This, however, cannot be substantiated in general. Precisely how long the straight line lasts on either form of buildup is a complex matter dependent on the validity of the necessary conditions for linearity which are different for the two types of plot. Besides, what does it matter how long a straight line lasts since it only requires a few points to define unambiguously. Therefore, in applying the above method of comparing the time derivatives no great effort need be extended in considering why the two plots should separate from one another. It is suffice to appreciate that they must do so and their early coincidence defines the correct straight line to use in the buildup analysis. Making this plot must surely warn the engineer of the theoretical error in seeking and extrapolating late straight lines on the Horner plot. In exercise 4.3, for instance, it would avoid the selection of the late straight line (Fig. 4.48), for which there is a flattening in the Horner time derivative (Fig. 4.70c), that is such a common mistake in analysing roll-over buildups.

The chapter is not intended to promulgate any form of competition between Horner and MDH analysis techniques, quite the contrary. It is instead recommended that all test interpretation should start by making *semi-log plots* of Horner and MDH buildups together with their time derivatives, as illustrated in Exercises 4.1–4.3, in which the early straight line is identified by comparison of the derivative plots, as described in this section. If the analysis has been performed correctly the results of Horner and MDH interpretations should be the same. This would give the engineer confidence in extending the analysis in any attempt at more sophisticated formation characterisation.

REFERENCES

[1] Gringarten, A.C.: Interpretation of Tests in Fissured and Multilayered Reservoirs with Double Porosity Behaviour: Theory and Practice, JPT, April 1984, 549.

[2] Varotsis, N. and Guleze, P.: On Site Reservoir Fluid Properties Evaluation, JPT, August 1990, 1046–1052.

[3] Dake, L.P.: Fundamentals of Reservoir Engineering, Elsevier, Amsterdam, 1978.

[4] Matthews, C.S. and Russell, D.G.: Pressure Buildup and Flow Tests in Wells, SPE Monograph, 1967.

[5] Earlougher, R.C.: Advances in Well Test Analysis, SPE Monograph, 1977.

[6] Ramey, H.J. Jr., Kumar, A. and Gulati, M.S.: Gas Well Test Analysis, Under Waterdrive Conditions, American Gas Association, Arlington, Va., 1973.

[7] Kumar, A. and Ramey, H.J. Jr.: Well-Test Analysis for a Well in a Constant Pressure Square, Soc. Pet. Eng. J., April 1974, 107–116.

[8] Larsen, L.: Wellbore Pressures in Reservoirs with Constant-Pressure or Mixed No-Flow/Constant Pressure Outer Boundary, JPT, September 1984.

[9] Krueger, R.F.: An Overview of Formation Damage and Well Productivity in Oilfield Operations, JPT, February 1986, 131–152.

[10] Howard, G.C. and Fast, C.R.: Hydraulic Fracturing, SPE Monograph 1970.

[11] Cinco-Ley, H. Samaniego, F.V. and Dominguez, N.A.: Transient Pressure Behaviour for a Well with a Finite- Conductivity Vertical Fracture, Soc. Pet. Eng. J., August 1978, 253–264.

[12] Cinco-Ley, H. and Samaniego, F.V.: Transient Pressure Analysis for Fractured Wells, JPT, September 1981.

[13] Dranchuk, P.M. and Quon, D.: Analysis of the Darcy Continuity Equation, Producers Monthly, October 1967.

[14] Al-Hussainy, R., Ramey, H.J., Jr. and Crawford, P.B.: The Flow of Real Gases Through Porous Media, JPT, May 1966, 624–636.

[15] Agarwal, R.G.: Real Gas Pseudo Time — A New Function for Pressure Buildup Analysis of MHF Gas Wells, SPE Fall Meeting, Las Vegas, Nev., September 1979.

[16] Lee, J.W. and Holditch, S.A.: Application of Pseudo Time to Buildup Test Analysis of Low-Permeability Gas Wells with Long Duration Wellbore Storage Distortion, JPT, December 1982, 2878–2887.

[17] Raghavan, R., Scorer, J.D.T. and Miller, F.G.: An Investigation by Numerical Methods of the Effect of Pressure Dependent Rock and Fluid Properties on Well Flow Tests, Soc. Pet. Eng. J., June 1972.

[18] Samaniego, F.V., Brigham, W.E. and Miller, F.G.: Performance Prediction Procedure for Transient Flow of Fluids Through Pressure Sensitive Formations, JPT, June 1979, 779–786.

[19] Cobb, W.M. and Dowdle, W.L.: A Simple Method for Determining Well Pressures in Closed Rectangular Reservoirs, JPT, November 1973, 1305–1306.

[20] Matthews, C.S., Brons, F. and Hazebroek, P.: A Method for the Determination of Average Pressure in a Bounded Reservoir, Trans. AIME, 1954, Vol. 201, 182–191.

[21] Nisle, R.G.: How to Use the Exponential Integral, Pet. Eng., August 1956, 171–173.

[22] Miller, C.C., Dyes, A.B. and Hutchinson, C.A., Jr.: The Estimation of Permeability and Reservoir Pressure from Bottom Hole Pressures Buildup Characteristics, Trans., AIME 1950, Vol. 189, 91–104.

[23] Bourdet, D., Ayoub, J.A. and Pirard, Y.M.: Use of Pressure Derivative in Well Test Interpretation, SPE Paper 12777 presented at the SPE California Regional Meeting, April 1984.

[24] Clark, D.G. and van Golf Racht, T.D.: Pressure Derivative Approach to Transient Test Analysis: A High Permeability North Sea Example, JPT, November 1985, 2023–2040.

[25] Theis, C.V.: The Relationship Between the Lowering of the Piezometric Surface and the Rate and Duration of Discharge Using Groundwater Storage, Trans. AGU, 1935, 519.

[26] Horner, D.R.: Pressure Buildup in Wells, Proc., Third World Pet. Congr., Leiden, 1951, 503.

[27] Kazemi, H. et al.: Complexities of the Analysis of Surface Shut-In Drillstem Tests in an Offshore Volatile Oil Reservoir, JPT, January 1983, 173–177.

[28] Fair, W.B.: Pressure Buildup Analysis with Wellbore Phase Redistribution, Soc. Pet. Eng. J., April 1981, 259–270.

[29] Ramey, H.J., Jr. and Cobb, W.M.: A General Pressure Buildup Theory for a Well in a Closed Drainage Area, JPT, December 1971, 1493–1505.

[30] Earlougher, R.C., Jr.: Practicalities of Detecting Faults from Buildup Testing, JPT, January 1980, 18–20.

[31] Davis, E.G., Jr. and Hawkins, M.F., Jr.: Linear Fluid-Barrier Detection by Well Pressure Measurements, JPT, October 1963, 1077–1079.

[32] Tiab, D. and Kumar, A.: Detection and Location of Two Parallel Faults Around a Well, JPT, October 1980, 1701–1708.

[33] Ehlig-Economides, C. and Economides, M.J.: Pressure Transient Analysis in an Elongated Linear Flow System , Soc. Pet. Eng. J., December 1985, 839–847.

[34] Prasad, R.K.: Pressure Transient Analysis in the Presence of Two Intersecting Boundaries, JPT, January 1975, 89–96.

[35] Yaxley, L.M.: Effect of a Partially Communicating Fault on Transient Pressure Behaviour, SPE Formation Evaluation, December 1987, 590–598.

[36] Ershaghi, I. and Woodbury, J.J.: Examples of Pitfalls in Well Test Analysis, JPT, February 1985, 335–341.

[37] Streltsova-Adams, T.D.: Pressure Transient Analysis for Afterflow — Dominated Wells Producing from a Reservoir with a Gas Cap, JPT, April 1981, 743–754.

[38] Brown, M.E. and Ming-Lung, M.: Pressure Buildup Analysis of Prudhoe Bay Wells, JPT, February, 1982, 387–396.

[39] McKinley, R.M. and Streltsova, T.D.: Early Time Pressure Buildup Analysis for Prudhoe Bay Wells, JPT, February 1984, 311–319.

[40] Larsen, L. and Kviljo, K.: Variable Skin and Cleanup Effects in Well Test Data, SPE Formation Evaluation, September 1990, 272–276.

[41] Dietz, D.N.: Determination of Average Reservoir Pressures from Build Up Surveys, JPT, August 1965, 955–959.

[42] Earlougher, R.C., Jr., Ramey, H.J., Jr. et al.: Pressure Distribution in Rectangular Reservoirs, JPT, February 1968, 199–208.

[43] Cobb, W.M. and Dowdle, W.L.: A Simple Method for Determining Well Pressures in Closed Rectangular Reservoirs, JPT, November 1973, 1305–1306.

[44] Peaceman, D.W.: Interpretation of Well-Block Pressures in Numerical Reservoir Simulation, Soc. Pet. Eng. J., June 1978, 183–194.

[45] Peaceman, D.W.: Interpretation of Well-Block Pressures in Numerical Simulation with Nonsquare Grid Blocks and Anisotropic Permeability, Soc. Pet. Eng. J., June 1983, 531–552.

[46] van Poollen, H.K., Breitenbach, E.A. and Thurnau, D.H.: Treatment of Individual Wells and Grids in Reservoir Modelling, Soc. Pet. Eng. J., December 1968, 341–346.

[47] van Poollen, H.K.: Radius of Drainage and Stabilization — Time Equations, Oil and Gas J., September 1964, 138–146.

[48] Johnson, P.W.: The Relationship Between Radius of Drainage and Cumulative Production, S.P.E., Formation Evaluation, March 1988, 267–270.

[49] Ramey, H.J., Jr.: Short Time Well Test Data Interpretation in the Presence of Skin Effect and Wellbore Storage, JPT, January 1970, 97–104.

[50] Earlougher, R.C., Jr. and Kersch, K.M.: Analysis of Short-Time Transient Test Data by Type Curve Matching, JPT, July 1974, 793–800.

[51] McKinley, R.M.: Wellbore Transmissibility from Afterflow-Dominated Pressure Buildup Data, JPT, July 1971, 863–872.

[52] Agarwal, R.G., Al-Hussainy, R. and Ramey, H.J., Jr.: An Investigation of Wellbore Storage and Skin Effect in Unsteady Liquid Flow: I. Analytical Treatment, Soc. Pet. Eng. J., September 1970, 279–290.

[53] Gringarten, A.C. et al.: A Comparison Between Different Skin and Wellbore Storage Type Curves for Early-Time Transient Analysis, SPE Paper 8205, 1979, 54th Fall Conference, Las Vegas, Nev.

[54] Slider, H.C.: A Simplified Method of Pressure Buildup Analysis for a Stabilized Well, JPT, September 1971, 1155.

[55] Agarwal, R.G.: A New Method to Account for Producing Time Effect When Drawdown Type Curves are Used to Analyse Pressure Buildup and Other Test Data, paper SPE 9289, 1980 Annual Fall Conference, Dallas, Texas.

[56] Bourdet, D. et al.: A New Set of Type Curves Simplifies Well Test Analysis, World Oil, May 1983, 95–106.

[57] Ehlig-Economides, C.A., Joseph, J.A., Ambrose, R.W., Jr. and Norwood, C.: A Modern Approach to Reservoir Testing, JPT, December 1990, 1554–1563.

Chapter 5

WATERDRIVE

5.1. INTRODUCTION

This chapter focuses on the purpose and practice of engineered waterdrive which is applied to enhance and accelerate oil recovery. It opens with a description of the factors most influencing the successful application of the process. The area studied is the North Sea, one of the first major provinces where waterdrive was elected as the principal recovery mechanism from the outset. There is a particular relevance in considering offshore projects in which the responsibilities and constraints faced by engineers are much more demanding than for land developments.

The main method of studying waterdrive is through the application of numerical simulation models and there is no attempt in the chapter to suggest a return to analytical techniques. Instead, the basic mechanics of waterflooding is studied with a view to enhancing the engineer's appreciation of how simulators function in development studies. In this respect, the one plea that is made is for the revival of the concept of the fractional flow of water which seems to have "gone missing" since the advent of simulation and, it is argued, is the key to understanding any form of displacement process. In fact, the whole purpose in performing waterdrive efficiency calculations is the generation of a relationship between fractional flow and oil recovery, whether achieved by analytical means or simulation.

Waterdrive in macroscopic reservoir sections occurs on the scale of flooding in hillsides rather than core plugs. At this level, there are three factors which govern the oil recovery efficiency: mobility ratio, heterogeneity and gravity. Precisely how they interact requires careful consideration and, as demonstrated, can sometimes provide surprising results in enhancing or downgrading oil recovery by waterdrive. Correct evaluation of the influence of the three factors amounts to paying rigorous attention to detail in determining the vertical sweep efficiency across sand sections, E_v, and there is a lengthy description of how the input data to studies, incorporating all the observed heterogeneity, must be handled to assure accuracy in its determination. The vertical sweep is implicity input to areal simulation models in the form of thickness averaged or pseudo-relative permeabilities. The simulator is then required to solve for the areal sweep, E_A, through the successful tracking of areal fluid movement. The chapter concludes with a description of how best to operate simulation models in the study of waterdrive and examines the development of some difficult waterdrive fields.

5.2. PLANNING A WATERFLOOD

(a) Purpose

Engineered waterdrive is the principal form of secondary recovery practised by the industry for the obvious reason that in many producing areas water is in plentiful supply and is therefore inexpensive. Not only that, but it is usually more stable than the alternative form of secondary flooding-gas drive, which is described in Chapter 6. Waterdrive serves two purposes in maintaining the reservoir pressure which energises the system and in displacing the oil towards the production wells.

Waterdrive has a lengthy history of application in the oil industry [1] but in many cases as an afterthought in mature fields with significant production history: applied once the pressure had fallen below the bubble point to improve upon the solution gas drive process. It is in offshore developments where engineered waterdrive has come to the fore and particularly in environments such as the North Sea which is the first major development area in which operators elected to apply the technique from the outset and platforms were designed accordingly. One of the main reasons for this choice of recovery mechanism was a straightforward matter of insurance against the failure event.

An operator may suspect that the oil reservoir has a well connected aquifer as depicted in Fig. 5.1a which, considering the general high level of permeabilities in most North Sea fields, could supply natural water influx that could dispense with the need and expense of water injection. What operators also realised, however, was that there was a risk that the oil column and aquifer may not be continuous as illustrated in Fig. 5.1b, in which the two are separated by a sealing fault. The likelihood of such segregation being undetected at the appraisal stage of development is heightened by the fact that data are collected under static conditions, as described in Chapter 1, which gives the minimum of information on areal communication. In a land development, the operator would have the opportunity of observing the field's performance, perhaps for a period of years, before having to finally decide on whether water injection was necessary or not. Offshore, however, the decision must be taken at the start of the development and the platform designed for waterdrive.

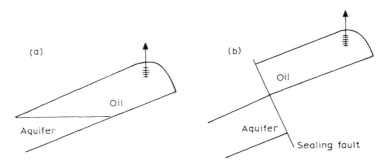

Fig. 5.1. Oil reservoir–aquifer configurations.

The huge steel and concrete structures standing in 500 ft of water could not have had their facilities upgraded with all the necessary equipment for injection following their installation. Lack of natural water influx would therefore have meant reliance on depletion drive which, for the generally undersaturated North Sea oilfields, would have resulted in greatly reduced oil recoveries, failure to achieve target oil production rates and economic failure. Consequently, with few exceptions, North Sea project design incorporated planing for engineered waterdrive.

An example of how things might have gone wrong if such design had not been catered for is provided by the initial production performance of the Thistle Field in the East Shetland Basin which came on stream in 1978 [2]. An early structural map on the top of the massive sand section is shown in Fig. 5.2.

The field is enclosed by a massive fault to the west and further faults to the north and south but following the appraisal stage it was believed that the central field area was open and would receive natural pressure support from an extensive aquifer to the east. The development started with a group of high-rate production wells in the western crestal area of the field but before long it was noticed that well pressures and rates were falling dramatically, meaning a lack of the anticipated natural water influx. The presence of a fault was inferred just to the west of wells 02A and 03A but its position was never exactly defined even following refined seismic surveys. The centrally located Thistle platform was fully equipped for water injection but the whole strategy for waterdrive had to be accelerated and the injection well locations altered. The field rate was reduced to arrest the serious pressure decline while injection wells were drilled in the down-flank areas of the new central fault block. This meant that instead of drilling a few injection wells far to the east to support the expected water influx, these had to be drawn back towards the centre of the field and supplemented to achieve the desired level of pressure support. Such revisions in plans give the drilling engineers severe headaches in offshore projects where slots are pre-assigned to wells deviating in different directions to try and reduce the complexity of the spaghetti of conductors immediately beneath the platform. Far from being a disaster, however, this incident at the start of the Thistle Field development proved beneficial to the partnership since the cut-back in production in 1978–1979 coincided with the period when the oil price took a significant leap towards $30 per barrel so the deferred production was worth very much more money when the oil eventually reached the surface.

Such adventures are not uncommon at the start of field developments but cause added trauma when they occur in offshore projects for which all the flexibility in design has to be anticipated and catered for. In particular, events in the Thistle Field demonstrate the need to consider engineered waterdrive as an insurance against failure in costly offshore developments. Sometimes the early production performance of fields establishes the existence of strong natural pressure support but even so operators usually opt for supplementing the water influx by injection with the aim of gaining a degree of engineering control over the flood.

The unprecedented commitment to waterdrive in the North Sea and the sheer scale of activities makes it an interesting area to study. Therefore, the remainder of this section will be devoted to a description of the conditions prevailing in this

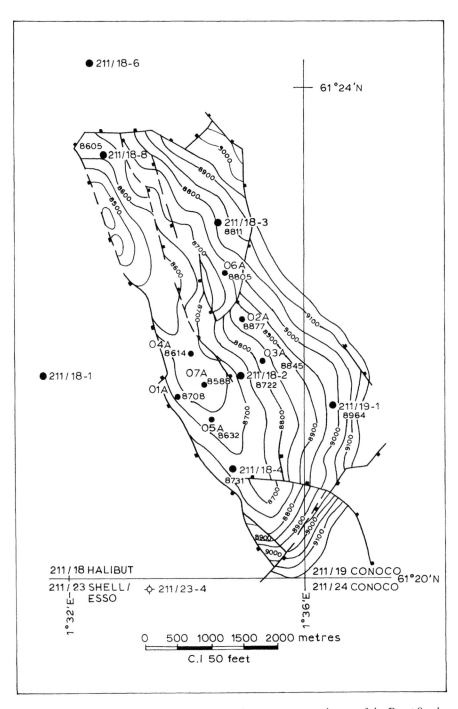

Fig. 5.2. Thistle Field, U.K. North Sea. Structural contour map on the top of the Brent Sands.

important province which made waterdrive the obvious choice as the dominant form of recovery method. Some of the comments relate strictly to offshore operations but nevertheless the list should provide points of guidance in vetting waterdrive project viability in general.

(b) Permeability

Most North Sea fields that have been selected for waterdrive have moderate to high permeabilities, Darcy levels being quite normal. Consequently, considering Darcy's law:

$$q = -\frac{kA}{\mu}\frac{dp}{dl} \tag{5.1}$$

the rates of production and injection wells are high; not quite so as extraordinary as in many Middle East Fields but typically 20,000 stb/d can be expected as an initial rate of an oilwell and 50,000 b/d for a successful injector. This means that quite large oil accumulations can be developed with relatively few wells which more importantly implies the requirement of few production platforms. It is these massive structures standing in 400–500 ft of water and carrying a price tag of one billion dollars plus that are by far the major cost item in projects and obviously their number must be kept to an absolute minimum in any field. The number of wells that can be drilled from any one platform is limited and therefore only high flow capacity fields were originally selected for economic development. For instance, the Thistle Field, described earlier, had a generous number of 60 well slots to develop the initially estimated STOIIP of one billion barrels and, in fact, only 40–45 may be required on account of the generally high permeabilities. There are many accumulations in the North Sea with permeabilities less than 50 mD that would be perfectly acceptable for development onshore but the requirement of perhaps a hundred low-rate production and injection wells in a dense pattern precludes their economic development offshore, at least from fixed production platforms in deep water.

The high flow rates also have the favourable effect of leading to relatively short project duration. Considering that platforms have a finite lifetime this leads to the maximum recovery of oil before any significant mechanical deterioration occurs and the operator is faced with evaluating:

value of remaining recoverable reserves vs. cost of platform refurbishment

which if unfavourably balanced would lead to field abandonment. Therefore, in the development of North Sea fields operators tend to produce at full potential keeping little in reserve and subtleties connected with the rate dependence of recovery are overlooked in comparison to optimising recovery within the project lifetime. In this respect, the British Government's guidelines on field developments [3] are quite consistent in stating that any production cut-backs, if ever deemed necessary, would be by the deferment of new projects rather than rate reductions in existing fields.

(c) Oil viscosity

North Sea fields selected for development by waterdrive invariably have low oil viscosity, typically less than one centipoise. Again, considering Darcy's law, equation 5.1, it can be seen that this condition has the same effect as high permeabilities in promoting high flow rates thereby accelerating field developments. Perhaps of greater significance, however, is the favourable influence of the low viscosity in enhancing the efficiency of water–oil displacement, at least on the microscopic scale. The parametric group which dictates the efficiency is known as the end-point mobility ratio (or simply the mobility ratio, as used in this text)

$$M = \frac{k'_{rw}}{\mu_w} / \frac{k'_{ro}}{\mu_o} \tag{5.2}$$

The incorporation of the maximum end-point relative permeabilities means that, by direct application of Darcy's law, the mobility ratio represents

$$M = \frac{\text{maximum velocity of the displacing phase (water)}}{\text{maximum velocity of the displaced phase (oil)}} \tag{5.3}$$

Using typical parameters for North Sea fields ($k'_{rw} = 0.3$, $k'_{ro} = 1$, $\mu_o = 0.8$ cp, $\mu_w = 0.4$ cp) gives a value of $M = 0.6$. The significance of the fact that $M < 1$ is that, in a one-dimensional flooding experiment in a homogeneous core plug, the displacement will be completely stable. That is, it is the water that is being injected and pushes the oil but since $M < 1$ the water cannot travel faster than the oil and therefore displaces it in a perfect piston-like manner (Fig. 5.3a). This is the most favourable form of displacement and means that the total volume of movable oil:

$$MOV = PV(1 - S_{or} - S_{we}) \tag{5.4}$$

can be recovered by the injection of an equivalent volume of water. Consequently, the flooding is both rapid and efficient.

Conversely, if $M > 1$, on account of high oil viscosity, then as shown in Fig. 5.3b the waterflood is inefficient. The water can now travel faster than the oil and, since it is the water that is doing the pushing, it does so and channels through the oil in an unstable fashion. Water breaks through prematurely at the end of the plug and if $M = 40$, for instance, it might require the circulation of about 100 pore volumes of water to recover the movable oil volume.

It is therefore natural that operators choose to conduct waterdrive in fields in which the oil viscosity is low so that the mobility ratio is less than unity. This is

Fig. 5.3. One-dimensional waterdrive experiment in a homogeneous core plug.

not just the case in the North Sea but everywhere, and the majority of the world's waterdrive fields operate under this favourable flooding condition.

It is particularly important that the mobility ratio be favourable in offshore developments. The circulation of large volumes of water to attain a high recovery in unfavourable mobility ratio fields would greatly prolong the project lifetimes which is an intolerable situation on account of the steady mechanical deterioration of the offshore facilities and the very high operating costs. The alternative would be premature abandonment with attendant loss in recovery. It is, however, worth making the point now, which is substantiated in section 5.4e, that provided an operator is prepared to circulate the requisite number of pore volumes of water through the reservoir it is always possible to recover all the movable oil. It's just that this can take so long and the amount of water production accompanying the oil become so large as to make the continued development of high-viscosity oilfields quite impracticable.

A further advantage in selecting low oil viscosity reservoirs for waterflooding ($M < 1$) is that at abandonment, the areal sweep of water will be very high. Finally, it must be stressed that the piston-like displacement associated with favourable mobility ratio displacement only occurs on the microscopic, one-dimensional scale of a core flooding experiment. In the flooding of macroscopic reservoir sections account must also be taken of the heterogeneity and gravity in accounting for the overall flooding efficiency.

(d) Oil volatility

Most of the waterdrive fields in the North Sea contain oil of moderate to low volatility. An obvious advantage in this is that gas oil ratios are of a tolerable level making gas disposal, which is always a problem offshore, fairly straightforward. Few of the reservoirs have natural gas caps and they often display a high degree of undersaturation sometimes amounting to thousands of psi. This condition is a distinct advantage at the start of an offshore development for it means that operators can afford to allow the reservoir pressure to decline by a significant amount initially without risking falling below the bubble point. And, while there is no strict regulation about producing below this pressure there is, at least, an "understanding" that operators must provide a very sound justification for doing so. The initial drop in pressure permits observation of the degree of areal and vertical communication in the reservoirs, which is knowledge denied to operators at the appraisal stage viewing data under static conditions. Pressure surveys conducted in each new development well, preferably using the RFT tool, permit the direct inspection of the degree of communication under dynamic conditions. This is essential information to facilitate the final design of the waterflood and ensure that injection and production wells are completed in reservoir units that are in direct communication.

A pioneering paper on the subject of using the RFT in the design of a waterflood was published by Amoco in 1980 [4] relating to the development of their Montrose Field in the U.K. sector of the North Sea. Production from this moderate sized field commenced in mid-1976 and the operator carried out a systematic programme

of running RFT surveys across the massive Palaeocene sand section in each new
development well, prior to setting the production casing. As can be seen from the
water saturation profile in Fig. 5.4 the oil column is confined to a 75 ft interval
in Layer I, only the top 45 ft of which is perforated. Beneath is a massive basal
aquifer whose degree of connection with the reservoir through the tight interval
of Layer II was uncertain. Also shown in Fig. 5.4 is the result of a static RFT
survey conducted in an appraisal well or the first development well. This reveals the
condition of hydrostatic equilibrium across the oil column and aquifer established
since oil migrated into the trap. In a sequential drilling project such as this, the

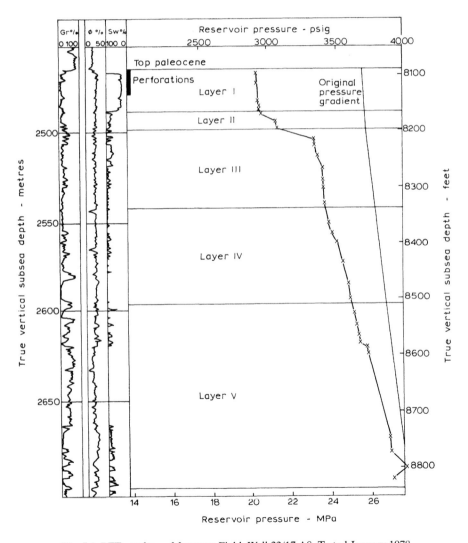

Fig. 5.4. RFT results — Montrose Field: Well 22/17-A8. Tested January 1978.

first and each successive well are put on a high-rate production once completed to encourage a large pressure drop at the selected locations of new wells. RFT surveys run in each reveal the reservoirs under dynamic conditions, as demonstrated by the second survey plot in Fig. 5.4. This well was completed some eighteen months after the start of continuous field production and demonstrates a good degree of both areal and vertical pressure communication. That is, the pressure drop caused by the production of earlier wells is transmitted to the new location via Layer I on which all wells have been perforated and then downwards across the entire aquifer. Layer II acts as a partial restriction to vertical fluid movement and the attainment of complete pressure equilibrium but nevertheless it does not act as a complete barrier to flow as the depletion in the lower aquifer sands demonstrates.

Similar dynamic pressure profiles were obtained in each new development well, as illustrated in Fig. 5.5, in which RFT surveys in six wells show the steady depletion

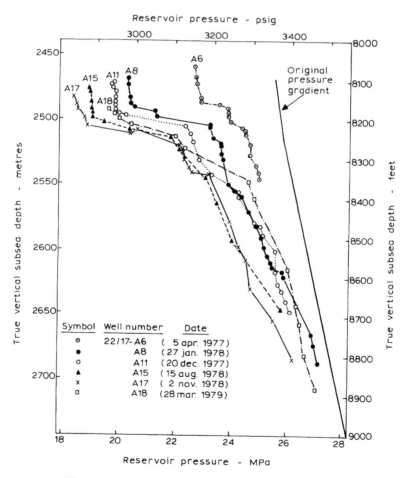

Fig. 5.5. RFT pressure data — Montrose Field (east flank).

resulting from continuous field production and also similar pressure profiles across the entire sand section. From inspection of these dynamic pressure data and their incorporation in a detailed numerical simulation model of the field, the operator was in a position to make a definite development decision concerning the nature of the waterdrive to be engineered namely, that it should be basal rather than from the edge. Had Layer II proven to be a complete barrier to vertical flow, it would have been necessary to drill long reach, highly deviated wells out to the aquifer on the periphery of the accumulation to act as injectors. The pressure profiles, backed-up by the simulation, demonstrated, however, that the injectors could be better drilled and perforated deep in the aquifer, with limited deviation, to perform a basal waterdrive. And, as described in section 5.4c, this is usually preferable to edge drive on account of its increased stability.

Since its inception in the mid seventies, the RFT has proven an invaluable tool in providing engineers with dynamic pressure profiles across reservoir sections. These data have proven essential in the planning of engineered floods and particularly in offshore developments in which, until the start of continuous production, only static reservoir data are available. Furthermore, the RFT provides the most reliable set of pressures for calibrating large 3D numerical simulation models. If such a model can be accurately history matched on a layer by layer basis using pressure profiles, as shown in Fig. 5.5, then a reliable tool should be obtained for predicting reservoir performance. The history matching is usually accomplished by varying vertical permeabilities between layers and areally by adjusting permeabilities and the position or sealing nature of faults.

To take full advantage of the RFT, however, some sensible, practical precautions must be taken, as demonstrated in the Montrose Field example. In this, the operator drilled the production wells much deeper than necessary across the massive basal aquifer in order to establish the degree of vertical communication. Furthermore, the high density of pressure survey points (Figs. 5.4 and 5.5) is necessary for accurate simulation model calibration.

(e) Overpressures

Many of the larger, deeper oil accumulations in the North Sea were significantly overpressured at initial conditions. In the main producing area of the East Shetland Basin, for instance, reservoirs/aquifers had fluid pressures of 1000–2000 psi in excess of the normal hydrostatic pressure regime. This proves to be a distinct advantage in waterdrive operations since the overpressure itself serves as a source of considerable free energy.

In the pressure–depth diagram (Fig. 5.6), it can be seen that if a hydrostatic aquifer (point C) were connected to the surface by tubing the static wellhead pressure would be zero. Alternatively, an aquifer with an initial overpressure of 1500 psi (point A) would display the same excess static pressure at the surface and, for the fluid gradients indicated, a static column of overpressured oil would have an excess wellhead pressure of 3000 psi for a 10,000 ft deep reservoir. Coupled with the usual high flow capacities, the excess surface pressure gives many North Sea wells

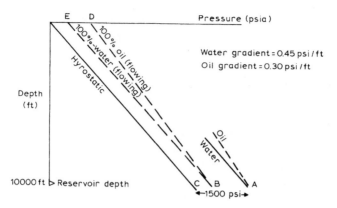

Fig. 5.6. Influence of overpressures on waterdrive.

a significant surge in production at the start of their lives and rates above 50,000 stb/d have been recorded. This is precisely what operators require to enable them to reach the plateau rates early with few wells and obtain a healthy cash flow.

As production continues the pressure drops as operators deliberately "feel their way round" the reservoirs using the RFT, as described above, before finally deciding on the optimum injection policy. Usually, pressure in maintained at some level (point *B*) between the initial overpressure and hydrostatic by sea water injection. Obviously, the lower this pressure the easier it is to inject but there is usually more to be gained by operating the flood at high pressure and certainly above hydrostatic. Initially, wells will produce only oil still with a high flowing wellhead pressure (point *D* in Fig. 5.6) but as the flood continues water breaks through to the producers and as their watercuts increase from 0 to 100% the flowing wellhead pressures decrease from point *D* to *E*. Nevertheless, provided the operator has maintained the average reservoir pressure above hydrostatic through injection, then wells will still be able to produce under natural flow conditions even at extremely high watercuts. Alternatively, if the average pressure is allowed to fall below the hydrostatic level then in order to maintain production it is necessary to install artificial lift (gas lift or pumping) which not only adds to the operating costs but also, in offshore projects, the equipment consumes precious space on the platforms which is always at a premium.

Sometimes, operators deliberately allowed the pressure to fall below hydrostatic by drilling an excess of production to injection wells or simply by deferring injection for too long. This was in the belief that the situation was quite controllable and that, as soon as considered necessary, the pressure could be raised commensurate with increasing watercuts to keep wells flowing naturally. Unfortunately, this is easier said than done and quite often merely increasing the injection rate does not necessarily raise the pressure. This is particularly the case in the high-permeability reservoirs of the North Sea fields in which the circulating system consists of three pipelines, the first being injection well tubing, the second the high flow capacity conduit between the injector and producer and the third the tubing in the production well. In such a system the only way to raise pressure is to install a choke somewhere and the only place that this can be done is by choking back production wells. This, of course, is anathema to many and they would prefer to install artificial lift. In bounded fault blocks it is usually possible to increase reservoir pressure by cutting back on production but in parts of fields open to aquifers even this measure may have little effect in raising the pressure and in such fields it is expedient not to let the pressure fall below hydrostatic in the first place.

The reason why people flirt with unnecessarily low pressures seems to be because of a mistaken idea that the way to produce more oil quickly and thereby improve their discounted cash flows is to keep on drilling producers at the expense of injectors. But each new producer drilled leads to an overall drop in reservoir pressure: therefore the potential of all wells in the field steadily declines and with it the field production rate. Considering that one good injection well could probably support several producers by maintaining their flow potentials, it is usually much better to advance their drilling in the field development plan and this will also improve the project economics.

Finally, considering the question of the optimum pressure at which to conduct a waterflood, there has been a substantial body of literature, over the years, directed at answering the questions

– is it preferable to flood above or below the bubble point pressure?
– if below, what is the optimum free gas saturation in the reservoir that will promote maximum oil recovery?

Naturally, the majority of the papers relate to laboratory experiments for it is difficult to answer them from direct observation of field performance for the simple reason that a field can only be developed once, affording no means of comparing different strategies, whereas laboratory experiments can be repeated as often as required.

Concerning the first question, there appears to be a consensus of opinion that operating a waterflood below the bubble point does enhance the ultimate oil recovery and one of the main reasons for this is because the presence of a free gas saturation reduces the residual oil saturation thus increasing the target movable oil volume, equation 5.4. Opinion is more divided, however, on the question of what the optimum free gas saturation should be. This is hardly surprising considering the range of microscopic phenomenon to be studied that might influence experimental results. These include rock characteristics (pore structure, permeability, etc.) rock–fluid properties (wettability, capillary effects, two- and three-phase relative permeabilities) and the fluid PVT properties themselves. In one rather convincing paper on the subject Arnold et al. [5] performed a series of solution gas drive/waterflooding experiments using a single core plug with four fluids displaying markedly different PVT properties. Since these can be accurately measured, whereas such effects as wettability and three-phase relative permeabilities cannot, it was hoped that by using a single core plug many of the microscopic effects and their variation would be negligible in comparison to the significant and quantifiable variations in fluid volatility. The results clearly demonstrated a range of free gas saturations between 7 and 35% PV for which the recovery could be optimised for the different fluid types. Compared to flooding at the bubble point, the increase was as high as 10% in the best case. Although it was not possible to quantify the results of the experiments in any generalised form, the authors were sufficiently encouraged to recommend that if an operator considers flooding below saturation pressure, then making a similar study using the actual reservoir rock and fluid would be worth the expenditure.

While such a recommendation can only be endorsed, it is also necessary to consider scaling-up the experimental results to the reservoir itself. That is, it is all very well to determine that in a microscopic core flooding experiment the optimum free gas saturation is, say, 25% PV but is there any way of engineering the flood in the macroscopic reservoir section so that such an average saturation could be attained in practice? In the reservoir, the effects of heterogeneity and gravity might lead to significant gas–oil segregation that could overwhelm the laboratory results in importance. Therefore, the suggested procedure in evaluating such a complex problem is as follows:

- Perform the laboratory experiments deemed necessary
- History match the experiments using a fine gridded numerical simulation model
- Construct a detailed cross-sectional simulation model of the reservoir incorporating all the observed heterogeneity and use it to study displacement efficiency on the macroscopic scale.

The second step is necessary to empirically evaluate microscopic phenomena, such as relative permeabilities, that are required as essential input to the reservoir model.

In most North Sea fields, waterdrive below saturation pressure was never seriously considered. In many, it would require flooding at pressures that were too low to sustain economic production rates in such costly developments.

(f) Reservoir depth

The majority of North Sea fields are deep, lying between 8000 and 12,000 ft below sea level. This has the advantage that large hydrocarbon accumulations can be developed by deviated drilling from a single platform. In a project, such as depicted in Fig. 5.7, the cost of development wells, usually the major expenditure in land fields, pales into insignificance in comparison to the price of the massive deep water platforms, consequently the number of these must be kept to an absolute minimum. Fortunately, by the mid-seventies, when many of the major North Sea projects were in the design or early development stage, the art of deviated drilling was already well established thus facilitating their development with few platforms.

The advantage of deep reservoirs in such an environment cannot be overemphasised. If, for instance, the average angle of deviation was 45º and the reservoir sketched below was at half the depth, it would require four times as many platforms to develop the accumulation which would be intolerable in the North Sea. A field such as Thistle, described earlier, has an areal extent of 6.5 × 2 km and with a datum depth of 9200 ft could easily be developed from one platform. It is only the much

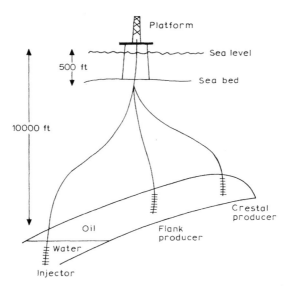

Fig. 5.7. Field development from a fixed production platform using deviated wells.

larger fields such as Ninian, Statfjord and Brent (see Fig. 5.12) that require three or four platforms for complete areal coverage. Although the advent of horizontal wells during the eighties has perhaps enlarged the scope for the development of shallower accumulations, the nature of many of the North Sea fields with massive overall sand sections subdivided into discrete reservoirs is more suitable for development with vertical or deviated wells which penetrate the full section and provide flexibility for recompletions. In such environments, depth will always be an advantage in deep water offshore developments.

5.3. ENGINEERING DESIGN OF WATERDRIVE PROJECTS

The reservoir engineer has many responsibilities in the basic design of waterdrive projects and these are heightened offshore by the fact that the most important decisions relating to platform design have to be made "up-front" during the appraisal stage with little or no knowledge of dynamic reservoir performance. Therefore, in describing the various stages, the concentration of theme will once again be on offshore developments and in particular on the North Sea where much has been learned about the subject — sometimes at high cost. The reservoir engineering responsibilities may be summarised as follows.

(a) Production plateau rate

This choice is not at the sole discretion of the reservoir engineer — far from it. Economists and management tend to dominate in the decision making, especially offshore where, once the platform production facilities are commissioned, there

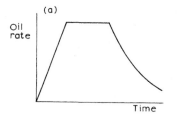
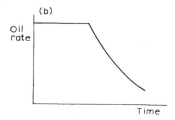

Fig. 5.8. Types of oilfield development. (a) Sequential drilling; (b) Pre-drilling.

is a need to produce at high rate in an attempt to optimise the discounted cash flow and in some areas to take advantage of early tax breaks. Nevertheless, the reservoir engineer is involved in making sure that the initial rates will not cause any reservoir damage, although this is extremely difficult to prove, and ensure that the plateau rate chosen is commensurate with the well numbers and injection/production facilities installed.

Major offshore projects usually have the type of production profile shown in schematic (Fig. 5.8a) resulting from sequential drilling. That is, few, if any, production wells are completed prior to platform installation. Instead, wells are drilled and brought on-stream sequentially and there is a buildup of oil production rate to the plateau level. In large North Sea projects, the plateau rates aimed at were between 10 and 16% of the recoverable reserves per annum and, in fact, it has proven difficult to achieve or exceed the larger figure. The limitation is that, on plateau, additional oil from newly drilled production wells is offset by increasing amounts of water production resulting from the water injection so that an equilibrium state is achieved consistent with the fluid production capacities of the equipment installed. The decline starts when the rate of increase of water production exceeds any gains in oil production.

In smaller more marginal fields, however, it is physically possible to produce at much higher recovery rates. For instance, in the case of the waterdrive development of a small, high flow capacity accumulation, it may be possible to produce above 100% of the reserves per annum and then move on rapidly to develop other, similar fields. Unfortunately, in some countries, there is a reluctance on the part of the regulatory authorities to permit such high rates of depletion which are seen as "exploitation". It is frequently claimed that ultimate recovery of oil is rate sensitive and that the slower it is produced, the higher the recovery. This seems to be a relic from the days of solution gas drive, in which recovery can be highly rate dependent, but is not so relevant in waterdrive fields, particularly if the mobility ratio is low ($M \leq 1$). Furthermore, rate dependence is very difficult to prove or disprove in reservoir engineering studies since none of the input data, such as relative permeabilities and capillary pressures, are themselves input as rate sensitive dependent, therefore the lack of rate dependence as the outcome is hardly surprising. In waterdrive the combined effects of heterogeneity and gravity can sometimes lead to rate sensitivity but it is rare that this should occur on account of the small gravity difference between water and oil. Gas drive is different, however, and as demonstrated in Chapter 6, the combination of unfavourable mobility ratio ($M > 1$) and large gravity difference between gas and oil do make the process rate dependent; not so much in terms of ultimate recovery but rather in the efficiency with which the oil is recovered.

The alternative form of production plateau shown in Fig. 5.8b results from the pre-drilling of development wells and is frequently practised in smaller offshore fields. That is, between appraisal and platform installation a template is set on the sea bed through which a number (possibly all) of the development wells are drilled. Then, when the platform is eventually positioned, the wells can be rapidly tied-back into the production facilities so that when switch-on occurs the plateau rate is achieved immediately. The advantage in terms of accelerated cash flow associated with this practice is obvious but, from the reservoir engineering point of view, the situation is far from satisfactory. Although each of the wells may be thoroughly evaluated, the data are acquired under static reservoir conditions whereas in the case of sequential drilling dynamic data are collected. In the latter case, as illustrated in section 5.2d, in connection with the Montrose Field, the acquirement of dynamic pressure profiles across the reservoir using the RFT enables the engineer to refine the plans for water injection before its implementation. The danger with pre-drilling is that the whole development proceeds based on the initial concepts of the geologists and reservoir engineers and, with the best will in the world, the chance of these being correct is slight, there is hardly ever an oilfield in which the eventual development plan follows the original. In particular, there is the risk with pre-drilling that production and injection wells may not be suitably located within communicating formations which is an essential requirement for waterflooding. With sequential drilling, if the RFT survey indicates that an injector and producer are incorrectly located, the former can be immediately sidetracked to a more suitable location.

In many pre-drilling projects, however, all the wells are drilled in advance and the platform usually has only a lightweight workover rig making sidetracking impossible if it is found after the start of production that the wells have not been correctly located. The above is not intended as a sermon against pre-drilling because for many small, marginal fields it is the only means of assuring economic viability, instead it is merely intended to warn engineers of the more obvious dangers associated with the practice. Ideally in such a project, the production wells can be pre-drilled, thus guaranteeing the attainment of the plateau at production start-up, but drilling of the injectors should be deferred. This, of course, would require a drilling rig on the platform but would provide the degree of flexibility associated with sequential drilling projects.

(b) Number of production/injection wells

The choice of the number of wells required for a project is the responsibility of the reservoir engineer. Onshore, there is no great pressure in making the decision: wells can usually be drilled, as required. Offshore, however, the correct decision must be made in advance and, for fields developed from a fixed platform, the well deck containing the drilling slots and wellhead equipment designed accordingly.

Fig. 5.9. Field average pressure profile at the start of a sequential drilling project.

In pre-drilling projects there is an equal if not greater need to anticipate well requirements but, in either case, if the numbers are underestimated there is always the possibility of drilling additional sub-sea completed wells with a mobile rig and tying them back to the fixed production facilities. This is an expensive alternative, however, and it would be much more satisfactory if the correct number of wells could be determined in advance.

Well numbers are selected based on the production/injection rates required and on the results of well tests conducted at the field appraisal stage. It is important to assess from such tests the ideal productivity and injectivity indexes that may be anticipated for the average development well and, therefore, as stressed in Chapter 4, section 4.4a, it is essential that in testing appraisal wells they should be perforated across the same completion intervals as planned for development wells.

In deciding the number of producers required it is a question of where should they be drilled and when, in order to achieve and maintain the production plateau. In the case of injection wells, however, the order of priorities is reversed with the first decision being when is it necessary to drill them followed by where should they be located. That is, in sequential drilling projects, as described in section 5.2d, the initial step is to produce at high rate without pressure support, while determining the degree of areal and vertical pressure communication by running the RFT in each new development well. As the rapidly declining pressure (Fig. 5.9) approaches the level at which required production levels are in danger of not being achieved — that determines when to drill injectors and initiate the waterflood at a suitable operating pressure. Where the wells should be drilled depends on the initial results of the RFT surveys which enable producers and injectors to be appropriately located in communicating regions of the field.

(c) Surface production/injection facilities

This is the most crucial aspect in the design of a waterdrive, especially offshore, and requires close liaison between the reservoir engineers, who specify the required capacities of the production/injection facilities, and the project engineers who design and install the equipment. It is not necessary for the reservoir engineer to fully appreciate the subtleties of design or operation of the complex surface facilities, although a broad understanding would obviously be helpful. What is vital is that the engineer is fully aware of the need to specify the capacities of the individual components and ensure that they are compatible with one another and with the reservoirs being produced. This section concentrates on the importance of sizing the equipment correctly and illustrates the consequences of making errors in this matter which can seriously jeopardise the commercial viability of a project.

The capacities that concern the reservoir engineer are:

– water injection $(q_{wi}, \text{b/d})$
– separator train $(q_o + q_{wp}, \text{stb/d})$
– water disposal $(q_{wp}, \text{b/d})$

and the significance of these is described in conjunction with Fig. 5.10.

The seawater to be injected is pumped to the surface from a depth of usually 100–200 ft, at which its plankton content is measured to be at an acceptably low level. After removal of corrosive oxygen (deaerator) the water is filtered to remove solids (organic debris) that might otherwise plug the formation and reduce injectivity. What constitutes an acceptable minimum size of solids particles depends largely on the nature of the formation and, in particular, the pore throat size. In one informative paper on the subject [6], it was assessed that for BP's massive Forties, North Sea field, the reduction of solids to a size of 5 μm would cause no significant impairment to seawater injection in a formation in which the average pore throat diameter was 15 μm. Standards, of course, will vary from field to field and must be investigated separately. The water is then injected into the reservoir by a battery of injection pumps.

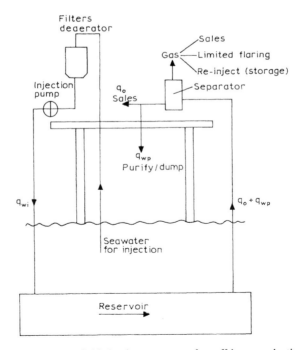

Fig. 5.10. Schematic of the fluid circulatory system of an offshore production platform.

Once underground, the water serves the dual purpose of maintaining pressure and displacing oil towards the production wells. To begin with only oil is produced but sooner or later (and unfortunately it is invariably sooner rather than later) water breaks through to the producers. Wells then exhibit an ever increasing watercut (fractional flow of water) defined as:

$$f_{ws} = \frac{q_{wp}}{q_o + q_{wp}} \tag{5.5}$$

in which the subscript "s" denotes the fact that the expression is here evaluated at surface conditions.

On the production side of the platform, the capacities of interest are those of the separator train and the oily water disposal equipment. Water, oil and gas are separated in several stages at pressures and temperatures designed to optimise the volume of stabilized crude oil. The water leaving the separators still has a fairly high content of dispersed oil which must be reduced to some low, environmentally acceptable level, usually about 50 ppm, before dumping the water back into the sea. Earlier equipment to remove the oil from the water was rather bulky and temperamental but now many platforms are equipped with compact and efficient hydro cyclones to perform this task.

The interdependence of the capacities of the component parts of the topsides facilities can be appreciated by considering the underground material balance for waterdrive under pressure maintenance conditions, that is

$$q_{wi} = q_o B_o + q_{wp} B_w \quad \text{(rb/d)} \tag{5.6}$$

in which the rates are those measured at the surface and it is assumed that the injected water contains no gas/air ($B_w = 1.0$ rb/stb). Although very simple, this is the most fundamental equation in designing a waterdrive and, before studying its application, it is worthwhile considering three important points about the equation itself, namely:

(1) It always works. That is, the engineer need not be concerned, as with normal application of material balance (Chapter 3, section 3.3) about whether there is a high degree of pressure equilibrium in the system — in waterdrive the pressure is being maintained at a constant level. The equation merely states that what goes in at one end of a core plug or reservoir must come out at the other.

(2) The equation is dominated by the left-hand side, the water injection rate. This is the *drive* in waterdrive. The left-hand side is under strict engineering control, the right is not, at least in the ratio of water to oil production. Therefore, for a successful waterdrive project, the engineer must concentrate on the injection — the production will then sort itself out.

(3) It is not just a reservoir material balance, it should also be regarded as the "platform equation" since it contains all the topsides equipment capacities:

q_{wi}	injection
$q_o + q_{wp}$	separators
q_{wp}	water disposal

Effectively, it is the equation that marries the reservoir to the platform.

The second of these points may seem fairly obvious but, in fact, it is not always fully appreciated, as described in section 5.10. It is the third point that will now be examined in connection with surface facilities design.

Consider what happens if the injected water breaks through to the producing wells prematurely and the subsequent watercut development is much more severe than anticipated at the platform design stage. That is, q_{wp} on the right-hand side of equation 5.6 is too large, too soon. Then, since q_{wi} is fixed by design, the oil rate, q_o, must eventually suffer and the production profile be adversely affected once the separator or water disposal capacities are exceeded. The manner in which the capacities affect the production profile is illustrated in Exercise 5.1, in which the material balance is expressed in a slightly different form. That is, recognising from equation 5.5 that:

$$q_{wp} = q_o \frac{f_{ws}}{1 - f_{ws}} \tag{5.7}$$

and substituting this in equation 5.6, gives:

$$q_{wi} = q_o \left(B_o + \frac{B_w f_{ws}}{1 - f_{ws}} \right) \tag{5.8}$$

which is a more convenient expression for studying production profiles.

Exercise 5.1: Topsides facilities design for an offshore waterdrive field

Introduction

This exercise illustrates the possible consequences of not anticipating correctly the harshness with which the watercut develops in an offshore waterdrive field. Although the data may appear "improbable" they are, in fact, based on a real example from the North Sea in which a last minute change in facilities design averted a serious dimunition in the profitability of the project.

Question

It is proposed to develop an 800 MMstb oilfield aiming at the production/injection profiles listed in Table 5.1 over the first six years of the project lifetime. These were generated from a detailed numerical simulation study conducted towards the end of the appraisal stage of development. Inadvertently, in the study, the operator modelled the reservoirs as being far too homogeneous: a mistake that is very easy to make, as will be demonstrated later in the chapter. The result was the evaluation of a watercut development trend referred to as "original" in Table 5.1 to which the production/injection figures relate. Several other fields in the area, producing from the same geological formations, had by this time a considerable amount of production history and their aggregate watercut development trend, listed in Table 5.1 as "observed", was noted as being considerably harsher. Appreciating that an error had possibly been made and that the design capacities of:

TABLE 5.1

Production/injection profiles and watercut development

Time (years)	Oil rate (Mstb/d)	Fractional oil recovery	Injection rate (Mb/d)	Average watercut	
				original (%)	observed (%)
0.5	40	0.009	–		
1	88	0.029	123		1.7
1.5	100	0.052	145	5.0	4.8
2	100	0.075	145	5.0	12.8
2.5	100	0.098	145	5.0	22.8
3	100	0.120	145	5.0	32.6
3.5	100	0.143	145	6.0	41.8
4	96	0.165	145	10.0	50.2
4.5	84	0.184	145	24.6	56.8
5	76	0.201	145	34.5	61.7
5.5	68	0.217	145	42.7	65.4
6	57	0.230	145	48.0	68.0

water injection = 145,000 b/d
separator train = 120,000 stb/d
water disposal   = 60,000 b/d

may be inadequate to sustain the required production profile, the operator decided to switch to the "observed" watercut development instead. Adopting the same trend (Fig. 5.11), calculate:

- the oil production profile if the above capacities were unchanged
- the required upgrading of capacities if the oil production profile listed in Table 5.1 is to be achieved.

The flood will occur at constant pressure for which B_o = 1.4 rb/stb and B_w = 1.0 rb/stb.

Solution

All the calculations are performed using the waterdrive material balance, as described in the text.

$$q_{wi} = q_o \left(B_o + \frac{B_w f_{ws}}{1 - f_{ws}} \right)$$

(5.8)

The first part of the question requires solving the equation for q_o using the original design figures for the capacities of the surface facilities. This is necessarily an iterative process. That is, in any time step an estimate is made of the oil rate which gives the cumulative and fractional recoveries at the end of the time step. From the latter, the watercut is read from the "observed" function in Fig. 5.11 and this value is returned to the material balance. The iteration proceeds until an oil

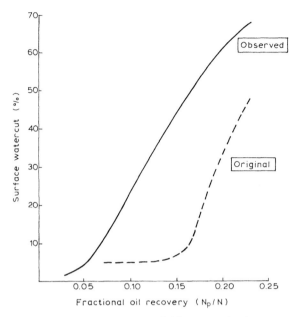

Fig. 5.11. Original and observed field watercut developments.

rate is achieved that is consistent with the capacities, which act as constraints. The results are listed in Table 5.2.

As can be seen, using the harsher watercut trend and the original capacities, the required oil rate can only be achieved for the first eighteen months. Thereafter, the injection capacity acts as a constraint until the fifth year when the water clean-up capacity takes over. As a result, after six years the cumulative recovery is 153 MMstb instead of the anticipated value of 184 MMstb (0.23 × 800 MMstb), a loss of 31 MMstb. The calculations have been continued beyond six years and, as can be seen, the required recovery of 184 MMstb is not achieved until almost three years later.

The second part of the question requires the removal of the surface capacity constraints and solution of the material balance directly keeping the oil rate at the required levels stated in Table 5.1. The results are listed in Table 5.3.

As can be seen, the required capacities and their percentage increase are:

water injection = 230,000 b/d (+59%)
separator train = 200,000 stb/d (+67%)
water disposal   = 130,000 b/d (+117%)

The exercise illustrates the seriousness of mismatching the capacities of the surface facilities to the reservoir performance. The deferment of 31 MMstb of oil over a three year period is not the sort of error than any engineer would want on his record. It is, unfortunately, a common mistake that occurs because there has been no tradition of reservoir engineers being concerned about equipment

TABLE 5.2

Production profiles using the original capacities of the surface facilities and the "observed" watercut development trend

Time (years)	q_o (Mstb/d)	N_p (MMstb)	N_p/N	f_{ws} (%)	q_{wi} (Mb/d)	q_{wp} (Mb/d)	q_{sep} (Mb/d)
0.5	40	7.30	0.009	–	–	–	40
1	88	23.36	0.029	1.7	126	2	90
1.5	100	41.61	0.052	4.8	145	5	105
2	94	58.77	0.073	12.2	145[a]	13	107
2.5	87	74.64	0.093	20.6	144	23	110
3	80	89.24	0.112	29.0	145	33	113
3.5	74	102.75	0.128	36.0	145	42	116
4	68	115.16	0.143	42.8	144	49	117
4.5	63	126.66	0.158	47.3	145	57	120
5	54	136.52	0.171	52.3	135	59[a]	113
5.5	47	145.09	0.181	55.9	125	60	107
6	42	152.76	0.191	58.8	119	60	102
6.5	38	159.69	0.200	61.5	113	61	99
7	34	165.90	0.207	63.4	106	59	93
7.5	32	171.74	0.215	65.0	104	59	91
8	30	177.58	0.222	66.4	101	59	89
8.5	29	182.87	0.229	67.8	102	61	90
9	27	187.80	0.235	68.8	97	60	87

[a] Denotes equipment capacity acting as a constraint.

capacities. Most of our history relates to the development of onshore fields where the mismatch of capacities in Exercise 5.1 would not be regarded as a mistake at all. At the end of the second year, the separator train capacity would be suitably increased and that of the water disposal plant during the fifth year. The only cost

TABLE 5.3

Modified profiles for the "observed" watercut development with no capacity constraints

Time (years)	q_o (Mstb/d)	N_p (MMstb)	N_p/N	f_{ws} (%)	q_{wi} (Mb/d)	q_{wp} (Mb/d)	q_{sep} (Mb/d)
0.5	40	7.30	0.009	–	–	–	40
1	88	23.36	0.029	0.017	125	2	90
1.5	100	41.61	0.052	0.048	145	5	105
2	100	59.86	0.075	0.128	155	15	115
2.5	100	78.11	0.098	0.227	169	29	129
3	100	96.36	0.120	0.324	188	48	148
3.5	100	114.61	0.143	0.417	212	72	172
4	96	132.13	0.165	0.500	230	96	192
4.5	84	147.46	0.184	0.567	228	110	194
5	76	161.33	0.202	0.620	230	124	200
5.5	68	173.74	0.217	0.654	224	129	197
6	57	184.14	0.230	0.680	201	121	178

would be that of the equipment itself. Offshore, however, the situation is quite different, and the engineer is directly confronted by space constraint. If ill-designed to begin with, there simply is not the space available on many platforms to upgrade the capacities to the required levels, the equipment is far too bulky. The additional injection capacity, for instance, is not just a matter of adding more pumps but also the very bulky equipment for oxygen removal and filtration all of which demand additional power generation and while upgrading water disposal capacity is now relatively straightforward, increasing the separator capacity runs into the space constraint once again. Paradoxically, as the structural engineers become ever more

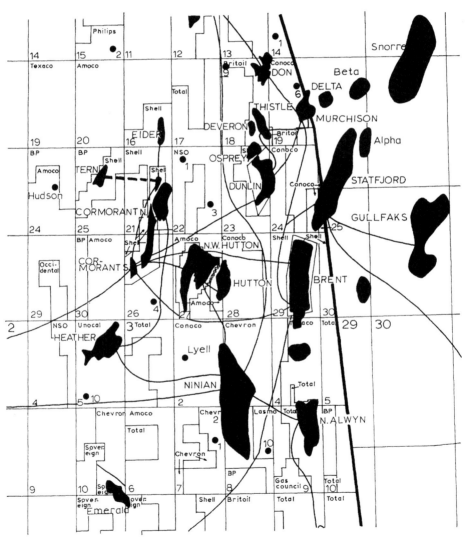

Fig. 5.12. Locations of the main fields in East Shetland Basin of the North Sea (1984).

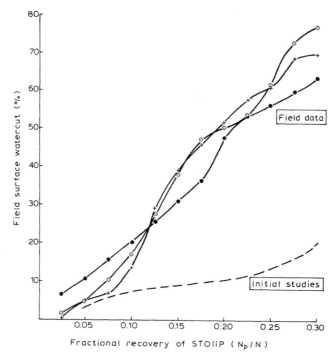

Fig. 5.13. Watercut development trends of three North Sea (East Shetland Basin) fields compared to their initial prediction.

brilliant in their design work and platforms become progressively more compact, the scope for placing additional equipment on board diminishes.

The error highlighted in Exercise 5.1 has afflicted quite a few North Sea fields, the worst place being the East Shetland Basin, the most prolific producing area lying between the United Kingdom and Norway (Fig. 5.12). The watercut development trends of three of the fields in the centre of the Basin are plotted in Fig. 5.13. Also plotted is the aggregate trend predicted from initial simulation studies and, as can be seen, the pattern is the same as revealed in Exercise 5.1. As already mentioned, this type of error arises from failure to account for reservoir heterogeneity in a realistic manner in studies and, in particular, the severity of permeability distributions across the reservoir sections. Fortunately, in the three fields whose watercuts are plotted in Fig. 5.13, there was sufficient space on the platforms to permit limited upgrading of facilities but if there is not then the penalties are:

– failure to attain the required production plateau or, if reached, failure to maintain it for as long as planned
– extension of project lifetimes while circulating large volumes of water at high watercut.

In a review paper "North Sea Scorecard" written by banker G.R. Castle in 1986 [7], of the fourteen North Sea waterdrive fields he considered, six never attained

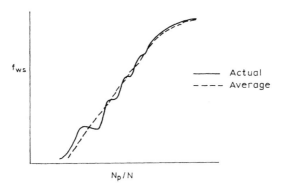

Fig. 5.14. Actual watercut trend, affected by field operations, and the average trend.

their planned production plateaus and on average by the end of 1983 the fourteen were 36% below their planned oil recovery targets. While there are many other factors behind these statistics, the main one is the production of much more water than anticipated and much earlier. Fortunately, it was not all bad news since, on average, the cash flows of the fields were 68% higher than expected at the end of 1983 on account of the unanticipated increase in oil prices during 1970's and early 1980's.

The primary aim of any reservoir engineering study of waterdrive, whether analytical or by numerical simulation, must be the establishment of a watercut development trend:

$$f_{\text{ws}} \text{ versus } \frac{N_{\text{p}}}{N} \tag{5.9}$$

for use in predictive calculations and particularly in sizing the capacities of surface facilities for offshore projects. There is no clear theoretical reason for the existence of any such relationship for a reservoir or field but empirically it appears justified. Early in the lifetime of a development it is possible to affect the trend by performing workovers or drilling new wells, as shown in Fig. 5.14. But the effect of such operations tends to diminish with time and generally there is an inevitability in the manner of watercut development imposed by the nature of the waterdrive, basal or edge, and the degree and type of reservoir heterogeneity. It is therefore the primary aim in the remainder of the chapter to investigate methods of generating watercut development trends for a variety of different reservoir types.

5.4. THE BASIC THEORY OF WATERDRIVE IN ONE DIMENSION

There are three levels on which the phenomenon of water–oil displacement can be viewed:

– the electron microscope scale (EMS)
– the microscopic, one-dimensional scale of a core flooding experiment
– flooding in hillsides

The first of these is a fascinating study observing the flow of liquids from one individual pore space to the next and considers such fundamentals as the trapping of residual oil droplets and the concept of wettability: which fluid, oil or water, preferentially clings to the rock particles and to what extent. But while this activity may lead to interesting still photographs and movies, the difficulty is that it does not lend itself to quantification. When, for instance, is the "wettability equation" going to be developed which could be used in a practical manner in describing flooding in macroscopic reservoir sections.

Flooding experiments in core plugs to determine such basic functions as relative permeabilities are conducted on the microscopic scale compared to flooding in hillsides which is the reality of practical reservoir engineering. Since the core plugs are small and of negligible thickness compared to the reservoir itself the experiment is to be regarded as one-dimensional in nature. The perennial difficulty in the description of waterdrive has always been how to relate the core flooding results, which are little affected by such complications as heterogeneity and gravity, to the hillside in which these same factors are usually dominant. The scaling-up of laboratory results for use in a meaningful fashion in field studies is one of the main topics in this and subsequent sections and particularly the sensitive matter of how the results of relative permeability experiments are input to numerical simulation models in such a way as to honour the basic laws of physics. A sensible linkage can be made between flooding in core plugs and hillsides but not, at present, between the EMS scale of observation and the hillside which remains a challenge for the future.

The basic theory of waterdrive, in fact, the one and only theory of waterdrive, is that of Buckley and Leverett which, at the time of writing, is celebrating its fiftieth anniversary. Before describing this elegant theory, however, it is first necessary to examine such fundamental concepts as relative permeabilities mobility ratio and fractional flow that are essential ingredients in Buckley-Leverett displacement mechanics.

(a) Rock relative permeabilities

The so-called rock relative permeability curves are measured in one-dimensional core flooding experiments. After cleaning the core plug and flooding it with oil, so that at initial conditions it contains oil and irreducible water, one of two types of experiment is usually performed. The most common is the viscous displacement of oil by injected water and the second is the steady-state type of experiment in which both oil and water are simultaneously injected into the plug at a succession of different volume ratios (water flow rate increasing, oil rate decreasing).

There has been considerable debate in the industry concerning which of the two experimental procedures is the more realistic in matching water-oil displacement in the reservoir and the debate is extended further in section 5.4g. For the moment, however, it will be assumed that experiments are performed using the viscous displacement technique since this honours the basic displacement theory of Buckley and Leverett. Even so, there is a lack of reality in simulating reservoir flooding conditions in core flooding experiments, the main drawbacks being:

- the flooding is often conducted at flow rates that may be orders of magnitude higher than in the reservoir to overcome capillary end effects and ensure that the experiment is concluded in a reasonable period of time.
- usually a synthetic, high-viscosity crude oil is used to deliberately encourage unstable displacement and therefore obtain relative permeability functions that are continuous across the entire movable saturation range.
- it is difficult to ascertain the wetting condition in the reservoir for duplication in laboratory experiments.

It is not the intention in this text to dwell on these difficulties, other than the second, which is returned to in section 5.4f. The practising engineer simply has to live with the conditions which are largely outwith his power to control.
Relative permeabilities are always plotted as functions of the increasing displacing phase saturation, in this case water, and are regarded as being functions of the saturation alone.
During the viscous displacement flood the water saturation increases from its irreducible value (Fig. 5.15), S_{wc}, at which it is immobile to the maximum or flood-out saturation, $S_w = 1 - S_{or}$, at which the oil ceases to flow. S_{or} is the residual oil saturation representing the unconnected oil droplets trapped in each pore space by surface tension forces at the end of the waterflood. This occurs in any flood in which the fluids are immiscible, that is they do not physically or chemically mix. All saturations in this text are expressed as fractions of the pore volume, PV. Consequently the maximum amount of oil than can be displaced from the core plug during a waterflood is:

$$MOV = PV(1 - S_{or} - S_{wc}) \qquad (5.4)$$

which is known as the movable oil volume while $1 - S_{or} - S_{wc}$ is termed the movable saturation range.
Relative permeabilities are used to modify Darcy's equation for two phase flow, as follows:

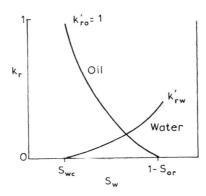

Fig. 5.15. Water–oil rock relative permeability curves.

$$q_w = -\frac{kk_{rw}}{\mu_w} A \frac{\Delta p_w}{\Delta L} \tag{5.10}$$

$$q_o = -\frac{kk_{ro}}{\mu_o} A \frac{\Delta p_o}{\Delta L} \tag{5.11}$$

In which, for each phase, the absolute permeability, k, is reduced through multiplication by the relative permeabilities k_{rw} or k_{ro} which are fractions between zero and unity and are dependent on the increasing water saturation. As the flood continues, the water finds it progressively easier to flow as its saturation increases until at flood-out it achieves its maximum or end-point relative permeability k'_{rw}. At the same time, the oil which has its end-point relative permeability, k'_{ro}, at the irreducible water saturation finds it more difficult to flow as the flood progresses until it eventually ceases flowing at the flood-out saturation when it becomes the discontinuous phase. The end-points are usually normalised, as shown in Fig. 5.15, so that $k'_{ro} = 1$.

(b) Mobility ratio

The significance of this ratio:

$$M = \frac{k'_{rw}}{\mu_w} \Big/ \frac{k'_{ro}}{\mu_o} \tag{5.2}$$

which represents the maximum velocity of water flow over that of oil, was described in section 5.2c, in that it dictates the efficiency of waterdrive on the microscopic scale. If $M \leq 1$ resulting from low oil viscosity, the displacement is piston-like and highly efficient such that all the movable oil is recovered by the injection of an equivalent volume of water. Alternatively, if the oil is viscous so that $M > 1$, the flood is inefficient and it can take the circulation of many MOVs of water to recover the single MOV of oil.

Another aspect of the mobility ratio worth investigating at this stage is its influence on the ease with which water can be injected into a reservoir. Consider the case of water–oil displacement in a thin core plug as depicted in Fig 5.16.

If the total pressure drop across the core plug, Δp, remains constant throughout the flood, then applying Darcy's law for piston-like displacement:

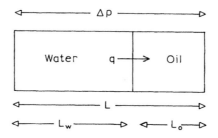

Fig. 5.16. Water–oil piston-like displacement.

$$\Delta p = \Delta p_w + \Delta p_o$$

$$\frac{q\mu_{av}L}{k_{av}A} = \frac{q\mu_w L_w}{kk'_{rw}A} + \frac{q\mu_o L_o}{kk'_{ro}A} \tag{5.12}$$

in which L_w and L_o are the lengths of the water and oil in the core plug at any stage of the flood. This may be expressed in terms of the average velocity:

$$v = \frac{k_{av}\Delta p}{\mu_{av}L} = \frac{\Delta p}{\dfrac{\mu_w L_w}{kk'_{rw}} + \dfrac{\mu_o L_o}{kk'_{ro}}} \tag{5.13}$$

or:

$$v = \frac{\dfrac{kk'_{rw}}{\mu_w}\dfrac{\Delta p}{L}}{\dfrac{L_w}{L} + M\dfrac{L_o}{L}} \tag{5.14}$$

But $L_w/L = R$, the fractional length of the water and therefore $L_o/L = 1 - R$; and since at the start of the flood when the core plug is full of oil:

$$v_i = \frac{kk'_{ro}}{\mu_o}\frac{\Delta p}{L} \tag{5.15}$$

then the ratio of the average velocity to the initial is:

$$\frac{v}{v_i} = \frac{M}{R + M(1 - R)} \tag{5.16}$$

This relationship implies that as the flood progresses such that $R \to 1$ then $v/v_i \to M$. The situation is depicted in Fig. 5.17. In the case that $M = 1$ the velocity of frontal advance remains constant throughout the flood. For $M > 1$, since the injected water is the more mobile fluid it becomes easier to inject as the flood

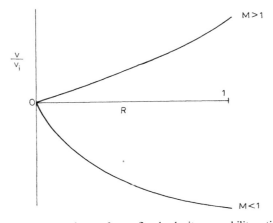

Fig. 5.17. Dependence of waterflood velocity on mobility ratio.

continues and therefore the velocity increases. Conversely, for $M < 1$ the injected water is less mobile and the flood slows as it progresses.

It will be appreciated, of course, that while equation 5.16 is perfectly correct for piston-like displacement for which it was derived ($M \leq 1$), it is inappropriate for $M > 1$ under which condition the displacement is unstable. Nevertheless, it is qualitatively quite correct in predicting that for $M > 1$ the flood will speed up as it progresses.

It is preferable in waterdrive operations to inject in the peripheral or basal aquifer, especially if $M < 1$, since the relative permeability to the injected water is near unity and therefore it is offered little resistance. In fields which are splintered into separate fault blocks not in communication with an aquifer or in low-permeability reservoirs where close pattern drive is required, it is necessary to inject water directly into the oil column. In this case, if $M < 1$ there is a very definite impedance to the injection which is exacerbated by the fact that the pressure falls off in a logarithmic fashion away from the injection wellbore.

Operators of some North Sea fields have experienced initial difficulties with the injection of cold seawater into the oil column, although it is often difficult to isolate the phenomenon from others such as movable fines blocking pore throats and scaling which can also cause injectivity impairment. Pre-heating the injected water to reduce its viscosity can help to alleviate the problem and yet another, although costly, approach is to inject a slug of surfactant at the start of the injection. This reduces the residual oil saturation immediately around the well to near zero and thus raises the end-point relative permeability to water towards unity which increases the mobility ratio and has a favourable effect on the injectivity. If the permeability is very high, the reduction in injectivity due to low mobility ratio is negligible while in low-permeability reservoirs it is overcome by fracturing the formation, which naturally occurs when attempting to inject water at practical rates. It is in the intermediate permeability range (hundreds of millidarcies), in which fracturing does not occur, that the difficulty seems to be at its worst.

(c) Fractional flow

The fractional flow of water at any point in a core plug or reservoir is defined as:

$$f_w = \frac{q_w}{q_w + q_o} \tag{5.17}$$

and is synonymous with the term watercut, equation 5.5, which specifically refers to the water produced in an experiment or from a well. Substituting for the rates using Darcy's law, equations 5.10 and 5.11, gives:

$$f_w = \frac{\dfrac{kk_{rw}}{\mu_w} A \dfrac{\Delta p_w}{\Delta L}}{\dfrac{kk_{rw}}{\mu_w} A \dfrac{\Delta p_w}{\Delta L} + \dfrac{kk_{ro}}{\mu_o} A \dfrac{\Delta p_o}{\Delta L}}$$

And assuming that the pressure gradients in the water and oil are similar, that is, neglecting capillary pressure effects, cancelling terms and dividing numerator and denominator by k_{rw}/μ_w gives:

$$f_w = \cfrac{1}{1 + \cfrac{\mu_w}{\mu_o}\cfrac{k_{ro}}{k_{rw}}} \tag{5.18}$$

which is the fractional flow equation for horizontal displacement. The neglect of capillary pressures is realistic in high-rate flooding experiments and is justified for displacement in macroscopic reservoir sections as described in section 5.6. In a waterflood either in the laboratory or the field, since pressure is usually maintained then the viscosity ratio μ_w/μ_o is constant; meaning that the fractional flow is strictly a function of the water saturation upon which the ratio k_{ro}/k_{rw} depends.

It is argued in this chapter that fractional flow is the most fundamental concept in the whole subject of waterdrive, much more so that relative permeabilities, because:

- it is a single function, the shape of which reveals all about the efficiency of the flood whereas the two relative permeability curves do not
- when applied to the description of flooding in the reservoir, it incorporates the correct, *in situ* oil and water viscosities, which is not the case in most relative permeability measurements.

The significance of the shape of the fractional flow is a subject that must be deferred until after the description of the Buckley-Leverett displacement theory (section 5.4d).

For waterdrive in an inclined reservoir, if the displacement is in the more gravity stable updip direction then the fractional flow of water is modified as:

$$f_w = \cfrac{1 - G}{1 + \cfrac{\mu_w}{\mu_o}\cfrac{k_{ro}}{k_{rw}}} \tag{5.19}$$

in which G is a positive gravity number derived in reference 9, chapter 10, as:

$$G = 4.886 \times 10^{-4}\,\frac{kk_{ro}A\Delta\gamma\sin\theta}{q\mu_o}$$

which can be more conveniently expressed as:

$$G = 2.743 \times 10^{-3}\,\frac{kk_{ro}\Delta\gamma\sin\theta}{\upsilon\mu_o} \tag{5.20}$$

where $\upsilon = 5.615\,q/A$ ft/day, is the average Darcy velocity of the flood. The larger the value of G the more gravity stable is the advance of the water and the watercut development is suppressed. The physical significance of the more important parameters in the group (θ, k, $\Delta\gamma$ and υ) is described below in connection with displacement in the reservoir.

Advancement angle (θº)

This is one of the most important parameters of all for its magnitude dictates whether the waterdrive is "edge" or "basal" which, in turn, affects the value of the average permeability, k, and the frontal velocity, v.

If the reservoir has a definite seal at its base then the waterdrive is from downdip and is edgewise. In this case the water moves between the bedding planes and the angle θ is equal to the angle of dip. If there is no seal beneath the reservoir, however, then the water will enter from the base at an angle whose maximum value can be $\theta = 90º$ (Fig. 5.18).

Edge waterdrive is the more complex of the two and therefore receives most attention in this chapter. Not only is it intrinsically less stable because of its smaller gravity number but also the efficiency of the waterdrive is critically dependent on the degree of heterogeneity, especially the permeability distribution, across the reservoir section. In the case of basal waterdrive, however, the gravity number is larger and the advancing water is little affected by the reservoir heterogeneity. That is, the rising water is confronted by "sheets" of sand, which even though they have different rock properties will appear uniform to the waterfront as it moves upwards.

Sometimes, as in the case of the Montrose Field described in section 5.2d, it is not possible to decide whether the drive is from the edge or base until the field has been produced on a continuous basis and dynamic RFT pressure profiles can be inspected. In the North Sea, most of the large fields in areas like the East Shetland Basin contain stacked reservoirs separated by shales and fall into the edge drive category. These are of Jurassic age. Conversely, the shallower, Palaeocene fields such as Montrose and BP's Forties Field consist of massive sand sections with no definite barriers to vertical fluid movement and therefore are predominantly affected by basal waterdrive. These fields generally, but not always, perform better and achieve higher ultimate recovery than if they were edge drive. This is illustrated by the contrasting watercut developments indicated in Fig. 5.19.

Watercuts have risen in a much harsher fashion in the edge drive fields than in the basal. This, however, as described in section 5.10, is due more to the adverse influence of the reservoir heterogeneity than the actual value of the gravity number, G, which may be insignificant in comparison. In basal drive fields the influence of heterogeneity is much less severe and water production is deferred. Basal water

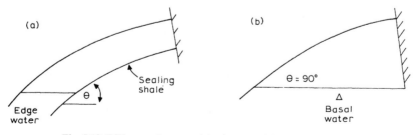

Fig. 5.18. Difference between (a) edge, and (b) basal waterdrive.

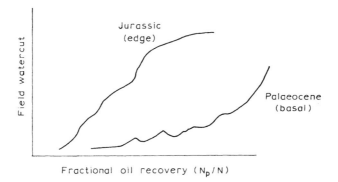

Fig. 5.19. Contrasting watercut developments in edge and basal drive fields.

coning will occur but if the reservoir is of substantial thickness then wells can be successively plugged back to defer water production.

Permeability (k)

The larger the permeability, the more the reservoir acts as a "tank" and therefore the more uniform and stable is the advance of water whether it is from the edge or base. The average permeability is usually higher for edge than basal waterdrive. In the first place, permeabilities tend to be greater parallel to the bedding planes rather than normal to them. Secondly, the average permeability for edge drive is usually evaluated as the arithmetic average, whereas for basal drive it is the harmonic average that is applied; meaning that the value is dictated by the layers with the lower vertical permeabilities.

Gravity difference ($\Delta \gamma$)

This is the difference in specific gravities in the reservoir between the water and oil and typically has a value of 0.30, which is slight. As mentioned in Chapter 6, however, for gas drive the gas–oil gravity difference is usually several times larger and therefore the parameter assumes greater significance in this process.

Average frontal velocity (v)

It should be noted that the average frontal velocity in equation 5.20 is the Darcy velocity: rate divided by the total cross-sectional area, A. As described in section 5.7b, however, the actual velocity may be evaluated from material balance considerations (equation 5.52) by replacing the area by $A\phi(1 - S_{or} - S_{wc})$ making it considerably larger than the Darcy value used in the equation. The velocity introduces a rate dependence in waterdrive calculations although usually the effect is rather small and does not necessarily affect the ultimate recovery but rather the efficiency with which oil is recovered. On account of the reduced vertical permeability and greater area of frontal advance, the velocity of basal waterdrive is less than for edge drive.

Taking parameters typical for the North Sea, the contrast in values of the gravity number, G, can be evaluated for edge and basal waterdrive by evaluating equation 5.20 for $\Delta\gamma = 0.30$, $\mu_o = 1$ cp:

Edge drive: $\theta = 6^\circ$, $k = 500$ mD, $v = 0.2$ ft/day $\qquad G = 0.22\,k_{ro}$

and

Basal drive: $\theta = 90^\circ$, $k = 50$ mD, $v = 0.004$ ft/day $\quad G = 10.29\,k_{ro}$

The latter is almost fifty times greater than for edge drive and leads to such stability in the upward movement of water that piston-like displacement is almost guaranteed even if the mobility ratio is unfavourable. The main problem with basal drive, as mentioned previously, is the localized wellbore coning of water.

For edge waterdrive, even at fairly high inclination, the gravity term in the fractional flow is often negligible and it is therefore omitted. The engineer should never assume this to be the case, however, and the matter should be checked prior to displacement efficiency calculations, as illustrated in Exercise 5.2. In this respect, it often makes little difference in terms of intrinsic stability whether the waterdrive is conducted in the updip or downdip direction. The former is by far the more common practice, however, for to displace oil downdip carries the attendant risk of displacing some of it into the aquifer where it becomes trapped.

(d) The Buckley-Leverett displacement theory

This is the basic theory of waterdrive governing all calculations in the subject whether performed using analytical or numerical simulation techniques. The theory, dating from 1942 [10], was derived for the following physical conditions:

– the displacement is one dimensional
– pressure is maintained
– the fluids are immiscible

The first is relevant for a core flooding experiment although as demonstrated in section 5.9, the theory can be applied in a much more general sense. To move fluids through the core plug there must be a positive pressure differential between the injection and production ends but the pressure difference remains constant throughout the flood and the water and oil viscosities are evaluated at the average value and are constant. The term immiscible means that the fluids do not mix physically or chemically. Consequently, there is a finite surface tension between them; meaning that at the end of the flood a finite saturation of the displaced phase, the oil, remains disconnected and trapped in each pore space. This is the residual oil saturation, S_{or}, and typically has a value of 0.25–0.35 PV in a waterdrive.

It was the aim of Buckley and Leverett to determine an expression for the velocity of a plane of constant water saturation passing through a core plug. This they did applying the physical principle of mass conservation for displacement at constant pressure. The mathematics is described in detail in references 1 and 9 and will not be repeated here; instead the concentration in theme will be on understanding the physics of the process. The resulting equation of Buckley-Leverett, expressed in absolute units, is:

$$v = \frac{q_i}{A\phi} \frac{df_w}{dS_w} \tag{5.21}$$

in which q_i is the constant injection rate and A the area of cross-section of the core plug. The equation states that the velocity of a plane of constant water saturation is directly proportional to the derivative of the fractional flow evaluated for the same saturation. Whether the fractional flow equation contains the gravity term or not (equation 5.18 or 5.19), it is strictly a function of the increasing water saturation through its dependence on the rock relative permeabilities, hence the full differential term in equation 5.21.

Consider water–oil displacement for a mobility ratio somewhat greater than unity. The fractional flow has the shape shown in Fig. 5.20a, displaying a point of inflexion for intermediate saturations. On account of this, the velocity distribution of saturations across the movable range must have a maximum point as depicted in Fig. 5.20b. This produces a physically unrealistic result for it suggests that saturations with low and high values can travel at the same velocity and may therefore occupy simultaneously the same position in the core plug. Looking through the plug, the water saturation distribution along its length would be as shown in Fig. 5.21.

In an excellent paper on Buckley-Leverett displacement [11], Cardwell referred to this anomaly as a physically absurdity — which indeed it is. The saturation distribution reveals that it should be possible for three saturations, S_{wc}, S_{w1} and S_{w2} to coexist at a single point in the linear displacement path. The literature reveals that the triple value in saturations posed quite an intellectual problem to engineers for many years. Buckley-Leverett themselves presented the argument that a triple value would not have arisen had they been able to incorporate capillarity in their theory. Others suggested that the anomaly would not have arisen had they used a method of solving differential equations, such as the ones they were dealing with, known as the method of characteristics. This, if used in conjunction with the concept of shock-front displacement, eradicates the triple value saturation.

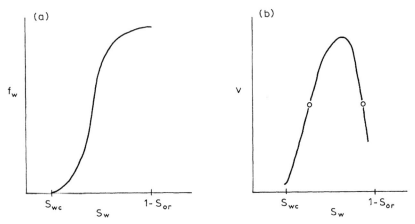

Fig. 5.20. (a) Fractional flow for an unfavourable mobility ratio ($M > 1$) and (b) corresponding velocity distribution.

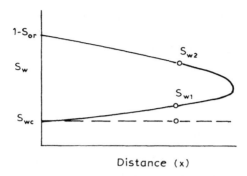

Fig. 5.21. Triple value saturations in Buckley-Leverett displacement.

That is, the higher values of saturation catch up with the lower leading to the development of a shock-front saturation discontinuity, S_{wf}. The magnitude of this saturation is dependent on the mobility ratio which is unfavourable ($M > 1$) for the displacement shown in Fig. 5.22. As noted in section 5.2c, however, for a favourable mobility ratio, $M < 1$, piston-like displacement occurs such that $S_{wf} = 1 - S_{or}$, the flood-out saturation, as shown in Fig. 5.23.

The reader must not harbour the impression that the Buckley-Leverett theory is somehow flawed — it is not, and in the following section a perfectly acceptable means of applying the theory is presented which caters fully for the phenomenon of shock-front development. The debate about "what went wrong with Buckley-Lever-

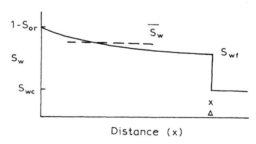

Fig. 5.22. Shock front development in immiscible displacement ($M > 1$).

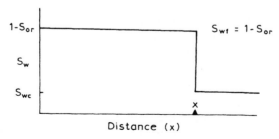

Fig. 5.23. Saturation distribution for piston-like displacement ($M \leq 1$).

ett" has raged now for fifty years but to this author, the most logical explanation of how the triple value saturation arose is because Buckley and Leverett pinned absolute faith in rock relative permeability curves defined across the entire movable saturation range and had they been a bit more circumspect in their use of these functions, as explained in section 5.4f, the anomaly would never have arisen in the first place.

(e) Welge displacement efficiency calculations

Welge's paper published in 1952 [12] provided engineers with a simple method of applying Buckley-Leverett's theory, including the shock-front effect, in a simple fashion to calculate oil recovery as a function of the cumulative water injected. Welge's aim was to calculate the average water saturation, $\overline{S}_w$ (Fig. 5.22), in the core plug as the flood progressed. The reason for doing so was that the difference between this and the initial saturation must be equal to the oil recovered from the plug, that is

$$N_{pd} = \overline{S}_w - S_{wc} \quad (PV) \tag{5.22}$$

Since saturations are always expressed in pore volumes (fractions of the pore volume) then so too is the oil recovery, N_{pd} — dimensionless pore volumes. In fact, in waterdrive recovery calculations it is conventional to work in pore volumes irrespective of whether they are being performed for a core flooding experiment or a reservoir problem since the recovery is always related directly to saturation changes. In the case of a reservoir flood, the pore volume recovery can be readily expressed as a real volume provided the dimensions of the system are known, as described in section 5.5b.

Welge determined the average saturation behind the flood using the simple one-dimensional integral:

$$\overline{S}_w = \frac{\int_0^x S_w \, dx}{x} \tag{5.23}$$

in which S_w is the water saturation at any point along the profile behind the front and x is the distance the front has travelled (Fig. 5.22). The integral appears quite simple but, in fact, as demonstrated in references 1 and 9 in which it is mathematically evaluated, it is a fairly complex integration by parts consuming several pages of text. The mathematics will not be repeated here but instead attention will be focused in the simple method presented by Welge [12] and its importance in calculating oil recovery.

The main interest in displacement calculations is when water first breaks through at the producing end of the core plug ($x = L$) and subsequently. The first step in applying the method is to draw the fractional flow relationship (equation 5.18 or 5.19). This is the vital step which incorporates the oil and water viscosities relevant

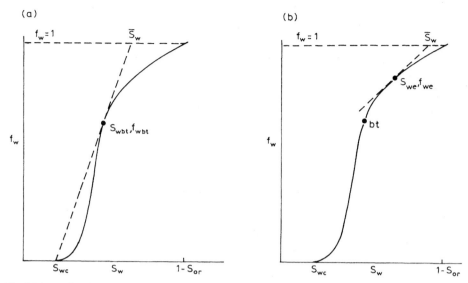

Fig. 5.24. Application of the Welge technique: (a) at breakthrough, (b) from breakthrough to flood-out $(M > 1)$.

for the actual flood with the saturation dependent relative permeability functions. Such a relationship is shown in Fig. 5.24, once again for unfavourable mobility ratio displacement, for which it is easiest to describe the technique. Drawing the tangent from $S_w = S_{wc}$ to the fractional flow curve determines, at the point of tangency, the shock-front or breakthrough saturation, $S_{wf} = S_{wbt}$, as the water first reaches the producing end of the core plug. The corresponding fractional flow of water is read from the ordinate. Furthermore, extending the tangent to intersect the line $f_w = 1$ gives the average water saturation in the plug at breakthrough, $\overline{S}_w$. At this instant the oil recovery calculation is trivial: since no water has been produced, all the injected water must have displaced an equal volume of oil as production, that is, at breakthrough:

$$N_{pd} = \overline{S}_w - S_{wc} = W_{id} \quad (PV) \tag{5.24}$$

in which W_{id} is the cumulative water injected in pore volumes.

Following breakthrough, the procedure is to move around the fractional flow curve from $S_w = S_{wbt}$ to $S_w = 1 - S_{or}$ (Fig. 5.24b) selecting values of S_{we}, the ever increasing water saturation at the end of the plug as the flood progresses and reading the corresponding values of f_{we} from the ordinate. It is suggested that the increments in saturation be no greater than 0.05 PV. Each time a new value of S_{we} is selected, extrapolation of the tangent at that point to the line $f_w = 1$ gives the increasing value of the average saturation in the core plug, $\overline{S}_w$, from which the recovery can always be evaluated using equation 5.22. This graphical technique is a little cumbersome, however, and the preferred approach is to use Welge's equation directly, as derived in chapter 10 of reference 9, which is equivalent to drawing the tangent:

$$N_{pd} = (S_{we} - S_{wc}) + (1 - f_{we})W_{id} \quad (PV) \tag{5.25}$$

All the parameters on the right can be read directly from the fractional flow curve except the cumulative water influx, W_{id} (PV), which comes directly from the Buckley-Leverett theory. That is, integrating equation 5.21 with respect to time gives:

$$x A\phi = W_i \frac{df_w}{dS_w}$$

and, after breakthrough when $x = L$:

$$W_{id} = \frac{W_i}{LA\phi} = \frac{1}{\dfrac{df_w}{dS_w}} \quad (PV) \tag{5.26}$$

Before illustrating the use of this important equation, it is worthwhile considering the significance of drawing the tangent from S_{wc} to the fractional flow, as in Fig. 5.24a: in doing so it seems to neglect the first part of the fractional flow function. What it signifies is that saturations in the range:

$$S_{wc} < S_w < S_{wbt}$$

are not free to move independently through the core plug, instead they are all caught up in the shock-front saturation discontinuity at the leading edge of the flood (Fig. 5.22). For ease of presentation, the theory of Buckley and Leverett has been described in terms of unfavourable mobility ratio displacement ($M > 1$). As pointed out in section 5.2c, however, by choice operators worldwide elect to perform engineered waterdrive in reservoirs that have rock and fluid properties that assure a favourable mobility ratio ($M < 1$), providing piston-like displacement on the microscopic scale.

For this important type of displacement, the fractional flow is concave upwards across the entire movable saturation range as illustrated in Fig. 5.25. It is not possible to draw a tangent to this curve, instead, the equivalent construction is a chord from S_{wc} which intersects at the flood-out saturation, $1 - S_{or}$. This, however, is the breakthrough saturation manifesting the phenomenon of piston-like displacement (Fig. 5.23). The intersection also occurs on the line $f_w = 1$ implying an average saturation behind the front of $\overline{S}_w = 1 - S_{or}$ which on substitution into the oil recovery equation 5.22 gives

$$N_{pd} = 1 - S_{or} - S_{we} \quad (PV) \tag{5.27}$$

meaning that the maximum oil recovery of 1 MOV is achieved at breakthrough in this most efficient form of displacement.

The continual upward curvature of the fractional flow also means that none of the saturations in the movable range, $1 - S_{or} - S_{wc}$, is able to move independently through the core plug, they are all caught-up in the complete shock-front discontinuity.

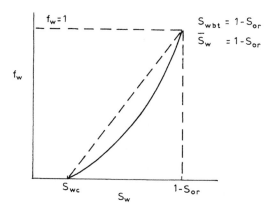

Fig. 5.25. Fractional flow curve for piston-like displacement ($M < 1$).

To illustrate the application of Welge's equation to a one-dimensional flood, consider the single set of rock relative permeabilities plotted in Fig. 5.26a applied to waterflooding for three different mobility ratios. The data common to each case are:

$$k'_{rw} = 0.2; \qquad k'_{ro} = 1.0; \qquad \mu_w = 0.4 \text{ cp}$$
$$S_{wc} = 0.20 \text{ PV}; \quad S_{or} = 0.30 \text{ PV}; \quad MOV = 0.50 \text{ PV}$$

and the oil viscosities and mobility ratios evaluated using equation 5.2 are as follows:

Case 1: $\mu_o = 0.5$ cp, $M = 0.25$
Case 2: $\mu_o = 5.0$ cp, $M = 2.5$
Case 3: $\mu_o = 50$ cp, $M = 25$

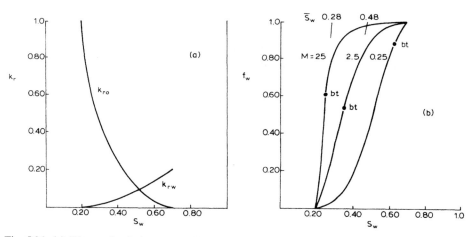

Fig. 5.26. (a) Water–oil relative permeability curves. (b) Fractional flow curves for three different mobility ratios.

TABLE 5.4

Fractional flow relationships for three different mobility ratios

S_w	k_{rw}	k_{ro}	Case 1 f_w	Case 2 f_w	Case 3 f_w
0.20	0	1	0	0	0
0.25	0.008	0.650	0.015	0.133	0.606
0.30	0.020	0.470	0.051	0.347	0.842
0.35	0.031	0.350	0.101	0.529	0.918
0.40	0.046	0.250	0.187	0.697	0.958
0.45	0.062	0.175	0.307	0.816	0.978
0.50	0.082	0.114	0.473	0.900	0.989
0.55	0.110	0.070	0.663	0.952	0.995
0.60	0.138	0.037	0.823	0.979	0.998
0.65	0.170	0.014	0.938	0.993	0.999
0.70	0.200	0	1	1	1

The three fractional flow relationships are listed in Table 5.4 and plotted in Fig. 5.26b. These have been evaluated for horizontal displacement using equation 5.18 for the single set of relative permeabilities and the three different water/oil viscosity ratios. That is, the relative permeabilities are somewhat arbitrary in themselves, being generated using fluid viscosities that usually bear no relation to the *in situ* reservoir values (refer to section 5.4f). It is only when the vital step is taken of generating the fractional flow curve using the relevant viscosities that the efficiency of the displacement becomes apparent, as in the present three cases.

Case 1 ($\mu_o = 0.5$ cp, $M = 0.25$)

According to the theory, such a low mobility ratio should guarantee piston-like displacement in a homogeneous core flooding experiment which is manifest by a totally concave upward fractional flow curve, as shown in Fig. 5.25. As can be seen in Fig. 5.26b, however, this situation is not quite realised and there is a point of tangency at $S_w = 0.625$ PV, $f_w = 0.880$. This is frequently observed when plotting fractional flow curves for low mobility ratio displacement. It results for a variety of reasons such as inhomogeneities in the core plug but is more particularly due to the experimental difficulty in measuring accurately the very low oil flow rates as the flood-out saturation is approached towards the end of the experiment. Under such circumstances, the engineer should simply accept the fact that if $M \leq 1$, the displacement is necessarily piston-like and that at breakthrough all the movable oil will be recovered, which in this case means that:

$$N_{pd} = 1 - S_{or} - S_{we} = 0.50 \text{ PV}$$

Case 2 ($\mu_o = 5$ cp, $M = 2.5$)

The inflexion in the fractional flow is not particularly evident in this case which results from the fact that the relative permeability functions themselves have fairly limited curvature (Figs. 5.26a and b). Nevertheless, a tangent to the fractional flow identifies the breakthrough parameters as

$$S_{wbt} = 0.35 \text{ PV}, \quad f_{wbt} = 0.529$$

while extrapolation of the tangent to the line $f_w = 1$ gives a value of $\overline{S}_w = 0.48$ PV and an oil recovery at breakthrough, in accordance with equation 5.24 of 0.28 PV, which is 56% of the MOV of 0.50 PV for this unfavourable mobility ratio displacement.

The Welge equation, equation 5.25, has been evaluated for saturations between S_{wbt} and flood-out, $1 - S_{or}$, as detailed in Table 5.5. In this, it should be noted that the values of S_{we} and f_{we} (columns 6 and 7) which are used directly in equation 5.25 are the averages of the values in columns 1 and 2 at which the derivative:

$$\frac{1}{W_{id}} = \frac{\Delta f_w}{\Delta S_w} \quad \text{(PV)}$$

is evaluated (the same applies in all Welge displacement calculations presented later in the chapter). As can be seen, to recover the MOV of 0.50 PV for this mobility ratio requires the circulation of 7.1 PV of injected water. In a reservoir development situation, especially offshore as described in section 5.3c, it can become intolerable to circulate such large volumes of water on account of the protraction of the project

TABLE 5.5

Welge water–oil displacement calculations for different mobility ratios

S_{we}	f_{we}	ΔS_{we}	Δf_{we}	$\Delta f_{we}/\Delta S_{we}$	S_{we}	f_{we}	W_{id}	N_{pd}
Case 2: $\mu_o = 5$ cP, M = 2.5								
0.35 (bt)	0.529							
0.40	0.697	0.05	0.168	3.360	0.375	0.613	0.298	0.290
0.45	0.816	0.05	0.119	2.380	0.425	0.757	0.420	0.327
0.50	0.900	0.05	0.084	1.680	0.475	0.858	0.595	0.359
0.55	0.952	0.05	0.052	1.040	0.525	0.926	0.962	0.396
0.60	0.979	0.05	0.027	0.540	0.575	0.966	1.852	0.438
0.65	0.993	0.05	0.014	0.280	0.625	0.986	3.571	0.475
0.70	1	0.05	0.007	0.140	0.675	0.9965	7.143	0.500
Case 3: $\mu_o = 50$ cP, M = 25								
0.25 (bt)	0.606							
0.30	0.842	0.05	0.236	4.720	0.275	0.724	0.212	0.134
0.35	0.918	0.05	0.076	1.520	0.325	0.880	0.658	0.204
0.40	0.958	0.05	0.040	0.800	0.375	0.938	1.250	0.253
0.45	0.978	0.05	0.020	0.400	0.425	0.968	2.500	0.305
0.50	0.989	0.05	0.011	0.220	0.475	0.984	4.545	0.348
0.55	0.995	0.05	0.006	0.120	0.525	0.992	8.333	0.392
0.60	0.998	0.05	0.003	0.060	0.575	0.997	16.667	0.425
0.65	0.999	0.05	0.001	0.020	0.625	0.999	50.000	0.475
0.70	1	0.05	0.001	0.020	0.675	0.9995	>50.000	0.500

lifetime and the difficulties in purifying the water prior to disposal. Therefore, an operator may be obliged to abandon the waterdrive at a watercut of, say, 90%, which would limit the oil recovery to 0.38 PV representing a loss of 24% of the MOV.

Case 3: (μ_o = 50 cP, M = 25)

For this highly unfavourable mobility ratio, water breakthrough occurs prematurely (Fig. 5.26b) for values of

$$S_{wbt} = 0.25 \text{ PV}, \quad f_{wbt} = 0.60$$

when the saturation behind the front is $\overline{S}_w$ = 0.28 PV. Consequently, the oil recovery at breakthrough is only 0.08 PV or 16% of the MOV. Thereafter, it would require the circulation of 50 PVs of water to recover the single MOV (Table 5.5). Abandoning at a 90% watercut would in this case result in an oil recovery of 0.22 PV which is only 44% of the MOV. This statistic explains why operators hesitate when faced with the prospect of developing a viscous oilfield by waterdrive.

Although the foregoing example relates to a one-dimensional core flooding experiment, the same method of Welge for interpreting Buckley-Leverett displacement mechanics applies also for describing waterflooding in macroscopic reservoirs (sections 5.6–5.7). Therefore, it is worthwhile considering some of the basics of waterdrive that emerge from the example and will be applied in a more practical form later. These are:

- The shape of fractional flow curve (Fig. 5.26b) is such that for favourable displacement conditions ($M \leq 1$) it moves to the right and is concave upwards; whereas for unfavourable displacement ($M > 1$) it moves to the left and is concave downwards. In the latter case, the small slope as the flood-out saturation is approached implies the circulation of large volumes of water since, from the Buckley-Leverett material balance

$$W_{id} = \frac{1}{\dfrac{\mathrm{d}f_w}{\mathrm{d}S_w}} \tag{5.26}$$

- The movable oil volume, MOV = $PV(1 - S_{or} - S_{wc})$, can always, by definition, be moved (recovered) in an engineered waterdrive: provided a sufficient number of PVs of water is circulated through the system. If the displacement is inefficient, however, it can become impracticable to handle such large volumes of water and projects may have to be terminated prematurely with attendant loss of oil recovery.

- The Welge equation, equation 5.25, provides a relationship between oil recovery, N_{pd} (PV), and cumulative water injected, W_{id} (PV). Both are equal at water breakthrough but if the mobility ratio is unfavourable ($M > 1$) then the injection steadily increases with respect to the oil production throughout the remainder of the flood. The Welge equation contains an even more significant relationship between f_{we} and N_{pd} which as described in section 5.3c is the fundamental requirement in planning a successful waterdrive. In equation 5.25, the relationship is expressed in PVs but a more practical formulation required in field calculations is presented in section 5.5b.

(f) Input of rock relative permeabilities to numerical simulation and analytical reservoir models

The reader should be in a position by now to appreciate some of the subtleties in the use of relative permeabilities in waterdrive studies. In the first place, as mentioned in section 5.4a, unless laboratories are instructed to do otherwise, they measure water–oil relative permeabilities using a synthetic oil of high viscosity (typically 17 cp). The sequence of events in generating relative permeabilities from an experiment performed with such a viscous oil are depicted in Fig. 5.27.

The high mobility ratio in the viscous drive flood precipitates premature breakthrough of water at the end of the core plug following which large volumes of water, W_{id} (PV), must be circulated to recover all the oil (N_{pd} = 1 MOV). The Welge equation, 5.25, is then solved to determine the fractional flow, as described by Jones and Roszelle [8], which results in a function that displays little inflexion (no shockfront development) and is concave downwards across practically the whole movable saturation range: indicating that the point of Welge tangency is at $S_{wbt} = S_{wc}$ and all saturations are free to move independently through the core plug. The function is similar to that shown in Fig. 5.26b for M = 25. In a final reverse step, a set of relative permeability curves is generated which must also be defined by points across the entire movable saturation range implying independent mobility of all saturations.

Since the majority of oilfields operated by engineered waterdrive contain, by choice, low-viscosity oil having $M < 1$ (section 5.2c), for which displacement is stable and piston-like on the microscopic scale — the questions must be asked:

- What relevance do conventional, high-viscosity experimental results have when applied to low-viscosity displacement which predominates under field conditions?
- What results would be obtained if the laboratory experiments were performed using *in situ*, low oil viscosity?

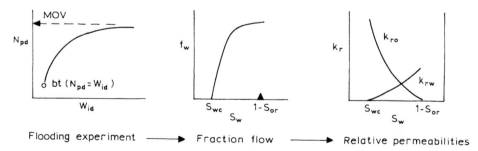

Fig. 5.27. Generation of relative permeabilities from a viscous displacement experiment using a high-viscosity oil ($M > 1$).

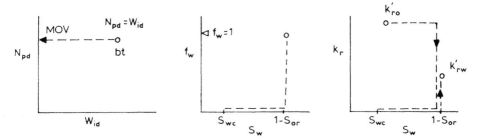

Fig. 5.28. Generation of relative permeabilities from a viscous displacement experiment using a low-viscosity oil ($M < 1$).

To address the latter question first; if the sequence of steps in Fig. 5.27 were repeated for a low oil viscosity flood ($M < 1$) in a homogeneous core plug, the result would be as shown in Fig. 5.28.

Under these circumstances, the experimental flood results in a single point such that all the movable oil is recovered by the injection of an equivalent volume of water, which manifests complete piston-like displacement. Consequently, the fractional flow consists of a single point also. The dashed line step-function implies that no water flow can be observed ($f_w = 0$) until the water piston reaches the end of the plug when complete flood-out occurs. Similarly in the final step, relative permeability "curves" are not obtained merely the end-points, k'_{rw} and k'_{ro}. In this case, the step functions mean that only oil flows at its maximum relative permeability until flood-out occurs when it stops abruptly to be replaced by the flow of water alone at its maximum relative permeability. It is an either–or situation reflecting the piston-like nature of the displacement. If such realistic experiments are performed it is possible to observe the development and movement of the water piston by looking through the core plug using a "brain scanner" or similar industrial device. There may be some "fuziness" at the front on account of microscopic heterogeneities but essentially the piston-like effect predominates.

It is therefore interesting to note that when you enquire of Service Company laboratory staff why they conduct relative permeability experiments with an artificial high-viscosity oil, the answer you receive is that if they did not adopt this practice they would not obtain "curves" only "points" — and that would not satisfy the customer! This has always seemed a perfectly reasonable if somewhat commercial reply.

To answer the first of the above questions concerning the relevance of the use of high-viscosity ($M > 1$) relative permeabilities in low-viscosity displacement problems ($M \leq 1$); the answer is that provided the engineer applies analytical techniques incorporating the concept of fractional flow then it really does not matter whether high- or low-viscosity experimental results are used — no error will be made. It is only when using numerical simulation models, which do not cater for fractional flow, that the engineer must be careful to distinguish between the different experiments when inputting relative permeabilities (saturation tables),

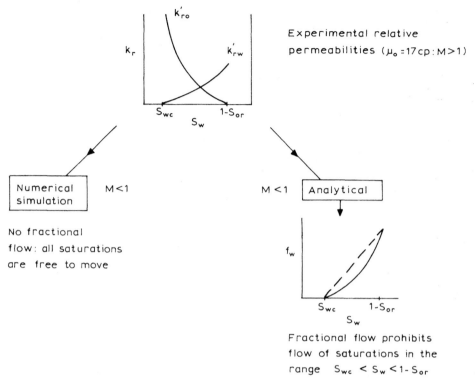

Fig. 5.29. Comparison of the input of rock relative permeabilities to numerical simulation and analytical models.

otherwise significant errors can occur. The situation is depicted in Fig. 5.29, in which a set of fully defined relative permeabilities, measured in a high oil viscosity experiment ($M > 1$), is input to analytical and numerical simulation models for a reservoir in which the mobility ratio is favourable ($M \leq 1$).

In adopting the analytical approach, the opening move is to plot the fractional flow: a step which incorporates the *in situ* oil and water viscosities relevant for the reservoir flooding problem under consideration. If the displacement is favourable ($M < 1$) then the curve will be concave upwards, or almost so, which manifests piston-like displacement and prohibits the independent movement of saturations in the movable range. The only water saturation allowed to move is the end-point value $S_w = 1 - S_{or}$ and the flow of oil and water is governed by their end-point relative permeabilities, k'_{ro} and k'_{rw} — the full curves are redundant and can be discarded. Precisely how the end-points are used in practical waterflooding problems in macroscopic reservoir sections will be described in section 5.5b.

If full rock relative permeabilities, measured using a high-viscosity oil ($\mu_o = 17$ cp), are input directly to a numerical simulation model, then since the concept of fractional flow is completely disregarded, the simulator grants mobility to all satur-ations in the movable range. In a low mobility ratio oil reservoir this is an unrealistic physical situation commonly referred to as "numerical dispersion" and means that while simulators may conserve mass to numerous places of decimal they do so in the wrong place and at the wrong time. The input of the continuous functions would also appear to violate Newton's third law of motion at the flood-front: in a low mobility ratio displacement there is a distinct shock-front in which the advancing water is held back (action and reaction) which is simply not catered for. Furthermore, the very input of continuous curves represents the paradoxical situation that in the input of PVT data the engineer may have specified that the oil viscosity is, say, 0.5 cp, yet by inputting full rock curves the simulator is being implicitly instructed that for fluid dynamics calculations the PVT is to be disregarded and an oil viscosity of 17 cp used instead!

It may be thought that there has been a basic oversight in the development of finite difference simulation models in their failure to appreciate the subtleties of fractional flow but this is not really the case. Simulators apply the fundamental principles of:

– mass conservation
– Darcy's law
– isothermal compressibility

to the description of immiscible water–oil displacement. In continuous space–time, the differential equations describing and linking these physical principles would lead to shock-front displacement just as does the theory of Buckley-Leverett and analytical method of Welge. The problem is that simulators work in discrete space and time (grid blocks, time steps) which, again considering a low mobility ratio flood, has the effect illustrated in Fig. 5.30.

The schematics represent various attempts to study oil–water displacement in a one dimensional, linear model for favourable conditions ($M < 1$) using numerical simulation. The diagrams show saturation distributions along the length of the linear system. In case (a), full relative permeabilities measured in a viscous oil displacement experiment are input to a one-dimensional model structured with a series of discrete grid blocks of individual length Δx aligned in the flood direction. The input of the full relative permeability curves permits all the movable saturations to flow. Therefore, as soon as the water saturation in the injection grid block exceeds the irreducible level then water is immediately dispersed to all the downstream grid blocks. In similar models of actual reservoirs the dispersion can be observed over distances of thousands of feet and is a total distortion of the shock-front displacement that the basic physics predicts should occur.

An obvious way to overcome this inaccuracy would be to structure the linear model with a large number of grid blocks in both the x and the z directions, as shown in case (b). The greater the number of grid blocks the more closely the model approaches continuum in space in which Buckley-Leverett shock-front displacement is honoured. In fact what is happening is that the input of full relative permeability curves still encourages dispersion in the x-direction but this is compensated by dispersion from block to block in the z-direction leading to a sharp frontal development. The slight spillage of water downstream of the front, shown

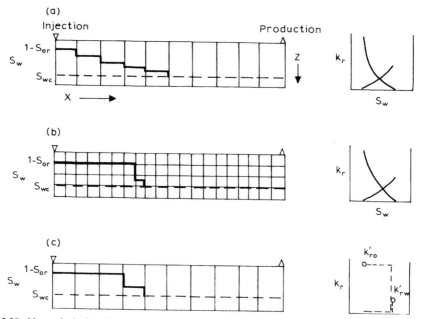

Fig. 5.30. Numerical simulation of a core flooding experiment for low mobility ratio waterdrive ($M \leq 1$).

in case (b), is due to viewing matters at the end of a discrete time step when the column of grid blocks has not filled to flood-out.

Yet another means of overcoming the numerical dispersion resulting from the input of continuous relative permeabilities for low mobility ratio displacement would be — not to input the functions to the model in the first place. Instead, the end-point relative permeabilities alone are input, connected to their respective end-point saturations by step functions, as shown in Figs. 5.28 and 5.30c. Then, even though the model is structured with coarse grid blocks in the x-dimension, as in case (a), the effect of these functions will be to inhibit movement of water from one block to the next until the upstream block has filled to flood-out. The situation is shown as case (c) and again indicates partial flooding of the first downstream block from the front due to freezing matters at the end of a discrete time step. The problem is, however, that running a detailed two-dimensional model, case (b), or a coarse model with step function relative permeabilities, case (c), are both expensive means of modelling shock-front displacement. In the former, the added number of grid blocks naturally increases the running cost while in the latter it is because numerical simulation models do not appreciate the input of step functions. Invariably they are constructed using the finite difference analogue and as such the input of step-functions can give the machines indigestion, to the extent that they can come to a grinding halt. Therefore, the functions should be input with some slight yet continuous curvature but even so the simulation runs can be slow and expensive, although matters are improving rapidly.

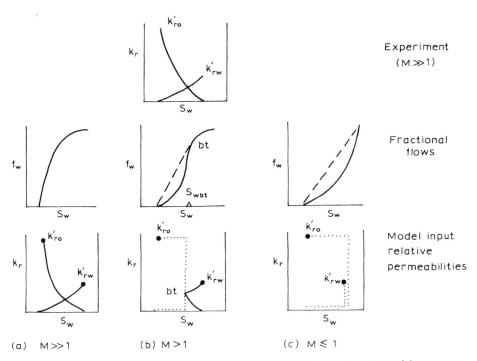

Fig. 5.31. Input of rock relative permeabilities to numerical simulation models.

The situation concerning the input of relative permeabilities to simulation models to overcome lack of consideration of the concept of fractional flow is summarised in Fig. 5.31. Starting with a single set of relative permeability curves measured using a high-viscosity oil and therefore continuous across the entire movable saturation range; consideration is given to how these data should be modified for use in simulation models for three different *in situ* mobility ratios. In case (a), for which $M \gg 1$, the fractional flow is strongly concave downward indicating premature water breakthrough and independent mobility of all saturations. Consequently, it is quite legitimate to input the rock curves directly into the model and the resulting saturation distributions will be quite realistic. Case (c) for $M \leq 1$ has already been described and requires the input of step functions to describe piston-like displacement. By analogy, the intermediate case (b) ($M > 1$), in which there is shock-front displacement such that $S_{wbt} < 1 - S_{or}$ requires the input of truncated relative permeabilities so that saturations in the range

$$S_{wc} < S_w < S_{wbt}$$

are denied independent movement. Unfortunately, as can be seen from these cases, simulators will function correctly on their standard diet of rock relative permeabilities only in the relatively rare cases of studying waterdrive in high oil viscosity fields. As the mobility ratio decreases the lack of reality in simulating immiscible displacement using relative permeability curves directly steadily increases until, for the most common flooding condition, $M \leq 1$, the degree of numerical dispersion is complete and numerical simulation will be at its least accurate.

The most unfortunate aspect concerning relative permeabilities and their use in coarse gridded numerical simulation is not that the models do not respect the concept of fractional flow but rather — that because they ignore it, then so too do reservoir engineers. Since the inception of simulation modelling in the mid 1960's the use of fractional flow by the industry has steadily diminished until now it is almost extinct; just like the application of Material Balance described in Chapter 3. Instead, the current interest seems to be devoted entirely to relative permeability measurements — almost to the exclusion of everything else. Yet, as described in this section, it is the relative permeabilities measured with some artificially high and often unreported oil viscosity, that are the arbitrary functions and must be regarded with great circumspection in reservoir engineering studies.

Full rock relative permeabilities always seem to have been treated with great veneration throughout the history of reservoir engineering. They are assumed to be intrinsically correct and all theory and practice is geared to accommodate this commonly held view. In fact, as argued throughout the chapter, this is a questionable attitude and full rock curves are never used directly: unless the problem in hand is that of flooding a reservoir which has the dimensions of a core plug and is full of 17 cp oil, which is a condition seldom encountered in practice.

To illustrate how the assumption that full rock curves are necessarily correct has influenced theoretical judgement, consider once again the dilemma of Buckley and Leverett and their triple value saturation (section 5.4d). Suppose that back in 1942 they decided to relate their theory to a particular oil of, say, viscosity 5 cp ($M > 1$) so that (as is now known) there would be some degree of shock-front displacement that was less than piston-like. If they had then performed relative permeability measurements using such oil and a perfectly homogeneous core plug, the results would have been as depicted in Fig. 5.32a.

The experiment would have demonstrated the existence of relative permeability step functions at the breakthrough saturation, S_{wbt}. The resulting fractional, flow calculated using equation 5.25, would only have finite values for $S_w > S_{wbt}$, diagram (b), manifesting the degree of shock-front displacement. Finally, calculating the

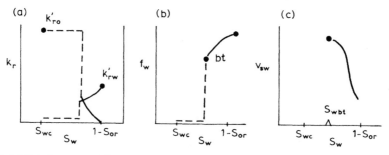

Fig. 5.32. (a) Relative permeabilities. (b) Fractional flow. (c) Saturation velocity distribution ($M > 1$).

water saturation velocity distribution using the original Buckley-Leverett equation, 5.21, the result would have been as shown in diagram (c). This, it will be noted, is single valued and would have led immediately to the construction of a realistic saturation distribution (Fig. 5.22) rather than the bulbous anomaly displayed in Fig. 5.21. Over the years, the Buckley-Leverett triple value saturation has been attributed to the neglect of capillarity, an over-simplified mathematical approach and various other causes. In fact, it would appear that the original derivation of the equation was quite correct and that the fault lay in trying to match a perfectly good theory to inappropriate laboratory measurements of relative permeabilities, which has given rise to over 50 years of intellectual bewilderment.

The foregoing does not imply that the engineer must specify to the laboratory that the relative permeabilities must be measured with a water/oil viscosity ratio compatible with that in the reservoir — that would be unnecessary and expensive. Instead, high-viscosity experiments are perfectly acceptable provided the engineer uses them with the *in situ* viscosities to construct a fractional flow and discriminate between those saturations that can and cannot move independently using the method of Welge.

(g) Laboratory experiments

Some further comments are necessary to describe the laboratory techniques applied in performing relative permeability experiments and, in particular, how the results should be modified for use in fields studies. As described in section 5.4a, the two most popular methods of measuring relative permeabilities are by the viscous displacement and steady-state processes. In the former (which has been implicitly the only type of experiment described in the chapter so far), water alone is injected directly into the core plug displacing the oil in accordance with Buckley-Leverett mechanics: a shock-front developing the magnitude of which is dependent on the mobility ratio. Usually the experiment is conducted using a high-viscosity oil to deliberately encourage unstable displacement ($M \gg 1$). Because of this it requires the circulation of many pore volumes of water to recover all the movable oil. Consequently, unrealistically high displacement rates are used to ensure that the experiment is concluded in a reasonable period of time. The relative permeabilities are generated applying the Welge equation, equation 5.25 in reverse [8] to the basic flood results.

By contrast, the steady-state experiment is conducted by simultaneously injecting both oil and water into one end of the core plug at a series of different ratios. At each, stability of oil and water saturations is attained (which is the time consuming part of the experiment) and the relative permeabilities are calculated by applying Darcy's law to the separate fluids (equations 5.10 and 11). Although rather slow, and therefore more expensive than viscous displacement, steady-state experiments are growing in popularity and one of the reasons for this is because of their contrived nature: it is possible to obtain full rock curves across the entire movable saturation range even if low-viscosity oil is used for which $M < 1$. The reason for this is because by injecting oil and water until stability is reached at different ratios there is no shock-front development in the experiment. Therefore, the fundamental displacement theory of Buckley-Leverett is conveniently thwarted and replaced by Darcy's empirical law which does not happen to be the basic theory of waterdrive. Fortunately, the distinction between the two types of experiment is more academic than real. If, for instance, full rock curves are obtained in a steady-state experiment using low-viscosity oil ($M < 1$), then plotting the more fundamental fractional flow will usually reveal a high degree of shock-front displacement indicating that the additional expense in performing the steady-state experiment was unwarranted.

There is a growing tendency for laboratories to perform viscous displacement experiments using low-viscosity oils and this can lead to anomalies in the reporting of results, two not uncommon examples being illustrated in Fig. 5.33. It is also becoming increasingly popular to plot the experimental results on a log-k_r scale, ostensibly to permit the engineer to better inspect the lower values of the relative permeabilities, Fig. 5.33a. If the experiment is performed with a moderately viscous oil so that the mobility ratio is, say, $M = 3$, then there will be an element of shock-front displacement so that the functions are physically defined by the solid lines. Next come the extrapolations from S_{wbt} to S_{wc}. As described in section 5.4f, these must necessarily be in the form of step functions (dotted lines Fig. 5.33a) but left to their own devices, laboratories will invariably perform the extrapolations in a non-linear fashion (dashed lines). It will be appreciated, however, that using the latter extrapolations in numerical simulation modelling will lead to unrealistic dispersion of saturations in the range $S_{wc} < S_w < S_{wbt}$. Even more extreme; on several occasions the author has observed viscous displacement experiments conducted under clinical conditions using low (*in situ*) viscosity oil and at reservoir pressure and temperature. Naturally, if the core plug is reasonably homogeneous, then under these circumstances rock curves will not be obtained at all, only the end-points. Yet the results have been presented as shown in Fig. 5.33b applying non-linear, non-scientific extrapolations across the entire movable saturation range instead of complete step functions (Fig. 5.31c). On enquiring about the physical justification for making such extrapolations the answer invariably received is that — "relative permeabilities always look like that" — to which there is no sensible reply.

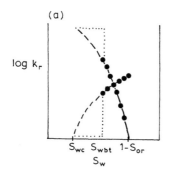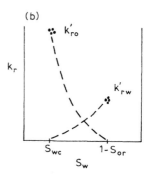

Fig. 5.33. Rock relative permeabilities: (a) log-k_r scale ($M = 3$), (b) linear-k_r scale ($M < 1$).

During the nineteen eighties, in particular, there were significant technological improvements leading to much more accurate experimental measurements of relative permeabilities and associated phenomena. While any such technical innovations are to be welcomed they sometimes impose a greater responsibility on the engineer in deciding on what is practical and what is not in manipulating the results of such highly accurate experiments for use in fields studies. As an example of this, consider the related phenomena of residual oil saturation, S_{or}, and rock wettability and how they are functions of experimental refinement. Figure 5.34a represents the closing stages of a relative permeability experiment as the residual oil saturation is approached. In earlier experiments the flood had to be terminated after circulating a limited number of pore volumes of water, W_{id}, because it was simply not possible to measure very low oil flow rates with any degree of accuracy. This resulted in a fairly high residual oil saturation, S_{or} (A). Now, however, there appears to be no such limitation and the author has noted relative oil flow rates as low as $k_{ro} = 0.000003$ being reported. This, of course, requires the circulation of unrealistically large volumes of water but has the effect of reducing the residual oil saturation to S_{or} (B). The reality of the experimental result can always be checked by drawing the fractional flow (Fig. 5.34b), which for a contemporary experiment would reveal an impracticably long tail at water cuts approaching 100% and would naturally lead to the selection of a higher, more realistic value of S_{or}. Another means of selecting a practical value of S_{or} from laboratory flooding experiments is to calculate the PVs of injected water required to attain different values of S_{or} and "cutting-of" at a reasonable number. W_{id} can be determined for reservoir, as opposed to laboratory, conditions in the following manner:

- For any set of rock curves plot the corresponding fractional flow using reservoir oil and water viscosities
- The reciprocal of the tangent to the fractional flow (after water breakthrough) determines W_{id} (equation 5.26)
- Plot the calculated values of W_{id} as a function of the increasing water saturation (Fig. 5.34c).

The plot normally displays an asymptotic approach towards a practical value

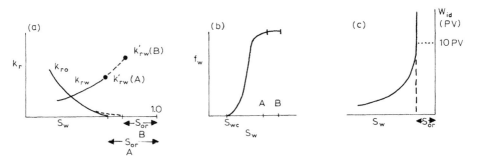

Fig. 5.34. (a) Termination of a relative permeability experiment. (b) Corresponding fractional flow. (c) Determination of a practical residual oil saturation.

of S_{or} for some terminal value of W_{id} of, say, 10 PVs and this procedure should be repeated for all the relative permeability experimental results available for a reservoir to seek some uniformity in the "cut-off" values of W_{id} and the resulting residual oil saturation. In practice, a reservoir will seldom (if ever) experience flooding with 10 PVs of water but regions close to injection wells will and this must be catered for in selecting a practical flood-out condition for the relative permeability functions.

The reservoir wettability condition is also (apparently) influenced by the degree of sophistication applied in laboratory experiments. As more and more water is circulated in a modern experiment then as illustrated in Fig. 5.34a the end-point relative permeability to water, $k'_{rw}(B)$, rises in comparison to the value, $k'_{rw}(A)$, that would have been attained in earlier experiments. Once its value increases above what is commonly regarded as the "magic number" of $k'_{rw} = 0.50$ (page 20 of reference 1) the reservoir is declared to be "oil-wet" which is not conducive to efficient oil recovery. In this respect the author has noticed an ever increasing number of oil-wet reservoirs in recent years and one that even experienced a metamorphosis between being classified as water-wet in 1979 and oil-wet in 1991. These observations are largely related to refinements in laboratory measuring techniques outwith their applicability or requirement in practical reservoir engineering.

It will be apparent from reading this section that practically all of the problems and misunderstandings associated with the use of rock relative permeability curves can be overcome simply by drawing and inspecting the fractional flow which incorporates the *in situ* oil and water viscosities and correctly accounts for the basic physics of immiscible displacement. It should therefore be regarded as a rule never to attempt to consider the meaning of a set of relative permeabilities without paying due regard to their fractional flow and the same maxim should apply to the treatment of pseudo-relative permeabilities to account for displacement in macroscopic reservoir intervals which are described in the following sections.

When applied to low mobility ratio floods ($M \leq 1$) the fractional flow demonstrates that only the end-point relative permeabilities are required; but the same is true in most macroscopic flooding problems for any mobility ratio (even using the mythical 17 cp crude). The reason for this is because when flooding in "hillsides" there is usually a strong element of fluid segregation meaning that the oil and water never mix intimately as in small-scale laboratory flooding experiments (section 5.6). Built into the end-points are all manner of "microscopic" complexities which are difficult to isolate and define. Potentially, the most significant of these is the wettability condition in the reservoir. The relative permeabilities plotted so far in this chapter have low end-points values to water, k'_{rw}, implying the favourable condition that water is the wetting phase. If the wetting phase is oil, however, the flow of water is encouraged to such an extent that the end-point relative permeabilities can be almost reversed in magnitude [13]. Under these circumstances, the favourable mobility ratio of $M = 0.25$ in case 1 of the example considered in section 5.4e would be raised to $M = 6.25$ which would seriously downgrade the efficiency of waterdrive and the timing of oil recovery. Fortunately, the influence of wettability is not necessarily so severe but the point being made is that effects such as this lie outwith the control of the field reservoir engineer who can only accept (while questioning) the basic experimental results. For wettability, of course, the main question relates to how the *in situ* reservoir condition can be established for duplication in laboratory experiments.

5.5. THE DESCRIPTION OF WATERDRIVE IN HETEROGENEOUS RESERVOIR SECTIONS

The previous section described the basic theory of waterdrive on the scale of a one-dimensional core flooding experiment. The same theory of Buckley-Leverett and the practical application technique of Welge will now be extended to the description of waterflooding in macroscopic, heterogeneous reservoir sections, which is a two-dimensional problem. In this respect, the engineer must be aware that in practice, waterflooding is conducted in "hillsides", not core plugs and the efficiency of the process is governed by three physical factors, namely:

– mobility ratio (M)
– heterogeneity
– gravity

The first of these encompasses all the experimental work performed in laboratories on such fundamental topics as relative permeabilities and wettability which are manifest in altering the mobility ratio; the influence of which on one-dimensional displacement has been described in section 5.4. Heterogeneity and gravity are closely interrelated and consideration of their combined influence on waterdrive efficiency is mandatory in reservoir sections of finite thickness. Without performing detailed waterdrive calculations, it is difficult to anticipate which of the three factors will predominate in influencing oil recovery calculations. The engineer must therefore be careful in making a priori assumptions about which is the more important to the exclusion of consideration of the others. The reason for that comment is because there is, quite naturally, a tendency for too much attention to be paid to the results of small-scale laboratory experiments rather than the macroscopic features of the reservoir itself. This may result from an in-balance in the literature concerning the subjects, there being many more papers on the microscopic aspects of waterflooding.

(a) Reservoir heterogeneity

Heterogeneity affects the oil recovery through its influence on both the vertical and areal sweep efficiencies according to the following simple equation

$$\frac{N_p}{N} = E_v \times E_A \tag{5.28}$$

in which:

N_p/N = fractional recovery of the STOIIP

E_v = vertical sweep efficiency — the fraction of the oil recovered in the reservoir cross-section by waterdrive

E_A = areal sweep efficiency — the fraction of the oil recovered areally by waterdrive.

The types of heterogeneity affecting the two components of the overall sweep efficiency are:

Vertical heterogeneity

By far the most significant parameter influencing the vertical sweep is the permeability and in particular its degree of variation across the reservoir section. Permeabilities can be observed to vary by several orders of magnitude within a matter of a few feet and when any parameter which plays a role in the physical description of a natural process is capable of such a high degree of variation its influence tends to "swamp" those of all other parameters — as is the case with permeability. In the description of vertical sweep, variations in porosity and water saturation are naturally catered for also but generally their variation, although weakly linked to the permeability, is slight in comparison (Fig. 5.35).

Areal heterogeneity

This includes areal variation in formation properties (h, k, ϕ, S_{wc}), geometrical factors such as the position and sealing nature of faults and boundary conditions due to the presence of an aquifer or gas cap.

Operators spend millions of dollars coring, logging and testing appraisal wells, all of which permits direct observation of the vertical heterogeneity. Therefore, if the data are interpreted correctly, it should be possible to quantify the vertical sweep, E_v, quite accurately. Areally, of course, matters are much more uncertain since methods of defining heterogeneity are indirect, such as attempting to locate faults from well test analysis and in many cases the results achieved lack uniqueness (Chapter 4, section 4.16). Consequently, the areal sweep efficiency is to be regarded as *the* unknown in reservoir development studies. To illustrate this point, consider the most important phase of such a study, the history matching of reservoir performance in a field which has been under waterdrive development for some years (Fig. 5.36). Unless a reliable history match of the production — pressure behaviour can be obtained the simulation model will have little value in predicting field perform-

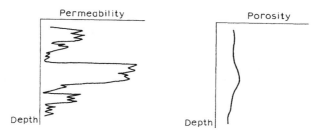

Fig. 5.35. Reservoir heterogeneity in the vertical cross-section.

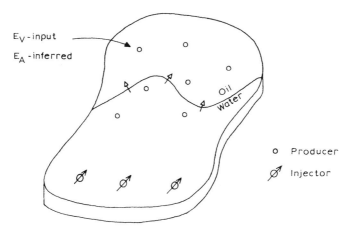

Fig. 5.36. Simulation model applied to history matching a waterdrive field.

ance, which is the basic aim in the study. Once pressure is maintained in a reservoir by waterdrive, there is little to be learned from matching a constant pressure in the model. Instead, a meaningful history match can only be attempted once water has broken through to the producing wells when the model can be "adjusted" to duplicate the timing of breakthroughs and subsequent watercut developments.

In doing so, the model is essentially being used to solve equation 5.28 in which the fractional oil recovery, N_p/N, is known and the vertical sweep, E_v, must also be treated as a "known". The model is then used to solve for the areal sweep, E_A, the principal unknown. This is accomplished by altering the formation properties from well to well, introducing faults and adjusting their throw and sealing qualities; all in an effort to manoeuvre the injected water around the reservoir and match its observed production spatially and with respect to time. Once accomplished, the model can be used as a tool for optimisation of the development by running it in the predictive mode. In particular the engineer is trying to assess the necessity for and location of additional production and injection wells, as described further in section 5.8.

The necessity of placing such a high degree of confidence in the vertical sweep, E_v, means that the main focus of attention in reservoir engineering must be the careful scrutiny of the "vertical" data collected in wells and its correct interpretation and processing for input to the model. The latter step usually involves the generation of pseudo-relative permeabilities on a well by well basis, as described in this section. The engineer may not always be entirely satisfied with the validity of some of the data acquired in wells but nevertheless it is the highest quality data collected and once input to the model the temptation to change it must be resisted in preference to altering the areal description (refer section 5.8). Concentration in this section is therefore devoted to the correct evaluation of vertical sweep efficiency.

(b) Recipe for evaluating vertical sweep efficiency in heterogeneous reservoirs

No matter what the nature of the vertical heterogeneity, the following recipe is applied to assess the sweep efficiency in edge waterdrive reservoirs.

- Divide the section in to N layers, each characterised by the following parameters: h_i, k_i, ϕ_i, S_{wc_i}, S_{or_i}, k'_{rw_i}, k'_{ro_i} (the subscript "i" relates to the ith layer).
- Decide whether there is vertical pressure communication between the layers or not.
- Decide upon the flooding order of the N layers and generate pseudo-relative permeabilities to reduce the description of the macroscopic displacement to one dimension.
- Use the pseudos to generate a fractional flow relationship which is used in the Welge equation to calculate the oil recovery, N_{pd} (PV), as a function of cumulative water influx, W_{id} (PV).
- Convert the oil volume to a fractional oil recovery, N_p/N, and relate this to the surface watercut, f_{ws}.

There are some subtleties attached to layering the section which are described later. Since the variation in permeability is usually the dominant factor, however, it is common to select individual layers based primarily on this parameter.

Residual oil saturations and end-point relative permeabilities are not usually available for each separate layer and if not input of these data should be handled as described in the following section. The most reliable method of determining the degree of pressure communication across the layers, which controls the cross-flow of fluids, is by running the RFT in each new development well after the start of continuous production, as described in section 5.2d. Two such survey results are shown in Fig. 5.37 which demonstrate the opposite extremes of complete pressure equilibrium and a total lack of it. It will be appreciated, of course, that there can be many states of partial equilibrium between these extremes, that is, complete equilibrium between several adjoining layers with no communication to the layers above or below. For the present, however, attention will be focused on the two conditions depicted below; partial equilibrium is described in section 5.10. The

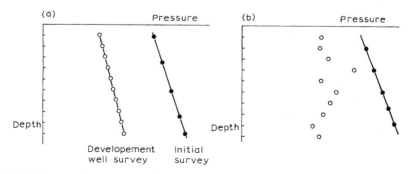

Fig. 5.37. RFT surveys demonstrating: (a) complete pressure equilibrium and cross-flow, (b) total lack of equilibrium with no cross-flow.

fact that dynamic pressure profiles can only be obtained some time after the start of production poses difficulties for the engineer especially when planning the development of offshore fields. As will be seen, a knowledge of the degree of pressure equilibrium is essential to quantify correctly the vertical sweep and in particular the rate of development of the field watercut upon which the surface topsides facilities are designed. Yet in many cases crucial decisions have to be made at the end of the appraisal stage when only static reservoir data are available (section 5.3c). At the start of production of a field it is recommended that an initial pressure drop be encouraged to gain information on the degree of communication as rapidly as possible which will help in the final planning of the water injection scheme (section 5.2d). But this will not alleviate the problems associated with predicting the necessary capacities of the surface equipment which will have already been installed by that stage. Since the RFT has only been available to the industry since the mid 1970's, the reader may wonder how engineers prior to that time managed to ascertain the degree of pressure equilibrium across reservoirs? So does the author.

Based upon the observed degree of pressure equilibrium across the sand section, the engineers must decide upon the order in which the selected layers will flood with the advancing water. If there is a total of N layers then, in principle, there are $N!$ ways in which they could successively flood. At present, however, interest is confined to the extreme conditions of complete equilibrium or a total lack of it. The former condition is referred to as vertical equilibrium [14,15] (section 5.6) in which it is considered that the oil and water segregate instantaneously with the latter falling to the base of the section under the influence of gravity. Consequently, the flooding order is dictated from the basal layer to the top. If the selected layers turn out to be isolated from one another, so that there is a total lack of cross-flow, then the order in which the layers flood is determined by the actual velocity of the frontal advance of water in each (section 5.7b), that is

$$v_i \propto \frac{k_i k'_{rw_i}}{\phi_i (1 - S_{or_i} - S_{wc_i})} \tag{5.29}$$

in which the dependence on permeability is obvious. Concerning the terms in the denominator, low porosity rock, having a smaller capacity, will flood faster than high and similarly the smaller the movable saturation the more rapid the advance of water. The N layers will therefore flood in decreasing sequence of their calculated values of v_i. Such displacement is described in section 5.7.

Having determined the flooding order of the layers the next step is to generate thickness averaged or pseudo-relative permeabilities across the section for edge waterdrive. That is, when the nth layer out of a total of N has flooded with water:

$$\overline{S}_{w_n} = \frac{\displaystyle\sum_{i=1}^{n} h_i \phi_i (1 - S_{or_i}) + \sum_{i=n+1}^{N} h_i \phi_i S_{wc_i}}{\displaystyle\sum_{i=1}^{N} h_i \phi_i} \tag{5.30}$$

$$\bar{k}_{rw_n} = \frac{\displaystyle\sum_{i=1}^{n} h_i k_i k'_{rw_i}}{\displaystyle\sum_{i=1}^{N} h_i k_i} \qquad (5.31)$$

$$\bar{k}_{ro_n} = \frac{\displaystyle\sum_{i=n+1}^{N} h_i k_i k'_{ro_i}}{\displaystyle\sum_{i=1}^{N} h_i k_i} \qquad (5.32)$$

The first expression calculates the ever increasing thickness averaged water saturation as the layers successively flood. It is naturally a thickness–porosity weighted average; the first term in the numerator relating to the n layers that have flooded and the second to the unflooded layers containing irreducible water. The thickness averaged relative permeabilities are permeability–thickness weighted averages of the end-point values. For water, equation 5.31 is evaluated over the flooded layers while for oil, equation 5.32 is summed over the unflooded layers. Detail of how the individual parameters in the relationships are derived from the basic data is deferred until section 5.6b. It will be noted that these averaging procedures cater for the "either–or" situation. That is, a layer is either at flood-out saturation, $1 - S_{or_i}$, for which only the end-point value of k'_{rw} is required, or it is completely unflooded: $S_w = S_{wc}$, $k_{ro} = k'_{ro}$. There is no in-between state. This assumption clearly requires justification which will be provided in sections 5.6 and 5.7 relating to vertical equilibrium and a total lack of it, respectively.

The significance of the averaging procedures is that they effectively reduce the complex two-dimensional description of waterdrive back to one-dimension in which the theory of Buckley-Leverett is formulated. If ten layers have been defined then equations 5.30–5.32 are evaluated ten times as the layers successively flood. The resulting points $\bar{k}_{rw}(\bar{S}_w)$ and $\bar{k}_{ro}(\bar{S}_w)$ define what are commonly referred to as pseudo-relative permeabilities, although as Thomas [16] has so succinctly pointed out — there is nothing "pseudo" about them at all. They are the realistic functions required to describe the complex flood in one dimension. If anything, it should be the rock relative permeabilities measured on thin core plugs that ought to be referred to as "pseudos" since they are hardly ever used directly in reservoir engineering calculations without some form of modification. Having generated the pseudos the next step is to use them in either equation 5.18 or 5.19 to obtain the fractional flow relationship. This important step incorporates the *in situ* oil and water viscosities relevant for the reservoir flooding condition. The function is then used in the one-dimensional Welge equation, equation 5.25, to evaluate the oil recovery in the vertical cross-section. As presented in section 5.4e, the oil produced and water injected in Welge's equation are expressed in pore volumes and therefore require modification to relate to actual volumes in field calculations. In the first place, it is more convenient to evaluate the equation in terms of hydrocarbon pore volumes (HCPV), that is:

$$N_{pD} = \frac{(\overline{S}_{we} - S_{wc}) + (1 - \overline{f}_{we})W_{id}}{(1 - \overline{S}_{we})} \quad (\text{HCPV}) \tag{5.33}$$

in which $\overline{S}_{we}$ and $\overline{f}_{we}$ are the thickness averaged saturation and fractional flow at the producing end of the cross-section which is the wellbore. If the flood is conducted at an average pressure for which the formation volume factor is B_o, then the oil recovered from the section by waterdrive is:

$$N_{pD} = \frac{N_p B_o}{N B_{oi}} \quad (\text{HCPV})$$

where $N B_{oi}$ is the initial HCPV of the undersaturated reservoir. Allowing for an initial phase of oil recovery due to depletion prior to commencing the flood and the overall areal sweep, E_A, the total recovery factor may be expressed as

$$\frac{N_p}{N} = \frac{B_{oi}}{B_o} \left(c_{\text{eff}} \Delta p + E_A N_{pD} \right) \tag{5.34}$$

The areal sweep, of course, is usually the outcome of detailed simulation studies but for the purposes of simple analytical calculations may be estimated as described in Exercise 5.2. The cumulative water injected can also be expressed in PVs as

$$W_{id} = \frac{5.615 q_i t}{LA\phi} \quad (\text{PV}) \quad (t, \text{days}) \tag{5.35}$$

in which $LA\phi = 1$ PV (ft^3) and q_i is the constant injection rate (b/d). It is the water injection that controls the flood and through which a time scale can be attached to the recovery using equation 5.35. Finally, the watercut, $\overline{f}_{we}$, appearing in equation 5.33 is at reservoir conditions and requires correction for shrinkage to obtain the equivalent expression at the surface, f_{ws}, that is

$$f_{ws} = \frac{q_w / B_w}{(q_w / B_w) + (q_o / B_o)}$$

in which the rates are expressed at reservoir conditions. Combining this with the reservoir fractional flow

$$f_{we} = \frac{q_w}{q_w + q_o}$$

gives

$$f_{ws} = \frac{1}{1 + \dfrac{B_w}{B_o} \left(\dfrac{1}{\overline{f}_{we}} - 1 \right)} \tag{5.36}$$

Equations 5.34 and 5.36 then provide the all important relationship between f_{ws} and N_p/N which, as described in section 5.3c, is so essential in the prediction of waterdrive performance and in particular for sizing the capacities of surface injection and production facilities.

The overall method described in this section is quite general and will be illustrated with examples in the remainder of the chapter. The only exception is that equations 5.30–5.32 cannot be applied to generate pseudos in reservoir sections in which there is no cross-flow between the layers and the mobility ratio is different from unity. This requires the generation of a thickness averaged fractional flow using the analytical technique of Dykstra and Parsons, as described in section 5.7d.

5.6. WATERDRIVE UNDER SEGREGATED FLOW CONDITIONS (VERTICAL EQUILIBRIUM)

(a) Basic description

This is the most common flooding condition encountered in nature and is characterised by the following physical conditions:

- There is a high degree of pressure equilibrium across the reservoir section which encourages cross-flow of fluids under the influence of gravity.
- The displacement occurs under strictly segregated conditions with a sharp interface between the water and oil.

The first implies an absence of vertical flow restrictions across the section and can be recognised by the type of RFT survey shown in Fig. 5.37a. Vertical permeabilities tend to be lower than horizontal but it is surprising to what extent they have to be reduced in a continuous reservoir section to remove or seriously reduce the vertical equilibrium (VE) condition of rapid fluid segregation.

Segregated flow implies that there is a negligible capillary transition zone between the oil and water resulting in a sharp interface at their boundary. The condition is satisfied in most engineered waterdrive projects where, by choice, operators select candidate reservoirs which have moderate to high average permeability (section 5.2b), in which case the extent of the dynamic capillary transition zone is usually negligible in comparison to the reservoir thickness. This has been confirmed both theoretically and experimentally [17,18] but more significantly the sharp interface between oil and water is observed repeatedly in the field: logging wells that have been drilled through partially flooded reservoirs. If a capillary transition zone must be accounted for in analytical calculations for lower permeability reservoirs then the method has been provided in references 9 and 19; although it is suggested that since the technique is rather cumbersome the engineer should use numerical simulation modelling from the outset. Even so, there is an inevitable lack of reality in attempting to quantify the phenomenon since no time dependence in capillary rise is catered for. Instead, capillary pressure data are input to models under the assumption that the water saturation distribution as a function of height above the free water level is attained instantaneously which is optimistic in providing an accelerated sweep of the upper parts of the reservoir. Therefore, the neglect of capillary phenomena, which is applied throughout the chapter, is not only realistic

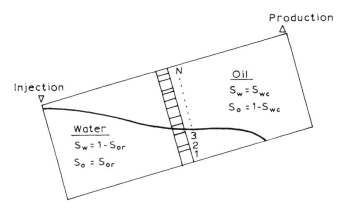

Fig. 5.38. Cross-sectional waterflooding under the VE condition.

for the majority of practical reservoir flooding projects but, if anything, tends to produce slightly pessimistic results which is a healthy state in dealing the any form of reservoir engineering problem.

In terms of saturations, the segregated flow condition described by Dietz [14] and Coats et al. [15] represents an either–or situation, as shown in Fig. 5.38. That is, behind the front water alone is flowing in the presence of residual oil, $S_w = 1 - S_{or}$, whereas ahead of the front only oil is flowing in the presence of irreducible water. There is no in-between situation and therefore only the end-point relative permeabilities are required to describe the dynamics of displacement. It should also be noted that this exclusive use of end-points applies irrespective of the mobility ratio. That is, even if the reservoir contained a viscous crude oil, the rapid segregation of the oil and water induced by the gravity difference in VE displacement would assure that the fluids never mix together as they do in a microscopic laboratory experiment: there will still be a sharp interface between them. Consequently, the end-point relative permeabilities are sufficient to account for water–oil displacement in a macroscopic reservoir section provided segregation occurs and for this most common flooding condition the laboratory "curves" are redundant.

The VE condition means that the order in which the N selected layers flood is from the base to the top of the section. It will be noted that although VE implies a dominance of gravity, at no stage does a vertical gravity term appear in any of the displacement equations. Instead, all that is necessary is recognition of the fact that since water is heavier than oil it will naturally slump to the base of the reservoir and this dictates the flooding order of the selected layers (Fig. 5.38). It is this order which is applied in the evaluation of the averaging procedures, equations 5.30–5.32, for generating pseudos.

Before illustrating segregated displacement with examples, it is first necessary to describe in some detail the significance of the individual parameters appearing in the averaging procedures and how they may best be evaluated from the basic data collected in appraisal and development wells.

(b) Data requirements and interpretation for input to the generation of pseudo-relative permeabilities

Whether evaluated analytically or by cross-sectional numerical simulation (section 5.8), equations 5.30–5.32 must be regarded as the most important in dictating the vertical sweep efficiency of waterdrive. Therefore it is necessary to consider carefully how the individual parameters in the relationships are input to calculations. In the first place, it will be noted that three of them, k_i, ϕ_i and h_i, appear both in the numerator and denominator meaning that they act as weighting factors and therefore their distributions across the sand section are of greater significance than their absolute values.

Permeability distribution

The absolute permeability of rock is a much overrated number in describing displacement mechanics. It will be noted, for instance, that it does not appear in the Buckly-Leverett equation, 5.21, which is so fundamental to the subject. The theory relies solely on the principle of mass conservation — what goes in at one end of a core plug must come out at the other regardless of the permeability. Similarly, in deriving the all important fractional flow relationship for horizontal flow, equation 5.18, the absolute permeability cancels and does not appear in the final result (one of its few appearances, however, is in the gravity term, G, in the fractional flow for dipping reservoirs, equation 5.20). The same is true for the averaging procedures used to generate pseudos, equations 5.30–5.32, in which the permeability acts as a weighting factor. Therefore, systematically doubling or halving the permeabilities will have no effect whatsoever on the vertical sweep efficiency derived using the pseudos. Occasionally, time is wasted in sensitivity studies in arbitrarily varying the permeabilities, while injecting at a fixed rate, only to find that the results are the same. The only point at which the absolute permeability will affect results is when it becomes so low that the water injection rate cannot be maintained at the maximum permissible injection pressure. This will increase the time required to flood the system but still not affect the displacement efficiency of the process.

It is, therefore, the distribution of permeabilities both vertically and areally that mainly influences the displacement efficiency and particularly the former for which, as already noted, variations of several orders of magnitude over slight depth increments are common. The effect of different permeability distributions across a continuous reservoir section is illustrated in Fig. 5.39. Case (a), in which there is a coarsening upwards in permeability represents what might be described as the "super homogeneous" reservoir. At the injection well, the bulk of the water enters the top of the section, in accordance with Darcy's law. But the viscous, driving force from the injection pumping decreases logarithmically in the radial direction and before the water has travelled far into the formation it diminishes to the extent that gravity takes over and dominates. The water, which is continually replenished at the top of the formation, then slumps to the base and the overall effect is the development of a sharp front and perfect, piston-like displacement across the macroscopic section. Case (b) demonstrates the inverse situation in which

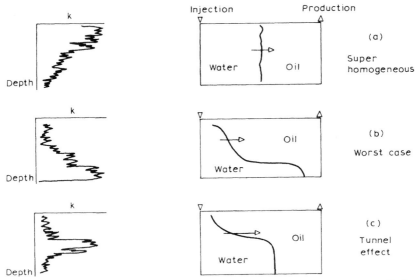

Fig. 5.39. Influence of permeability distributions across a continuous reservoir section on displacement efficiency (average permeability, $\bar{k}$, is the same in each reservoir).

the permeabilities increase with depth. The majority of the injected water now enters at the base of the section at the injection wellbore and being heavier it stays there. This leads to premature breakthrough and the circulation of large volumes of water to recover all the oil trapped at the top of the section. The third case, (c), is intermediate between the two. There is piston-like displacement across the lower part of the section but a slow recovery of oil from the top.

Considering the three factors which influence vertical sweep-mobility ratio, heterogeneity and gravity; in case (a) the latter two complement each other and can produce a favourable displacement efficiency even for an unfavourable mobility ratio. In case (b), the opposite is the case and gravity and heterogeneity conspire against each other to produce unstable displacement even for a favourable mobility ratio. Case (c) lies between the two. Watercut developments as a function of time for the three cases are depicted in Fig. 5.40.

The rapid watering out of production wells with a favourable permeability distribution (within a matter of weeks) is a bit alarming when observed in the field and can cause a significant drop in the overall production. Nevertheless, this represents the ideal situation in which practically all of the movable oil in the vertical cross-section is recovered. The case (b) watercut development is the unhealthy one with premature breakthrough and the requirement to dispose of large volumes of produced water during the lifetime of the project. In such reservoirs the attainment of the equivalent recovery of case (a) is possible by circulating enough pore volumes of water through the system but in practical terms recovery is reduced by premature curtailment of operations, especially in the offshore environment, since much of the movable oil can only be recovered at excessively high watercut.

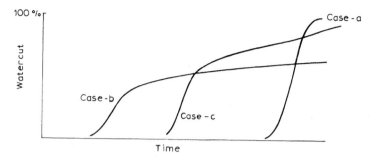

Fig. 5.40. Watercut developments for the three permeability distributions shown in Fig. 5.39.

The nature of permeability distributions in clastic rocks is to a large extent governed by the depositional environment and the reservoir engineer must seek advice from geologists concerning the likely areal extent of permeability distribution patterns observed in any particular well. In a marine environment, for instance, cycles of regression (sea receding from a land area) and transgression (sea advancing over a land area) give rise to coarsening upward and downward trends in permeability as illustrated in Fig. 5.41. In case (a), point *A* is originally in deep water and only fine sediment is transported from the land to be deposited at this distance. As the sea recedes, point *A* is located closer to the shore line resulting in the deposition of coarser material which is not transported as far as the fine sediment. In geological time, the regressive cycle is relatively short and the overall result is a coarsening

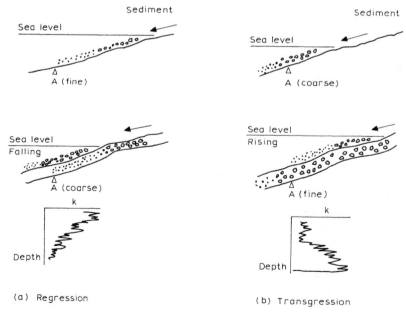

Fig. 5.41. The influence of regressive and transgressive cycles on permeability distributions.

upward in the size of the rock particles and the flow channels between them which is reflected in the permeability. During a period of transgression, Fig. 5.41b, the opposite happens and as the sea level rises finer and finer material is deposited at point *A* resulting in a coarsening downwards.

Considering the importance of permeability distributions on the vertical sweep efficiency, it is essential that permeability data be collected, inspected and incorporated in reservoir models in a realistic fashion. The only reliable way of observing permeability distributions is by coring wells. This opinion is not shared by everyone, however, for at one seminar on Reservoir Environments, a geologist is reported as saying:

"The only definite thing we know about core is that it is not in the reservoir
any more".

Nevertheless, it might be asked — what can replace it? In depletion type fields core data is useful but when planning any form of secondary recovery flood in which one fluid displaces another of different density, the acquisition of core data is mandatory since it provides the most pertinent data of all for quantifying flooding efficiency. Because of this, all appraisal wells should be fully cored across reservoir sections and the campaign extended into the development drilling phase until the geologists and engineers are quite satisfied of the areal control of permeability distributions observed in individual wells. In particular, appraisal wells which are inadvertently drilled into an edge aquifer must also be fully cored. This is where the peripheral injection wells will be located and it is of equal importance to view their permeability distributions as it is for the updip producers. Furthermore, it sometimes happens that aquifer permeability can be partially destroyed resulting from diagenesis meaning that the required injection rate may not be achieved through wells located in the aquifer. If so, it is important to determine this at an early stage in the development and plan for injection into the oil column. This will not only diminish the areal sweep but if the mobility ratio is low ($M < 1$) can cause the injection difficulties described in section 5.4b.

Another, indirect method of generating permeability distributions is through the use of "petrophysical correlations". This seems to be growing in popularity largely because it is linked to the statistical manipulation of large data sets which is continually improving through advances in computer software. Originally, such correlations sought to relate permeability to porosity, as illustrated in Fig. 5.42.

Having cored one well a plot is made of the core measured permeabilities, k_{core}, versus the core porosities, ϕ_{core}. This is usually done for each separate reservoir or for selected rock types. Allowance for compaction between surface measured core porosities and their *in situ* reservoir values then permits a relationship to be established between the core and petrophysical log derived porosities, ϕ_{log}, which enables k_{core} to be directly related to ϕ_{log}. Thereafter, permeability distributions can be generated in wells that have not been cored using the correlation and, if it is regarded as satisfactory, the coring programme itself may be restricted and reliance placed instead on this indirect method of generating permeabilities from log interpretation. More recently, the industry has accepted and started to apply

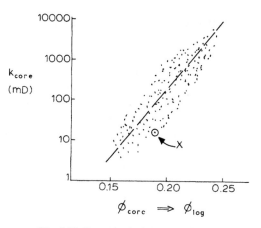

Fig. 5.42. Petrophysical k/ϕ correlation.

much more complex forms of petrophysical correlation which seek to relate the permeability to the porosity, water saturation, gamma ray response grain density ..., each of which is multiplied by a constant coefficient and raised to some non-integer power whose values are established by the application of regression analysis, for instance

$$\log k = A\,(\phi_{\log})^{l} + B\,(S_{w})^{m} + C\,(\gamma\text{-ray})^{n} + D\,(\text{density})^{o}\,\dots$$

Worthy though these attempts may be they suffer from the common fault that they simply are not accurate enough for the requirements of reservoir engineers. The statistical manipulation of the data has a smoothing effect which systematically underrates the true severity of the permeability distribution. That is, the full distribution is contained within the scatter plot shown in Fig. 5.42 but simply drawing a trend line to best match the k/ϕ scatter can never reveal it. Such petrophysical correlations are capable of providing a reasonable value of the average permeability of a reservoir but, as described earlier in this section, this does not happen to be a particularly useful number in reservoir engineering. Instead, what is often required is the particular detail of a permeability at a certain depth in the section (for instance, point X in Fig. 5.42), which is simply not catered for by the correlation. An example of a major reservoir whose performance was critically dependent on one such point is provided in section 5.10b. Furthermore, concerning the ever increasing application of highly complex transforms, one has to question the physical reality in their use. That is, is it physically reasonable to attempt to correlate a dependent variable, the permeability, which can vary rapidly over several orders of magnitude, against a group of parameters whose variation is only weakly linked? From a strictly scientific point of view the answer must be — no. In the past, engineers were protected from the errors associated with "statistical smoothing" of permeability data by their inability to apply complex regression analysis. Now, however, this can be performed with ease but the modern engineer is confronted with the added problem of differentiating between what appears technically sophisticated and what is scientific.

No matter how petrophysical transforms may develop in the future they will never be able to unscramble the true heterogeneity implicit in the scatter of points in Fig. 5.42 by attempting to represent it by a single function for the reason that Nature is just not that simple. Like it or not, the only way to view permeability correctly through foot-by-foot measurements of the parameter on plugs cut from the core across the interval or through use of the mini-permeameter across the section. There is no substitute for coring. Even so, experience has shown, especially in areas like the North Sea, that even honouring the core data fully in displacement efficiency calculations tends to produce optimistic results. That is, there is often a struggle to match the full severity of permeability variation occurring in Nature. The reason for this is probably because the missing intervals in any cored section are likely to be the higher permeability zones which, being more friable, reach the surface as loose sand which precludes any measurement of their rock properties and it is these intervals that exert the greatest influence on displacement efficiency in the reservoir.

While on the subject of the dangers in attempting to quantify reservoir heterogeneity using statistical methods, it is worthwhile drawing attention to one of the most popular and often recommended methods [1] in the industry — and yet one which contains a severe deficiency in its treatment of the basic physics of displacement. The method requires the plotting of core permeabilities on log-probability paper and such plots have the salving feature of smoothing the data into a straight line. The value of the average permeability (Fig. 5.43), $\bar{k}$, is determined at the 50% probability and the value of k_σ at one standard deviation from the mean. The permeability variation is then calculated as:

$$V = \frac{\bar{k} - k_\sigma}{\bar{k}}$$

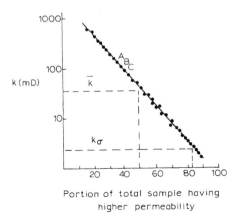

Fig. 5.43. Log-probability plot of core permeabilities.

and the sweep efficiency computed by entering a set of charts relating V to the mobility ratio, M, for varying water-oil ratios. The oversight in this method is that it completely dismisses the effect of gravity and incorporates only the two remaining factors which influence waterdrive: mobility ratio and heterogeneity. That is, measurements A, B and C (Fig. 5.43) could be obtained from core plugs cut from the top, bottom and middle of the reservoir — there is no regard to position. Consequently, the method treats each core plug as a separate reservoir which is isolated from all others. This, of course is the antithesis of the vertical equilibrium condition being described in this section and if the method were applied to the two reservoirs (a) and (b) depicted in Fig. 5.39, it would produce the bizarre result that the vertical sweep efficiency in each was the same, whereas, as illustrated in Exercise 5.3, (a) should provide a much superior sweep, dependent on the abandonment watercut imposed. There may be some justification in using the method in reservoir sections, as described in section 5.7, in which there is no cross-flow between the defined layers but even so the alternative analytical methods presented in that section are preferable.

Considerable effort has been expended in the industry by studying permeability heterogeneity strictly in terms of probability distributions [1] (the log-normal being of common occurrence) and service companies performing routine core analysis frequently present their results in this manner. But who actually uses such data in studies is unclear; it certainly must not be the reservoir engineer. The statistical study of heterogeneity is invalid unless the results are coupled with some statement concerning the location of the high and low permeabilities in the reservoir section. Without it, the force of gravity and therefore Newton's second law of motion are totally disregarded: the most unfavourable probabilistic permeability distribution can provide excellent sweep efficiency provided the higher permeabilities are towards the top of a continuous sand section.

Concerning a related topic: great care must be exercised in applying "cut-offs" for reservoir rock in displacement studies. Petrophysicists will exclude sections with low k and ϕ and high S_w but in doing so dot not always specify where the exclusion of such rock has occurred: top, middle or base of the reservoir and to incorporate gravity in the waterdrive calculations it is necessary that such information be provided along with their net pay figures.

In using the core data directly in vertical sweep efficiency calculations, the opening move is to plot the permeabilities on a linear scale versus depth or thickness across the reservoir section. By tradition, the popular method of making such plots is of log-k versus depth which inevitably leads to a complete visual distortion of the harshness of the permeability variation. The reason for choice of the log scale is to enable the several orders of magnitude variation in permeability to be fitted on a reasonable sized piece of paper, while at the same time "blowing-up" the lower permeabilities to a level at which they can be inspected without risking eye strain. Unfortunately, the log scale also has the effect of visually depressing the high permeabilities and the overall effect is to make the reservoir section appear very much more homogeneous than it is in reality. If Darcy's law stated that:

$v \propto \log k$

there would be justification in using the log-k scale — but it doesn't and therefore it is strongly recommended that engineers should not make any decisions concerning the layering of systems without first inspecting the permeability distribution plotted on a linear-k versus depth plot. Failure to do so can and has led to very serious errors of judgement in accounting for heterogeneity in the vertical section which in turn has resulted in the significant overestimation of the sweep efficiency. There are several examples illustrating the difference between log-k and linear-k plots across reservoirs provided in this chapter (Figs. 5.59, 5.77 and 5.81), which will illustrate the point.

Having viewed the heterogeneity correctly, the next step is to divide the reservoir into N discrete layers. Since for the VE condition there is complete cross-flow in the reservoir there is no great subtlety in this exercise. The layers are usually selected based on the permeability variation, as illustrated in Exercises 5.2 and 5.3. There should be a sufficient number, however, to provide a reasonable degree of smoothness in the pseudo-relative permeabilities and fractional flow. Quite often when interpreting the core report engineers correct the air-measured permeabilities for the Klinkenberg effect (slippage of gas (air) molecules at low flow rate) and compaction in an attempt to determine the *in situ* permeabilities in the reservoir for liquid flow. Such corrections are usually unnecessary, however, for if the Klinkenberg and compaction corrections are uniform for a reservoir, which is normally the case, then since the permeability only appears as a weighting factor in equations 5.30–5.32 for generating pseudos, the application of the uniform corrections will have no influence on the resulting functions. Therefore, in many cases it is permissible to use the air measured permeabilities directly for each layer. Appreciation of this point not only saves time but is preferable since the fewer corrections applied to any basic data — the better. Within any selected layer there may be, say, 10 or 15 individual core measured permeabilities which are averaged arithmetically as

$$\bar{k}_i = \frac{\sum\limits_j h_j k_j}{\sum\limits_j h_j} \tag{5.37}$$

to give the average permeability of the ith layer out of the N selected; the subscript "j" referring to the individual core measured permeabilities within the layer.

Porosity distribution

As with the permeability, the porosity appears in the averaging procedures, equations 5.30–5.32 as a weighting factor. Consequently, provided the compaction correction applied to core porosities to determine their *in situ* values is uniform, the helium measured core porosity can be used directly in calculations. Again, this not only saves time but uses the absolute uncorrected data. Within each layer the average porosity is calculated as

$$\bar{\phi}_i = \frac{\sum\limits_{j} h_j \phi_j}{\sum\limits_{j} h_j} \tag{5.38}$$

Layer thickness

Since the thickness of each layer also acts as a weighting factor in the averaging procedures, it removes the necessity for correcting along-hole to vertical depths in deviated wells, provided the angle of deviation is reasonably constant across the reservoir, which is frequently the case. Therefore, the core measured depths can be used directly, having first made any core shift corrections necessary to correlate the core with the petrophysical logs.

Relative permeability data

These include the end-point relative permeabilities, k'_{rw} and k'_{ro}, and saturations, S_{wc} and S_{or}, which are all that are necessary to generate pseudos for segregated flow using the averaging procedures, equations 5.30–5.32. The end-point water saturations seldom match those determined by petrophysical log analysis for a given layer. Consequently, the laboratory measured relative permeabilities are usually "normalized" by plotting the increase in water saturation as a fraction of the movable saturation range:

$$S_w^* = \frac{S_w - S_{wc}}{1 - S_{or} - S_{wc}} \tag{5.39}$$

which enables them to be matched to the movable saturation range determined for each separate layer.

Usually, there are far fewer rock relative permeability measurements than there are layers selected in the section, nevertheless the averaging procedures dictate that if an experiment has been performed on a core sample from the ith layer then the results must naturally be input to that layer. If interpolation is necessary then it is usually based on different rock types. If, for instance, there are 20 layers and only three relative permeability measurements for low, medium and high-permeability samples then the results should be input to layers in the section with similar rock properties. Thus the sparse relative permeability data may be assigned to layers within the reservoirs contained in a single well and this can be applied for each individual well: relating the experimental results determined from cores in the well to exactly the layer (rock type) from which the core samples were cut. If the layers are continuous from one well to another and have similar rock properties then the experimental results from one well may be extrapolated to the same layer in another for which data are not available.

Under no circumstances, however, must the common practice of "averaging" relative permeability functions be attempted. In this, families of normalized curves, possibly from similar rock types throughout the field, are plotted together, as shown in Fig. 5.44a. An average set of functions is then sought for water and oil either by "eye-balling" or using more sophisticated computer averaging procedures. The

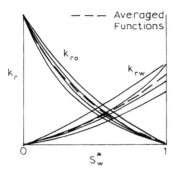
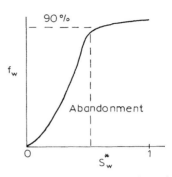

Fig. 5.44. (a) Averaging normalized relative permeabilities ($M < 1$), (b) resulting physically inconsistent fractional flow.

resulting functions are then input to all similar rock types in simulation models. Unfortunately, there is no physical principle that suggests that such a practice has any validity whatsoever. If any form of averaging were considered necessary (which it is not) then it would seem more appropriate to perform it on the fractional flow relationships which have a greater degree of physical reality — but even this is hard to justify. The sort of error the practice can give rise to is illustrated in Fig. 5.44. A single set of averaged rock curves was generated for a low mobility ratio flood ($M < 1$) and these were used directly in a numerical simulation study producing a dismal result in terms of waterdrive efficiency, which caused considerable delay in financing the project. Had the engineer checked the physical reality of the averaged functions by plotting the corresponding fractional flow, Fig. 5.44b, the inconsistency would have been at once evident. Instead of producing a curve that was totally concave upwards (Fig. 5.25: $M < 1$) representing piston-like displacement on the microscopic scale, the physically invalid averaged rock curves produced a fractional flow with a long tail at excessive watercut. Welge calculations using this function indicated that it would take the circulation of over 20 PVs of water to recover all the movable oil and at an abandonment watercut of 90%, the effect was to stop half the defined movable oil from "moving".

(c) Catering for the presence of edge water in VE flooding

Provided there is a reasonable degree of pressure communication throughout an accumulation, the water injection wells are located, by preference, in the aquifer rather than the oil column. The reasons are because there is less resistance to injection in the aquifer (section 5.4b) and in edge waterdrive fields the areal sweep is increased. The initial condition in such a VE waterdrive is depicted in Fig. 5.45, in which the injection well has been ideally located so that the entire reservoir section is just in the water. The pseudo-relative permeability curves for the updip producer are generated using the averaging procedures, equations 5.30–5.32, and are quite unaffected by the wedge of water present in the reservoir at initial conditions. That is, as the advancing water reaches the producer, the layers defined in its section

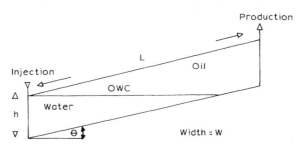

Fig. 5.45. Initial conditions in an edge waterdrive field.

must still flood from the base to the top as dictated by the VE condition. The Welge equation, equation 5.33, is then applied to calculate the oil recovery (N_{pD}, HCPV) and water injection (W_{id}, PV) ignoring the presence of the initial volume of water in the reservoir: assuming that the total HCPV is filled with oil. These results are then modified to account for the initial condition, as follows.

The volume of the downdip water, expressed as a fraction of the total pore volume is:

$$\frac{\dfrac{h^2}{2 \tan \theta} W \phi}{L h W \phi} = \frac{h}{2L \tan \theta}$$

which fills the total pore space below the oil–water contact. But if, instead of being originally present, this volume of water had been injected into an oil filled reservoir, it would only fill the movable oil volume. Therefore, the initial condition is the same as having injected a volume of water equal to:

$$W_{Id} = \frac{h}{2L \tan \theta}(1 - S_{or} - S_{wc}) \quad \text{(PV)} \tag{5.40}$$

which is referred to as the equivalent volume of injection.

Similarly, the fraction of the total HCPV of the section initially flooded by water is

$$\Delta N_{pD} = \frac{h}{2L \tan \theta} \tag{5.41}$$

which represents the oil volume that cannot be recovered on account of the initial condition. Therefore, having calculated N_{pD} and W_{id} for the entire system, all that is necessary is to reduce their values in Welge calculations to

$$N'_{pD} = N_{pD}(1 - \Delta N_{pD}) \quad \text{(HCPV)} \tag{5.42}$$

and

$$W'_{id} = W_{id} - W_{Id} \quad \text{(PV)} \tag{5.43}$$

where both are evaluated as fractions of the volume of the total section, excluding the initial water. Since the total pore volume is

$$PV_T = \frac{LhW\phi}{5.615} \quad \text{(rb)}$$

then the actual volumes of oil production and water injection can be calculated as

$$N_p = N'_{pD} \times PV_T \times \frac{(1 - S_{wc})}{B_o} \quad \text{(stb)} \tag{5.44}$$

and

$$W_i = W'_{id} \times \frac{PV_T}{B_w} \quad \text{(stb)} \tag{5.45}$$

in which the formation volume factors are evaluated at the average flooding pressure. The method which provides a "reasonable approximation" for accounting for the presence of edge water is illustrated in Exercise 5.2. The correction naturally decreases as the dip angle of the reservoir increases.

(d) VE displacement in a homogeneous acting reservoir

Waterdrive in homogeneous reservoirs is worth considering separately because it may be described in a very simple manner which is illustrative of the VE-flooding condition. Surprisingly, homogeneous acting reservoirs are not as uncommon as may be imagined. Apart from uniform slabs of rock, which admittedly are of rare occurrence, any reservoir section that displays a randomness in permeability and porosity distributions (with no definite trends such as coarsening upwards or downwards) may, as illustrated in Exercise 5.2, be treated as being homogeneous. In fact, the greater the degree of randomness in rock properties, the more appropriate is this simple description. Flooding in such a reservoir is illustrated in Fig. 5.46. Consider viewing the displacement at a fixed time and position, X, in the section for which a and b represent the fractional thickness of the reservoir occupied by the original edge water and the rise of the dynamic oil water contact. In generating pseudo-relative permeabilities for such a system, the thickness averaged water saturation at point X may be evaluated as

$$\overline{S}_w = a + b(1 - S_{or}) + (1 - a - b)S_{wc}$$

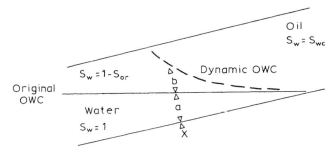

Fig. 5.46. Water–oil displacement in a homogeneous acting reservoir.

which allows for 100% water saturation below the original oil–water contact (OOWC) and the either–or situation in the original oil column dictated by the VE condition, for any mobility ratio. That is, behind the dynamic contact the water saturation is everywhere the flood-out value, $1 - S_{or}$, while ahead of the front it is at the irreducible level, S_{wc} — there is no in-between. The extremes in the thickness averaged water saturation are therefore

Original $(b = 0)$: $\overline{S}_w = S_{wc} + a(1 - S_{wc})$

Flood-Out $(b = 1 - a)$: $\overline{S}_w = 1 - S_{or} + aS_{or}$

And the difference between these gives a movable average saturation range of

$$(1 - a)(1 - S_{or} - S_{wc})$$

The corresponding thickness averaged relative permeabilities to water and oil at point X are

$$\overline{k}_{rw} = a + bk'_{rw}$$

and

$$\overline{k}_{ro} = (1 - a - b)k'_{ro}$$

which are for flow of the fluids at their maximum or end-point values, k'_{rw} and k'_{ro}, within the original oil column, as required under the VE condition, while below the OOWC, $k'_{rw} = 1$. Then, re-expressing the equation for $\overline{S}_w$ as:

$$b = \frac{(\overline{S}_w - S_{wc}) - a(1 - S_{wc})}{(1 - S_{or} - S_{wc})}$$

which may be used to evaluate the pseudo-relative permeabilities as

$$\overline{k}_{rw} = a \left(1 - \frac{(1 - S_{wc})k'_{rw}}{1 - S_{or} - S_{wc}} \right) + \left(\frac{\overline{S}_w - S_{wc}}{1 - S_{or} - S_{wc}} \right) k'_{rw} \tag{5.46}$$

and

$$\overline{k}_{ro} = \frac{aS_{or}k'_{ro}}{1 - S_{or} - S_{wc}} + \left(\frac{1 - S_{or} - \overline{S}_w}{1 - S_{or} - S_{wc}} \right) k'_{ro} \tag{5.47}$$

In these expressions it will be noticed that every term on the right-hand side is a constant except $\overline{S}_w$. Consequently, the pseudos are simply linear relationships between $\overline{k}_{rw}$ and $\overline{k}_{ro}$ and the increasing average water saturation, $\overline{S}_w$, which is contained in the second term of equations. In particular, for updip regions of the reservoir far removed from the OOWC ($a = 0$) the pseudos are reduced to

$$\overline{k}_{rw} = \left(\frac{\overline{S}_w - S_{wc}}{1 - S_{or} - S_{wc}} \right) k'_{rw} \tag{5.48}$$

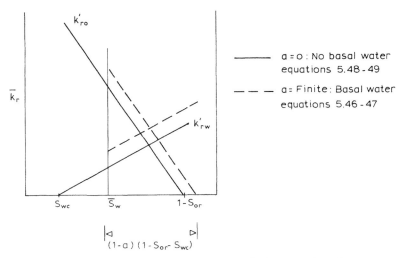

Fig. 5.47. Linear pseudo-relative permeabilities for VE displacement in a homogeneous acting reservoir.

and

$$\bar{k}_{ro} = \left(\frac{1 - S_{or} - \bar{S}_w}{1 - S_{or} - S_{wc}} \right) k'_{ro} \tag{5.49}$$

which are simple linear functions across the entire movable saturation range, $1 - S_{or} - S_{wc}$. The pseudos are plotted in Fig. 5.47.

This type of displacement was originally described by Dietz [14] and has been explained in some detail in chapter 10 of reference 9, in which, because the pseudos are linear, it was demonstrated that an analytical expression could be readily derived for calculating the oil recovery as a function of water injected (for $M > 1$). In this chapter, however, the preferred method is to apply the general "recipe" for performing vertical sweep efficiency calculations presented in section 5.5b. That is, generating the pseudos effectively reduces the problem from a two-dimensional macroscopic flood back to one dimension, making them appropriate for use in Buckley-Leverett displacement using the Welge method. The principle is demonstrated in Exercise 5.2 and is appropriate for any value of the mobility ratio.

Exercise 5.2: Water–oil displacement under the vertical equilibrium condition

Introduction

The exercise demonstrates that a reservoir whose vertical permeability distribution is completely random may be treated as homogeneous, provided there is pressure equilibrium across the section so that the VE condition prevails. Following this, the displacement efficiency is evaluated for three different mobility ratios and recovery calculations performed which illustrate the application of the equations presented in section 5.5b.

Question

The core measured permeability distribution across a sand of net thickness 109 ft is plotted in Fig. 5.48. It will be noted that the distribution is random in nature in that it contains no recognisable pattern: it appears the same whether viewed normally or upside down. Regular RFT surveys conducted under dynamic conditions in each new development well revealed complete hydrostatic equilibrium across the sand from which it was inferred that the VE condition would govern the displacement. The waterflood is characterised by a single set of normalised rock relative permeability curves with end-points $k'_{rw} = 0.3$, $k'_{ro} = 1.0$ and residual oil saturation of $S_{or} = 0.270$ PV. The section has been subdivided into eleven layers, ordered from the base to the top, which is the sequence in which they will flood with water under the VE condition. The formation properties and irreducible water saturations are listed in Table 5.6. A symmetry element from the line drive waterflood is shown in Fig. 5.49 and the remainder of the data required for problem solution are:

$B_o \approx B_{oi}$ = 1.475 rb/stb at 4500 psia: flooding at near initial pressure to avoid falling below the bubble point

B_w = 1.03 rb/stb

$\Delta \gamma$ = 0.32 (difference in water–oil specific gravities in the reservoir)

μ_w = 0.5 cp

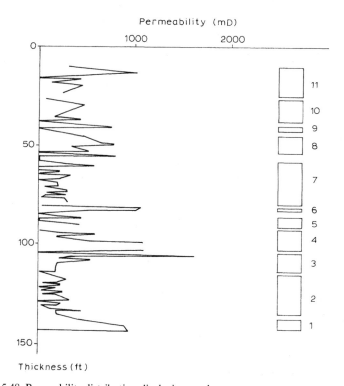

Fig. 5.48. Permeability distribution displaying randomness across a reservoir section.

TABLE 5.6

Formation properties and thickness averaged parameters

Layer No.	h_i (ft)	k_i (mD)	ϕ_i	S_{wc_i}
11	15	350	0.21	0.25
10	11	250	0.20	0.28
9	2	500	0.23	0.24
8	9	450	0.23	0.24
7	22	150	0.18	0.27
6	1	1000	0.24	0.24
5	5	300	0.21	0.27
4	10	600	0.23	0.24
3	9	250	0.20	0.27
2	20	150	0.19	0.28
1	5	650	0.24	0.25

$\sum h_i = 109$ ft
$\sum h_i k_i = 33350$ mD.ft: $\bar{k} = 306$ mD
$\sum h_i \phi_i = 22.23$ ft: $\bar{\phi} = 0.204$
$\sum h_i \phi_i S_{wc_i} = 5.823$ ft: $\bar{S}_{wc} = 0.262$ PV

MOV $= (1 - S_{or} - \bar{S}_{wc}) = 0.468$ PV $= 0.468/(1 - 0.262) = 0.634$ HCPV.

- (a) Generate pseudo-relative permeabilities and fractional flow relationships for three oil viscosities of $\mu_o = 50, 5$ and 0.8 cp (for which the mobility ratios may be calculated using equation 5.2 as $M = 30, 3$ and 0.48).
- (b) Calculate the oil recoveries as a function of the cumulative water injected and time and determine the relationships between surface watercut development and oil recovery.

Solution
(a) Pseudo-relative permeabilities/fractional flow: The pseudo-relative permeabilities for VE displacement are listed in Table 5.7. These have been generated applying the averaging procedures, equations 5.30–5.32 to the formation data listed in Table

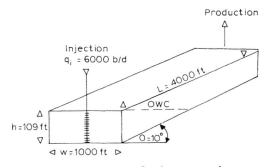

Fig. 5.49. Line drive waterflood symmetry element.

TABLE 5.7

Generation of pseudo-relative permeabilities for VE flooding

Flooding order	h_i (ft)	k_i (mD)	ϕ_i	S_{wc_i}	$\overline{S}_w$	$\overline{k}_{rw}$	$\overline{k}_{ro}$
11	15	350	0.21	0.25	0.730	0.300	0
10	11	250	0.20	0.28	0.662	0.253	0.157
9	2	500	0.23	0.24	0.617	0.228	0.240
8	9	450	0.23	0.24	0.607	0.219	0.270
7	22	150	0.18	0.27	0.562	0.183	0.391
6	1	1000	0.24	0.24	0.480	0.153	0.490
5	5	300	0.21	0.27	0.474	0.144	0.520
4	10	600	0.23	0.24	0.453	0.130	0.565
3	9	250	0.20	0.27	0.402	0.076	0.745
2	20	150	0.19	0.28	0.365	0.056	0.813
1	5	650	0.24	0.25	0.288	0.029	0.903
0					0.262	0	1.000

5.6, assuming the flooding order of the eleven layers to be from the base to the top of the section. The functions are plotted in Fig. 5.50 as the dots (•) and crosses (+) from which it can be seen that they correspond very closely to linear pseudos required to describe displacement in a macroscopic, homogeneous acting reservoir. This confirms the assertion made in section 5.6d, that a reservoir section with a random distribution of permeability can be described as homogeneous.

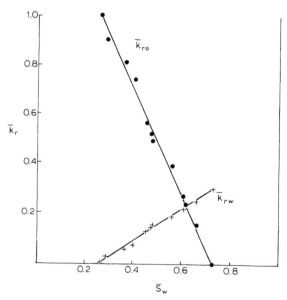

Fig. 5.50. Linear oil–water pseudo-relative permeabilities for VE displacement in a homogeneous acting reservoir.

TABLE 5.8

Pseudo-fractional flow functions for three mobility ratios (Fig. 5.51)

$\overline{S}_w$	$\overline{k}_{rw}$	$\overline{k}_{ro}$	Fractional flow		
			$M = 30$	$M = 3$	$M = 0.48$
0.262	0	1	0	0	0
0.300	0.024	0.919	0.723	0.207	0.033
0.350	0.056	0.812	0.873	0.408	0.083
0.400	0.088	0.705	0.926	0.555	0.143
0.450	0.120	0.598	0.953	0.667	0.215
0.500	0.153	0.492	0.969	0.757	0.300
0.550	0.185	0.385	0.980	0.828	0.402
0.600	0.217	0.278	0.987	0.886	0.525
0.650	0.249	0.171	0.993	0.936	0.677
0.700	0.281	0.064	0.998	0.978	0.864
0.730	0.300	0	1	1	1

In generating the fractional flow relationship for the three mobility ratios ($M = 30$, 3 and 0.48), it is necessary to examine the magnitude of the gravity term (equation 5.20)

$$G = 2.743 \times 10^{-3} \frac{\overline{k} k_{ro} \Delta \gamma \sin \theta}{v \mu_o} \tag{5.20}$$

acting in the downdip direction to stabilize the displacement (section 5.4c). Using the values of $\overline{k} = 306$ mD, $\Delta \gamma = 0.32$, $\theta = 10°$ then G may be evaluated as

$$G = \frac{0.047 \,\overline{k}_{ro}}{v \mu_o} \tag{5.50}$$

And, considering the low mobility ratio case ($\mu_o = 0.8$ cp: $M = 0.48$) for which the displacement should be piston-like, it may be calculated that injecting at 6000 b/d into the symmetry element shown in Fig. 5.48 leads to the calculation of an average flood velocity of approximately $v' = 3$ ft/day using equation 5.52 (section 5.7b). The corresponding Darcy velocity required for use in equation 5.50 is therefore

$$v = v' \phi (1 - S_{or} - S_{wc}) = 0.3 \text{ ft/day} \tag{5.52}$$

Thus, evaluating the gravity term, G, for the maximum oil relative permeability of $\overline{k}_{ro} = 1$ would yield a value of $G = 0.20$ which is of significant magnitude. The term has therefore been included as $G = 0.196 \,\overline{k}_{ro}$ in the results listed in Table 5.8. For the higher mobility ratios, however, not only are the oil viscosities much larger but so too are the velocities due to underrunning of the oil by water (Fig. 5.52) which justifies the omission of the gravity term as negligible in these cases. The fractional flows have been calculated using equations 5.18 and 5.19 and are listed in Table 5.8 and plotted in Fig. 5.51. The calculations have been performed using the linear pseudo-relative permeabilities (Fig. 5.50) for water and oil expressed as

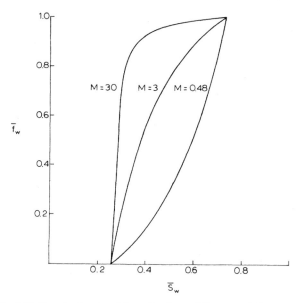

Fig. 5.51. Pseudo-fractional flows for three different mobility ratios.

$$\overline{k}_{rw} = 0.641\,\overline{S}_w - 0.168$$

$$\overline{k}_{ro} = 1.560 - 2.137\,\overline{S}_w$$

Inspecting the fractional flows (Fig. 5.51), it can be seen that for the cases in which $M > 1$, it is not possible to draw a conventional tangent to the curves as required in Welge calculations (section 5.4e). Instead, the point of tangent coincides with the origin of the two curves $(\overline{S}_w = \overline{S}_{wc}, \overline{f}_w = 0)$ which means that there is no shock front development $(S_{wbt} = S_{wc})$ and all saturations in the movable range, $1 - S_{or} - S_{wc}$, have independent mobility. The nature of the displacement is depicted in Fig. 5.52. In the least favourable case, for which $M = 30$, the injected water can travel 30 times faster than the oil it is displacing and, under the influence of gravity, it falls to the base of the reservoir and does just that: forming an extensive tongue of water which moves rapidly towards the producer.

Viewing matters at the production well, a slow increase in water saturations across the entire movable saturation range would be observed as oil at the top of the reservoir was slowly recovered. For $M = 3$, the situation is similar although less dramatic but both cases ($M = 30$, $M = 3$) illustrate the point that for unfavourable mobility ratio displacement ($M > 1$) under the VE condition in a homogeneous acting reservoir, there is no shock-front development across the macroscopic section as there is in a one-dimensional core flooding experiment. Instead, all saturations are free to move independently. In the favourable case that $M = 0.48$, the shape of the fractional flow curve (Fig. 5.51) is concave upwards across the entire movable saturation range which, as described in section 5.4e (Fig. 5.25) is manifest of the condition of piston-like displacement across the reservoir section. In this situation

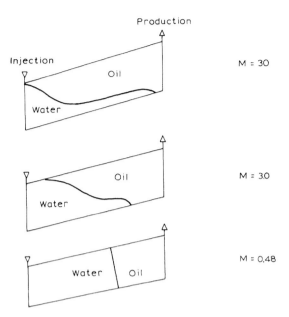

Fig. 5.52. Water–oil displacement for VE flooding in a homogeneous acting reservoir.

wells will flood-out in a short period, a matter of weeks, indicating both rapid and highly efficient recovery.

(b) Recovery calculations These have been performed for $M = 30$ and $M = 3$ (the case $M = 0.48$ being trivial) using the Welge equation:

$$N_{pd} = \frac{(\overline{S}_{we} - \overline{S}_{wc}) + (1 - \overline{f}_{we})W_{id}}{(1 - \overline{S}_{wc})} \quad \text{(HCPV)} \tag{5.33}$$

in which the presence of edge water in the reservoir is initially disregarded. The values of $\overline{S}_{we}$ and $\overline{f}_{we}$ at the end of the reservoir section which are used in the equation are listed in columns 5 and 6 of Table 5.9, in which the results are presented. These are the average values at which the cumulative water influx of

$$W_{id} = \frac{\frac{1}{\Delta \overline{f}_{we}}}{\Delta \overline{S}_{we}} \quad \text{(PV)}$$

has been evaluated. Allowance is then made for the edge water present in the reservoir, as described in section 5.6c, by calculating the "equivalent" initial water volume as:

$$\begin{aligned}
W_{Id} &= \frac{h}{2L\tan\theta}(1 - \overline{S}_{or} - \overline{S}_{wc}) \quad \text{(PV)} \tag{5.40} \\
&= \frac{109(1 - 0.27 - 0.0262)}{2 \times 4000 \times 0.176} = 0.036 \text{ PV}
\end{aligned}$$

and the fraction of the total HCPV $LhW\bar{\phi}(1 - \bar{S}_{wc})$ initially flooded with water as:

$$\Delta N_{pd} = \frac{h}{2L \tan \theta} = 0.077 \tag{5.41}$$

The correction for initial edge water is then made by applying the equations

$$N'_{pD} = N_{pD}(1 - \Delta N_{pD}) = 0.923 N_{pD} \quad \text{(HCPV)} \tag{5.42}$$

$$W'_{id} = W_{id} - W_{Id} = W_{id} - 0.036 \quad \text{(PV)} \tag{5.43}$$

which both are expressed as fractions of the total volume, excluding the edge water (columns 9 and 10, Table 5.9) and, as can be seen, for this geometry the corrections are not particularly significant.

Since the pore volume of the total system is

$$PV_T = \frac{LhW\phi}{5.615} = \frac{4000 \times 109 \times 1000 \times 0.204}{5.615} = 15.84 \text{ MMrb}$$

then applying equations 5.44 and 5.45 the actual oil recovery and water injection can be expressed as

$$N_p = \frac{15.84(1 - 0.262)}{1.475} N'_{pD} = 7.925 N'_{pD} \quad \text{(MMstb)}$$

$$W_i = \frac{15.84}{1.03} W'_{id} = 15.38 W'_{id} \quad \text{(MMstb)}$$

At an injection rate of 6000 stb/d (6180 rb/d: 2.256 MMrb/year) it would take 7.02 years to inject one total pore volume of 15.84 MMrb ($W'_{id} = 1$). The time scale can therefore be attached to the project as

$$t = 7.02 W'_{id} \quad \text{(years)} \tag{5.51}$$

Finally, the surface watercut can be evaluated using equation 5.36 as

$$f_{ws} = \frac{1}{1 + 0.698 \left(\dfrac{1}{f_{we}} - 1 \right)}$$

N_p, W_i, t and f_{ws} are listed in the final four columns of Table 5.9.

The plots in Fig. 5.53 show the nature of the watercut development for the three cases as a function of the oil recovery from the vertical reservoir section: the calculations taking no account of the areal sweep which is assumed to be perfect throughout the flood ($E_A = 1$). In this respect the results are equivalent to those that would be obtained from a two-dimensional cross-sectional numerical simulation study of the flood. The three cases may be described as follows.

$M = 0.48$, $\mu_o = 0.8$ cp. This demonstrates perfect piston-like displacement across the macroscopic sand section (Fig. 5.52) with all the movable oil being recovered by the injection of the same volume of water. In this case, for a symmetry element of a regular line drive pattern, Fig. 5.49, the areal sweep might be expected to be

TABLE 5.9

Displacement calculations for VE flooding ($M = 30$, $M = 3$)

$\bar{S}_{we}$	$\bar{f}_{we}$	$\Delta\bar{S}_{we}$	$\Delta\bar{f}_{we}$	$\bar{S}_{we}$	$\bar{f}_{we}$	W_{id} (PV)	N_{pD} (HCPV)	W'_{id} (PV)	N'_{pD} (HCPV)	N_p (MMstb)	W_i (MMstb)	Time (years)	f_{ws}
$M = 30$													
0.262	0												
0.300	0.723	0.038	0.723	0.281	0.362	0.053	0.072	0.017	0.066	0.52	0.26	0.12	0.448
0.350	0.873	0.050	0.150	0.325	0.798	0.333	0.177	0.297	0.163	1.29	4.57	2.08	0.850
0.400	0.926	0.050	0.053	0.375	0.900	0.943	0.281	0.907	0.259	2.05	13.95	6.37	0.928
0.450	0.953	0.050	0.027	0.425	0.940	1.852	0.371	1.816	0.342	2.71	27.93	12.75	0.957
0.500	0.969	0.050	0.016	0.475	0.961	3.125	0.454	3.089	0.419	3.32	47.51	21.68	0.972
0.550	0.980	0.050	0.011	0.525	0.975	4.545	0.510	4.509	0.471	3.73	69.35	31.65	0.982
0.600	0.9874	0.050	0.0074	0.575	0.984	6.757	0.571	6.721	0.527	4.18	103.37	47.18	0.989
0.650	0.9932	0.050	0.0058	0.625	0.990	8.621	0.605	8.585	0.558	4.42	132.04	60.27	0.993
0.700	0.9977	0.050	0.0045	0.675	0.996	11.111	0.627	11.075	0.579	4.59	170.33	77.75	0.997
0.730	1	0.030	0.0023	0.715	0.999	13.043	0.634	13.007	0.585	4.64	200.05	91.31	0.999
$M = 3$													
0.262	0												
0.300	0.207	0.038	0.207	0.281	0.104	0.184	0.249	0.148	0.230	1.82	2.28	1.04	0.143
0.350	0.408	0.050	0.201	0.325	0.308	0.249	0.319	0.213	0.294	2.33	3.28	1.50	0.389
0.400	0.555	0.050	0.147	0.375	0.482	0.340	0.392	0.304	0.362	2.87	4.68	2.13	0.571
0.450	0.667	0.050	0.112	0.425	0.611	0.446	0.456	0.410	0.421	3.34	6.31	2.88	0.692
0.500	0.757	0.050	0.090	0.475	0.712	0.556	0.506	0.520	0.467	3.70	8.00	3.65	0.780
0.550	0.828	0.050	0.071	0.525	0.793	0.704	0.554	0.668	0.511	4.05	10.27	4.69	0.846
0.600	0.886	0.050	0.058	0.575	0.857	0.862	0.591	0.826	0.545	4.32	12.70	5.80	0.896
0.650	0.936	0.050	0.050	0.625	0.911	1.000	0.612	0.964	0.565	4.48	14.83	6.77	0.936
0.700	0.978	0.050	0.042	0.675	0.957	1.190	0.629	1.154	0.581	4.60	17.75	8.10	0.970
0.730	1	0.030	0.022	0.715	0.989	1.364	0.634	1.328	0.585	4.64	20.42	9.32	0.992

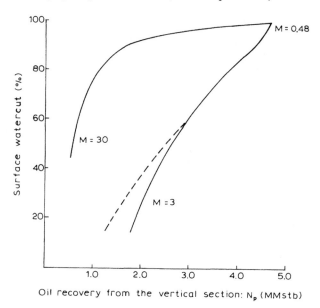

Fig. 5.53. Watercut developments (M = 30, 3, 0.48).

close to unity throughout the flood. In practice, the producing well would water-out completely within a very short period of perhaps a few weeks. The total movable oil (Table 5.9) is 0.585 (HCPV) = 0.585(1 – 0.262) (PV) (expressed as a fraction of the total volume: 15.84 MMrb) Therefore, for piston-like displacement the volume of injected water must be the same: W'_{id} = 0.432 PV which, applying equation 5.51, establishes the duration of the flood as just over three years.

M = 3, μ_o = 5 cp. For this slightly unfavourable mobility ratio there is a degree of underrunning of the oil leading to premature breakthrough of water after producing approximately 40% of the STOIIP in the section. At breakthrough the areal sweep will be less than perfect but for a line drive pattern it should rise towards unity quite rapidly. Although areal sweep calculations are not catered for in this chapter, being best handled by numerical simulation, some notional allowance for the effect [1] could be made by accelerating the breakthrough time, as illustrated by the dashed line in Fig. 5.53. The early watercut trend is not of great importance, however, because at this stage there is usually adequate capacity in the topsides facilities to handle the total gross fluid production. It is only later when the watercut rises to higher values that constraints are imposed by the capacities (section 5.3c); then it is important to accurately predict the watercut trend and this should be possible using the relationship plotted in Fig. 5.53. The trend is used in material balance calculations, as illustrated in Exercises 5.1 and 5.4, to determine a practical abandonment condition commensurate with the surface topsides capacities. Abandoning at a watercut of 90% (a reasonable value to choose since above that the watercut rises quite sharply) would yield a recovery of 4.3 MMstb, 93% of the total movable volume, in almost six years.

$M = 30$, $\mu_o = 50$ cp. In this case the watercut development is so harsh that downgrading it further to account for unfavourable areal sweep would not make any appreciable difference. The 90% watercut level is reached within five years for an oil recovery of 1.7 MMstb. Thereafter, it would require the circulation of 200 MMstb of water, taking 91 years, to recover the remainder of the recoverable oil and while this may appear to offer the engineer job security for life, it is hardly a practical proposition. In situations like this where the flooding characteristics are so poor it pays to drill more injection wells, at the expense of producers, to circulate the large volumes of water necessary to attain a higher recovery in a reasonable period of time.

Exercise 5.3: The influence of distinctive permeability distributions on the vertical sweep efficiency for the VE-flooding condition

Introduction

Waterdrive is first studied in a reservoir in which there is a natural coarsening downward in rock properties, which is unfavourable for VE flooding. The sand section is then inverted so that it coarsens upwards and the effect on the fractional flow relations is compared. Displacement efficiency is studied for several different mobility ratios.

Question

The natural permeability distribution across the reservoir being studied is plotted in Fig. 5.54a and displays a fairly strong coarsening downward trend in rock properties. The 94 ft thick sand section has been divided in to 10 layers as shown in Fig. 5.54b and detailed in Table 5.10. Dynamic RFT surveys across the formation indicate that the flooding is likely to occur under the VE condition. There is only one set or rock relative permeability curves for the reservoir with end-point values of $k'_{rw} = 0.330$, $k'_{ro} = 1.0$ and residual oil saturation of $S_{or} = 0.330$ PV.

- (a) Generate pseudo-relative permeabilities for VE water–oil displacement for the normal reservoir section (Fig. 5.54b — coarsening downward) and for the inverted system (Fig. 5.54c — coarsening upward).
- (b) Determine reservoir fractional flow relationships for the following oil viscosities:

  A: $\mu_o = 1.24$ cp ($M = 1.00$: equation 5.2)
  B: $\mu_o = 20$ cp ($M = 16.1$: equation 5.2)

  The water viscosity is in $\mu_w = 0.41$ cP for both cases.
- (c) Calculate the oil recovery (N_{pD}, HCPV) and relate it to the reservoir fractional flow of water for both oil viscosities.

Solution

(a) Pseudo-relative permeabilities. These have been generated for VE flooding using the averaging procedures, equations 5.30–5.32. The results are listed in Table 5.11 for both the natural coarsening downward sequence of permeabilities and its inversion. In both cases the flooding order of the 10 layers is from base to top.

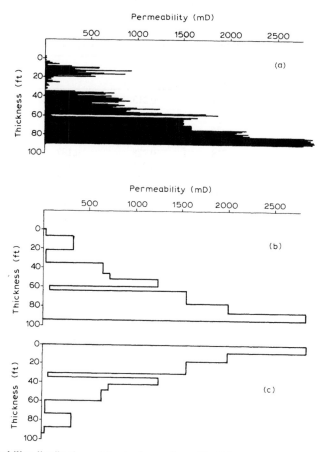

Fig. 5.54. Permeability distributions: (a) actual core data; (b) 10-layer equivalent model; (c) inverted 10-layer model.

The pseudos are plotted for both cases in Fig. 5.55 and have markedly different shapes the significance of which can be best appreciated by determining their fractional flow functions for the two different mobility ratios.

These have been calculated for horizontal flow (equation 5.18) for the two oil viscosities of $\mu_o = 1.24$ cp ($M = 1$) and $\mu_o = 20$ cp ($M = 16.1$) and for the unfavourable and favourable permeability distributions. The results are listed in Table 5.12. The values of $\bar{k}_{rw}$ and $\bar{k}_{ro}$ used have been read from the smoothed curves in Fig. 5.55 for saturation increments of 5%. The functions are plotted in Fig. 5.56 and demonstrate a considerable difference dependent on the nature of the heterogeneity and mobility ratio. For the unfavourable coarsening downward trend in permeabilities, Fig. 5.54b, both fractional flow curves are concave downwards across the entire movable saturation range, $1 - \bar{S}_{or} - \bar{S}_{wc} = 0.483$ PV, indicating separate mobility of all these saturations. For $M = 16.1$ the displacement efficiency

TABLE 5.10

Formation properties (coarsening downward) and thickness-averaged parameters

Layer No.	h_i (ft)	k_i (mD)	ϕ_i	S_{wc_i}
1	7	34	0.213	0.210
2	14	320	0.220	0.196
3	14	32	0.215	0.205
4	11	650	0.227	0.195
5	6	718	0.228	0.187
6	8	1244	0.235	0.180
7	4	74	0.220	0.192
8	13	1560	0.253	0.175
9	9	2000	0.250	0.165
10	8	2840	0.259	0.168

$\sum h_i = 94$ ft
$\sum h_i k_i = 87872$ mD.ft: $\quad \bar{k} = 935$ mD
$\sum h_i \phi_i = 21.817$ ft: $\quad \bar{\phi} = 0.232$
$\sum h_i \phi_i S_{wc_i} = 4.079$ ft: $\quad \bar{S}_{wc} = 0.187$ PV

MOV $= (1 - S_{or} - \bar{S}_{wc}) = 0.483$ PV $= 0.483/(1 - 0.187) = 0.594$ HCPV.

TABLE 5.11

Generation of pseudo-relative permeability curves for coarsening downward and upward sequences of permeability (Fig. 5.54)

Layer No.	h_i (ft)	k_i (mD)	ϕ_i	S_{wc_i}	flood order	$\bar{S}_w$	$\bar{k}_{rw}$	$\bar{k}_{ro}$	flood order	$\bar{S}_w$	$\bar{k}_{rw}$	$\bar{k}_{ro}$
						Coarsening downward				Coarsening upward		
									0	0.187	0	1
1	7	34	0.213	0.210	10	0.670	0.330	0	1	0.218	0.001	0.997
2	14	320	0.220	0.196	9	0.639	0.329	0.003	2	0.285	0.018	0.946
3	14	32	0.215	0.205	8	0.572	0.312	0.054	3	0.349	0.019	0.941
4	11	650	0.227	0.195	7	0.507	0.311	0.059	4	0.404	0.046	0.860
5	6	718	0.228	0.187	6	0.453	0.284	0.140	5	0.434	0.062	0.811
6	8	1244	0.235	0.180	5	0.423	0.268	0.189	6	0.476	0.100	0.698
7	4	74	0.220	0.192	4	0.381	0.230	0.302	7	0.496	0.101	0.694
8	13	1560	0.253	0.175	3	0.361	0.229	0.306	8	0.570	0.177	0.463
9	9	2000	0.250	0.165	2	0.287	0.153	0.537	9	0.622	0.245	0.259
10	8	2840	0.259	0.168	1	0.235	0.085	0.741	10	0.670	0.330	0
					0	0.187	0	1				

is extremely poor while even for $M = 1$ the heterogeneity predominates over the favourable mobility ratio giving a fairly harsh development in the fractional flow. The results for the inverted, coarsening upward permeability distribution gives much more satisfactory fractional flows. For $M = 1$, as might be expected, the curve, Fig. 5.56b, is concave upward across the movable saturation range implying piston-like

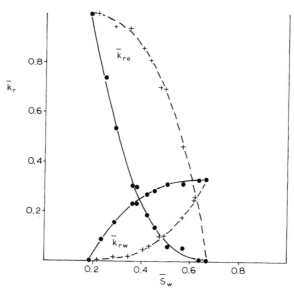

Fig. 5.55. Water–oil VE-pseudos: ●—●—●—● = coarsening downward, +—+—+—+ = coarsening upward.

TABLE 5.12

Generation of fractional flows for $M = 1.0$ and 16.1

$\overline{S}_w$	Coarsening downward				Coarsening upward			
	$\overline{k}_{rw}$	$\overline{k}_{ro}$	fractional flow		$\overline{k}_{rw}$	$\overline{k}_{ro}$	fractional flow	
			$M = 1.0$	$M = 16.1$			$M = 1.0$	$M = 16.1$
0.187	0	1	0	0	0	1	0	0
0.20	0.031	0.908	0.094	0.625	0.001	0.998	0.003	0.047
0.25	0.109	0.700	0.320	0.884	0.005	0.982	0.015	0.199
0.30	0.169	0.507	0.502	0.942	0.013	0.961	0.039	0.398
0.35	0.220	0.350	0.655	0.968	0.021	0.923	0.064	0.526
0.40	0.258	0.230	0.772	0.982	0.040	0.868	0.122	0.692
0.45	0.283	0.140	0.859	0.990	0.070	0.780	0.213	0.814
0.50	0.307	0.083	0.918	0.994	0.106	0.680	0.320	0.884
0.55	0.319	0.046	0.954	0.997	0.150	0.540	0.457	0.931
0.60	0.327	0.021	0.979	0.999	0.213	0.374	0.633	0.965
0.65	0.329	0.005	0.995	1	0.200	0.143	0.860	0.990
0.67	0.330	0	1	1	0.330	0	1	1

displacement over the entire sand (section 5.4e). Even for $M = 16.1$, however, there is a slight degree of inflexion and a breakthrough point can be identified for $\overline{S}_w = 0.40$ PV, $\overline{f}_w = 0.70$, making it more favourable than for $M = 1$ for the coarsening downward in permeabilities.

TABLE 5.13

Welge calculations for the fractional flow functions plotted in Fig. 5.56

$\bar{S}_{we}$	Coarsening down: $M = 1.0$				Coarsening down: $M = 16.1$				Coarsening up: $M = 16.1$			
	$\bar{f}_{we}$	W_{id} (PV)	N_{pD} (HCPV)	$\bar{f}_{we}$	$\bar{f}_{we}$	W_{id} (PV)	N_{pD} (HCPV)	$\bar{f}_{we}$	$\bar{f}_{we}$	W_{id} (PV)	N_{pD} (HCPV)	$\bar{f}_{we}$
0.187	0				0							
0.20	0.094	0.138	0.170	0.047	0.625	0.021	0.026	0.313				
0.25	0.320	0.221	0.263	0.207	0.884	0.193	0.105	0.755				
0.30	0.502	0.275	0.307	0.411	0.942	0.862	0.200	0.913				
0.35	0.655	0.327	0.339	0.579	0.968	1.923	0.276	0.955		breakthrough		
0.40	0.772	0.427	0.382	0.714	0.982	3.571	0.341	0.975	0.692	0.308	0.379	0.692
0.45	0.859	0.575	0.423	0.816	0.990	6.250	0.400	0.986	0.814	0.410	0.417	0.753
0.50	0.918	0.847	0.470	0.889	0.994	12.500	0.477	0.992	0.884	0.714	0.487	0.849
0.55	0.954	1.389	0.525	0.936	0.997	16.667	0.508	0.996	0.931	1.064	0.537	0.908
0.60	0.979	2.000	0.560	0.967	0.999	25.000	0.539	0.998	0.965	1.471	0.571	0.948
0.65	0.995	3.125	0.589	0.987	1	50.000	0.569	1.000	0.990	2.000	0.594	0.978
0.67	1	10.000	0.594	0.998	1	>50.000	0.594	1.000	1	5.000	0.594	0.995

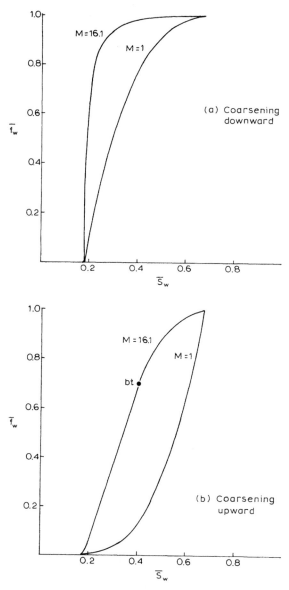

Fig. 5.56. Fractional flow relationships for different forms of heterogeneity and mobility ratios.

(c) Welge displacement calculations. These have been performed for the four fractional flow curves (Fig. 5.56) applying equation 5.33 to determine the relationships between $\overline{f}_{we}$ and the vertical sweep, N_{pD} (HCPV), both evaluated at *in situ* reservoir conditions. The results are listed in Table 5.13 and plotted in Fig. 5.57. In the table, for each case, the first column is the value of $\overline{f}_{we}$ read from the plots in Fig. 5.56 while the figure listed in column four is the average value used in the Welge

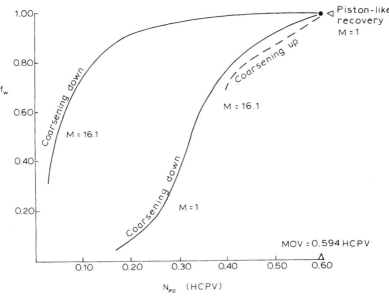

Fig. 5.57. Reservoir fractional flow of water as a function of the vertical sweep efficiency (N_{pD}, HCPV).

equation to evaluate the cumulative water injected, W_{id} (PV), using equation 5.26. In viewing the results, consideration should be given to the separate influence of the three factors — mobility ratio, heterogeneity and gravity which, as described in section 5.5a, govern the efficiency of displacement in macroscopic reservoir sections.

Coarsening downward

M = 1.0. In a core flooding experiment or for displacement in a homogeneous reservoir, such a favourable mobility ratio would provide perfect, piston-like displacement but in this particular example the effects of heterogeneity and gravity work together to downgrade the sweep efficiency as the heavier water is channelled preferentially through the high-permeability layers at the base of the section. Even so, the situation is quite tolerable and abandoning at a reservoir watercut of f_{we} = 0.95 would give a vertical sweep of N_{pD} = 0.54 HCPV (Fig. 5.57) which is in 91% of the total movable oil of 0.594 HCPV. This would require the circulation of W_{id} = 1.67 PV of injected water (Table 5.13).

M = 16.1. In this case, mobility ratio, heterogeneity and gravity all exert an adverse influence on the efficiency of the waterdrive. Again the heavier water finds its way to the high flow capacity basal sands where, on account of its unfavourable mobility ratio, it moves 16 times faster than the oil leading to premature breakthrough and production with sustained high watercut. As can be appreciated from Fig. 5.57, almost 70% of the movable oil remains to be recovered once the reservoir watercut has risen above 90%. Inefficient waterdrives such as this place a great strain on the water injection and production facilities, especially offshore (section 5.3c). If, however, the surface equipment capacities are adequately sized and sufficient injection wells drilled to enable large volumes of water to be circulated quickly then there are significant rewards in terms of oil recovery. For instance, (Table 5.13), abandoning at a watercut of 98.6% would recover 0.400 HCPV which is 67% of the movable volume. This would require, however, the circulation of W_{id} = 6.25 PV of water which may be physically realistic but the economics of doing so, which are highly dependent on the environment, may prove so unfavourable as to limit the recovery to a lower level.

Coarsening upward

$M = 1.0$. This is the ideal case in which mobility ratio, heterogeneity and gravity all work together to enhance the recovery. As described in section 5.6b, the water preferentially enters the high-permeability layers at the top of the section then falls under the influence of gravity forming a complete water piston across the entire reservoir section. In this case, the recovery calculation is trivial and is therefore not listed in Table 5.13. That is, the total movable oil of 0.594 HCPV is recovered by the injection of precisely the same volume of water: W_{id} = 0.594 HCPV = 0.483 PV.

$M = 16.1$. This provides the most surprising result of all the cases for it establishes that a favourable sweep efficiency can be achieved even for an intrinsically high mobility ratio flood, provided the permeability distribution is favourable, as in the present case. Here, the combined effects of gravity and heterogeneity overcome the adverse mobility ratio to produce results that are better than for $M = 1$ in the coarsening downward reservoir environment. At an abandonment watercut of 95%, the vertical sweep would be 0.57 HCPV which is 96% of the movable oil in the cross-section.

The reader may accept as an obvious fact that in a coarsening downward sequence of permeabilities, the reservoir will flood from the base to the top with water, which is the order in which the averaging procedures 5.30–5.32 are evaluated — but feel more suspicious that this same order should apply in a coarsening upward environment. Nevertheless, as frequent comparison with cross-sectional numerical simulation models has demonstrated, the vertical permeabilities between the model layers have to be severely reduced to remove the VE condition. In models, of course, one has the freedom to alter permeabilities in the vertical direction whereas in applying the principle of VE flooding analytically there is no possibility of doing so. Instead, it is merely assumed that in such displacement, the flooding must naturally occur from the base to the top of the section with water since it is the heavier of the two fluids.

5.7. WATERDRIVE IN SECTIONS ACROSS WHICH THERE IS A TOTAL LACK OF PRESSURE EQUILIBRIUM

(a) Reservoir environment

Reservoirs in which the VE-flooding condition prevails are usually associated with either a marine or beach type depositional environment in which the sands are

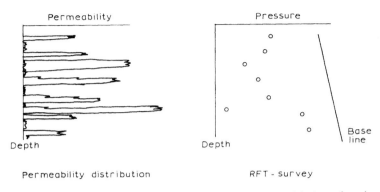

Fig. 5.58. Results of a dynamic RFT survey conducted across a deltaic sand section.

relatively clean and largely free from vertical flow barriers of significant areal extent. The type of section being described now, however, is one containing many (usually) thin sands that are physically separated from one another by impermeable shales or tight sands which precludes pressure communication between them. Therefore, there is no vertical cross-flow of fluids: apart from where layers may be juxtaposed across non-sealing faults. This type of section normally results from deposition in a non-marine, deltaic environment in which there are numerous channel sands of distinct flow capacity separated both vertically and areally from each other by shales and silts. It is hardly surprising that the set of sands observed in one well may be absent in others, except for the main distributary channels, since the probability of drilling through the less significant, meandering channels in different wells is slight. Such sections can be recognised by running RFT surveys under dynamic conditions in each new development well, as described in sections 5.2d and 5.5b. As shown in Fig. 5.58 the surveys usually reveal a distinct lack of pressure equilibrium across the section, the lower pressures being usually associated with the higher permeability, better connected sands.

The permeability distribution across a deltaic sand section of 200 ft thickness is shown in Fig. 5.59. The two plots on linear and more conventional logarithmic permeability scales reveal the significant distortion in the latter, referred to in section 5.6b, in visually exaggerating the importance of the tighter intervals while apparently diminishing the significance of the higher permeability sands. Permeability distributions must always be plotted using a linear permeability scale arranged so that the higher values are fully represented since these, which act as thief zones in a waterdrive, are the most significant sands in the section. If, in doing so, the tighter sections do not show-up, then this is the correct physical interpretation for it means that effectively they do not exist and will contribute little or nothing to the oil recovery under waterdrive conditions. For the section shown in Fig. 5.59, for instance, the operator merely observed the heterogeneity on the conventional logarithmic scale, (Fig. 5.59b), and as a result assigned a uniform permeability to all sands in the section whereas, in reality, there is at least a three order of magnitude difference in permeabilities between sands in the section. Making full allowance for

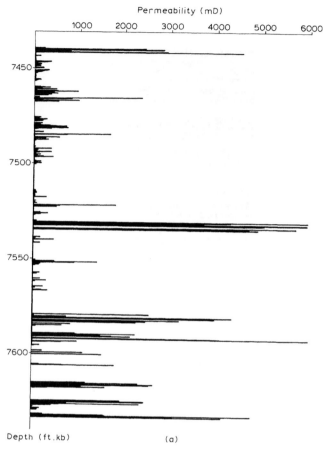

Fig. 5.59. (a) Permeability distribution across a deltaic reservoir plotted on a linear permeability scale.

the heterogeneity in recovery calculations, as described later in this section, halved the vertical sweep efficiency in comparison to the uniform description — no small downward adjustment.

While most of the flow directly into the wellbore is through the higher permeability channels, there can sometimes be a degree of vertical cross-flow of oil from low to high flow capacity sands in the reservoir which is encouraged by the large permeability × area product for vertical fluid movement. In this respect, the section acts as a dual porosity system, as defined in Chapter 4, section 4.2b. Of course, recovery from the tighter intervals increases with the pressure differential between these and the higher flow capacity channels. Therefore, in developing such fields the difficult choice has to be made between straightforward depletion, which encourages large pressure differentials and some cross-flow, and waterdrive in which the maintenance of pressure inhibits the process and often leads to flooding of only the higher permeability intervals. For onshore developments the operator has the opportunity to experiment by allowing significant pressure depletion, even below the

Fig. 5.59 (continued). (b) Permeability distribution across a deltaic reservoir plotted on a logarithmic permeability scale.

bubble point, followed by a low pressure waterflood. Offshore, however, the luxury of choice is not usually available and operators are obliged to opt for waterdrive from the outset. There simply is not time to cater for depletion recovery from the poorer sections and normally it is preferable to obtain accelerated oil recovery from the higher permeability sands alone. It follows that in this type of reservoir environment great care must be exercised in applying petrophysical cut-offs on k and ϕ for they will depend on the recovery mechanism, depletion or waterdrive, and if such cut-offs are imposed, it is necessary to state precisely where in the section the sand has been excluded.

Use of the RFT under dynamic conditions has proven invaluable in establishing the degree of "connectivity" in these complex sand sections. Since there is little or no pressure communication vertically, what is reflected in surveys such as depicted in Fig. 5.58 is the effect of areal communication between wells. Unfortunately, since the RFT has only been available since the mid-seventies, there are not many recorded cases of the application of the tool in these sand sections to date. There is one such deltaic section, however, that is common to many North Sea fields in which the technique has been routinely applied. This is the Ness Section * which is a part of the overall Brent, Middle Jurassic reservoirs in the prolific production area of the East Shetland Basin between the Shetland Islands and Norway (Fig. 5.12). In some fields in this area, when dynamic RFT surveys have been run, they demonstrate a degree of depletion in all sands across the Ness Section, including those of very low flow capacity: even though the permeability contrast is in some places almost as severe as shown in Fig. 5.59a. This is a most encouraging observation for it means that even though the sands do not appear correlatable from well to well they are somehow connected — no matter how tortuous the routes. Therefore, waterdrive in that particular sand section is feasible and is practised.

The remainder of the section concentrates on methods for determining the vertical sweep efficiency in these complex sand sections which again amounts to generating pseudo-relative permeabilities. The manner in which these are used in numerical simulation modelling to match and predict fluid movement in deltaic environments will then be considered in section 5.8c.

(b) Data requirements and interpretation for input in the generation of pseudo-relative permeabilities

Most of the comments on this subject made in section 5.6b, for flooding in more continuous sand sections under the VE condition, are equally relevant to waterdrive in deltaic formations. There are, however, two notable differences as described below.

Layering/permeabilities

The choice of layers and handling of layer data has to be approached with a little more discretion than in the case of VE flooding. In a deltaic section, as shown in Fig. 5.59a, for instance, the selection of the numerous layers is itself fairly straightforward and is based largely on the permeability values. Each sand that is separated from its neighbours by a significant tight interval is initially designated as a layer. But some of these may themselves have a reasonable thickness and characteristic permeability distribution which, if the sands are clean so that the VE condition may be assumed to prevail within each, will influence the sweep efficiency in the separate layers and must be catered for as illustrated in Fig. 5.60. In the upper sand, the coarsening upward trend in permeabilities will result in a perfect piston-like displacement across its section and the sand may therefore be described as a single layer with thickness averaged permeability and porosity. The lower sand is the inverse, however, and will effectively flood as three separate layers and must be included as such in the overall layer count: each represented by its individual value of k_i and ϕ_i. In this way the two initial layers would eventually be represented by four which are effectively physically separate from

---

* Christened after Loch Ness in Scotland, home of the famous but elusive "Monster" — the sand section has been aptly named.

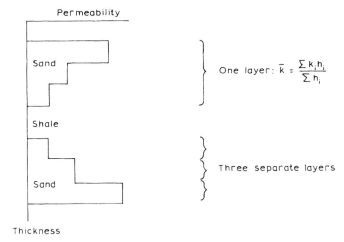

Fig. 5.60. Catering for VE flooding within clean, individual sands in a deltaic section.

one another with no cross-flow between them. Applying this approach, using the uncorrected core data directly, as described in section 5.6b, the section shown in Fig. 5.59a was originally represented by over 50 non-communicating layers which were subsequently reduced after consideration of their separate flooding order, as described below.

Flooding order of the layers

Having divided the entire section into a set of non-communicating layers, it is necessary to predict in which sequence they will flood in order to generate pseudo-relative permeabilities. The sequence will be dependent on the velocity of water frontal advance which was presented as equation 5.29 and may be readily derived by combining material balance with Darcy's law. The former applied to the cumulative volume of water injected at constant rate in any layer is

$$W_i = q_i t = LA\phi(1 - S_{or} - S_{wc})$$

from which the velocity may be determined as

$$v' = \frac{L}{t} = \frac{q_i}{A\phi(1 - S_{or} - S_{wc})}$$

It should be appreciated that this is the real velocity of frontal advance compared to the Darcy velocity which is defined as simply $v = q_i/A$. The relationship between the two is therefore

$$v = v'\phi(1 - S_{or} - S_{wc}) \tag{5.52}$$

meaning that the Darcy velocity, which is used in the gravity term in the fractional flow equation, equation 5.20, is typically an order of magnitude smaller than the actual velocity.

Substituting for the rate in the above expression, using Darcy's law gives

$$v' = \frac{k k'_{rw} \dfrac{\Delta p}{\Delta L}}{\mu_w \phi (1 - S_{or} - S_{wc})} \tag{5.53}$$

In applying this expression, it is conventionally accepted that the pressure differential between the injection and production wells is practically the same in each layer which means that

$$v'_i \propto \frac{k_i k'_{rwi}}{\phi_i (1 - S_{or_i} - S_{wc_i})} = \frac{k_i k'_{rw_i}}{\phi_i \Delta S_i} \tag{5.29}$$

in which the subscript denotes the ith layer and is attached to the parameters that can vary from layer to layer and ΔS_i is the movable saturation range. It will be appreciated that this expression has been derived under the assumption of piston-like displacement in each layer but in spite of this approximation is usually perfectly appropriate for determing the flooding order of the layers; which is in decreasing sequence of the velocities, v'_i. In fact, if variation in k'_{rw} and S_{or} is not catered for through lack of sufficient experimental data, as is often the case, then the flooding order is dictated by

$$v_i \propto \frac{k_i}{\phi_i} \text{ or even } k_i$$

and in reservoirs with a strong variation in permeability, quite often it is this parameter alone that predominates over all others in deciding the flooding order.

In applying this method to the deltaic section shown in Fig. 5.59a, the 50 separate layers, selected as described above, were reduced to 16 by grouping them in sets with like-values of k_i/ϕ_i. This is quite acceptable because they are physically separated and therefore members within each group will flood simultaneously irrespective of their position in the section. The flooding order of the 16 composite layers was then arranged in decreasing sequence of their average k_i/ϕ_i values. Having first reduced the number of layers and then determined their flood order, the section data are suitably prepared for calculations to generate pseudos, as described below.

(c) Stiles method

This dates from 1949 [20] and was originally presented in a form that led to direct oil recovery calculations following water breakthrough in each of the non-communicating layers. The flooding order was dictated by the method described in the previous section which was originally devised by Stiles and is, in fact, the key feature in his approach to the problem. In this text, the method has been modified so that its application results in the generation of pseudo-relative permeabilities and a fractional flow which can then be used in the general "recipe" for evaluating vertical sweep presented in section 5.5b.

The Stiles method is restricted to reservoirs in which the mobility ratio is unity, or close to unity. The significance of this has been described in section 5.4b in that the velocity of frontal advance of water in each separate layer will remain constant during the flood. That is, the velocities will be different in each layer, as dictated by equation 5.53, but as the flood progresses the differences will remain constant: there is no velocity dispersion. This is the simplest condition to consider but it should not be regarded as specialised because as mentioned previously, the majority of the world's waterdrive fields satisfy, by choice, the condition that $M \sim 1$ and the method therefore finds broad application. In using Stiles method, the procedure is as follows:

- Inspect the core and log data and divide the section into a total of N separate layers, as described in the previous section.
- Order the N layers in the sequence in which they will successively flood-out with water. This should be done by applying equation 5.29 but often ordering in sequence of decreasing values of k_i/ϕ_i or k_i is all that is necessary.
- Generate pseudo-relative permeabilities by applying the averaging procedures, equations 5.30–5.32. Since $M \sim 1$, the displacement in each layer is piston-like, thus only end-point saturations and relative permeabilities are required.

Application of the averaging procedures is justified on account of the lack of velocity dispersion during the flood. That is, it does not matter at which point between the injection and production well they are applied, the results will be the same. Each time a new layer floods the procedures are evaluated producing eventually a total of N values of $\overline{S}_w$ and the pseudos, $\overline{k}_{rw}$ and $\overline{k}_{ro}$. The only difference between Stiles and VE pseudo-generation is in the (flooding) order in which the procedures are evaluated. In the latter case, it is dictated by gravity from the base to the top of the continuous sand section whereas for Stiles the flooding order is established by application of equation 5.29 to the non-communicating sands. It will also be appreciated that, with the exception of the allowance for gravity within each sand layer (Fig. 5.60), gravity plays no part in Stiles type displacement on account of the vertical separation of the layers. Therefore, their flooding order is independent of their position in the section. This means that the displacement efficiency is entirely dictated by the mobility ratio and heterogeneity. Stiles type pseudos and fractional flow have the typical shapes depicted in Fig. 5.61. In fact, it should be noted that they ought to be step functions: once a layer has flooded, there will be plateau levels in $\overline{k}_r$ and $\overline{f}_w$ until the next layer floods which is accompanied by a step-like change to a new level. If sufficient layers are used, however, the points may be joined as smoothed functions which is required for applying analytical techniques. The pseudo-relative permeability to water has the characteristic concave-downward shape which is normal for this type of displacement and indicates that a slight increase in the water saturation, as the first few high-permeability layers flood, is accompanied by a high average relative flow of water — which is an unfavourable state of affairs. In particular, the fractional flow has also a characteristic concave-downward shape which, as described in section 5.4f, implies that all saturations in the movable range, $1 - S_{or} - S_{wc}$, are capable of independent mobility. In this respect, while for $M \sim 1$ there is piston-like displacement in each separate layer, there is no shock-front development across

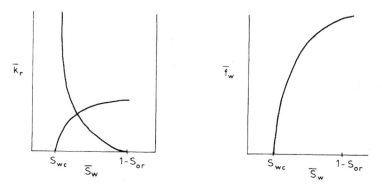

Fig. 5.61. Typical shapes of Stiles pseudo relative permeabilities and fractional flow.

the macroscopic sand section. This relates back to the comment made at the beginning of section 5.5 that, in reality, flooding in reservoirs occurs on the scale of "hillsides" not core plugs (individual layers). Stiles-type displacement is illustrated in Exercise 5.4.

(d) Dykstra-Parsons method

This is the general approach [21] to calculating the vertical sweep efficiency in deltaic sand sections since it is appropriate for all values of the mobility ratio. It therefore caters for velocity dispersion of the flood front between the individual layers. That is, as described in section 5.4b, if:

$M < 1$: the velocity of frontal advance in each layer will be reduced as the flood progresses which tends to stabilize the macroscopic flood front

$M > 1$: the velocity of frontal advance in each layer increases as the flood progresses which promotes instability in the macroscopic flood front.

Consider the flooding of an individual layer in which it is assumed that piston-like displacement occurs. This will be valid for $M \leq 1$ but an approximation for $M > 1$. It is, however, quite acceptable considering the scale of the overall system being flooded. Then, applying Darcy's law at the flood front:

$$\frac{k'_{rw}}{\mu_w} \frac{dp_w}{dx} = \frac{k'_{ro}}{\mu_o} \frac{dp_o}{dx}$$

or

$$\frac{dp_w}{dx} = \frac{1}{M} \frac{dp_o}{dx}$$

If x is the fraction of the total length of the system flooded by water and Δp the total pressure drop across the length, which remains constant and is assumed the same for all layers, then

$$\frac{\mathrm{d}p_{\mathrm{w}}}{\mathrm{d}x} = \frac{\Delta p - x\dfrac{\mathrm{d}p_{\mathrm{w}}}{\mathrm{d}x}}{M(1-x)}$$

which may be expressed as

$$\frac{\mathrm{d}p_{\mathrm{w}}}{\mathrm{d}x} = \frac{\Delta p/M}{Ax+1} \tag{5.54}$$

where

$$A = \frac{1}{M} - 1 \tag{5.55}$$

Then applying equation 5.53 to the particular layer

$$v' = \frac{\mathrm{d}x}{\mathrm{d}t} = \frac{kk'_{\mathrm{rw}}}{\mu_{\mathrm{w}}\phi(1 - S_{\mathrm{or}} - S_{\mathrm{wc}})} \frac{\mathrm{d}p_{\mathrm{w}}}{\mathrm{d}x}$$

$$\frac{\mathrm{d}x}{\mathrm{d}t} = \text{constant} \times \frac{kk'_{\mathrm{rw}}}{\phi \Delta S_{\mathrm{w}}} \times \frac{1}{Ax+1} \tag{5.56}$$

which on integration gives:

$$\text{constant} \times t = \frac{\phi \Delta S_{\mathrm{w}}}{kk'_{\mathrm{rw}}} \left(\frac{Ax^2}{2} + x \right)$$

which is an expression that is valid for all layers. Consequently, when the ith layer of the section has just flooded with water, the position of the front in the jth layer, still to flood, can be calculated as

$$\tfrac{1}{2}Ax_j^2 + x_j = \frac{\lambda_j}{\lambda_i}\left(\tfrac{1}{2}A + 1\right) \tag{5.57}$$

where

$$\lambda = \frac{kk'_{\mathrm{rw}}}{\phi \Delta S_{\mathrm{w}}}$$

As each layer floods, the frontal positions in all the remaining unflooded layers can be calculated by solution of the quadratic equation, 5.57; the order in which the layers flood being predicted in decreasing sequence of λ, as for Stiles type displacement. If $M = 1$, then $A = 0$ and equation 5.57 reduces to $x_j = \lambda_j/\lambda_i$ which is appropriate for Stiles constant velocity displacement. The final step is to determine directly an expression for the fractional flow, which for a section of fixed width, w, and individual layer thickness, h_i, may be evaluated using equation 5.56 as:

$$f_w = \frac{\displaystyle\sum_{i=1}^{n} \frac{\lambda_i h_i}{A+1}}{\displaystyle\sum_{i=1}^{N} \frac{\lambda_i h_i}{Ax_i+1}} \sim \frac{\displaystyle\sum_{i=1}^{n} \frac{k_i h_i}{A+1}}{\displaystyle\sum_{i=1}^{N} \frac{k_i h_i}{Ax_i+1}} \tag{5.58}$$

which is the condition pertaining after the nth layer out of a total of N has flooded: the numerator representing flow in the flooded layers and the denominator across the total section. Considering the predominance of the kh-product over other parameters, the second expression is usually a perfectly acceptable approximation. The corresponding thickness averaged water saturation can be calculated using the averaging procedure, equation 5.30. This interpretation does not yield pseudo-relative permeabilities but if these are required as input to a numerical simulation model, for instance, they may be obtained from the fractional flow. That is, a concave downward shaped water curve is assumed from which the oil curve may be obtained by solution of equation 5.18. Exactness is not required since they are relative functions. The resulting fractional flow relationship is used in vertical sweep calculations, as described in section 5.5b. The Dykstra-Parsons method is illustrated in Exercise 5.5.

Refinements to the basic method of Dykstra-Parsons have been described [22,23] but it is suggested that in attempting to deal with more complex flooding problems, such as when there is suspected partial cross-flow between layers, the engineer should use cross-sectional numerical simulation modelling directly. In spite of the generally unfavourable displacement efficiency in deltaic type reservoir environments, the vertical sweep calculated by methods such as Stiles and Dykstra-Parsons are invariably optimistic. This is because in their derivation it is assumed that the pressure drop across each of the sands is the same and remains constant. But if a reservoir section is depleted prior to initiating the water injection, as is usually the case, then there will be differential depletion between the layers with those of high flow capacity having the lower pressures. This can be directly observed in RFT surveys conducted in development wells and, in fact, is the best indicator that there is little or no cross-flow in the section (Fig. 5.58). The effect of injecting water into such a differentially depleted section is that most of the water will be preferentially channelled into the lower pressure, high flow capacity sands at the expense of the poorer sands in the section. The velocity of water entry into a sand will be proportional to $k\Delta p = k(p_{wf} - \overline{p})$, where p_{wf} is the wellbore injection pressure and $\overline{p}$ is the average pressure in each, which will vary. The overall effect is to exacerbate the severity of the permeability distribution leading to accelerated breakthrough and, in general, a more severe fractional flow of water. Nor is equilibrium necessarily attained as the flood progresses for, as mentioned in section 5.2e, it is very difficult to restore pressures in high-permeability sands. Therefore, to add the requisite harshness to the fractional flow, it is suggested, as an approximation, that the layer permeabilities, k_i, be multiplied by the ratio $\Delta p_i / \Delta p_{max}$, in which Δp_{max} is the maximum pressure drop observed in any sand. This measure will adversely affect the permeability distribution and consequently the fractional flow. It is not worthwhile trying to be more exact since the pressure distribution observed in the initial RFT will vary during the flood in a manner which is difficult to determine with accuracy. Multi-rate PLT surveys, described in Chapter 4, section 4.20c, will prove helpful in monitoring pressure changes in sands but lack the accuracy and resolution of RFT surveys. The flooding performance of such complex sand sections is often not even amenable to accurate description using numerical simulation modelling: the problem being that you cannot model that which cannot be seen. A simple means of history matching such a field is described in section 5.9b.

(e) Well workovers

Following water breakthrough in the higher flow capacity sands in a deltaic type section it is always possible to isolate these zones if it proves economically viable to do so. This may be affected by cement squeeze, casing patches or straddle packing across the offending intervals. If the last of these, then safe packer seats should be left between the individual sands when initially perforating. Such remedial workovers are usually performed only in production wells, not the injectors. These may each be supporting several wells in which water has not yet broken through in the higher flow capacity sands and therefore to exclude them in the injection well would be counter productive. Besides, one is never sure whether or not water entering the better sands in an injection well somehow connects with and sweeps oil in the poorer sands remote from the well. In this respect, there is a degree of randomness in sand connections which dictates that the completion policy should be one of blanket perforation in both injection and production wells to maximise the probability of water finding its way, by whatever tortuous route, into the lower permeability sands. Attempting to be cautious and not perforate the better sands initially, to defer water production, can lead to incalculable loss of recovery. If a workover is successfully accomplished in a production well, then the pseudos and fractional flow will have to be re-evaluated, excluding the isolated sands.

Workovers in this type of reservoir environment stand every chance of success for the simple reason that the sands are physically separated from one another, meaning that they can also be mechanically isolated in the wellbore. In the case of continuous sand bodies subjected to edge waterdrive under the VE condition, however, there is no guarantee of success associated with well workovers. Premature water breakthrough may be anticipated, for instance, in a coarsening downward sand section, as depicted in Fig. 5.39b, but if the lower part of the sand is isolated in the wellbore then production will be from the upper, lower permeability interval. Consequently, the producing pressure drawdown will be higher than from the whole section and this will induce water coning upwards from the flooded basal zone into the restricted perforations. This occurs quite rapidly and frequently the net oil rate is not enhanced by such workovers. It is only in thick, continuous sand sections, and especially those subjected to basal waterdrive, that systematic plugging back of watered out perforated intervals can significantly inhibit water production.

Exercise 5.4: History matching and prediction of a waterdrive field performance using the method of Stiles

Introduction

The watercut development of a field is history matched over a six and a half year period. Extrapolation of the trend in the watercut development then permits the generation of a production profile over a ten year period applying the material balance method described in section 5.3c.

Question

The production history of an offshore waterdrive field with a STOIIP of $N = 190$ MMstb is listed in Table 5.14 and the watercut development plotted in Fig. 5.62. As can be seen, it is very harsh, rising to almost 80% for a fractional oil recovery of less than 30% STOIIP. This was attributed to the deltaic depositional environment

TABLE 5.14

Six and a half years of field production history

Time (years)	Oil rate (stb/d)	Cumulative oil recovery (MMstb)	Fractional recovery	Watercut (fraction)
1	8800	1.606	0.008	–
	25560	6.271	0.033	0.020
2	23400	10.541	0.055	0.172
	18350	13.890	0.073	0.260
3	35780	20.420	0.107	0.292
	41820	28.052	0.148	0.366
4	36200	34.659	0.182	0.497
	26730	39.537	0.208	0.595
5	24240	43.961	0.231	0.657
	20530	47.707	0.251	0.683
6	15790	50.589	0.266	0.708
	15370	53.394	0.281	0.742
7	12590	55.691	0.293	0.778

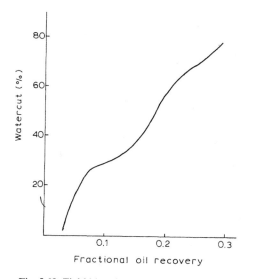

Fig. 5.62. Field historic watercut development.

and the harshness of the permeability distribution observed in the only cored well in the field which is plotted in Fig. 5.63. The total section has been divided into 8 layers, as indicated, and the core and petrophysical layer data are listed in Table 5.15. Although no RFT surveys were run to provide dynamic pressure profiles, it is strongly suspected that there is a lack of pressure communication between the layers based on both direct observation of the section and the severity of the watercut development. It is intended to continue with the waterdrive at near initial pressure using the following equipment capacities:

Injection:	60000–65000 b/d
Water disposal:	60000 b/d
Separator train:	70000 stb/d

Other data required in the exercise are:

Relative permeabilities:	$k'_{ro} = 1$,	$S_{or} = 0.28$ PV
(one normalized set)	$k'_{rw} = 0.39$	
PVT (at initial pressure):	$B_o = 1.255$ rb/stb,	$B_w = 1.0$ rb/stb
	$\mu_o = 1.0$ cp,	$\mu_w = 0.3$ cp

- Generate a 10 year oil production forecast for the field: keeping the rate constant until the end of the seventh year and applying full injection capacity thereafter.

Solution

If it is assumed that there is no cross-flow of fluids between the 8 defined layers, then it is perfectly acceptable to apply the method of Stiles to this problem since the mobility ratio may be calculated from the input data as $M = 1.3$. The first step is to re-order the 8 layers in the sequence that they will flood which is arranged in terms of decreasing values of $k/\phi \Delta S$ (equation 5.29), in which $S_{or} = 0.280$ PV, as assessed in the laboratory flooding experiment. The flooding order is listed in Table 5.15 and as can be seen it is not very different from that obtained by merely arranging the layers in decreasing sequence of their permeabilities. The flooding order has been transferred to Table 5.16, in which the pseudos and fractional flow are generated using equations 5.30–5.32 and 5.18 for horizontal displacement.

The functions are plotted in Fig. 5.64 and have the characteristic shapes referred to in section 5.7c. In particular, the fractional flow reveals an extremely harsh initial increase: precisely as observed in the field. In fact, the point marked "X" on the fractional flow corresponds to the state of flooding in the field, indicating that only a few of the higher permeability layers have flooded and most of the movable oil in the tighter sections still remains to be recovered. The fractional flow is concave downwards across the full movable saturation range implying that all saturations can move independently and must be included in the Welge recovery calculations (section 5.5b). These are presented in Table 5.17 in which small saturation increments of 1% have been used over most of the movable saturation range for increased accuracy. Values of $\overline{S}_{we}$ and $\overline{f}_{we}$ in columns 6 and 7 are the averages at which W_{id} (PV) has been evaluated and are used in the Welge equation,

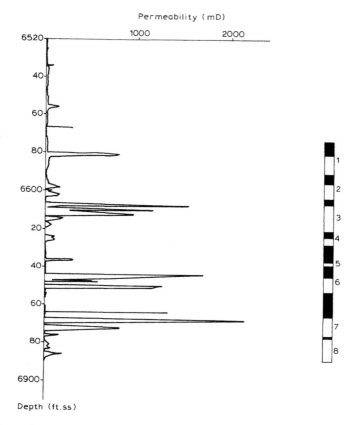

Fig. 5.63. Permeability distribution across the deltaic sand section in the only cored well in the field.

TABLE 5.15

Division of the section (Fig. 5.6) into 8 layers and assignment of a flooding order

Layer No.	h (ft)	k (mD)	ϕ	S_{wc}	$k/\phi/\Delta S$	Flood order
1	10	174	0.22	0.37	2260	5
2	8	103	0.18	0.41	1845	6
3	14	487	0.21	0.35	6268	3
4	4	73	0.22	0.42	1106	8
5	2	141	0.18	0.39	2374	4
6	8	904	0.21	0.34	11328	2
7	10	1223	0.20	0.32	15288	1
8	13	70	0.20	0.42	1167	7

$\sum h_i = 69$ ft
$\sum h_i k_i = 30328$ mD·ft: $\bar{k} = 440$ mD
$\sum h_i \phi_i = 14.1$ ft: $\bar{\phi} = 0.204$
$\sum h_i \phi_i S_{we_i} = 5.247$ ft: $\bar{S}_{wc} = 0.372$ PV

MOV $= (1 - S_{or} - \bar{S}_{wc}) = 0.348$ PV $= 0.348/(1 - 0.372) = 0.554$ HCPV.

TABLE 5.16

Generation of pseudos and fractional flow

Flood order	h (ft)	k (mD)	ϕ	S_{wc}	$\overline{S}_w$	$\overline{k}_{rw}$	$\overline{k}_{ro}$	$\overline{f}_w$
					0.372	0	1	0
1	10	1223	0.20	0.32	0.429	0.157	0.597	0.467
2	8	904	0.21	0.34	0.474	0.250	0.358	0.699
3	14	487	0.21	0.35	0.551	0.338	0.133	0.894
4	2	141	0.18	0.39	0.560	0.342	0.124	0.902
5	10	174	0.22	0.37	0.614	0.364	0.067	0.948
6	8	103	0.18	0.41	0.646	0.375	0.040	0.969
7	13	70	0.20	0.42	0.701	0.386	0.010	0.992
8	4	73	0.22	0.42	0.720	0.390	0	1

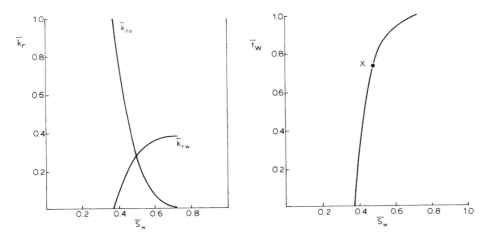

Fig. 5.64. Pseudo-relative permeabilities/fractional flow: 8-layer section.

equation 5.33. Since the flooding occurs close to the initial pressure, then from equation 5.34 it can be seen that $N_{pD} = N_p/N$ (HCPV): the vertical sweep. The final column is the surface watercut evaluated using equation 5.36.

The relationship f_{ws} versus N_p/N is plotted in Fig. 5.65 in comparison to the actual field watercut development. Initially, there is a mismatch between the two which reflects the fact that for the field data

$$\frac{N_p}{N} = E_v \times E_A \tag{5.28}$$

whereas for the theoretical calculations, $N_p/N = E_v$ ($E_A = 1$). From the disparity in the functions it appears that at breakthrough the areal sweep efficiency for the field as a whole was only about 20% but increased thereafter until at the end of the history matching period, when the actual and theoretical curves meet, it had risen to 100%. This results from the symmetrical nature of the flood pattern and the

TABLE 5.17

Welge calculations, history and prediction

$\bar{S}_{we}$	$\bar{f}_{we}$	$\Delta \bar{S}_{we}$	$\Delta \bar{f}_{we}$	W_{id} (PV)	$\bar{S}_{we}$	$\bar{f}_{we}$	$N_{pD} = N_p/N$ (HCPV)	f_{ws}
0.372	0							
0.400	0.242	0.028	0.242	0.116	0.386	0.121	0.185	0.147
0.425	0.445	0.025	0.203	0.123	0.413	0.344	0.194	0.396
0.440	0.542	0.015	0.097	0.155	0.433	0.494	0.222	0.551
0.450	0.593	0.010	0.051	0.196	0.445	0.568	0.251	0.623
0.460	0.640	0.010	0.047	0.213	0.455	0.617	0.262	0.669
0.470	0.681	0.010	0.041	0.244	0.465	0.661	0.280	0.710
0.480	0.721	0.010	0.040	0.250	0.475	0.701	0.283	0.746
0.490	0.757	0.010	0.036	0.278	0.485	0.739	0.295	0.780
0.500	0.786	0.010	0.029	0.345	0.495	0.772	0.321	0.810
0.510	0.813	0.010	0.027	0.370	0.505	0.800	0.330	0.834
0.520	0.837	0.010	0.024	0.417	0.515	0.825	0.344	0.855
0.530	0.857	0.010	0.020	0.500	0.525	0.847	0.365	0.874
0.540	0.874	0.010	0.017	0.588	0.535	0.866	0.385	0.890
0.560	0.902	0.020	0.028	0.714	0.550	0.888	0.411	0.909
0.580	0.923	0.020	0.021	0.952	0.570	0.913	0.447	0.929
0.600	0.940	0.020	0.017	1.176	0.590	0.932	0.474	0.945
0.620	0.952	0.020	0.012	1.667	0.610	0.946	0.522	0.956

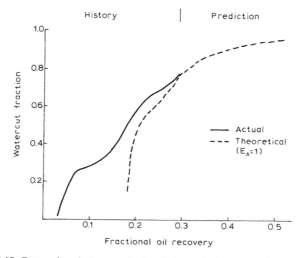

Fig. 5.65. Comparison between actual and theoretical watercut developments.

favourable mobility ratio. Altogether, the theoretical watercut trend is considered quite convincing in matching the field history and certainly explains why the field watercut had risen so sharply. Therefore, its theoretical extrapolation (dashed line, Fig. 5.65) is considered realistic to use in a predictive manner to generate a field production profile. This is done using the waterdrive material balance, equation 5.8, which, since $B_w = 1.0$ rb/stb is reduced to the form

$$q_{wi} = q_o \left(B_o + \frac{f_{ws}}{1 - f_{ws}} \right)$$

in which $B_o = 1.255$ rb/stb at the initial reservoir pressure.

Taking annual time increments the equation is solved iteratively, as described in Exercise 5.1, to predict the oil rate: subject to any capacity constraints in the topsides equipment. The results are listed in Table 5.18 and plotted in Fig. 5.66. The collapsing production profile at the end of the third year, due to excess water production, is reminiscent of the production profile for a solution gas drive field. This is followed by a lengthy period of slow decline which results from the flattening

TABLE 5.18

Ten years prediction: oil production forecast

Time (years)	q_o (stb/d)	N_p (MMstb)	N_p/N	f_{ws}	q_{wi} (b/d)	q_{wp} (b/d)	q_{sep} (stb/d)
7.5	12600	57.991	0.305	0.783	61300	45500	58100
8	11800	62.298	0.328	0.810	65000	50200	62000
9	9600	65.802	0.346	0.846	64800	52700	62300
10	8400	68.868	0.362	0.866	65000	54300	62700
11	7700	71.679	0.377	0.878	65000	55400	63100
12	6900	74.198	0.391	0.884	65000	56400	63300
13	6300	76.498	0.403	0.900	64600	56700	63000
14	5900	78.652	0.414	0.907	64600	57200	63100
15	5300	80.587	0.424	0.916	64400	57800	63100
16	5000	82.412	0.434	0.921	64600	58300	63300
17	4700	84.128	0.443	0.926	64700	58800	63500

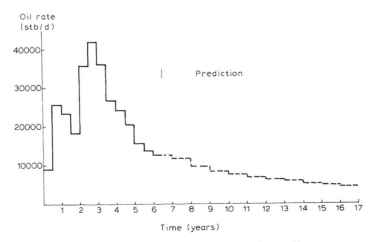

Fig. 5.66. Ten year prediction: oil production profile.

of the theoretical watercut development trend during the prediction period (Fig. 5.65). The production constraint is the maximum injection capacity of 65000 b/d but for the relatively small platform, there is little or no scope for increasing this capacity which would also affect the separator and water disposal equipment.

This practical example illustrates the main purpose in performing waterdrive calculations referred to in section 5.3c. That is, to generate a theoretical relationship between f_{ws} and N_p/N, preferably history match it against field performance, as in the present case, and use the relationship to predict future performance, subject to the constraints imposed by the surface equipment capacities.

Exercise 5.5: Dykstra-Parsons displacement calculations

Introduction

Water–oil displacement is considered in precisely the same sand section studied in the previous exercise but for mobility ratios different from unity. This necessitates the application of the Dykstra-Parsons method to allow for velocity dispersion of the fronts as the flood progresses.

Question
- Apply the Dykstra-Parsons method to generate fractional flow curves for the sand section whose permeability distribution is plotted in Fig. 5.63 and detailed in Table 5.15. Perform comparative calculations for favourable and unfavourable mobility ratios of $M = 0.2$ and $M = 5$, assuming that the relative permeability data from the previous exercise are relevant to both cases ($k'_{ro} = 1$, $k'_{rw} = 0.39$, $S_{or} = 0.280$ PV).

Solution

It is assumed, as before, that there is no pressure communication between the individual sands in the section and therefore, since the two mobility ratios of $M = 0.2$ and 5 are considerably different from unity, the Dykstra-Parsons method must be applied as described in section 5.7d, as follows.

$M = 0.2$ *(favourable).* The first step is to solve the quadratic

$$\tfrac{1}{2}Ax_j^2 + x_j = \frac{\lambda_j}{\lambda_i}\left(\tfrac{1}{2}A + 1\right) \tag{5.57}$$

in which, since $k'_{rw} = 0.39$ for all layers, $\lambda = k/\phi\Delta S_w$. The equation is solved each time water breakthrough occurs in a layer to calculate the fractional displacement of the flood fronts in the remaining unflooded layers. For example, if water has just broken through in layer 4 ($i = 4$, flooding order) then the quadratic must be solved for $j = 5, 6, 7, \ldots$ to determine the frontal positions in these layers. Since $M = 0.2$, then $A = 4$ (equation 5.55) which reduces the quadratic to

$$2x_j^2 + x_j - 3\frac{\lambda_j}{\lambda_i} = 0$$

TABLE 5.19

Calculation of relative frontal positions ($M = 0.2$)

Relative front position	λ (mD):	Layer flooding order						
		1	2	3	4	5	6	7
		15288	11328	6268	2317	1845	1167	1106
x_1		1						
x_2		0.833	1					
x_3		0.573	0.695	1				
x_4		0.288	0.358	0.535	1			
x_5		0.243	0.304	0.460	0.871	1		
x_6		0.171	0.216	0.335	0.645	0.756	1	
x_7		0.164	0.207	0.322	0.632	0.731	0.968	1

which may be solved for x as

$$x_j = \frac{-1 + [1 + 24(\lambda_j/\lambda_i)]^{\frac{1}{2}}}{4}$$

Values of λ have been taken from Table 5.15 and are presented in Table 5.19 in the correct flooding order. Since the values of λ in layers 1 and 5 (original layering) are similar, the two have been aggregated to give a single layer which is fourth in the flooding order: reducing the layers from 8 to 7.

As an example of how the elements in Table 5.19 are calculated; after breakthrough in layer 1 ($i = 1$) the frontal position in layer 4 ($j = 4$) is

$$x_4 = \frac{-1 + \left(1 + 24 \times \dfrac{2317}{15288}\right)^{\frac{1}{2}}}{4} = 0.288$$

The next step is to directly calculate the fractional flow of water as

$$\overline{f}_w = \frac{\displaystyle\sum_{i=1}^{n} \frac{k_i h_i}{A + 1}}{\displaystyle\sum_{i=1}^{N} \frac{k_i h_i}{A x_i + 1}} \tag{5.58}$$

in which the numerator is summed over the n layers that have flooded at any stage and the denominator over all the layers. The calculations are set out in Table 5.20, in which the elements in the matrix are values of $k_i h_i/(A x_i + 1)$ The flooded layers lie above the dividing line and the unflooded, for which $x < 1$ below. The fractional flow is then the sum of the elements above the line divided by the sum of all the elements in the column. The average saturation corresponding to the fractional flow is calculated using equation 5.30 as the layers successively flood and is the same as presented in Table 5.16. The resulting fractional flow curve is plotted in Fig. 5.67.

TABLE 5.20

Calculation of the Dykstra-Parsons fractional flow ($M = 0.2$)

Layer	kh (mD.ft)	Layer flooding order						
		1	2	3	4	5	6	7
1	12230	2446	2446	2446	2446	2446	2446	2446
2	7232	1669	1446	1446	1446	1446	1446	1446
3	6818	2071	1804	1364	1364	1364	1364	1364
4	2022	940	831	644	404	404	404	404
5	824	418	372	290	184	165	165	165
6	910	540	488	389	254	226	182	182
7	292	176	160	128	83	74	60	58
Fractional flow — $\overline{f}_{\mathrm{w}}$		0.296	0.516	0.784	0.916	0.951	0.990	1
$\overline{S}_{\mathrm{w}}$ ($\overline{S}_{\mathrm{wc}} = 0.372$ PV)		0.429	0.474	0.551	0.614	0.646	0.701	0.720

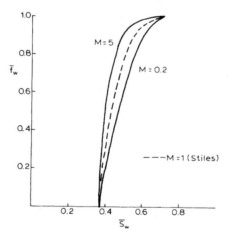

Fig. 5.67. Dykstra-Parsons ($M = 0.2, 5$) and Stiles ($M = 1$) fractional flow curves for the deltaic sand section Fig. 5.63.

$M = 5$ *(unfavourable)*. In this case, $A = -0.80$ and equation 5.57 is reduced to

$$-0.4x_j^2 + x_j - 0.6\frac{\lambda_j}{\lambda_i} = 0$$

which may be solved for x as

$$x_j = \frac{1 - [1 - 0.96(\lambda_j/\lambda_i)]^{\frac{1}{2}}}{0.80}$$

    The relative frontal positions as the flood proceeds are listed in Table 5.21 and the fractional flow is calculated in Table 5.22 and plotted in Fig. 5.67.

    The effect of velocity dispersion between the flood fronts in the isolated layers is evident by comparing the fractional flow curves with that of Stiles from the previous

TABLE 5.21

Calculation of relative frontal positions ($M = 5$)

Relative front position	λ (mD):	Layer flooding order						
		1	2	3	4	5	6	7
		15288	11328	6268	2317	1845	1167	1106
x_1		1						
x_2		0.578	1					
x_3		0.277	0.394	1				
x_4		0.095	0.129	0.246	1			
x_5		0.075	0.102	0.191	0.643	1		
x_6		0.047	0.063	0.117	0.352	0.467	1	
x_7		0.044	0.060	0.111	0.330	0.436	0.875	1

TABLE 5.22

Calculation of the Dykstra-Parsons fractional flow ($M = 5$)

Layer	kh (mD.ft)	Layer flooding order						
		1	2	3	4	5	6	7
1	12230	61150	61150	61150	61150	61150	61150	61150
2	7232	13452	36160	36160	36160	36160	36160	36160
3	6818	8759	9956	34090	34090	34090	34090	34090
4	2022	2188	2255	2517	10110	10110	10110	10110
5	824	877	897	973	1697	4120	4120	4120
6	910	946	958	1004	1267	1453	4550	4550
7	292	303	307	320	397	448	973	1460
Fractional flow — $\overline{f}_w$		0.697	0.871	0.965	0.977	0.987	0.994	1
$\overline{S}_w$ ($\overline{S}_{wc} = 0.372$ PV)		0.429	0.474	0.551	0.614	0.646	0.701	0.720

exercise in which there was no dispersion ($M \sim 1$). For $M = 0.2$, the flood fronts are retarded to a greater extent in the higher flow capacity sands than in the low which tends to stabilize the macroscopic frontal advance, providing a more efficient vertical sweep. In the case of the unfavourable mobility ratio ($M = 5$) the effect is reversed: the fronts in the high flow capacity sands are accelerated with respect to the low which decreases the efficiency of the vertical sweep. It should also be noted that all the fractional flows, even for $M = 0.2$, are concave downwards across the movable saturation range implying independence of movement of all saturations which requires the inclusion of the entire spectrum in Welge calculations. This is the most characteristic feature associated with water oil displacement in isolated layers within a sand section.

5.8. THE NUMERICAL SIMULATION OF WATERDRIVE

It is not intended in this section to duplicate the detail contained in the many fine books on the subject of numerical simulation [16,24–26] but rather to comment on the broader aspects of the purpose and practice of simulation applied to waterdrive fields.

(a) Purpose

Application of numerical simulation provides the engineer with the means of quantifying areal effects in reservoirs by attempting to answer the questions of where is the injected water going to spread and why. Therefore, to be able to construct a reliable model for this purpose it is first necessary that it be capable of reproducing the production-pressure performance of the field to date. This initial calibration of the model, known as history matching (section 5.5a), acquires credibility in proportion to the maturity of the field, especially in waterdrive projects.

Considering the various stages in the development of a large offshore waterdrive field, such as in the North Sea (Fig. 5.68) where accurate prediction of performance is crucial, the reliability of history matching may be summarised as follows.

Appraisal

During this period, of course, there is simply no history to match and the number of unknowns is at its maximum. As noted in Chapter 3, the single material balance equation can have over 10 unknowns and simulation modelling introduces more: such as the requirement for geometry and the need for relative permeabilities. Therefore, the attainment of a unique and correct result from a simulation study at this stage is a little too much to expect. Nor does it help that during appraisal, information is only gathered under static conditions giving no indication of areal or vertical communication within the reservoir sections penetrated; knowledge which is essential in the planning of a waterflood. It is therefore hardly surprising that mistakes were made in simulation studies at this early stage in several North Sea

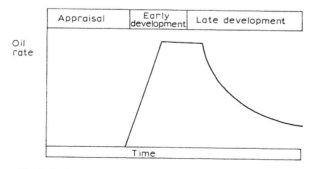

Fig. 5.68. Production profile for an offshore waterdrive field.

fields. These had nothing to do with simulation itself but resulted instead from
failure to observe correctly the severe degree of heterogeneity across many of the
sand sections and incorporate it realistically in models. Two examples of this type of
oversight are provided in section 5.10. The main aims in simulation at the appraisal
stage are to generate long term production profiles to assess project viability but of
even greater importance, as described in section 5.3c, is the realistic assessment of
the watercut development since this is the key to sizing the capacities of the topsides
production/injection equipment.

Early development

During the rate build up and early part of the plateau production period,
data are collected from the reservoir under dynamic conditions for the first time.
Unfortunately, quite often the rate of data collection exceeds the rate at which
it can be assimilated into numerical simulation models, which require rapid and
frequent updating. In this respect, it is the phase when engineers should be
able to "think in their feet" in making decisions, such as the locations of future
development wells, and when and where to initiate injection. Although simulation
modelling proves cumbersome at this frenetic stage to be of much assistance in
development decision making (but the situation improves with each passing year) it
is nevertheless the period when some of the most important data of all are being
collected to ensure the reliable history matching of a model. These are the dynamic
RFT survey results from each new development well revealing the degree of areal
and vertical communication in the reservoirs (section 5.2d). These are the highest
quality pressure data that will ever be recorded in the field and it is therefore
imperative that the engineer match the results from this initial pressure drop on a
layer by layer basis in each well that has been surveyed. Following this, pressure is
maintained at a near constant level by the water injection providing little of interest
to be of use in calibrating the model. The plateau production period represents an
equilibrium state in which any new oil is balanced by the onset of water production
which must be history matched in the model. Therefore, while simulation modelling
may not be at the fore-front in the decision making during the early development
stage, this is nevertheless the period when the important ground work can be done
to construct a reliable model for the final phase of development.

Late development

This consists of the latter half of the plateau period and the decline towards
eventual abandonment. Quite naturally, it is during this period when simulation
models should be at their most useful. By this time, water has spread to most parts
of the field and the simulator must be "adjusted" to match both the timing of
water breakthrough in individual wells and their subsequent watercut development
history. The greater the number of wells that can be accurately history matched on
water production the more reliable the model should be for predictive purposes.
The main objectives at this stage are the prolongation of the plateau production
for as long as possible and the attempt to arrest the rate of decline. This can
be accomplished by drilling additional wells together with the sidetracking and

recompletion of existing wells once they have watered-out. This may not be the most glamorous activity during the development of a field but it can prove to be very rewarding financially.

Implicitly, as described in section 5.5a, the simulator is being used to solve the equation

$$\frac{N_p}{N} = E_v \times E_A \qquad (5.28)$$

in which N_p/N represents the production history and is therefore a known quantity. On the right-hand side, E_v, the vertical sweep can be much more accurately calculated than the areal sweep E_A, for which the equation (simulation model) is solved. A persistent problem in modelling, however, is that if engineers do not take sufficient care in using all the core and log data correctly in the calculation of E_v, then the result must be that E_A will be incorrectly assessed to match the observed production history. This is simply a case of two wrongs equalling one right! There is a common tendency to overestimate the vertical sweep efficiency, E_v, for the reasons stated in section 5.6b namely, the use of petrophysical correlations to generate permeability distributions and the adherence to the tradition of plotting permeability distributions with a log-k rather than a linear scale, both of which can lead to an incorrect smoothing of the true heterogeneity. This, in turn, means that E_A must be somehow reduced to match the production history. The easiest way to accomplish this is by introducing fictitious sealing faults in the model to channel and steer the water around the field where the unrealistically smooth vertical permeability distribution has failed to do so. The consequences of this type of error can be quite expensive. It is not uncommon that operators will drill or sidetrack a well into a hypothetical fault block confined by such faults, only to find that the water has, in fact, already reached the location due to the lack of any real physical barrier and the new well commences production at a watercut over 80%. Because of this the engineer must be particularly careful in locating faults in simulation models, as a convenience, and it should only be done following expert geological advice. Numerical simulation modelling offers far too many degrees of freedom to the unwary and by far the best way of "tuning" the areal unknowns is to make sure that the vertical sweep, E_v, has been correctly accounted for in the first place. This usually amounts to the generation of physically realistic pseudo-relative permeabilities for input to models, as described in earlier sections of the chapter.

(b) Generation of pseudo-relative permeabilities using cross-sectional simulation modelling

The aim in generating pseudo-relative permeabilities for models is to reduce the vertical dimension from two (multi-layered) back to preferably one (single-layered) which will obviously reduce the running cost of the model. In doing so, however, the engineer must make quite sure that no lack or reality in the displacement mechanics is introduced into the simplified model (Fig. 5.69).

A two-dimensional cross-sectional model is structured with grid blocks of small

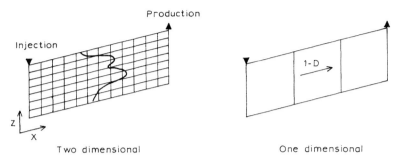

Fig. 5.69. Two-dimensional model run to generate pseudos for an equivalent one-dimensional model.

dimensions in both the x and z directions. The layering should be sufficiently refined to cater for all the vertical heterogeneity, particularly the permeability variation, in the cored wells which are being represented in the model. The width is restricted to one grid block of arbitrary dimension but the injection rate must be scaled to represent a fixed fraction of the total STOIIP in the section per annum, say 8%, which should be commensurate with the intended recovery rate of the field as a whole. In the simplest form of generating pseudos, at the end of each time step in the run the simulator post processor evaluates for each column of grid blocks the following averaging procedures to calculate the thickness averaged water saturation and pseudo-relative permeabilities across the N layers:

$$\overline{S}_{\mathrm{w}} = \frac{\displaystyle\sum_{1}^{N} h_i \phi_i S_{\mathrm{w}_i}}{\displaystyle\sum_{1}^{N} h_i \phi_i}$$

$$\overline{k}_{\mathrm{rw}} = \frac{\displaystyle\sum_{1}^{N} h_i k_i k_{\mathrm{rw}_i}(S_{\mathrm{w}_i})}{\displaystyle\sum_{1}^{N} h_i k_i}$$

$$\overline{k}_{\mathrm{ro}} = \frac{\displaystyle\sum_{1}^{N} h_i k_i k_{\mathrm{ro}_i}(S_{\mathrm{w}_i})}{\displaystyle\sum_{1}^{N} h_i k_i}$$

These, it will be recognised, are analogous to equations 5.30–5.32 used to generate analytical pseudos, except that they do not cater for the either–or situation that a layer is either completely flooded or it is not flooded at all, there being no intermediate condition. Thus in calculating the averaged water saturation, the numerator contains one summation across all the layers and S_{w_i} is simply the water saturation in the ith layer of the column whether the irreducible level, S_{wc}, the flood-out saturation, $1 - S_{or}$, or some intermediate value. It is the last that can lead to inaccuracy in a great many simulation studies. If the mobility ratio is low and favourable ($M \leq 1$), as is so often the case, then if full rock relative permeability curves defined across the total movable saturation range are input to the model there must inevitably be some numerical dispersion of water between grid blocks, as described in section 5.4f. Therefore, some of the grid block saturations, S_{w_i}, may include this unrealistic dispersed water. Not only that, but the individual grid block relative permeabilities, $k_{rw_i}(S_{w_i})$ and $k_{ro_i}(S_{w_i})$, are functions of the dispersed water saturation, related through the input rock relative permeability curves, which will lead to artificial estimates of water and oil mobilities. The degree of distortion resulting from numerical dispersion will be dependent on the physical nature of the flood: cross-flow (VE) or a total lack of it and the construction of the numerical simulation cross-sectional model itself.

Cross-flow (VE)

Two extreme reservoir conditions of coarsening upward and downward in rock properties are depicted in Fig. 5.70, together with the pseudos and fractional flow relationships that may be anticipated for each. Case (a), as described in section 5.6b (Fig. 5.39a), should provide a perfect, piston-like displacement across the entire sand. Provided that a reasonable number of grid blocks are structured in both the x and z directions, then even if full rock relative permeabilities are input to the model there will be very little numerical dispersion. This is for the reason explained in section 5.4f, that dispersion in the x direction will be compensated by dispersion in the z direction giving the overall effect of a piston front as the water slumps from the high-permeability layers to the base of the section. This will result in the generation of realistic pseudos in the simulation and a fractional flow that is concave upward across the entire movable saturation range. An example of such displacement is provided in Exercise 5.3 (Fig. 5.56b, $M = 1$). Alternatively, case (b), coarsening downward, can yield an unfavourable form of displacement, even if the mobility ratio is low (Exercise 5.3, Fig. 5.56a, $M = 1$). This is manifest in the concave downward shape of both the water pseudo-relative permeability and fractional flow across the entire movable saturation range. Numerical dispersion can be more significant in modelling this type of formation. There is a strong tendency for the water to slump to the base of the reservoir (Fig. 5.39b) where it is held by gravity. Therefore, at the base of the reservoir there can be a significant dispersion in the water tongue in the x-direction with no compensating vertical dispersion. Therefore, to generate realistic pseudos for this type of formation it is necessary to structure the model with smaller grid blocks vertically than for case (a) either for the entire cross-section or at least for the basal high-permeability layers. Naturally, this will only be required if the mobility ratio is reasonable favourable so that there is a degree of shock-front displacement on the microscopic scale. Alternatively, truncated rock relative permeabilities can be input to the simulation, as shown in Figs. 5.30 and 5.31: the degree of truncation being dependent on the mobility ratio.

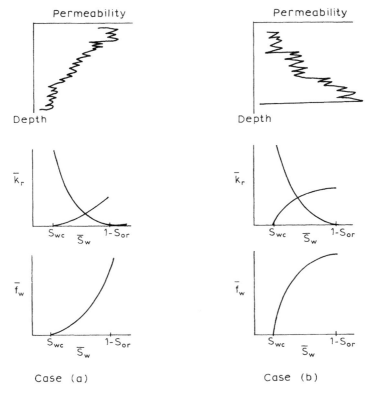

Fig. 5.70. VE pseudos and fractional flows resulting from displacement in (a) coarsening upward, (b) coarsening downward in rock properties.

No cross-flow

In modelling a reservoir section such as shown in Fig. 5.63, consisting of physically isolated layers, the numerical dispersion can be very significant for favourable mobility ratio displacement ($M \leq 1$) using full rock relative permeabilities: if each sand is modelled with just one layer in the model. This is again because there can only be numerical dispersion in the x-direction, there will be none vertically between the separate layers. It can be overcome by structuring the model so that each sand is modelled with fine grid blocks in the x and z directions or by using truncated rock curves as input to cater for the microscopic shock-front displacement. Either method, however, can prove costly in computer time.

It will be clear, therefore, that at the start of any cross-sectional simulation representing flooding in which there is some element of shock-front displacement on the microscopic scale, it is necessary for the engineer to consider carefully to what extent numerical dispersion will influence the results and take steps to reduce or eliminate the effect. Saturation distributions should be examined carefully to inspect whether forbidden saturations in the range $S_{wbt} < S_w < 1 - S_{or}$ are separately mobile in the model. Alternatively, the engineer may opt for the generation of pseudos analytically using the simple techniques described in sections 5.6 and 5.7. Experience shows that they invariably reproduce the results obtained using fine gridded models and even if the simulation approach is preferred the analytical pseudos should also be generated as a check and to increase the engineer's awareness of the physics of the displacement. Any significant difference in results obtained by the different approaches should be carefully scrutinized for it really should not occur.

For "in-between" cases, that is when there is neither vertical equilibrium nor a total lack of it, Jacks et al. [27] have presented a simulation method for generating dynamic pseudo-relative permeabilities to cater for the sluggish attainment of vertical equilibrium in reservoirs with restricted vertical permeability. This consists of treating the displacement as occurring as a sequence of VE states which vary with time. An analytical method of generating pseudos for this condition is presented in the two examples in section 5.10.

The next important step is the use of the pseudos generated by simulation or analytically: first of all in the reduced one-dimensional cross-sectional model (Fig. 5.69) and then in the two-dimensional areal simulation model structured with just one layer. Inputting the pseudos to the one-dimensional cross-section is used as a validation step. That is, for the same injection rate, the timing of water breakthrough in the production well and subsequent rate of watercut development should be the same for the fine and coarse gridded cross-sections. If they do not match, then the pseudos are often adjusted until there is correspondence.

From inspection of the two fractional flow curves in Fig. 5.70 and the comments made in section 5.4f, it will be apparent to the reader that it is not always possible to use the pseudos generated as input to a one layer model and attain physically realistic results. In the case that the displacement efficiency is poor, due either to unfavourable mobility ratio or the adverse effect of heterogeneity-gravity, then the use of pseudos in reduced dimensional models is perfectly acceptable. The act of generating pseudo-relative permeabilities is to reduce problems to one dimension, that is to the scale of a core flooding experiment in which Buckly-Leverett displacement mechanics can be applied directly. If the displacement is unfavourable, the one-dimensional fractional flow will be concave downwards across the movable saturation range (Fig. 5.70b) meaning that all saturations have mobility and the full spectrum is used in Welge calculations of vertical sweep (Exercise 5.2, $M = 30$, 3, Fig. 5.51; and Exercise 5.3, $M = 1$, 16.1, Fig. 5.56a). Therefore, inputting the pseudos to the model will produce realistic results because they too are defined across the full movable saturation range.

If the displacement is ideal on account of low mobility ratio and/or the helpful influence of heterogeneity-gravity then the situation is quite different and the pseudos cannot be used in a reduced one-dimensional model. This is because the one-dimensional fractional flow (Fig. 5.70a) is concave upwards across the entire movable saturation range, meaning that the saturations cannot move independently, they are all caught-up in the piston-like shock-front (section 5.4e). Examples of such displacement are presented in Exercise 5.2, $M = 0.48$, Fig. 5.51 and Exercise 5.3, $M = 1$, Fig. 5.56b. Therefore, inputting the pseudos, which are independently defined across the full saturation range (Fig. 5.70a), is simply informing the simulation model that all saturations are capable of individual movement which of course is a distortion of the physics of displacement and can lead to significant numerical dispersion. The paradoxical situation therefore arises that the worse the displacement efficiency the more capably will simulators model the physics and, conversely, the more ideal the flooding condition the poorer will be the simulated result using one-dimensional models.

If ideal pseudos, such as depicted in Fig. 5.70a, are input to a single layer simulation model without checking, then it can lead to engineers inserting artificial barriers, both areally and vertically, to inhibit the flow and production of dispersed water. Another means of reducing the dispersion, whether for ideal, piston-like or any degree of shock-front displacement, is to use a "hold-up" on the water pseudo-relative permeability, similar to that shown in Figs. 5.31b and c, which prevents its movement until a selected saturation is exceeded. But the best way to model a reservoir in which the displacement is favourable is not to use pseudos at all. Instead a full, multi-layered three-dimensional model should be structured. This will be expensive but since it is required to model the "perfect" condition then presumably there will be money in the bank to pay for the luxury.

Many errors associated with the input and use of pseudos in models would never arise if engineers would adopt the practice of always drawing the fractional flow before ever attempting to use a set of relative permeabilities, whether they are laboratory rock curves or pseudos. Numerical simulation models disregard the concept of fractional flow altogether and therefore the onus rests with the engineer to compensate for this deficiency, which amounts to honouring Newton's third law of motion.

Another difficulty associated with the transition from detailed two-dimensional modelling to a one-dimensional equivalent arises from the difference in length of the grid blocks in the x-direction. In Fig. 5.69, for instance, one block in the one-dimensional model is represented by four in the fine gridded model. The use of such large blocks exacerbates the numerical dispersion and a refined method of overcoming this difficulty has been described by Kyte and Berry [28]. In a simpler version, if four grid blocks in the x-direction are going to be replaced by one, then the pseudo-relative permeabilities computed in the detailed model in the fourth column of cells, that is the furthest downstream, should be input to the one-dimensional model as a function of the increasing average saturation in the four columns of cells in the fine gridded model. This will cause a hold-up in the transfer of water in the one-dimensional model thus reducing the numerical dispersion.

In determining average porosity/permeability for input to the one-dimensional model provided there a reasonable degree of sand continuity arithmetic averaged values are used. On the other hand if it is considered that the distribution of formation properties is quite random then there is justification in using a geometric average [1] permeability. It is not of critical importance which is used, however, since as described in section 5.6b, the absolute value of permeability seldom appears in the important equations of waterdrive. Its significance is as a weighting factor in calculations and therefore it is the permeability distribution, both vertically and areally that matters and this is not necessarily affected whether arithmetic or geometric averaging is applied.

(c) Areal numerical simulation modelling

Whether performed with a three-dimensional, multi-layered model or two-dimensional version using pseudos, the main aim in such modelling is first of all to track and subsequently to predict the areal movement of water. To do this it is essential to model the vertical sweep accurately, either by having sufficient layers to incorporate all the heterogeneity or by generating realistic pseudos which cater for the heterogeneity implicitly. When using pseudos, the initial phase of the study is usually to pick one, or possibly two, linear sections through the reservoir intersecting wells that have been thoroughly appraised by coring and logging. The pseudos generated in this exercise are then used throughout the entire reservoir. It is suggested, however, that this approach of selecting a particular section is imposing a bias on the direction of fluid movement which may not, in reality, coincide with the orientation of the selected section. Furthermore, the use of layer data from just the few wells excludes consideration of equally valid data collected in other parts of the reservoir. It is suggested as an alternative that if, for instance, seven wells have been cored and logged over the full areal extent of the reservoir, then pseudos should be generated for each of them either analytically or using detailed cross-sectional modelling using the layer data from the well being studied alone. These pseudos implicitly represent the efficiency of the flood as water passes the well location whose data was used in their generation. This approach not only utilizes all the data from wells throughout the field but is also free from directional bias. It will then be necessary, however, to assign areas of influence to pseudos generated in the individual wells. This should be done taking geological advice particularly on the depositional environment and its influence on permeability distributions.

The most difficult type of reservoir to model is that described in section 5.7 which usually results from deltaic deposition of sediments. Not only are the pseudos themselves difficult to generate but the main problem is the general lack of correlation of individual sands between wells which makes the construction of a detailed layered model extremely difficult. Yet in many fields of this type where dynamic RFT surveys have been run (section 5.7a) it is usually noted that all sands, even the tighter ones, display some degree of depletion, suggesting that they may be somehow connected, albeit in a random manner. In recent years, attention has been focused on the stochastic modelling of such reservoirs in which a degree of statistical randomness is attributed to sand-shale distributions between wells. Perhaps a simpler approach to the problem is not to attempt to construct a layered model in the first place. Instead, the above procedure is adopted of generating pseudos in all cored wells on a field wide basis. These are guaranteed to produce fractional flow curves which are concave downwards across the movable saturation range, even if $M < 1$ (Exercise 5.5, $M = 0.2$, Fig. 5.67), meaning that they can always be used in a reduced one layer model without reservation (section 5.8b): there will be no artificial dispersion of water. The main factor dictating the harshness of the watercut development in fields like this is invariably the severity of the permeability distributions which is catered for in the pseudos. This approach may seem an oversimplification but it is physically justified and often, the more complex the problem, the more convincing is the simple solution.

Finally, once pseudos of any type have been input to a model, the engineer should resist the temptation to alter them. In some texts on simulation, the suggestion is made that history matching be accomplished by varying the shapes and end-points of the pseudo-relative permeabilities. But built into the pseudos are all the expensively acquired well data which — like it or not, are the most reliable collected from the reservoir. The greatest uncertainty for both geologists and reservoir engineers is the variation in formation properties and presence of faults or discontinuities between wells. Therefore, in so far as it is possible, the history match should be achieved by varying the lesser known areal properties with the aim of solving equation 5.28 for the areal sweep, E_A.

5.9. THE EXAMINATION OF WATERDRIVE PERFORMANCE

As mentioned in the introductory chapter to the book, section 1.2d, the numerical simulation modelling of mature waterdrive fields: those that have reached the back end of their production plateaus or are in decline, should provide the most rewarding experience for the reservoir engineer in applying these sophisticated techniques. The reason is because by that stage of a field's development, water has usually broken through to most, if not all, of the producing wells. Therefore, if the timing of water breakthrough in each well can be areally matched together with their subsequent rates of watercut development then, in principle, it should be possible to obtain an extremely reliable history matched model to apply in the prediction of performance in order to determine what can best be done to bolster the production of the field during its later years of terminal decline. The possibilities include drilling new wells or sidetracking existing completions, recompletions, workovers, optimising the lift method and improving the efficiency of the surface facilities to both inject and produce fluids. The efficacy of these possibilities can be compared in a series of runs with the reliably calibrated numerical simulation model. So much for the principle, in practice it often does not quite work out like that. The two most common difficulties encountered being:

– There is simply too much historic data to be matched
– The study is too time consuming and, therefore, expensive.

Considering the first of these, the engineer can be completely overwhelmed by the sheer quantity of the input data collected over the years of production history even for a moderate sized field. This tends to provide too many degrees of freedom and consequently an apparent lack of uniqueness in solution: there may be several ways in which the historic water production can be matched for a given well. Because of this, the second difficulty arises in that it can easily take six months or more to achieve a satisfactory history matched model. This is a hopeless state of affairs because just as in the usual traumatic build up to the plateau production rate, section 1.2d, the decline can often be no less dramatic and significant development decisions have to be made within days rather than months. Under these circumstances the engineer must have at hand a reliable, physically sound method of assessing the history and future of the field which can be successfully applied in a matter of hours rather than months- thinking on one's feet. Furthermore, on account of the length of the study time taken with large simulation models they consume many engineering man hours and are therefore expensive. Upon this realisation the field operator often adopts the attitude that since the field is in decline anyway, then why bother with such elaborate and expensive studies and frequently the reservoir engineering A-team is moved on to work on the new field development, which is much more interesting. This overlooks the fact that there is often a lot of money to be made by bolstering the production from declining fields in which all the wells and surface facilities are in place.

Since reliable and physically sound methods of evaluating the performance of mature waterdrive fields are not abundant in the literature, operators usually resort to the application of what must surely be the most primitive technique ever introduced into the subject of reservoir engineering- decline curve analysis. In his now famous book on production engineering [*Principles of Oil Well Production*, McGraw-Hill, Inc. 1964], Professor T.E.W. Nind of the University of Saskatchewan wrote eight pages in chapter 1 on the subject of decline curve analysis, commencing with the following warning:

"When [performance] estimates are based on the mathematical or graphical techniques of production-rate-decline curve analysis, it should always be remembered that the analysis is merely a convenience, a method that is amenable to mathematical or graphical treatment, and it has *no basis in the physical laws* governing the flow of oil or gas through the formation."

Thereafter follow eight pages of learned description of the subject Concluding with the following remark.

"It must be reiterated that the production-rate-decline curves (exponential, harmonic or hyperbolic) are conveniences, enabling extrapolations of future well or field per-formances to be made. There is, however, *no physical basis* for these curves, and the production engineer must not feel surprised if his wells or pools do not follow the estimated production-rate-decline curves, no matter how carefully these may have been prepared.

So — no physics. Such comments are not very encouraging concerning the veracity of the decline curve analysis technique. When I first read the Professor's comments two thoughts occurred. Firstly, I need never apply this technique for the rest of my career and secondly, if the Professor felt such antipathy towards the application of decline curves, what were the intervening eight pages all about? Perhaps the worst aspect of all concerns professional pride. Have we reservoir engineers suffered all those years at college and university having had all that mathematics, physics chemistry and engineering painfully banged into our heads only to emerge into the oil industry merely to draw straight lines on semi-log plots- hopefully not. Finally, any kind of log or semi-log plot should *only* be used if the mathematics used to describe the physics of the situation itself contains logarithmic terms, as in semi-log pressure buildup plotting (Chapter 4). But there are no logarithmic terms in the physical description of immiscible displacement, consequently the use of semi-log and log-log plots in decline curve analysis will only lead to visual distortion, which in turn can cause serious errors in predicting waterdrive performance. As repeated throughout this book, if the reservoir engineer cannot identify basic physical principles associated with the task in hand, then danger threatens. Yet the technique of decline curve analysis is probably the most commonly applied throughout the industry in attempting to predict the performance of waterdrive fields and even major operators, in important waterdrive areas such as the North Sea, will opt for decline curve analysis having abandoned numerical simulation as impracticable.

The problem with any method such as decline curve analysis is that it relies on the extrapolation of trends in surface production statistics: oil rate, watercut, water-oil ratio etc, and, even though these are dampened by plotting them on logarithmic scales, they are still too directly affected by operational activity: sidetracks, addition of surface facilities etc, to provide any safe means of extrapolation to predict performance. What it proves necessary to do is to use the surface cumulative production and injection statistics to generate a reservoir fractional flow of water which, for the reasons described below, *must* provide a smooth, rational function which is amenable to extrapolation using conventional fractional flow mechanics. The situation is depicted in Fig. 5.71a and b. The aggregate water injection rate of all the wells, q_{wi}, is converted to cumulative injection as are the aggregate production rates of all the producers, q_o and q_{wp}, the latter two also provide the surface watercut, f_{ws}. The cumulative production and injection statistics are then expressed in reservoir PVs and used in the Welge equation

$$N_{pd} = (S_{we} - S_{wc}) + (1 - f_{we})W_{id} \tag{5.25}$$

This is solved for the single unknown, S_{we}, the varying water saturation at the *end*

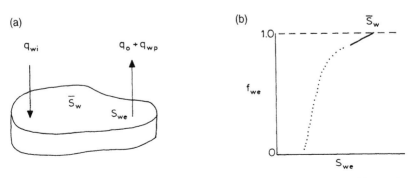

Fig. 5.71. (a) "Black Box" treatment of a reservoir or simulation model; (b) resulting underground fractional flow.

of the flooded system, thus establishing the reservoir fractional flow relationship f_{we} versus S_{we}. Since $1\,PV = N B_{oi}/(1 - S_{wc})$, the conversion from surface cumulative volumes to reservoir pore volumes is affected using the relationships

$$N_{pd} = \frac{N_p B_o}{N B_{oi}}(1 - S_{wc})$$

(5.59)

$$W_{id} = \frac{W_i}{N B_{oi}}(1 - S_{wc})$$

(5.60)

in which B_o and B_w are the FVFs for the oil and water at the average flooding pressure in the system [B_w is usually unity for injected water but not for aquifer water]. Finally, by transposing equation 5.36, the reservoir fractional flow becomes

$$f_{we} = \frac{1}{1 + \dfrac{B_o}{B_w}\left(\dfrac{1}{f_{ws}} - 1\right)}$$

(5.61)

which, on account of the oil shrinkage as it is produced to the surface, is slightly smaller than the surface watercut, f_{ws}.

Plotting the reservoir fractional flow f_{we} versus S_{we} *must always* lead to the generation of a smooth, rational fractional flow function Fig. 5.71b, the only exceptions to this being when the production/injection statistics are themselves faulty or the technique is applied to a reservoir which should, in fact, be subdivided into several discrete faulted blocks, in which case a reservoir fractional flow must be applied to each separately if those single blocks can be defined. The reservoir fractional flow must naturally be affected by operational activity but it is in a much more subtle and illuminating fashion than for the surface watercut, as will be demonstrated.

It will be recognised that the application of the Welge equation (5.25) to the production of a macroscopic reservoir, as described above, is precisely the same usage as that presented by Jones and Roszelle [8] for the measurement of viscous or unsteady state relative permeabilities described in section 5.4f. But how is it possible that flooding in such disparate geometrical systems as a core plug and what is effectively an underground hillside be described in exactly the same manner. The reasons are:

- If it proves possible to inject water into a reservoir and maintain some reasonable average pressure throughout, then it qualifies for description as a "zero dimensional" system.
- Zero dimensional systems are described by application of the concept of material balance and for waterdrive that is the basic theoretical statement of Buckley-Leverett

$$W_{id} = \frac{1}{\left(\dfrac{\partial f_{we}}{\partial S_{we}}\right)}$$

(5.26)

It should be noted that the injection of water need not maintain pressure in the reservoir, the necessary condition for the meaningful application of material balance as described in Chapter 3, section 3.3 is that it must be possible to somehow define an average pressure trend in the reservoir, whether there is pressure equilibrium between the individual wells or not. Under these circumstances the reservoir is effectively zero dimensional and displays tank-like behaviour. In the popular literature the Buckley-Leverett equation is usually derived for flooding in a one dimensional system such as a core plug, section 5.4d, and many believe it is restricted to one dimensional displacement. What this overlooks, however, is that the equation so derived is, in fact, simply a statement of material balance, which is basic to the whole subject of waterdrive and, of course, material balance just happens to be zero dimensional. Consequently, once water injection into a reservoir has been established it can be described using the Buckley-Leverett equation, quite irrespective of the system's size or shape and the practical application of the equation is through use of the Welge equation, section 5.4e. This means that the method being described is invariably successful and this author has only noted a few failures usually when the monitoring of production/pressure data has been inadequate.

Once a reservoir fractional flow relationship has been generated, its validity can be checked by extrapolating the tangent to the final points of the plot to the line $f_{we} = 1$. The point of intersection gives the average water saturation in the reservoir from which the final oil recovery at the end of the history can be evaluated as

$$N_{pd} = \overline{S}_w - S_{wc} \tag{5.22}$$

Similarly, the reciprocal of this tangent gives the cumulative water influx at the end of the history in accordance with equation (5.26). The values of N_{pd} and W_{id} so determined must coincide with the final values of the oil production/water injection statistics expressed in pore volumes. Furthermore, while methods for evaluating oil recovery using pseudo relative permeabilities and fractional flow functions (sections 5) only assess the vertical sweep efficiency, the reservoir fractional flow method accounts for the total volumetric sweep.

In applying the reservoir fractional flow technique to history matching and prediction, the following details must be considered.

(a) Starting point

In most waterdrive reservoirs there is an initial period of depletion prior to the commencement of water injection. This is a practice to be encouraged, if reservoir conditions permit (section 5.2d). The fractional flow calculations, however, only cater for the oil recovered by the water injection. Therefore, the starting point of the calculations occurs at the time of water breakthrough in the production wells when the unique condition prevails that

$$N_{pd(bt)} = W_{id(bt)}$$

which is exactly the same condition pertaining in a core flooding experiment, section 5.4e. Consequently, at the start of the calculations the oil recovered by waterdrive at breakthrough must be set equal the water injected and, therefore, the oil recovered by depletion is

$$\Delta N_{pd(depl)} = N_{pd(total)(bt)} - W_{id(bt)}$$

The net oil recovery by waterdrive used in the calculations is then

$$N'_{pd} = N_{pd(total)} - N'_{pd(depl)}$$

In the final calculations, $\Delta N_{pd(depl)}$ is added to the waterdrive recovery to give the total oil recovered.

Since the *starting point* represents the water breakthrough saturation and fractional flow, the initial value of S_w may be in excess of the average connate water saturation, S_{wc}, as illustrated, for instance, in Fig. 5.24a, which relates to a core flooding experiment representing an element of shock front displacement. In the field, such a distinct *step function* is not usually observed — but almost so. In this respect, the initial saturation is usually set as the average connate water saturation, S_{wc} and the invariably initial rapid rise in f_{we} after breakthrough is manifest of a near macroscopic shock front in the reservoir. In fact, results of the fractional flow calculations are quite independent of the value of S_{wc} used in equation 5.25 as the reader can easily verify by re-working either of the field examples in this section for a different value of the initial saturation, say, $S_{wc} = 0$. It will be found that the calculated cumulative oil recoveries and water injection will be identical to the values obtained using the finite values of S_{wc}, as in the examples. It is useful if the average connate water saturation in the production wells is used a starting point because then, the saturation to which the late-time fractional flow will aim is the *practical* flood-out water saturation in the reservoir, $1 - S_{or}$, as distinct from that determined in laboratory flooding experiments and it is usually found that this practical saturation is often considerably smaller than the experimental value, meaning that the residual oil saturation in the reservoir is much greater. The reason for this is because the combined effects of areal heterogeneity and vertical heterogeneity coupled with gravity result in considerable volumes of in situ oil being completely by-passed during the waterflood and while this is duly accounted for by the reservoir fractional flow it is not, of course, catered for in one dimensional flooding experiments. It is not uncommon in a heterogeneous reservoir, for instance, to determine a value of $_sor$ as high as 0.60 PV from the fractional flow calculations compared to 0.30 PV obtained in controlled flooding experiments in core plugs, as illustrated in the first example.

(b) Natural waterdrive

In the event that a reservoir has a strong natural waterdrive, to the extent that pressure is maintained by the water influx alone, making injection unnecessary, then the value of the effective injection, W_i used in the fractional flow calculations is simply expressed by the cumulative underground withdrawal which, for an undersaturated oil reservoir is

$$W_i = N_p B_o + W_p B_p$$

in which the FVFs are evaluated at the flooding pressure. This is illustrated in the second example in this section.

If there is an influx that is insufficient to maintain pressure so that supplementary water injection is required then

$$W_i = W_{i(\text{injection})} + W_e B_w$$

in which W_e is the cumulative natural water influx, evaluated by material balance aquifer fitting calculations, as described in Chapter 3, section 3.8. It is usually found that the onset of water injection tends to inhibit the influx from the aquifer.

(c) Prediction

This is the difficult part of the exercise. In order to predict, somehow the reservoir fractional flow established over the production history must be extrapolated around the *long corner* towards a practical flood-out water saturation and, of course, the extrapolation is decidedly non-linear. To complicate matters further the extrapolation is not just based on the physical situation but also on the intentions of the operator, as illustrated in the third example in this section. For the moment only the engineering aspects will be considered. If a plot is made of $1/W_{id} = \partial f_{we}/\partial S_{we}$ versus S_{we}, then at first there will be a steep decline in the function from high initial values, corresponding to the early steep rise of the fractional flow, which as it bends downwards has a progressively smaller slope (example b, Fig. 5.74b). Eventually the function tends to flatten although, provided the injection or water influx is maintained, it must continuously decrease. If conditions during latter part of the flood are reasonably stable, that is, the circulation of water is at constant rate, then the $1/W_{id}$ function tends to decline in a near linear fashion which can be fitted by an equation of the form

$$\frac{\partial f_{we}}{\partial S_{we}} = a S_{we} + b$$

in which a and b are constants. Integrating this equation yields a quadratic of the form

$$f_{we} = \frac{a^2}{2} S_{we}^2 + b S_{we} + c$$

which is the non-linear element entering the extrapolation. Predicted values of f_{we} and S_{we} obtained are then used in the Welge equation (5.25), which is applied in its more conventional manner to predict N_{pd} as a function of f_{we} described in section 5.4e and these can be converted to surface conditions N_p versus f_{ws} using the relationships described in section in section 5.5. Finally a prediction of the oil production profile can be obtained using the simple material balance equation

$$q_{wi} = q_o \left(B_o + \frac{B_w f_{ws}}{1 - f_{ws}} \right) \tag{5.8}$$

(d) Perturbations in the fractional flow

It is important to realise that the fractional flow and water saturations (f_{we}, S_{we}) appearing in Welge's equation (5.25) are the values at the *end* of the flooded system, hence the use of the subscript "e". This applies whether the equation is being applied to a core flooding relative permeability experiment in the laboratory or the flooding of a massive reservoir. Frequently, engineers have the impression that the saturation is the average saturation in the system- which it definitely is not. As mentioned earlier, it is difficult to work with surface production data directly in analytical calculations because the watercut, for instance, is too directly affected by operational activity to permit its meaningful extrapolation. But if the surface watercut is perturbed for any reason then so too must the reservoir fractional flow, and it is, although in a quite different manner. If, for instance, a successful workover is conducted in a producing well, then both f_{we} and S_{we}, representing the aggregate values for all the producers will both decrease. The points will temporarily plot in a backward trend on the function, or close to it, but eventually, once the positive effect of the workover has diminished, will simultaneously increase again but on a more favourable fractional flow trend with a smaller slope which will be adhered to until the next perturbation occurs. In this manner the points are nudged around the corner of the fractional flow. This is best illustrated in the second example, Fig. 5.74b, which demonstrates two such favourable perturbations.

(e) Example — North Sea Waterdrive Field

This example illustrates the generation of a reservoir fractional flow using production/injection data for an isolated fault block of an extremely complex North Sea field. It is of the delta top depositional environment type described in section 5.7, with little or no correlation between individual sands from one well to the next. Nevertheless RFT surveys run under dynamic conditions during the initial stage of depletion, prior to the pressure maintenance phase, indicated that all the numerous sands in the section were differentially depleted, indicating a degree of connectivity no matter how tortuous. Attempts to history match the field's performance using numerical simulation modelling systematically failed to produce a reliable predictive tool. Therefore, in attempting to understand the nature of the waterflood, there was little choice but to generate and examine the fractional flow of the system as a whole, as described below.

The technique is applied to a fault block containing one injection well with a capacity of 20,000 b/d which supported two producers. There was a brief period of depletion recovery before the injection was initiated. Water breakthrough occurred

TABLE 5.23

Production/injection statistics for the example North Sea field and calculation of the reservoir fractional flow

Month	Field data			Reservoir fractional flow			
	N_p (MMstb)	W_i (MMstb)	f_{ws} (fraction)	N'_{pd} (PV)	W_{id} (PV)	f_{we} (fraction)	S_{we} (PV)
1	9.091	5.928	0.002	0.058	0.058	0.002	0.190
2	9.555	6.559	0.027	0.064	0.064	0.021	0.191
3	9.970	7.127	0.050	0.070	0.069	0.038	0.194
4	10.367	7.700	0.091	0.075	0.075	0.071	0.195
5	10.764	8.259	0.111	0.080	0.080	0.086	0.197
6	10.921	8.492	0.092	0.082	0.082	0.071	0.196
7	11.225	9.066	0.279	0.086	0.088	0.227	0.208
8	11.532	9.621	0.417	0.090	0.093	0.352	0.220
9	11.833	10.111	0.416	0.093	0.098	0.351	0.219
10	12.088	10.691	0.464	0.097	0.104	0.396	0.224
11	12.330	11.725	0.494	0.100	0.109	0.425	0.227
12	12.578	11.859	0.529	0.103	0.115	0.460	0.231
13	12.800	12.419	0.565	0.106	0.120	0.496	0.236
14	13.020	12.964	0.606	0.109	0.126	0.538	0.241
15	13.189	13.484	0.647	0.111	0.131	0.582	0.246
16	13.341	14.011	0.668	0.113	0.136	0.604	0.249
17	13.480	14.518	0.691	0.115	0.141	0.629	0.253
18	13.584	14.934	0.717	0.116	0.145	0.658	0.256
19	13.693	15.346	0.756	0.117	0.149	0.701	0.262
20	13.793	15.784	0.777	0.119	0.153	0.725	0.267
21	13.896	16.260	0.771	0.120	0.158	0.719	0.266
22	13.972	16.695	0.808	0.121	0.162	0.761	0.272
23	14.051	17.130	0.807	0.122	0.166	0.760	0.272
24	14.172	17.130	0.721	0.123	0.166	0.662	0.257
25	14.306	17.130	0.639	0.125	0.166	0.573	0.244
26	14.409	17.130	0.642	0.126	0.166	0.576	0.246
27	14.475	17.130	0.677	0.127	0.166	0.614	0.253
28	14.557	17.249	0.720	0.128	0.167	0.661	0.261
29	14.610	17.561	0.783	0.129	0.170	0.732	0.273
30	14.640	17.811	0.815	0.129	0.173	0.770	0.279
31	14.661	18.011	0.841	0.130	0.175	0.800	0.285
32	14.667	18.011	0.814	0.130	0.175	0.768	0.279
33	14.707	18.131	0.802	0.130	0.176	0.754	0.277
34	14.755	18.466	0.796	0.131	0.179	0.747	0.276
35	14.801	18.850	0.808	0.131	0.183	0.761	0.277
36	14.819	18.985	0.822	0.132	0.184	0.778	0.281
37	14.864	19.208	0.825	0.132	0.186	0.781	0.281
38	14.900	19.430	0.813	0.133	0.188	0.767	0.279
39	14.946	19.954	0.835	0.133	0.193	0.793	0.283
40	14.992	20.321	0.838	0.134	0.197	0.797	0.284
41	15.034	20.656	0.839	0.134	0.200	0.798	0.284
42	15.077	20.971	0.849	0.135	0.203	0.810	0.286
43	15.108	21.213	0.852	0.135	0.206	0.814	0.287
44	15.135	21.325	0.849	0.136	0.207	0.810	0.287

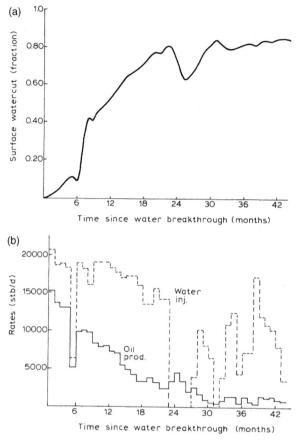

Fig. 5.72. (a) Aggregate watercut development of two producing wells. (b) Production and injection profiles for the field sector.

almost simultaneously in the producers, within months of the start of injection, and their cumulative production/injection statistics together with the surface watercut are listed in Table 5.23 over the 44 month period following breakthrough. The aggregate watercut development of the two producers is shown in Fig. 5.72a and the production and injection rate histories in Fig. 5.72b. In spite of the fact that the mobility ratio is favourable ($M < 1$), the rate of watercut development was extremely severe and this may be attributed to the adverse effects of reservoir heterogeneity mainly in the vertical cross section, there being little or no influence of gravity in such a flood (section 5.7). The production/injection rate profiles show that for the first 23 months there was reasonable continuity of injection but then the injector was closed-in for four months for a workover and thereafter the injection was generally more intermittent than before. Furthermore, one of the production wells was closed-in due to mechanical failure after 31 months and remained in that state until the end of the 44 month period under consideration. The data necessary

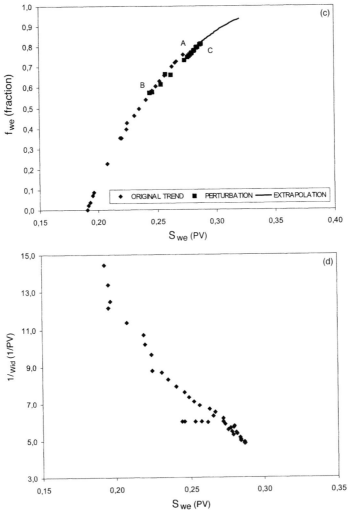

Fig. 5.72 (continued). (c) Reservoir fractional flow function. Heterogeneous North Sea reservoir. (d) $1/W_{id}$ versus S_{we} plot. Heterogeneous North Sea reservoir.

to generate a reservoir fractional flow are as follows:

N = 65 MMstb (STOIIP) B_{oi} = 1.284 rb/stb
S_{wc} = 0.190 PV B_o = 1.319 rb/stb (flooding pressure)
S_{or} = 0.28 PV (experimental determination) B_w = 1.0 rb/stb

Therefore

$$1\ PV = \frac{NB_o}{1 - S_{wi}} = \frac{65 \times 1.284}{1 - 0.19} = 103.04\ \text{MMrb}$$

and $N_{pd} = N_p B_o/103.04$ (PV), $W_{id} = W_i/103.04$ (PV) and the reservoir fractional flow of water is

$$f_{we} = \frac{1}{1 + 1.319 \left(\dfrac{1}{f_{ws}} - 1\right)}$$

Therefore, at breakthrough (row 1, Table 5.23) $N_{pd} = 0.116$ PV and $W_{id} = 0.058$ PV. Since these must be equal at this time, the oil recovery during the initial phase of depletion must be $0.116 - 0.58 = 0.58$ PV [It is a pure coincidence that the depletion recovery is the same as the water injected at breakthrough]. Therefore, the oil recovered by waterdrive used in the calculations is

$$N'_{pd} = N_p B_o/103.04 - 0.058$$

The depletion recovery of 0.058 PV = 5.98 MMrb = 4.53 MMstb is added to the waterdrive component to give the total oil recovery. Details of the fractional flow calculations are listed in Table 5.23 and the plot of the function is shown as Fig. 5.72c. The data plot in a regular manner up to point A, (diamond points) which corresponds to the time when the injection well was closed-in for repair for four months, 23 months after breakthrough of the injected water. During the closure both f_{we} and S_{we} simultaneously decrease and the points plot backwards towards point B. On resumption of the injection the points plot in the forward direction once again (square points) but there is a shift in the fractional flow which is bent slightly downwards, eventually moving to point C at the end of the production history. It is the positive action of repairing the injection well that causes the favourable reduction of slope in the fractional flow forcing the points around the long final corner and accessing more of the movable oil. Contributing to this effect also is that after 31 months, when one of the production wells was permanently closed-in, there was an excess of injection with respect to production which, in strict accordance with Buckley-Leverett's equation (5.26) leads to a reduction in slope of the fractional flow. To check on the validity of the function, extrapolation of the tangent to the final points on the curve to the line $f_{we} = 1$ gives a value of the average water saturation in the reservoir of 0.326 PV and subtracting the connate water saturation of 0.19 PV from this results in a value of $N'_{pd} = 0.136$ PV, while the reciprocal of the tangent gives $W_{id} = 0.207$ PV. Both figures correspond with the final values in Table 5.23.

To facilitate extrapolation of the fractional flow to higher water saturations the trend in values of $1/W_{id}$ ($= \partial f_{we}/\partial S_{we}$) versus S_{we} has been plotted as shown in Fig. 5.72d. Initially the values decline sharply corresponding to the rapid slope change after breakthrough. As can be seen, the trend following the perturbation is more favourable than that before on account of its steeper decline and the final 13 points have been extrapolated in a linear fashion to eventually provide a quadratic expression for the increasing fractional flow as

$$f_{we} = -37.81 S_{we}^2 + 26.57 S_{we} - 3.69$$

This function is shown as the solid line in Fig. 5.72c for which the final value of $f_{we} = 0.937$ corresponding to an abandonment surface watercut of 95%. As described in the text, the extrapolated reservoir fractional flow is used in the Welge equation (5.25), which is applied in the conventional manner to generate a surface watercut trend as a function of the fractional oil recovery. This, in turn, is used in the waterdrive material balance, equation 5.8, to obtain a production profile commensurate with the capacities of the surface equipment. Since this technique has already been illustrated in Exercises 5.1 and 5.4, it will not be considered further.

At abandonment, the oil recovery by waterdrive can be calculated using the Welge equation as 0.156 PV = 12.187 MMstb obtained by the circulation of 0.410 PV of water. Adding the depletion component gives a total oil recovery of 16.717 MMstb, which is 1.58 MMstb ($+10.4\%$) in excess of the recovery at the end of the 44 month history (Table 5.23). The operator must now decide whether it will prove economically viable to repair the damaged production well and increase the operating efficiency of the injection pumps and well to obtain this incremental oil. Since the waterdrive recovery at abandonment is 0.156 PV, the intercept of the extrapolated fractional flow on the line $f_{we} = 1$ will give a value of the average saturation in the reservoir of $0.156 + 0.19 = 0.346$ PV, which is the flood out water saturation $1 - S_{or}$. This implies a volume averaged residual oil saturation for the reservoir of $S_{or} = 0.654$ PV compared to the average value of 0.28 PV determined in a series of controlled flooding experiments in thin core plugs. The disparity between field and laboratory figures, as mentioned in the text results from the waterflood experiencing the full, adverse effect of heterogeneity in the field, which cannot be duplicated in the laboratory experiments. The total recovery factor is only 16.2% STOIIP at abandonment, which is a rather dismal figure for waterdrive but is still a considerably better value than would have been obtained by depletion leading to solution gas drive in such a complex reservoir environment.

(f) Example — the East Texas Field

The drilling of well Daisy Bradford Number 3 in October 1930 led to the discovery of the famous East Texas Field. In Daniel Yergin's book *The Prize* [5, chapter 1], there is an interesting and at times amusing account of the discovery and early development of the field which had such an important influence on the growth of the state of Texas and its emergence as the centre of gravity of the technical oil industry. In the early days of development, before the concept of field unitisation had been recognised, there was the usual mad scramble to claim, drill and produce as much oil as possible as rapidly as possible. It was this undisciplined approach that originally led to the evolution of Petroleum Engineering and Reservoir Engineering during the 1930s and, in particular, which saw the derivation of first means of quantifying oil recovery through application of material balance, presented through the AIME by Schilthuis in 1936.

This important oilfield in the history of the industry can be physically defined as follows

STOIIP $= 7034$ MMstb
B_{oi} $= 1.3118$ rb/stb (at $p_i = 1620$ psia)
S_{wc} $= 0.20$ PV

Therefore

$$1 \text{ PV} = \frac{7034 \times 1.3118}{1 - 0.20} = 11534 \text{ MMrb}$$

Cumulative oil and water production histories from 1930 to 1992 are listed in Table 5.24, together with the pressure and FVF records. Historic oil rates, watercut development and the pressure history are plotted in Figs. 5.73a–c. There has been a massive natural water influx into the field, to the extent that, following an early, rapid pressure decline during the 1930s, the pressure support from the aquifer increased and then maintained pressure at ± 600 psi below the initial level (Fig. 5.73c). No water injection was necessary and the *effective* value of W_{id} supplied by the aquifer (column 9 of Table 5.24) is simply the underground withdrawal

$$W_{id} = \frac{N_p B_o + W_p B_w}{11534}$$

While the cumulative oil in PVs (column 8) and reservoir fractional flow (column 7) are evaluated as

$$N_{pd} = \frac{N_p B_o}{11534}$$

and

$$f_{we} = \frac{1}{1 + \dfrac{B_o}{B_w}\left(\dfrac{1}{f_{ws}} - 1\right)}$$

Table 5.24 is a simple spreadsheet application in which S_{we} in column 11 is evaluated as the solution of equation (5.25) and the resulting reservoir fractional flow is plotted in Fig. 5.74a. Whereas the surface watercut trend (Fig. 5.73b) is directly affected by operational activities, the reservoir fractional flow provides a remarkably smooth function over the 62 years of production history. There are two significant periods of intense remedial activity in the field (drilling, workovers, etc.) but these appear as perfectly rational perturbations that in themselves can be interpreted to determine the efficacy of the remedial work.

Initially, there is a steep rise in the fractional flow (Fig. 5.74a) immediately after breakthrough and the function rises to point A when the first remedial campaign began in 1949. The favourable effect of this reduces both S_{we} and f_{we} and the points temporarily plot backwards and off the curve to point B. As the water production increases again, there is a second reversal and the points again plot in the forward direction, eventually towards point C. The overall positive effect of the initial remedial campaign is still being felt, however, since the slope of the fractional flow is smaller and therefore more favourable. A second remedial

TABLE 5.24

East Texas Field: Production/injection statistics, 1930–1992 and generation of the reservoir fractional flow

Year	Field data						Reservoir fractional flow				
	Pressure (psia)	B_o (rb/stb)	B_w (rb/stb)	N_p (MMstb)	W_p (MMstb)	f_{ws} (fraction)	N_{pd} (PV)	W_{id} (PV)	f_{we} (fraction)	S_{we} (PV)	$1/W_{id}$ (1/PV)
1930	1619.86	1.3118	1.0126	0.027	0						
1931	1415.73	1.3135	1.0132	109	0	0.0012	0.0124	0.0124	0.0009	0.200000003	80.488
1932	1280.07	1.3148	1.0136	254	1	0.0031	0.0290	0.0290	0.0024	0.2000	34.476
1933	1118.78	1.3169	1.0141	462	3	0.0102	0.0527	0.0530	0.0079	0.2002	18.872
1934	1110.65	1.317	1.0141	645	7	0.0202	0.0736	0.0742	0.0157	0.2006	13.473
1935	1096.34	1.3173	1.0142	820	12	0.0312	0.0937	0.0947	0.0242	0.2012	10.557
1936	1089.69	1.3174	1.0142	985	24	0.0663	0.1125	0.1146	0.0518	0.2038	8.726
1937	1011.74	1.3187	1.0144	1160	50	0.1306	0.1326	0.1370	0.1036	0.2098	7.297
1938	1017.51	1.3186	1.0144	1310	89	0.2063	0.1498	0.1576	0.1666	0.2184	6.345
1939	1016.05	1.3186	1.0144	1450	140	0.2665	0.1658	0.1781	0.2185	0.2266	5.615
1940	988.98	1.3191	1.0145	1590	202	0.3069	0.1818	0.1996	0.2541	0.2329	5.010
1941	902.38	1.3209	1.0148	1730	320	0.4574	0.1981	0.2263	0.3930	0.2608	4.419
1942	913.96	1.3206	1.0147	1850	445	0.5102	0.2118	0.2510	0.4446	0.2724	3.985
1943	933	1.3202	1.0147	1980	577	0.5038	0.2266	0.2774	0.4383	0.2708	3.605
1944	959.61	1.3197	1.0146	2110	734	0.5470	0.2414	0.3060	0.4815	0.2828	3.268
1945	945.4	1.3199	1.0146	2250	903	0.5469	0.2575	0.3369	0.4813	0.2827	2.968
1946	968.49	1.3195	1.0146	2370	1090	0.6091	0.2711	0.3670	0.5451	0.3042	2.725
1947	1022.86	1.3185	1.0144	2480	1290	0.6452	0.2835	0.3970	0.5831	0.3180	2.519
1948	974.47	1.3194	1.0145	2600	1500	0.6364	0.2974	0.4294	0.5737	0.3144	2.329
1949	1070.69	1.3177	1.0143	2690	1700	0.6897	0.3073	0.4568	0.6311	0.3388	2.189
1950	1015.86	1.3186	1.0144	2790	1880	0.6429	0.3190	0.4843	0.5807	0.3159	2.065
1951	1074.92	1.3176	1.0142	2890	2030	0.6000	0.3301	0.5086	0.5359	0.2941	1.966
1952	1062.47	1.3178	1.0143	2980	2180	0.6250	0.3405	0.5322	0.5619	0.3073	1.879
1953	1078.16	1.3175	1.0142	3070	2330	0.6250	0.3507	0.5556	0.5620	0.3073	1.800
1954	1047.31	1.318	1.0143	3160	2480	0.6250	0.3611	0.5792	0.5619	0.3074	1.727
1955	1146.5	1.3165	1.014	3230	2630	0.6818	0.3687	0.5999	0.6227	0.3423	1.667
1956	1109.87	1.3171	1.0141	3310	2790	0.6667	0.3780	0.6233	0.6063	0.3326	1.604
1957	1077.81	1.3175	1.0142	3380	2960	0.7083	0.3861	0.6464	0.6515	0.3608	1.547
1958	1166.75	1.3162	1.014	3430	3120	0.7619	0.3914	0.6657	0.7114	0.3993	1.502
1959	1142.82	1.3166	1.014	3490	3300	0.7500	0.3984	0.6885	0.6979	0.3904	1.452
1960	1165.09	1.3163	1.014	3530	3470	0.8095	0.4029	0.7079	0.7660	0.4372	1.413

TABLE 5.24 (continued)

Year	Field data						Reservoir fractional flow				
	Pressure (psia)	B_o (rb/stb)	B_w (rb/stb)	N_p (MMstb)	W_p (MMstb)	f_{ws} (fraction)	N_{pd} (PV)	W_{id} (PV)	f_{we} (fraction)	S_{we} (PV)	$1/W_{id}$ (1/PV)
1961	1139.5	1.3166	1.0141	3580	3630	0.7619	0.4087	0.7278	0.7114	0.3986	1.374
1962	1182.5	1.316	1.0139	3620	3790	0.8000	0.4130	0.7462	0.7550	0.4302	1.340
1963	1130.72	1.3167	1.0141	3670	3950	0.7619	0.4190	0.7663	0.7114	0.3978	1.305
1964	1186.45	1.316	1.0139	3710	4110	0.8000	0.4233	0.7846	0.7550	0.4311	1.275
1965	1175.39	1.3161	1.0139	3750	4260	0.7895	0.4279	0.8024	0.7429	0.4216	1.246
1966	1156.1	1.3164	1.014	3790	4430	0.8095	0.4326	0.8220	0.7660	0.4402	1.217
1967	1142.07	1.3166	1.014	3840	4590	0.7619	0.4383	0.8419	0.7114	0.3953	1.188
1968	1112.2	1.317	1.0141	3900	4750	0.7273	0.4453	0.8630	0.6725	0.3627	1.159
1969	1139.71	1.3166	1.0141	3950	4920	0.7727	0.4509	0.8835	0.7237	0.4068	1.132
1970	1083.44	1.3175	1.0142	4030	5110	0.7037	0.4603	0.9097	0.6464	0.3387	1.099
1971	1063.71	1.3178	1.0143	4100	5300	0.7308	0.4684	0.9345	0.6763	0.3659	1.070
1972	1039.46	1.3182	1.0143	4180	5500	0.7143	0.4777	0.9614	0.6580	0.3489	1.040
1973	1059.61	1.3178	1.0143	4250	5700	0.7407	0.4856	0.9868	0.6874	0.3771	1.013
1974	1036.05	1.3182	1.0144	4320	5910	0.7500	0.4937	1.0135	0.6978	0.3874	0.987
1975	1023.52	1.3185	1.0144	4390	6130	0.7586	0.5018	1.0410	0.7074	0.3973	0.961
1976	1055.96	1.3179	1.0143	4460	6340	0.7500	0.5096	1.0671	0.6978	0.3871	0.937
1977	1038.2	1.3182	1.0144	4520	6570	0.7931	0.5166	1.0944	0.7468	0.4395	0.914
1978	1023.41	1.3185	1.0144	4590	6810	0.7742	0.5247	1.1236	0.7251	0.4158	0.890
1979	1045.74	1.3181	1.0143	4650	7060	0.8065	0.5314	1.1523	0.7623	0.4575	0.868
1980	1033.94	1.3183	1.0144	4700	7340	0.8485	0.5372	1.1827	0.8116	0.5144	0.845
1981	1054.82	1.3179	1.0143	4760	7630	0.8286	0.5439	1.2149	0.7881	0.4865	0.823
1982	1045.14	1.3181	1.0143	4810	7940	0.8611	0.5497	1.2479	0.8267	0.5334	0.801
1983	1052.84	1.318	1.0143	4860	8270	0.8684	0.5554	1.2826	0.8355	0.5444	0.780
1984	1055.62	1.3179	1.0143	4910	8620	0.8750	0.5610	1.3191	0.8434	0.5545	0.758
1985	1012.53	1.3187	1.0144	4960	8990	0.8810	0.5671	1.3577	0.8506	0.5642	0.737
1986	1051.54	1.318	1.0143	5000	9330	0.8947	0.5714	1.3918	0.8674	0.5868	0.718
1987	1034.34	1.3183	1.0144	5050	9670	0.8718	0.5772	1.4277	0.8395	0.5481	0.700
1988	1046.49	1.3181	1.0143	5090	10000	0.8919	0.5817	1.4611	0.8639	0.5829	0.684
1989	985.33	1.3192	1.0145	5120	10400	0.9302	0.5856	1.5004	0.9111	0.6523	0.667
1990	1095.81	1.3173	1.0142	5160	10700	0.8824	0.5893	1.5302	0.8524	0.5634	0.654
1991	1090.84	1.3173	1.0142	5190	11100	0.9302	0.5928	1.5688	0.9112	0.6535	0.637
1992	1081.53	1.3175	1.0142	5220	11500	0.9302	0.5963	1.6075	0.9112	0.6536	0.622

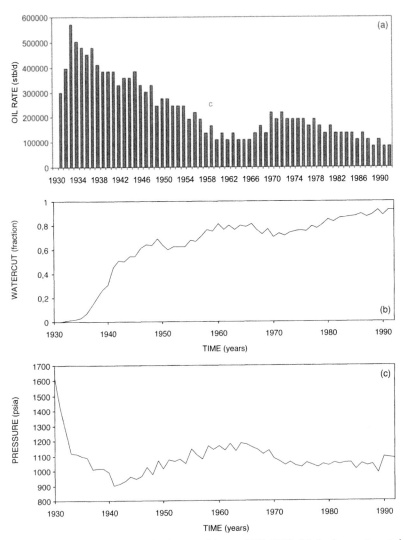

Fig. 5.73. East Texas Field. (a) Oil production rate history, 1930–1992. (b) Surface watercut development, 1930–1992. (c) Field average pressure history, 1930–1992.

campaign commenced in 1960 at point C and the cycle of reversal of the fractional flow points is repeated, this time to point D. Finally, the points move again in the forward direction around the remainder of the curve but once more with a smaller and more favourable slope. There has been ongoing well-activity since 1960 but not such intensive campaigns as to cause the magnitude of the two earlier perturbations. These can be seen by plotting the later trend in the fractional flow on a larger scale, which reveals four subsequent, lesser perturbations to the fractional flow but each having the positive effect of favourably decreasing the slope of the function. In fact, the day-to-day remedial work that is conducted in most fields is necessary just to

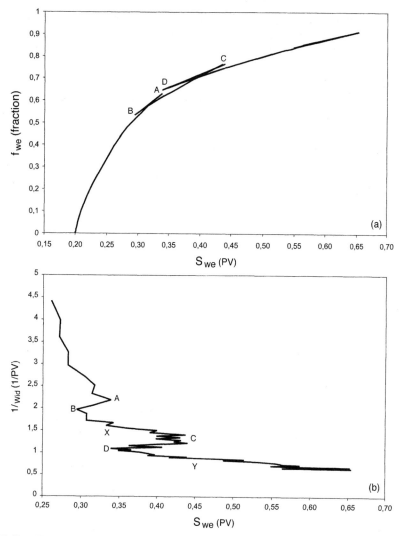

Fig. 5.74. East Texas Field. (a) Reservoir fractional flow curve, 1930–1992. (b) $1/W_{id}$ versus S_{we} plot, 1930–1992.

keep the fractional flow behaving itself properly. The oil recovery during each of the major perturbations could calculated by extrapolating the tangents to the fractional flow curve before and after the events. Extrapolation of these to the line $f_{we} = 1$ gives the average water saturations in the field before and after and their difference is the oil recovered in PVs. However, there is a simpler method of doing this, as described below.

Fig. 5.74b is a plot of the $1/W_{id}$ function which through the Buckley-Leverett equation (5.26) is proportional to the slope of the fractional flow, $\partial f_{we}/\partial S_{we}$. As such, the function must always decrease provided the natural water influx continues unabated, which it does. The initial decline in $1/W_{id}$ is high, corresponding to the rapid increase in the watercut, but eventually this trend is abated as the field struggles around the long final corner of the fractional flow. The perturbations due to remedial work show up much more clearly in this plot than in the fractional flow itself. Points A, B, C and D illustrating the reversals in Fig. 5.74a have been transferred to Fig. 5.74b, in which the effects appears more significant. For the two major remedial campaigns there is a sudden drop in f_{we} accompanied by a reversal in values of S_{we}. If the duration of the effect of the remedial work is defined as the time it takes for S_{we} to be restored to its original value prior to the activity, at A and C on Fig. 5.74b, for the 1949 and 1960 campaigns respectively, then these points are indicated as X and Y on the plot which occur at $S_{we} \sim 0.33$ PV in 1956 and 0.44 PV in 1977. The oil recovery during the period of influence of the two remedial campaigns is then

$$1949\text{–}1956: \text{ Oil recovery } = N_{pd(1956)} - N_{pd(1949)} = 0.3780 - 0.3073 = 0.0707 \text{ PV}$$
$$= 0.0707 \times 11534 = 815.5 \text{ MMrb} = 804.2 \text{ MMstb}$$

$$1960\text{–}1977: \text{ Oil recovery } = N_{pd(1977)} - N_{pd(1960)} = 0.5166 - 0.4029 = 0.1137 \text{ PV}$$
$$= 0.1137 \times 11534 = 1311.4 \text{ MMrb} = 1293.3 \text{ MMstb}$$

in which 1 PV $= 11534$ MMrb and the average value of the oil FVF for both periods is $B_o = 1.014$ rb/stb. The oil recoveries in PVs are simply read from Table 5.24 for the appropiate years. It should be stressed that these oil recoveries are the totals achieved during the separate periods of activity, unfortunately the method does not lend itself to calculation of the *incremental* recoveries during the same periods. The remedial work started in 1960 was much more protracted than the earlier effort and could better be described as sustained well work. During the final levelling off of the $1/W_{id}$ plot, four additional minor perturbations to the plot are evident, corresponding to ongoing, routine maintenance of wells.

The final, slight downward trend of the $1/W_{id}$ plot could be linearly extrapolated to obtain a prediction of f_{we} versus S_{we} which would be a quadratic function, as described in the main text. This in turn could be used in the Welge equation (5.25) to predict N_{pd} as a function of f_{we} and converting this to a surface relationship, N_p/N versus f_{ws} would enable an oil production profile to be generated. To do this in a meaniful fashion, however, requires a knowledge of the operator's intentions for the further development of the field and of the capacities to inject and produce fluids, information to which the author is not privy. The importance of the *intentions* is illustrated in the following example.

(g) The influence of operational activity

The third example of the use of the reservoir fractional flow relates to the examination of numerical simulation output rather than field production statistics. The field being studied is a large, tight, fractured limestone accumulation with various other complications including a significant initial gas cap. When production started in the early 1970s, using vertical well completions, it was difficult to prevent production of gas from the gas cap through the vertical fractures and individual well pressure drawdowns had to be restricted to inhibit this effect, resulting in a limited production rate from the field as a whole. During the 1980s with the advent of horizontal well completions the field took on a new lease of life and, in addition, a water injection scheme was initiated. The use of extended horizontal wells, some approaching 10,000 feet in length, became standard practice both for oil producers and water injectors and the performance of the field was greatly enhanced due to the reduced pressure drawdowns implicit in the use of such completions.

The numerical simulation study being examined was conducted during 1995 and considered two development plans. The first was the 1991 plan, still being implemented, which was a fairly low-key affair. The second was the newly devised and much more aggressive 1995 plan. This called for the drilling of 30 additional horizontal wells, most of them water injectors, and the upgrading of surface injection/production facilities, sufficient to enable the circulation rate of water through the reservoir to be doubled. Usually when making this sort of comparative study, the 1991 simulation model would be used for the first case and the 1995 model for the second. But in this study, the operator adopted a much more sensible approach of using the most recent 1995 model for both development cases. Therefore since both runs employed the same numerical chassis, this affords a direct comparison of the effectiveness of the two development plans isolated from other factors.

The cumulative injection and production statistics output from the model were used in the Welge equation (5.25), and the resulting fractional flow curves are plotted in Figs. 5.75a and b for the 1991 and 1992 development plans respectively. As can be seen for both, the movable oil volume in this tight, fractured reservoir is very small but it is nonetheless a viable project. The final points on the functions coincide in time and represent the relinquishment date of the licence. Linear extrapolations of the tangents to the final points on the curves to the line $f_{we} = 1$ give the average water saturations in the reservoir at that time, which are 0.281 PV for the 1991 plan and 0.381 PV for the 1995 plan. Subtracting the connate water saturation of $S_{wc} = 0.20$ PV from each gives ultimate recoveries for the two cases of $N_{pd(1991)} = 0.081$ PV and $N_{pd(1995)} = 0.156$ PV. Therefore, the updated plan practically doubles the oil recovery and the economics associated with the upgrade are very favourable. Physically what happens is perfectly consistent with the Buckley-Leverett equation (5.26): the circulation of more pore volumes of water in a given time in the 1995 plan has the effect of bending the fractional flow in the favourable, downward direction, thus increasing the target movable oil.

Therefore, while the fractional flow functions, Figs. 5.75a and b implicitly contain all the effects of both areal and vertical heterogeneity built into the numerical model plus the influence of pressure and fluid properties, it is evident that since these are the same for both simulation runs, the shape of the fractional flow and the ultimate oil recovery depend also, to a large extent, on the operator's intentions for further field development. Therefore it is necessary to be able to read the operator's mind to perform a meaningful prediction of waterdrive performance — merely studying the physics and mathematics of the situation is inadequate, which should be an obvious fact.

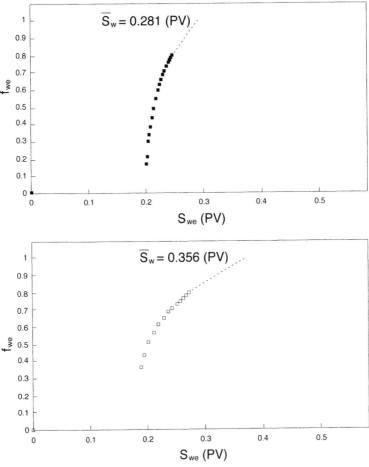

Fig. 5.75. (a) Example 5 — Reservoir fractional flow (1991 development). (b) Example 6 — Reservoir fractional flow (1995 development).

In this respect, successful prediction of waterdrive performance also depends on the operator's awareness of the true significance of the Buckley-Leverett theory that, to progress around the long final corner of the fractional flow accessing more and more of the movable oil, requires the circulation of a large volume of water. Certainly, the operator of the above field seems to appreciate this point but a great many simply do not, as is evident from inspection of the decline performance of some fields in major producing areas such as the North Sea. The reason for this is pointed to in the introduction to this chapter in the statement that, the very concept of fractional flow itself seems to have "gone missing" in reservoir engineering because it never has been considered in the construction of numerical simulation models. Therefore, since the Buckley-Leverett equation (5.26) is totally dependent on the fractional flow and its slope, it is small wonder that the basic theory of waterdrive is not always given due consideration in the development planning of waterdrive fields

(h) Comment

Key to understanding the reservoir fractional flow technique is the appreciation that the Buckley-Leverett theory is dimensionless and is simply a statement of material balance for waterdrive. Therefore, provided water can be successfully injected into any system, such that an average pressure trend can be defined, it qualifies for description as zero dimensional to which Buckley-Leverett's equation (5.26) can be directly applied, whether it is a core plug or complex reservoir. Nevertheless, it is a cause of some aesthetic satisfaction to see, repeatedly, how well the technique works, particularly in a massive field such as East Texas over a 62 year production history. With a STOIIP of over 7000 MMstb and dimensions of 72 km in length by 8–16 km in width — this is quite a large core plug!

Just as described for volumetric material balance in Chapter 3, because the description of waterdrive using this method is zero dimensional, it means that the macroscopic reservoir is treated as a black box. It may contain all sorts of complexities: areal and vertical heterogeneity, faults, horizontal wells etc, but the entire system will reveal its nature simply through consideration of the cumulative fluid input and output and definition of some average pressure. Having examined the overall physics of the system's history and, in particular, whether there appears any scope for bending the fractional flow further downwards in the favourable direction through some form of remedial operational activity, only then should the engineer explore the detailed innards of the black box to determine what in practice can best be done to improve oil recovery. This may require the construction of a detailed numerical simulation model to determine the optimum location of new wells, workover/recompletion possibilities, upgrading the surface facilities etc. It will be noted that this is the antithesis of the more modern approach to reservoir engineering in which physical phenomena in the reservoir are viewed on smaller and smaller scales, even to the level of displacement through individual pore spaces. The engineer is then confronted with the difficult process of up-scaling of observations to reservoir proportions, a subject which is currently at the forefront of reservoir engineering research. The modern approach is facilitated by the rapid advances that have been made in recent years in high-tech laboratory equipment to enable viewing on such a small scale. It is the "bottom-up" approach, whereas application of the reservoir fractional flow is the "top-down" and is probably the safest opening move to studying any field problem.

The fractional flow method is ideal to use in conjunction with numerical simulation modelling, once there has been significant water breakthrough, as illustrated in the final example in the section. The output from simulators is not very informative being simply tables and plots of production/injection statistics. Consequently, from a series of sensitivity runs, it is not always obvious why one run is better than others. But since the simulator also calculates the volume averaged pressure in the reservoir at each time step, then all the data are at hand to compute and plot the reservoir fractional flow for the simulated reservoir which again must necessarily be a smooth, rational function matching the simulated history perfectly. Comparison of the fractional flows can be revealing in differentiating between runs and invariably the better simulation runs are those in which a greater number of pore volumes of water were circulated in the system in a given time. The fractional flow should match the simulated history exactly and a series of analytical predictions can be made using the extrapolation technique described above, incorporating assumed perturbations due to remedial activity, which can give guidance in selecting cases for detailed simulation. But, while the numerical model may take months to history match and predict for a complex field with lengthy production history, the fractional flow only takes a matter of hours to apply using spread sheets programs.

5.10. DIFFICULT WATERDRIVE FIELDS

This section describes the development of two complex waterdrive reservoirs. The first is in a relatively small offshore field in South East Asia while the second describes what is perhaps the most problematical but widespread reservoir in the U.K. and Norwegian sectors of the North Sea: the Etive–Rannoch at the base of the Brent sand section which extends across the entire East Shetland Basin (Fig. 5.12). Both are characterised by the fact that the flooding is neither under the VE condition nor with a total lack of cross flow between the layers. Instead, it occurs at an intermediate condition between the two. Another common feature is that in both fields, the severity of the vertical permeability distributions was initially underrated.

(a) Field A

This is located offshore in a new development area and therefore there was no previous production experience to refer to for guidance. Not long after the start of injection water breakthrough occurred in four wells in what was regarded as the best area of the field (Fig. 5.76). Thereafter, the rate of watercut development was extremely harsh and in well A, for instance, (studied in this section) had risen to 80% for a fractional oil recovery of 19% whereas in the original development study breakthrough was not anticipated until the recovery had reached 34%. Such behaviour in four wells in an un-faulted area of the field can only be attributed to severe heterogeneity across the sand section and indeed this is confirmed by inspection of the permeability distribution across the 70 ft thick reservoir in well A, the only one fully cored in that area of the field (Fig. 5.77). Originally, the operator only viewed these data on an conventional log-k scale (dashed line) and while appreciating that there was a high-permeability tunnel at the centre of the reservoir, assumed that the flooding of the sands would be from the base to the top of the section under the VE condition, providing the type of sweep depicted in Fig. 5.39c. This assumption was incorporated in a numerical simulation study upon which the project was designed. The harsh watercut development clearly demonstrated that VE flooding was not occurring and that the flooding order of the layers must be

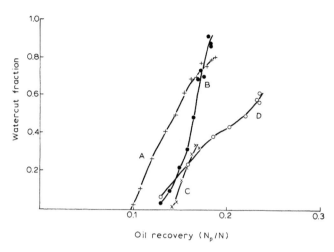

Fig. 5.76. Watercut development in four production wells.

somewhat more complex. So far in this chapter only two well defined flooding orders have been considered: flooding from base to top (VE) and the flooding of isolated sands following Stiles ordering. But if there are N layers in a sand section, this still leaves $N! - 2$ other possible flooding orders for which the averaging procedures, equations 5.30–5.32, may be evaluated to generate pseudos. In Fig. 5.77 the section has been divided into 8 layers implying more than 40,000 possible flooding orders — offering the engineer involved with the project job security for life evaluating the possibilities. In fact, inspection of the section reveals that the alternatives can be narrowed down to a few which can be used in an attempt to match the watercut development, the correct one being selected on a trial and error basis.

It is considered that the flooding is neither under the VE condition nor for Stiles ordering but somewhere between the two. That is, the very tight section between layers 5 and 6 removes the possibility of VE across the entire section but within and between the remaining sands there may be VE. The order which turns out to be the most appropriate for matching the history and its rationale for selection is as follows.

Layer
5   Whether it is separated from 4 or not it has the highest value of $k/\phi\Delta S$ and is at the base of the massive central section which dictates that it must flood first.
4   This high-permeability coarsening upward sand will flood as a single unit under the VE condition (Fig. 5.60).
6   Likely to be separated from the sands above but floods third on account of its high value of $k\phi\Delta S$.
3   Coarsening upward sand will flood as a single unit (VE).
7   Downward flood under some influence of gravity.
2   Floods next on account of $k/\phi\Delta S$ value.

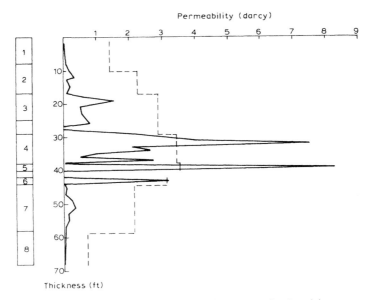

Fig. 5.77. Permeability distribution (— — — — = log-*k* scale).

8   Downward flood under the influence of gravity.
1   The only layer left.

The layer data are listed in Table 5.25, arranged in the above flooding order.
Other data required for displacement efficiency calculations are:

Relative permeabilities (single set): $k'_{ro} = 1.0$ $S_{or} = 0.40$ PV
                                          $k'_{rw} = 0.29$

PVT (flooding at near initial pressure): $\mu_o = 2.0$ cp, $\mu_w = 0.35$ cp
                                          $B_o = 1.1$ rb/stb, $B_w = 1.0$ rb/stb

from which the mobility ratio can be calculated as $M = 1.66$.

TABLE 5.25

Layer data arranged in the estimated flooding order

Layer No.	Flood order	h (ft)	k (mD)	ϕ	S_{wc}
5	1	2	3500	0.22	0.135
4	2	9	2933	0.21	0.140
6	3	2	1500	0.17	0.150
3	4	8	785	0.18	0.160
7	5	14	137	0.16	0.160
2	6	9	171	0.17	0.160
8	7	10	6	0.11	0.201
1	8	8	28	0.12	0.190

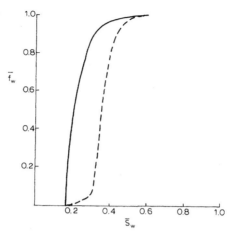

Fig. 5.78. Reservoir fractional flow relationships (———— = flooding order, Table 5.25; — — — — = VE flooding).

This is sufficiently close to unity to neglect any velocity dispersion that may occur between isolated layers and the calculations to generate the pseudos, fractional flow and the relationship f_{ws} versus N_p/N proceed as in Exercise 5.4. These will be left as an exercise for the reader. The reservoir fractional flow relationships, both for the flooding order listed in Table 5.25 and for the original assumption of VE displacement (flooding from the base) are plotted for comparison in Fig. 5.78. The disparity between the two is very significant. The function for VE flooding demonstrates a high degree of shock front development as the central and lower layers flood in a piston-like fashion, whereas for the flooding order presented in Table 5.25 the fractional flow development is very much more severe, as in the field, and the function displays the familiar downward curvature across the entire movable saturation range. The final step is to use the fractional flows in Welge displacement calculations to evaluate the vertical sweep and the relationship between the surface watercut development and fractional oil recovery (section 5.5b). The results are plotted in Fig. 5.79 in comparison with the observed watercut in well A, to which the core data relates. As can be seen, the results for the Table 5.25 flooding order history match the actual watercut development in a perfectly acceptable fashion except for a slight discrepancy at breakthrough. This is because the theoretical calculations are for the vertical sweep (assuming $E_A = 1$) whereas the field data incorporate an areal sweep which is less than unity. The geometry is that of a line drive, however, and since the mobility ratio is close to unity, the areal sweep increases rapidly to 100% where the theoretical and observed watercuts coincide. In comparison, the VE assumption produces a highly optimistic result in which breakthrough is deferred until 34% of the STOIIP has been recovered.

Production profiles have been generated for the two cases using the field watercut extrapolations (Fig. 5.79) aiming at a plateau oil rate of 4000 stb/d from the four wells for a maximum injection rate of 9000 b/d. The method employs the waterdrive

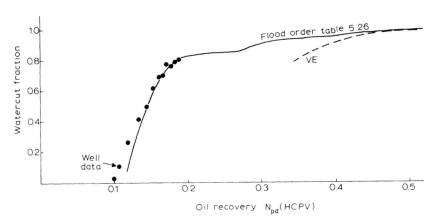

Fig. 5.79. Comparison of theoretical and actual watercut developments (well A).

material balance, equation 5.8, in the manner illustrated in Exercises 5.1 and 5.4. The profiles are compared in Fig. 5.80 in which it can be seen that the premature water breakthrough, allowing for the full effect of heterogeneity, causes a collapse in the plateau rate after eighteen months compared to three and a half years for the VE flood. Comparative recovery statistics at an abandonment rate of 400 stb/d are as follows:

	VE flooding	Full heterogeneity
Oil recovery (MMstb):	6.35	4.94
Recovery factor (%):	42	33
Water injected (MMb):	17	27
Abandonment watercut (%):	95	95
Time of abandonment (years)	7	9.5

The results confirm the conclusions reached in section 5.3c for heterogeneous fields: that in comparison to more homogeneous developments they suffer from loss of recovery and prolongation of their producing lifetime while circulating large volumes of water. Overall it requires 2.7 barrels of injected water to recover a barrel of oil in the favourable VE case compared to 5.5 b/b, allowing for the full degree of heterogeneity. In a situation like this, the operator should concentrate on circulating water through the complex reservoir section as rapidly as possible. Advantage should be taken of the fact that water production from this restricted area of the field occurred early in the lifetime of the project, prior to water breakthrough elsewhere. Therefore, the injection/production facilities are not being used to full capacity, making it possible to concentrate on this poorer field area and inject and produce at high rate and watercut before water production occurs in other field areas. Rather than drill additional producers, injectors should be given preference to facilitate the more rapid circulation of water.

There is no great technical sophistication in the calculations, nevertheless they are quite satisfactory in history matching the severe watercut development resulting

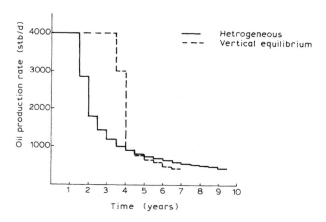

Fig. 5.80. Oil production profiles for full heterogeneity and VE flooding.

from the harshness of the permeability distribution. The reader may wonder why problems like this cannot be better handled by detailed cross-sectional modelling rather than the method adopted, which amounts to "good old fashioned guesswork" of the likely flooding order. It must be remembered, however, that for the 8 layers there are still more than 40,000 ways the engineer could set the vertical permeabilities between model layers to influence the flooding order. Simulation models do not teach engineers physics, they merely reflect the consequences of the input assumptions to the study. In this particular field, the VE-flooding condition resulted from a detailed cross-sectional simulation study in which the vertical permeabilities were incorrectly set.

(b) Field B

The field described may be considered as a prototype for the central area of the East Shetland Basin of the North Sea. In particular, attention is focused on the development of the complex Etive–Rannoch reservoir (named after Scottish Lochs) at the base of the Middle Jurassic Brent sand section. The difficulties in developing the reservoir have been reported in several technical papers for different fields [29–31] and have already been described in connection with RFT pulse testing in Chapter 2, section 2.8, in which a permeability distribution across the total sand section is included as Fig. 2.22. A "blocked" permeability distribution for the same prototype well is plotted in Fig. 5.81 for both log-k and linear scales. The Etive sand has extremely high permeabilities compared to the Rannoch beneath and separating the two is a tight micaceous sand interval (70 ft from the top of the section, Fig. 5.81b) which appears correlatable from field to field across the basin. The interval acts as a partial barrier to vertical fluid movement but nevertheless numerous dynamic RFT surveys across the section revealed the condition of apparent hydrostatic equilibrium: provided no disturbance was caused in neighbouring wells prior to the survey. Thus the two sands are hydraulically

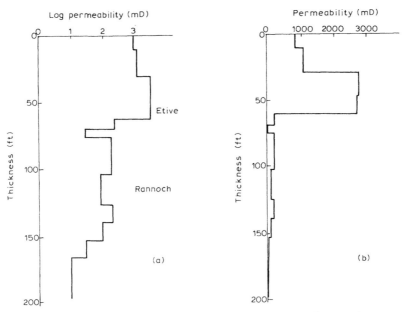

Fig. 5.81. Permeability distribution across the Etive–Rannoch reservoir.

connected and therefore comprise a single reservoir unit and must be treated as such — which is what causes the problem in their development. If the tight section between the sands was a complete barrier to flow, the reservoirs could be developed separately and, even though the permeability of the Rannoch appears to be low (Fig. 5.81), this is only in comparison to the overlying Etive. In fact, it has very respectable permeabilities in the hundreds of millidarcies range and could be flooded quite efficiently if in isolation.

Alternatively, if the vertical flow restriction were not present then a super efficient piston-like displacement might be anticipated across the entire section as water slumped from the high-permeability Etive downwards into the Rannoch. Annoyingly, the vertical permeability across the micaceous layer is of the order of 10–15 mD, for which neither of the above conditions is realized.

If water is injected across the entire section then it tends to rush across the base of the Etive with little downward drainage of water into the Rannoch on account of the relatively weak forces of gravity and capillary imbibition from the high to low-permeability sand. The downward flooding rate is only 2–3 ft/year and the oil displaced from the Rannoch rises into the Etive through which it is produced. It might be thought that the solution would be to restrict the perforations in both injection and production wells to the Rannoch, safely below the tight interval, in an attempt to develop this sand first and then proceed to the Etive. This has, of course, been tried but only with limited success for the reason shown in Fig. 5.82. Water injected into the lower permeability Rannoch (*A*) raises the fluid potential difference between the sands to such an extent (sometimes as high as 1000 psi)

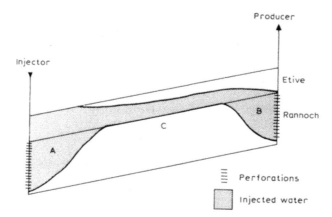

Fig. 5.82. Movement of injected water resulting from well completions on the Rannoch sand.

that the water moves upwards against gravity and for such a differential easily flows through the micaceous sand into the Etive through which it is channelled towards the producer. At point B, production from the low-permeability Rannoch at high-pressure drawdown pulls water down through the tight interval into the producing perforations in the form of a cone, typically extending about 800 ft from the wellbore. In between the wells, however, at point C, the gravity–capillary forces prove too weak to cause effective downward flooding through the restriction. The success in applying this completion policy varies from field to field and depends on the average permeability in the Rannoch and the vertical permeability across the micaceous interval. In some fields, for instance, the method was precluded because it would require far too many injection wells for which the necessary drilling slots were not available on the platform. Oil recovery from the Etive is high, over 50% of the STOIIP but, on account of the unfortunate flooding pattern shown in Fig. 5.82, will probably not exceed 25% in the Rannoch.

Original studies in some of the earlier fields to be developed in the area, upon which the platforms were designed, proved to be optimistic both in terms of ultimate oil recovery and the rate of recovery. Concerning the latter there was a failure to appreciate how rapidly the watercut would develop through what amounts to "short-circuiting" the water through the Etive. Operators suffered from the disadvantages of developing a new area where there was no production experience and the fact that at the appraisal stage the reservoirs were only viewed under static conditions giving no hint of the degree of communication between the Etive and Rannoch. In some cases, errors of observation were made again through the dangerous practice of only viewing the permeability distribution across the sands on the log-k scale. As can be seen in Fig. 5.81a, this makes the reservoir appear perfect: with a gently coarsening upward in properties which should have provided piston-like displacement across the section. On this scale, the tight interval appears as just a minor "blip" which would cause little hindrance to vertical fluid movement. This led to an over simplification in some of the early simulation studies as shown in Fig. 5.83.

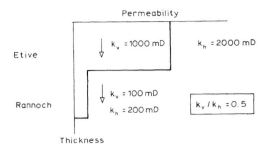

Fig. 5.83. Over simplified cross-sectional model for Etive–Rannoch flooding.

Typically, the two sands were modelled as homogeneous with thickness-averaged horizontal permeabilities, k_h, and a uniform vertical/horizontal permeability ratio, k_v/k_h, which sets the vertical permeabilities, k_v. In its initialization routines, the simulator then calculates the vertical permeability at the interface between the sands using a harmonic averaging procedure [9] which, if the layers are of equal thickness, has the form:

$$k_v = \frac{2 K_{ET} \times K_{RAN}}{K_{ET} + K_{RAN}} = \frac{2 \times 1000 \times 100}{1000 + 100}$$

which for the data presented in Fig. 5.83 would give a vertical permeability between the sands of 180 mD. This, in turn, proved large enough to permit total slumping of the water downwards from the Etive to form a piston frontal advance across both sands, as depicted in Fig. 5.84. As the front reached the flank wells, they watered out "overnight" and their simultaneous closure had the effect of controlling the overall watercut until the flood had advanced to the crestal wells, when the recovery was complete (Fig. 5.84). This pattern will be recognised as the favourable plot in Fig. 5.13, whereas the reality corresponds to the plots labelled "field data". This type of oversight resulted from a lack of attention to detail. The detail in this case being revealed by the linear permeability distribution, Fig. 5.81b, which shows that at the base of the Etive, there is a contrast in permeability of at least two orders of magnitude which must be catered for in any simulation study. That is, it is quite legitimate to model the sands as two layers, with pseudos input for each, but between the sands the vertical permeability must be set at a suitably

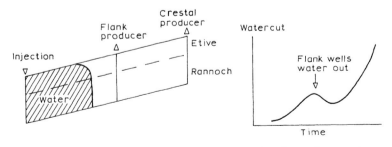

Fig. 5.84. Modelling of piston-like displacement, Etive–Rannoch reservoir.

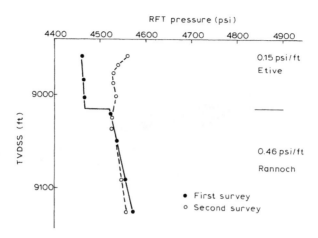

Fig. 5.85. Vertical pulse test using the RFT (Etive–Rannoch reservoir).

low level — but what is that "suitable" level and how should it be determined? Direct measurement of vertical permeabilities on cores cut from the micaceous zone is not altogether convincing because it represents such a small, and possibly unrepresentative, area of the total interface. One method has been described [32] which relies on a form of vertical pulse test performed with the RFT (Fig. 5.85). In this a disturbance was caused in the reservoir by closing in a high-rate injection well some 40 hours before running an RFT survey in a new development well at the start of a logging job. Injection was then resumed and at the end of the logging job, 40 hours later, a second RFT survey was run. The closure of the injector caused a negative pressure pulse laterally through the high-permeability Etive which was then transmitted downwards into the Rannoch while the re-injection during the logging job caused exactly the opposite effect. The pressure gradients in Fig. 5.85 are quite artificial since at this location the Etive was flooded with water and, due to poor downward drainage, the Rannoch contained oil. They reflect a lag in response in both sands due to the pulses which were propagated primarily through the extremely high permeability at the base of the Etive. For instance, in the lower part of the Rannoch pressures are still falling at the time of the second survey in response to the original negative pressure pulse. Dynamic profiles such as this can be evaluated analytically [33] or perhaps more accurately by history matching the test sequence with a three-dimensional numerical simulation model [34]. Results indicate that for some fields the vertical permeability between the two sands is about 10–15 mD. This type of test can only be performed in high-permeability reservoirs. The injector was located 2000 ft from the new development well but on account of the high permeability a significant pressure pulse of 70 psi could be detected 40 hours after creating the original perturbation in the injection well.

Knowledge of the vertical interface permeability enables a realistic cross-sectional model to be constructed to gain a better physical understanding of the displacement process. Such a model is shown in Fig. 5.86. It consists of 13 layers, including the

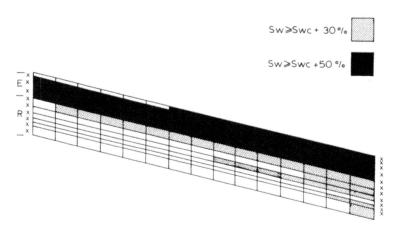

Fig. 5.86. Cross-sectional simulation of waterdrive in the Etive–Rannoch reservoir.

tight interval, and the producer and injector are perforated across the entire section. The flooding state displayed in Fig. 5.86 shows, after several years of injection, that the Etive has been almost completely flooded but the vertical and lateral sweep of the Rannoch is both irregular and poor. Pseudo-relative permeabilities can be generated directly from the model but it is also possible to do the same analytically. The method is precisely the same as described in the previous field example: make a reasonable "guess" at the flooding order of the 13 layers and evaluate the averaging procedures, equations 5.30–5.32, for this order to generate the pseudos. In this type of flood the guess is not particularly difficult but suppose it is made to coincide with the flooding order observed in the cross-section, Fig. 5.86. Then the analytical and simulated results, expressed in terms of the fractional flow across the formation, Fig. 5.87, correspond very closely. This merely demonstrates

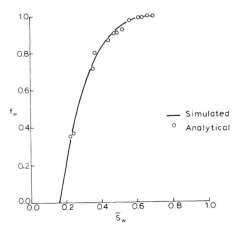

Fig. 5.87. Simulated versus analytical fractional flow relationships, Etive–Rannoch reservoir.

again the veracity of the simple analytical methods for generating pseudos/fractional flows presented in this chapter. Both are concave downwards across the movable saturation range displaying an unfavourable displacement efficiency for the reservoir as a whole: in spite of the fact that in all East Shetland Basin fields the mobility ratio is favourable ($M < 1$). It is the dominance, once again, of vertical heterogeneity with little assistance from gravity that predominates in dictating the vertical sweep. Manifested as a field watercut, the reservoir fractional flow produces the trends plotted in Fig. 5.13 for three of the fields in the area. The unanticipated severity of the watercut development put great strain on the capacities of the topsides injection and production facilities, and particularly the latter, which were underdesigned to handle so much water so soon. Much thought is still and will continue to be given to raising the oil recovery from the low-permeability Rannoch and any technical innovations such as horizontal well completions, to compensate for the lack of well slots on the platforms, is hastily seized upon. In spite of this, estimated recovery factors have tumbled since the initial evaluations and in some cases, continue to do so albeit in the usual quiet and dignified manner. As an operator once described the situation — "We shall approach the truth asymptotically".

(c) The overall management of waterdrive fields

There is no particular interest in the development of excellent waterdrive fields — they can look after themselves. Where the reservoir engineering is most urgently required is in difficult fields in which recovery can sometimes be sacrificed through misconceptions concerning the very basics of the subject. Foremost amongst these are the

- deferment in drilling injection wells
- drilling too few injectors in proportion to producers
- fear of producing water
- incorrect design and use of surface facilities.

All these seem to arise from a misunderstanding of the basic nature of the process in which it is the *drive* in the word waterdrive that must be emphasised and can best be appreciated by consideration of the basic statement of material balance:

$$q_{wi} = q_o B_o + q_{wp} B_w \tag{5.6}$$

As pointed out in section 5.3c, it is the left-hand side of this equation, the injection that controls and can be controlled and which exerts the *drive* — nature looks after the right-hand side. Yet it is in difficult fields: defined as having a severe concave downwards fractional flow because of unfavourable mobility ratio, adverse heterogeneity, or both, where the drive aspect of the process requires greatest consideration.

To proceed around the "long corner" of an unfavourable fractional flow (Fig. 5.88) requires the circulation of many pore volumes of water, accompanied by production at very high watercut, otherwise much of the movable oil will remain unrecovered. The sooner this is appreciated in a field and measures taken to accelerate the rate

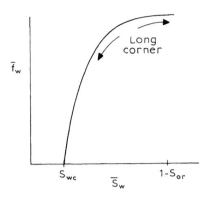

Fig. 5.88. Reservoir fractional flow in a "difficult" waterdrive field.

of circulation of water through the reservoirs, the better. Therefore, the deferment of drilling injectors and the policy of drilling too few are counter productive. In poor waterdrive fields, it is the injection wells that provide the greatest return on investment — not the producers. The oil must be driven out of the reservoirs — not sucked.

The fear of producing water will also lead to a reduction in oil recovery in a given time. It is necessary to produce large volumes of water to produce oil in a poor field. To get matters in perspective, once the watercut of a field exceeds 50%, which can happen with alarming rapidity in some cases, then it becomes — a water field with a producing oilcut. Management of the water becomes the primary concern and the oil is produced as a by-product. In this respect, as emphasised throughout the chapter, it is desirable, especially in offshore projects, to size the capacities of the water injection/production facilities correctly in the first place. This is often difficult but can be done if sufficient attention is paid to detail in generating realistic fractional flow relationships for the reservoir sections. But even if errors are made in the original design, the basic aim should be to utilize the facilities installed to the maximum throughout the project lifetime. That is, as described for Field A in this section, attention must be focused on the rapid circulation of water through the worst parts of the field where water production occurs prematurely. Additional injectors should be drilled and as many pore volumes of water as possible circulated while there is still spare capacity to inject and produce it: before breakthrough occurs in the field as a whole when the poorer areas will have to be abandoned prematurely.

REFERENCES

[1] Craig, F.F., Jr.: The Reservoir Engineering Aspects of Waterflooding, SPE Monograph, 1971.
[2] Nadir, F.T.: Thistle Field Development, SPE Europec Conference (EUR 165), October 1980.
[3] U.K. Government: North Sea Depletion Policy, H.M. Stationery Office, December 1992.
[4] Bishlawi, M. and Moore, R.L.: Montrose Field Reservoir Management, SPE Europec Conference, London, (EUR 166), October 1980.

[5] Arnold, D.M., Hall, P.C. and Crawford, P.B.: The Effect of Fluid Properties and Stage Depletion on Waterflood Oil Recovery, Trans. AIME, 1962: 1165–1168.

[6] Mitchell, R.W.: The Forties Field Seawater Injection System, JPT, June 1978.

[7] Castle, G.R.: North Sea Scorecard, SPE, 61st Annual Technical Conference, New Orleans, October 1986.

[8] Jones, S.C. and Roszelle, W.O.: Graphical Technique for Determining Relative Permeabilities from Displacement Experiments, JPT, May 1978.

[9] Dake, L.P.: Fundamentals of Reservoir Engineering, Elsevier, Amsterdam, 1978.

[10] Buckley, S.E. and Leverett, M.C.: Mechanism of Fluid Displacement in Sands, Trans. AIME, 1942, Vol. 146: 107.

[11] Cardwell, W.T.: The Meaning of the Triple Value in Noncapillary Buckley-Leverett Theory, Trans. AIME, 1959, Vol. 216: 271.

[12] Welge, H.J.: A Simplified Method for Computing Oil Recovery by Gas or Water Drive, Trans. AIME, 1952, Vol. 195: 91.

[13] Hagoort, J.: Measurement of Relative Permeability for Computer Modelling/Reservoir Simulation, Oil and Gas J., February 20, 1984: 62.

[14] Dietz, D.N.: A Theoretical Approach to the Problem of Encroaching and By-Passing Edge Water, Akad. van Wetenschappen, Amsterdam, 1953, Proc. Vol. 56B: 83.

[15] Coats, K.H., Dempsey, J.R. and Henderson, J.H.: The Use of Vertical Equilibrium in Two Dimensional Simulation of Three Dimensional Reservoir Performance, Soc. Pet. Eng. J., March 1971: 63.

[16] Thomas, G.W.: Principles of Hydrocarbon Reservoir Simulation, IHRDC Publishers, Boston, Mass., 1982: 160.

[17] Rapoport, L.A. and Leas, W.J.: Properties of Linear Waterfloods, Trans. AIME, 1953, Vol. 198: 139.

[18] Van Daalen, F. and van Domselaar, H.R.: Waterdrive in Inhomogeneous Reservoirs — Permeability Variations Perpendicular to the Layer, Soc. Pet. Eng. J., June 1972: 211.

[19] Coats, K.H., Nielsen, R.L., Terhune, Mary H. and Weber, A.G.: Simulation of Three Dimensional, Two Phase Flow in Oil and Gas Reservoirs, Soc. Pet. Eng. J., December 1967: 377.

[20] Stiles, W.E.: Use of Permability Distribution in Water Flood Calculations, Trans. AIME, 1949, Vol. 186: 9.

[21] Dykstra, H. and Parsons, R.L.: The Prediction of Oil Recovery by Waterflood, Secondary Recovery of Oil in U.S., API, 1950: 160.

[22] Osman, M.E.: Waterflooding Performance and Pressure Analysis of Heterogeneous Reservoirs, SPE. Middle East Oil Technical Conference, Bahrain, 1981 (SPE 9656).

[23] El-Khatib, N.: The Effect of Crossflow on Waterflooding of Stratified Reservoirs, Soc. Pet. Eng. J., April 1985: 291.

[24] Peaceman, D.W.: Fundamentals of Numerical Reservoir Simulation, Elsevier, Amsterdam, 1978.

[25] Aziz, K. and Settari, A.: Petroleum Reservoir Simulation, Applied Science Publishers, London, 1979.

[26] Crichlow, H.B.: Modern Reservoir Engineering — A Simulation Approach, Prentice-Hall Inc., Englewood Cliffs, N.J., 1977.

[27] Jacks, H.H., Smith, O.J. and Mattax, C.C.: The Modelling of a Three-Dimensional Reservoir with a Two-Dimensional Reservoir Simulator — The Use of Dynamic Pseudo Functions, Soc. Pet. Eng. J., June 1973: 175.

[28] Kyte, J.R. and Berry, D.W.: New Pseudo Functions to Control Numerical Dispersion, Soc. Pet. Eng. J., August 1975: 265.

[29] Massie, I., Beardall, T.J., Hemmens, P.D. and Fox, M.J.: Murchison: A Review of Reservoir Performance during the First Five Years, SPE Europec Conference.

[30] Stiles, J.H. and Valenti, N.P.: The Use of Detailed Reservoir Description and Simulation Studies in Investigating Completion Strategies, Cormorant — UK North Sea, SPE Europec Conference, London: 1984 (SPE-16553).

[31] Bayat, M.G. and Tehrani, D.H.: The Thistle Field-Analysis of its Past Performance and Optimisation of its Future Development, SPE Europec Conference, London: 1984 (SPE-13989).

[32] Dake, L.P.: Application of the Repeat Formation Tester in Vertical and Horizontal Pulse Testing in the Middle Jurassic Brent Sands, SPE Europec Conference, London, 1982 (EUR 270).

[33] Stewart, G.: The Interpretation of Distributed Pressure and Flow Measurements in Produced Reservoirs, SPE Europec Conference, London, 1982 (EUR 272).

[34] Lasseter, T., Karakas, M. and Schweitzer, J.: Interpreting an RFT-Measured Pulse Test with a Three Dimensional Simulator, SPE Formation Evaluation, March 1988: 139.

Chapter 6

GAS RESERVOIR ENGINEERING

6.1. INTRODUCTION

Following a brief description of the basic PVT for gas/gas-condensate systems, three main topics relating to the broader aspects of gas reservoir engineering are addressed. These are: the application of material balance, the immiscible displacement of oil by injected gas and dry gas recycling to enhance the liquid hydrocarbon recovery from gas-condensate reservoirs.

In applying the material balance equation to the production/pressure history of a gas reservoir the basic aims are to define the drive mechanism (natural waterdrive or volumetric depletion) and estimate the gas initially in place (GIIP), both being required in constructing a more detailed numerical simulation model for field performance predictions. Yet, as demonstrated, the most popular method in the Industry for applying material balance: the p/Z plot, can be extremely insensitive in practice leading to misjudgment of the drive mechanism and a serious overestimation of the GIIP. The more sensitive method of Havlena-Odeh is recommended as a means of checking the validity of the p/Z plot.

For gas displacing oil and dry gas displacing wet in recycling, the emphasis is on presenting useful analytical techniques for describing the sweep efficiency of both processes in macroscopic reservoir sections. In doing so the theory of waterdrive presented in the previous chapter is extended and modified to cater for these different forms of displacement.

6.2. PVT REQUIREMENTS FOR GAS-CONDENSATE SYSTEMS

Improvements in our understanding of the complex thermodynamics of hydrocarbon mixture phase behaviour have been so significant and extensive since the 1950's that by now it has emerged as a subject in its own right. The advances in knowledge have been encouraged by, and kept pace with, the rapid developments in computer software that are so essential for the manipulation of complex equations of state (EOS) and compositional numerical simulation modelling required to describe gas recycling operations and EOR miscible gas floods.

Presented in this section, however, are the essential PVT fluid property relations adequate to describe the macroscopic features of gas and gas-condensate reservoir development that must be considered preparatory to performing more detailed studies in which the PVT data may have to be handled in a more refined manner.

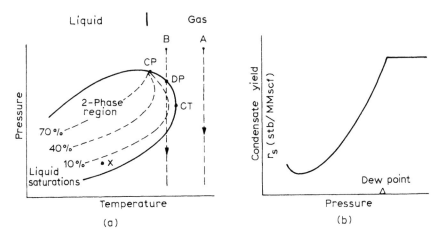

Fig. 6.1. (a) Hydrocarbon mixture phase envelope. (b) Condensate yield function.

The types of hydrocarbon system described in the chapter are illustrated by the phase envelope and have initial states at points *A* and *B* (Fig. 6.1a). Since these both lie to the right of the critical point (CP), they are initially in the gaseous phase in the reservoir. During reservoir pressure depletion, which is normally assumed to occur under isothermal conditions, the gas initially at point *A* will remain as a single phase single phase gas in the reservoir since its depletion path lies to the right of the cricondentherm (CT): the maximum temperature on the phase envelope. Thus it never crosses into the two-phase region. Nevertheless, in producing the gas to the surface, there will be a reduction in both pressure and temperature so that some liquid hydrocarbons will be collected in the surface separator operating at pressure and temperature represented by point *X* within the two-phase envelope.

A hydrocarbon mixture with initial condition at point *B* (Fig. 6.1a), which has a temperature intermediate between that of the critical point and the cricondentherm, is referred to as a retrograde gas-condensate system. Pressure depletion at constant temperature means that, at the dew point (DP), the path crosses into the two-phase region and liquid hydrocarbons (condensate) will be deposited in the reservoir. There are two unfortunate consequences associated with this occurrence. In the first place, the condensate is deposited at such low liquid saturation that it is usually trapped by surface tension forces and is therefore immobile and cannot be produced. Secondly, it is the heavier, richer hydrocarbon components that are condensed first and therefore their retention in the reservoir represents a serious loss of the more valuable hydrocarbons. The PVT parameter which is of greatest significance in quantifying the potential loss of liquids is the condensate yield (Fig. 6.1b). This represents the condensate recovered by the surface separators: r_s (stb/MMscf of dry gas). Above the dew point, all the liquid hydrocarbons contained in each MMscf of gas are recovered but below the dew point, since liquid is deposited in the reservoir, there is a growing deficiency in the volume of condensate recovered at the surface as the reservoir pressure continues to decline. Eventually, as the reservoir pressure decreases so that it approaches the base of the two-phase region some condensate in the reservoir will evaporate thus increasing the surface yield but in the field this is not always observed since abandonment may occur at a higher pressure. The condensate yield is measured in constant volume depletion experiments, as described in section 6.2c. The initial value of the yield affects decision making concerning the manner of field development. If it is low, say, less than 50 stb/MMscf, then it may not prove economically viable to do other than simply deplete the accumulation and suffer the inevitable loss in condensate recovery. If the yield is higher, however, and it can exceed 250 stb/MMscf, then measures can be taken to maintain the reservoir pressure above the dew point so that each MMscf of gas produced will contain its maximum condensate yield and no liquid will be deposited in the reservoir.

The most popular and effective means of maintaining pressure is by the process of dry gas recycling in which, after removal of the liquid condensate at the surface, the dry gas is reinjected into the reservoir where it helps to maintain pressure and displace the wet gas towards the producing wells, as described in section 6.5. Alternatively, pressure maintenance could be achieved by water injection but considering the potential wastage of gas associated with this process, described in section 6.3, it is not usually considered as a recovery method.

(a) Equation of state

An advantage in describing the PVT properties of gas, in comparison to oil, is that all three parameters: p, V and T can be related by a simple equation of state (EOS), that is

$$pV = ZnRT \tag{6.1}$$

in which, in field units

p = pressure (psia)
R = gas constant (10.732)
V = volume (cu.ft.)
T = absolute temperature (460 + ºF)
n = quantity of gas (lb moles)
Z = dimensionless, Z-factor.

At low pressure the Z-factor is close to unity but at higher pressure it demonstrates the typical sort of deviation shown in Fig. 6.2a. The Z-factor accounts for the fact that at higher pressures the gas molecules are so closely packed that they occupy a finite volume and exert a significant inter-molecular attraction upon one another. At the commencement of any gas field development study, the engineer must determine both the Z-factor and p/Z as functions of pressure (Fig. 6.2b), the latter being required in material balance calculations (section 6.3).

The Z-factor can be determined by direct laboratory experiment (section 6.2c) or by correlation, the basic one for hydrocarbon gases being that of Standing and Katz [1] as described in reference 2 and many other texts. Nowadays, however,

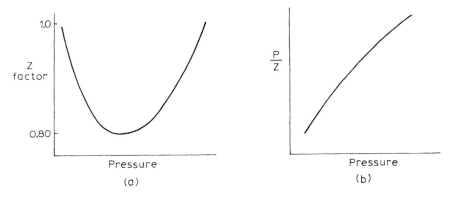

Fig. 6.2. (a) typical shape of Z-factor function. (b) Relationship between p/Z and pressure.

Z-factor calculations, making full allowance for hydrocarbon and non-hydrocarbon components, can be readily performed using pocket calculator programs such as contained in the Hewlett-Packard "Petroleum Fluids Pac" [3]. Programs such as this essentially use the basic Standing-Katz correlation, as best fitted analytically using a complex EOS; in this particular case being the equation of Benedict et al. [4]. Since it is assumed that such methods for calculating Z-factors are readily available to engineers, the correlation methods will not be described further in this text.

(b) Surface/reservoir volume relationships

There are two ways of relating surface volumes of gas to their equivalent in the reservoir. The first of these is the gas expansion factor, E (scf/rcf), which may be derived for a given quantity of gas using equation 6.1 as

$$E = \frac{V_{sc}}{V_{res}} = \left(\frac{Z_{sc}RT_{sc}}{p_{sc}}\right)\left(\frac{p}{ZRT}\right) = \left(\frac{1 \times R \times (460 + 60)}{14.7}\right)\left(\frac{p}{ZRT}\right)$$

$$E = 35.37\frac{p}{ZT}\left(\frac{\text{scf}}{\text{rcf}}\right) \tag{6.2}$$

The second form is the inverse relationship — B_g (rcf/scf) the gas formation volume factor (FVF), which is more commonly used in oilfield engineering applications such as material balance (Chapter 3). While for oilfield engineering the common unit of B_g is rb/scf for gas reservoir engineering it is more conventional to use the compatible units of rcf/scf — which is the practice adhered to in this chapter. The relationship between the two factors is obviously $E = 1/B_g$.

If anything, use of the gas expansion factor, E, seems more popular in gas reservoir engineering than the gas FVF, B_g, for the simple reason that the former is practically a linear function of pressure at constant temperature, as is evident from equation 6.2, whereas B_g is almost hyperbolic in nature (see Fig. 6.3b). Consequently, for computational ease and accuracy in interpolation and extrapolation the use of the gas expansion factor is often preferred.

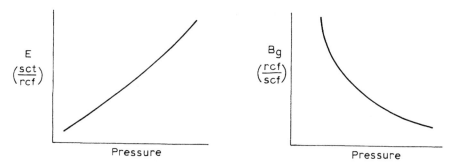

Fig. 6.3. (a) Gas expansion factor. (b) Gas FVF — as functions of pressure.

(c) Constant volume depletion (CVD) experiments

This is the basic form of laboratory experiment required to define the PVT properties of retrograde condensate systems [5]. A quantity of the reservoir fluid is charged to a visual PV cell which is maintained at reservoir temperature throughout the experiment. If the initial gas volume is G (scf) and the gas expansion factor is E_i (scf/rcf), then the cell's volume at this pressure, which is the starting point of the experiment, is G/E_i — the initial HCPV. Following the determination of the dew point, by visual inspection, the cell pressure is decreased in stages and for each step, the expanded fluids are withdrawn from the cell at the reduced pressure.

At each stage of depletion, the volume of liquid condensate deposited in the cell is measured and reported as a fraction of the initial HCPV and the volume of the gas expelled is measured at both cell and standard conditions. These permit the calculation by material balance of

- r_s (stb/MMscf), the diminishing condensate yield below the dew point pressure (Fig. 6.1b)
- the two-phase Z-factor

The latter, which accounts for the deposition of liquid condensate in the cell, is calculated from material balance as

Gas produced	=	GIIP	−	gas remaining in the cell
(scf)		(scf)		(scf)

$$G_p \quad = \quad G \quad - \quad \frac{G}{E_i}E$$

where G/E_i is the original HCPV of the cell. At constant temperature, E is directly proportional to p/Z (equation 6.2) and therefore the material balance may be conveniently solved for the Z-factor as

$$Z = \frac{p}{\dfrac{p_i}{Z_i}\left(1 - \dfrac{G_p}{G}\right)} \tag{6.3}$$

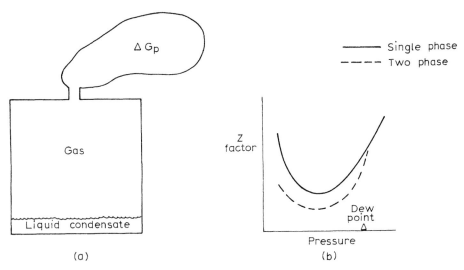

Fig. 6.4. CVD experiment and resulting single- and two-phase Z-factors.

If dealing with a dry gas, such that no condensate is deposited, equation 6.3 will determine the single-phase Z-factor. But for a retrograde condensate gas, liquid will be left behind so that the cumulative production at any stage of depletion, G_p, will be smaller than for a dry gas. Consequently, the so-called two-phase Z-factor for a retrograde condensate will be lower than the single-phase value once the pressure has fallen below the dew point, as shown in Fig. 6.4b. Also reported, even for retrograde condensates, is a single-phase Z-factor obtained by applying the ratio of equation 6.1 evaluated for the gas in the PV cell and at standard conditions.

For reservoir engineering calculation purposes, however, it is the two-phase Z-factor that is used in such applications as material balance (section 6.3), since in relating the gas recovery G_p to the initial value, G, for a given pressure, it makes allowance for the fact that liquid condensate is left behind in the cell, as it is in the reservoir. It is still necessary, however, to take account of the gas liquids that are condensed in the surface separators. The G_p used in material, balance calculations (section 6.3) must be the cumulative "wet" gas production whereas, what is measured are the volumes of dry gas and separated liquids. The latter, however, may be expressed as an equivalent gas volume, as described in references 2 and 6, using the expression, in field units

$$GE = 1.33 \times 10^5 \frac{\gamma_o N_p}{M} \quad \text{(scf)} \tag{6.4}$$

where γ_o is the specific gravity of the condensate, measured at each level of depletion, M its molecular weight and N_p is the cumulative condensate production (stb). Adding the GE volume to the cumulative dry gas production, which usually amounts to a fairly small correction, gives the cumulative wet gas production G_p for use in field development studies. In field examples in the remainder of the chapter it may be assumed that for dry gas reservoirs the *GE* correction has been applied and for retrograde condensate reservoirs the combination of the *GE* correction and use of the two-phase *Z*-factor.

(d) Gas compressibility and viscosity

These unrelated PVT properties are considered in conjunction because they do more to distinguish gas from oil reservoir engineering than any other parameters: the gas compressibility being very high and the viscosity extremely low, compared to liquids. The compressibility is the isothermal value which can be calculated by differentiation of the real gas EOS (equation 6.1) as [2]

$$c = -\frac{1}{V}\frac{\partial V}{\partial p}\bigg|_T = \frac{1}{p} - \frac{1}{Z}\frac{\partial Z}{\partial p} \quad \text{(psi)}^{-1} \tag{6.5}$$

which to a first approximation may be evaluated as the reciprocal of the pressure. Therefore, for an initial pressure of, say, 5000 psia the gas compressibility would be 200×10^{-6}/psi which is an order of magnitude greater than for a typical undersaturated oil.

Gas viscosity is usually calculated at each stage of depletion during CVD experiments. That is, chromatographic gas analysis establishes the mole fractions of the hydrocarbon and non-hydrocarbon constituents from which the viscosity may be reliably calculated using standard correlations [3]. A typical value of gas viscosity in the reservoir is $\mu = 0.025$ cp making it 40 times smaller than for even a favourable light oil viscosity of 1 cp. Gas viscosity increases with pressure while the compressibility is almost hyperbolic, as shown in Fig. 6.5. Thus the product of the two, μc, is usually fairly constant over quite large pressure differences.

(e) Semi-empirical equations of state (EOS)

An important application of CVD experiments is that the results are used to calibrate semi-empirical equations of state, whose development has received much attention in the literature in recent years. The basic idea is that if the EOS can be

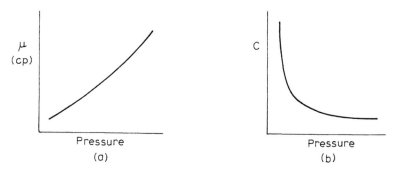

Fig. 6.5. (a) Gas viscosity and (b) compressibility as functions of pressure.

tuned to replicate the experimental results, then it may be used to calculate PVT properties, not just of the bulk gas, but catering from the individual hydrocarbon components and their phases, gas or liquid, at different states of pressure and temperature. In this way, it can be used to predict the outcome of separator flash calculations, for instance, instead of performing a costly and time consuming series of experiments. The EOS approach is to attempt to match modified forms of the basic equation of van der Waal, for a unit quantity of gas

$$p = \frac{RT}{V - b} - \frac{a}{V^2} \tag{6.6}$$

which is the fundamental way of expressing equation 6.1 and in which a and b were originally believed to be positive constants. The term a/V^2 accounts for the intermolecular attraction and b for the finite volume occupied by the gas molecules. Two of the most popular EOS modifications of equation 6.6 are those of Redlich and Kwong [7] (as adapted by Soave [8]) and Peng and Robinson [9]. The latter is based on an altered version of the EOS

$$p = \frac{RT}{V - b} - \frac{a(T)}{V(V + b) + b(V - b)} \tag{6.7}$$

which may be alternatively expressed in terms of the Z-factor (pV/RT) in cubic form as

$$Z^3 - Z^2(1 - B) + Z(A - 3B^2 - 2B) - (AB - B^2 - B^3) = 0$$

in which $A = ap/R^2T^2$ and $B = bp/RT$. The parameters a and b can be evaluated for each component of a gaseous mixture and for their composite; they may include dependence on critical pressures and temperatures, binary interaction coefficients between the non-hydrocarbon components and the hydrocarbons and in particular between methane and the heavier hydrocarbon elements — and many other factors. The art of "tuning" an equation of state to duplicate the results of CVD experiments [5,10] consists of applying non-linear regression analysis to evaluate the constants and interaction parameters until an acceptable match is achieved. Altogether, it is a complex and, to a certain extent, a subjective process and, although methods are continually improving, sometimes the degree of sophistication sought is difficult to justify in practice. It is recommended, for instance, that for useful application of an EOS to a gas-condensate system, that compositional analysis be available extending to at least C_{20} which is necessary because one of the most influential parameters in tuning an EOS (to match the dew-point pressure) is the interaction coefficient between methane and the heavier hydrocarbon components, whose mole fractions must therefore be well defined. And, although for a given fluid sample, sophisticated chromatographic analysis can readily define C_{20+} components, there will always be some doubt about whether the heavier components existing in the reservoir have been collected and are contained in the surface sample being used for EOS calibration. The most popular method of sampling condensate reservoirs is by surface recombination which, even under favourable circumstances, may lead to drop-out of some of the heavier fractions in the reservoir, close to the well, or during flow to the surface. This can only be overcome by taking great care in achieving stability of flow during testing [11]. Consequently, the basic limitation in EOS modelling must inevitably be the reliability of sampling so that the equation is calibrated using actual reservoir fluids rather than a laboratory sample which may well be deficient in some of the heavier components.

There are many applications of EOS techniques throughout the gas production/ transportation business [12]. In reservoir engineering, two of the main applications are

- Flash equilibrium calculations to determine conditions of surface separation that will optimise the stabilized volumes of condensate production from gases.
- Application in compositional numerical simulation modelling to simplify the handling of PVT relationships.

In the latter case, while it would be impracticable in terms of computer running time to cater for >20 separate components, groups of hydrocarbons and non-hydrocarbons can be aggregated to form sets of pseudo-components. These may vary between two and twelve, dependent on the complexity of the hydrocarbon system and the reservoir application. For example, a set of six pseudo-components may consist of: $C_1 + N_2$, C_2–C_5, C_6–C_9, C_{10}–C_{17}, C_{18}–C_{24} and C_{25+}. The equation of state is then adjusted to determine the pseudo-component properties subject to consistency checks suggested by Coats [10].

Even so, compositional modelling is a luxury and the engineer must be quite certain at the commencement of a study whether its use is justified. As described in section 6.5, for instance, it is not required for the majority of depletion fields nor for high-pressure dry gas recycling, above the dew point pressure, when simpler approaches can be applied without significant divergence from the results obtained by full compositional modelling.

6.3. GAS FIELD VOLUMETRIC MATERIAL BALANCE

(a) Appropriateness in application

Whether material balance can be applied to a hydrocarbon accumulation as a whole depends upon how rapidly any pressure disturbance is equilibrated in the reservoir so that it may be treated as zero dimensional. This, in turn, is dependent on the magnitude of the hydraulic diffusivity constant, $k/\phi\mu c$, as described in Chapter 3, section 3.3: the larger the value of this parametric group, the more rapidly is pressure equilibrium achieved. Considering a reservoir with average rock properties k and ϕ, the diffusivity constant will be several times larger if it contains gas rather than oil on account of the lower μc-product of the former. To take typical data at a pressure of 5000 psia

$$[\mu c]_{oil} = 1 \text{ cp} \times 20 \times 10^{-6}/\text{psi} = 20 \times 10^{-6} \text{ cp/psi}$$

$$[\mu c]_{gas} = 0.02 \text{ cp} \times \frac{1}{5000 \text{ psi}} = 4 \times 10^{-6} \text{ cp/psi}$$

in which the gas compressibility has been approximated as the reciprocal of the pressure using equation 6.5. Therefore, in spite of the high gas compressibility, its extremely low viscosity dominates in making the diffusivity constant five times greater than for oil, in this particular case, which enhances the prospect for meaningful application of material balance — even in tight gas reservoirs. Therefore, there has always been a tradition in the Industry to apply the technique to history match and predict reservoir performance and also to estimate the GIIP, which has survived to a greater extent than for oilfields. Nevertheless, the engineer, must never simply assume that the technique is appropriate in every situation but rather verify that pressure equilibrium is rapidly attained in a reservoir by plotting individual well pressures as a function of time or cumulative production, as illustrated in Fig. 3.2. Parts of the reservoir may be isolated by sealing faults, requiring the separate application of material balance to each compartment.

Even in a situation in which there is a distinct lack of pressure equilibrium throughout the reservoir, the necessary condition for application of material balance: that an "average" pressure decline be defined for the system, can be achieved by applying the averaging procedure

$$\frac{\bar{p}}{Z} = \frac{\sum_j (p/Z)_j (\Delta G_p)_j / (\Delta p/Z)_j}{\sum_j (\Delta G_p)_j / (\Delta p/Z)_j} \tag{6.8}$$

which is summed over all the individual well production/pressure histories. It is the analogy for gas fields of equation 3.19 applied to non-equilibrium oilfields and can be generated using precisely the same arguments as in Chapter 3, section 3.3[†]. In applying equation 6.8, the ΔG_p's are the increments in cumulative gas production over the selected time steps of, say, six months and the $\Delta p/Z$'s are the corresponding changes in this group which replaces the pressure p, as it often does in gas reservoir engineering, which naturally arises from the definition of E, equation 6.2. In principle, application of equation 6.8 enables an average pressure (p/Z) decline trend to be defined for any reservoir, irrespective of the lack of pressure uniformity between wells. And, although the technique may appear somewhat cumbersome, there are usually compelling reasons for applying material balance to a reservoir in advance of the construction of a more complex numerical simulation model.

Gas material balance is supposed to be one of the simplest subjects in the whole of reservoir engineering and, indeed, the physics and mathematics are quite trivial. Yet there are great subtleties attached to its application which have perhaps not been stressed sufficiently in the literature and, if unappreciated, can lead the engineer into serious error in assessing the reservoir drive mechanism (waterdrive or volumetric depletion) and the GIIP. Foremost amongst the sources of error is the use of the traditional p/Z versus cumulative production plot in attempting to history

---

[†] Personal communication: Dr. E. Balbinski, A.E.A., Winfrith, Dorset, 1991.

match the performance of gas reservoirs. It can lead to a complete misinterpretation of drive mechanism and a serious overestimation of the GIIP. In fact, it is the intention in writing this section to dissuade engineers from using this most popular technique in isolation but rather in conjunction with other methods. The two principal methods of applying material balance will be briefly described and their accuracy compared in analyzing the production–pressure history of a hypothetical, example field whose performance can be predicted exactly (section 6.3d).

(b) Havlena-Odeh interpretation

In this, the material balance is expressed in reservoir volumes of production, expansion and influx as

Underground withdrawal (rcf)	$=$	Gas expansion (rcf)	$+$	Water expansion/ pore compaction (rcf)	$+$	Water influx (rcf)

$$G_p B_g + W_p B_w = G(B_g - B_{gi}) + G B_{gi} \frac{(c_w S_{wc} + c_f)}{1 - S_{wc}} \Delta p + W_e B_w \qquad (6.9)$$

And, adopting the nomenclature of Havlena and Odeh, similar to that used in Chapter 3, section 3.2.

$$F = G_p B_g + W_p B_w = \text{total gas and water production (rcf)}$$
$$E_g = B_g - B_{gi} = \text{underground gas expansion (rcf/scf)}$$

$$E_{fw} = B_{gi} \frac{(c_w S_{wc} + c_f)}{1 - S_{wc}} \Delta p = \text{expansion of the connate water and reduction of the pore space (rcf/scf)}$$

This reduces equation 6.9 to the simple form

$$F = G(E_g + E_{fw}) + W_e B_w \qquad (6.10)$$

In most practical cases $E_{fw} \ll E_g$ and may be omitted but not before checking that this is a valid neglect of the term across the entire range of pressure depletion. The material balance then becomes

$$F = G E_g + W_e B_w \qquad (6.11)$$

Finally, dividing both sides of the equation by E_g gives:

$$\frac{F}{E_g} = G + \frac{W_e B_w}{E_g} \qquad (6.12)$$

Using the production, pressure and PVT data, the left-hand side of this expression should be plotted as a function of the cumulative gas production, G_p. This is simply for display purposes to inspect its variation during depletion. Plotting F/E_g versus production time or pressure decline, Δp, can be equally illustrative. The plot will

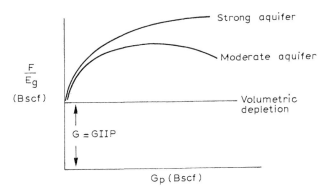

Fig. 6.6. Diagnostic gas material balance plot to determine the GIIP and define the drive mechanism.

have one of the three shapes depicted in Fig. 6.6. If the reservoir is of the volumetric depletion type, $W_e = 0$, then the values of F/E_g evaluated, say, at six monthly intervals, should plot as a straight line parallel to the abscissa — whose ordinate value is the GIIP. Alternatively, if the reservoir is affected by natural water influx then the plot of F/E_g will usually produce a concave downward shaped arc whose exact form is dependent upon the aquifer size and strength and the gas offtake rate (section 6.3d). Backward extrapolation of the F/E_g trend to the ordinate should nevertheless provide an estimate of the GIIP ($W_e \sim 0$) but, as illustrated in section 6.3d, the plot can be highly non-linear in this region yielding a rather uncertain result. The main advantage in the F/E_g versus G_p plot, however, is that it is much more sensitive than other methods in establishing whether the reservoir is being influenced by natural water influx or not. If it is, then common wisdom suggests that every effort should be made to accelerate the production of gas to the greatest extent that is practicably possible. The aim is to evacuate the gas before the less mobile water can catch-up and trap significant quantities of gas behind the advancing flood front. In principle, this is quite feasible since the mobility ratio for water–gas displacement is abnormally low, typically:

$$M = \frac{k'_{rw}}{\mu_w} \bigg/ \frac{k'_{rg}}{\mu_g} = \frac{0.2}{0.4} \bigg/ \frac{1.0}{0.02} = 0.010 \qquad (6.13)$$

meaning that under an imposed pressure differential, the gas can travel 100 times faster than water by which it is being displaced and can therefore be removed before the water has the opportunity to advance significantly. The amount of gas that is trapped may be determined from the equation of state, equation 6.1, as:

$$p \times (\text{pore volume flooded}) \times S_{gr} = ZnRT \qquad (6.14)$$

where S_{gr} is the residual or trapped gas saturation expressed as a fraction of the pore volume and is regarded as independent of the pressure at which the flood occurs. On account of this constancy, it can be seen that the quantity of trapped gas, n (lb-moles), is directly proportional to the pressure: the higher the pressure, the greater the quantity of trapped gas. Conversely, if the pressure is reduced by rapid gas evacuation the volume of gas trapped in each individual pore space, S_{gr}, remains unaltered but its quantity is reduced. The residual gas saturation is usually high, typically having a value of 30–40% PV [6,13]. It can be determined by water–gas imbibition flooding experiments but quite often such experiments are not performed because of a commonly held belief by many operators that gas fields are little affected by natural water influx (section 6.3e).

Considering the gas in place equation

$$\text{GIIP} = G = V\phi(1 - S_{wc})E_i$$

it can be seen that the pore volume of the reservoir, $V\phi$, and hydrocarbon pore volume, $V\phi(1 - S_{wc})$, may be expressed as

$$\text{PV} = \frac{G}{E_i}\frac{1}{1 - S_{wc}}, \quad \text{HCPV} = \frac{G}{E_i} \tag{6.15}$$

and the movable gas volume (MGV) by waterflooding as

$$\text{MGV} = \text{PV}(1 - S_{gr} - S_{wc}) = \frac{G}{E_i}\frac{(1 - S_{gr} - S_{wc})}{(1 - S_{wc})} \tag{6.16}$$

Consequently, after an influx of $W_e B_w$ into a waterdrive gas reservoir, the fractional volumetric sweep may be evaluated as

$$\alpha = W_e B_w \bigg/ \frac{G}{E_i}\frac{(1 - S_{gr} - S_{wc})}{(1 - S_{wc})} \tag{6.17}$$

and therefore the volume of gas trapped behind the advancing water at this stage of the flood can be expressed at standard conditions as

$$\alpha\,(\text{PV})\,S_{gr}E = \alpha\frac{G}{E_i}\frac{S_{gr}}{(1 - S_{wc})}E = \alpha\frac{G S_{gr}}{(1 - S_{wc})}\frac{p/Z}{p_i/Z_i} \tag{6.18}$$

(c) p/Z-interpretation technique

This is by far the most popular method of applying gas material balance in which the equation is formulated at standard conditions (scf) as

$$\begin{array}{ccc}
\text{Cumulative} \\
\text{gas production}
\end{array} = \text{GIIP} - \quad \text{Gas remaining in the reservoir}$$

$$G_p = G - \left(\frac{G}{E_i} - \frac{G}{E_i}\frac{(c_w S_{wc} + c_f)}{1 - S_{wc}}\Delta p - W_e B_w\right)E$$

in which E_i and E are the gas expansion factors at the initial and reduced average pressure. G/E_i is the original HCPV and the second term within the brackets caters for the expansion of the connate water and reduction of pore volume resulting from compaction. The term W_e represents, in this case, the cumulative net water influx, that is, influx minus production. The underground volume of remaining gas is multiplied by the gas expansion factor, E, at the reduced pressure to convert it to standard conditions. Usually, the water and pore compressibilities are negligible in comparison to that of gas and the second term within the brackets can be omitted — after checking its relative magnitude. This reduces the equation to

$$\frac{G_p}{G} = 1 - \frac{E}{E_i}\left(1 - \frac{W_e B_w E_i}{G}\right)$$

and, if the reservoir temperature remains constant, the gas expansion factors can be replaced by their corresponding values of p/Z to give

$$\frac{p}{Z} = \frac{p_i}{Z_i}\frac{\left(1 - \dfrac{G_p}{G}\right)}{\left(1 - \dfrac{W_e B_w E_i}{G}\right)} \qquad (6.19)$$

The term $W_e B_w/(G/E_i)$ represents the fraction of the HCPV invaded by water. Consequently, the greater the influx the higher the pressure for a given offtake of gas. In the event that there is no influx and the reservoir is of the volumetric depletion type, then the equation may be reduced to the form

$$\frac{p}{Z} = \frac{p_i}{Z_i}\left(1 - \frac{G_p}{G}\right) \qquad (6.20)$$

which is a simple linear relationship between p/Z and the fractional gas recovery and gives rise to the popular field technique of plotting the reservoir averaged values of p/Z, in which the pressures are referred to some common datum level, as a function of the cumulative gas production G_p. If the reservoir is of the volumetric depletion type, then the plot must necessarily be linear, as shown in Fig. 6.7, and its extrapolation to the abscissa ($p/Z = 0$) enables the effective GIIP to be determined as $G_p = G$.

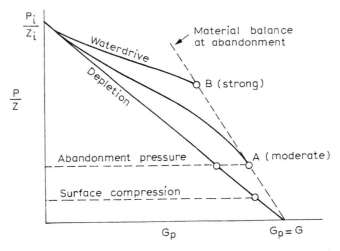

Fig. 6.7. Gas material balance plots for depletion and waterdrive reservoirs.

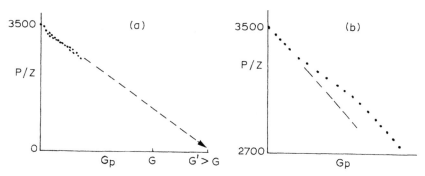

Fig. 6.8. p/Z plots for a waterdrive gas reservoir. (a) Across the total range of p/Z (0–3500). (b) Over a reduced range of p/Z (2700–3500).

Alternatively, if there is natural water influx from an adjoining aquifer, the p/Z plot is, in principle, non-linear. The technique may seem fairly straightforward but this is where the potential danger lies in application of the p/Z plot: deciding what is and what is not a straight line. In a great many cases the plot for a waterdrive field will *appear* to be linear until a very advanced stage of depletion when, in fact, it is not. Then, as shown in Fig. 6.8a, extrapolation of the apparent linear trend to the abscissa will yield a value of the GIIP which is too large ($G' > G$). The error is twofold. In the first place, following inspection of the p/Z plot and noting its apparent linearity, the engineer *assumes* the reservoir to be of the volumetric depletion type. This is followed by the erroneous extrapolation of the trend to the abscissa which determines too large a value of the GIIP. In many cases, the error arises simply from plotting the data across the total range of p/Z ($0-p_i/Z_i$) to demonstrate the full extrapolation (Fig. 6.8a); whereas, if the plot is made with an enlarged p/Z scale, over the range of depletion, then subtle curves appear in the plot as shown in Fig. 6.8b. From this, it can be seen that the only linear portion of the plot occurs very early in the lifetime of the field, before the water influx is significant and extrapolation of this early trend will give a more reliable value of the GIIP, although it is still likely to be too large.

The dual error in mistaking both the drive mechanism and the GIIP is one that has been pointed to frequently in the literature [2,14,15] but obviously the warning has not been stated forcibly enough because the error is still of frequent occurrence in the Industry. Perhaps the most succinct statement concerning the practical application of the p/Z plot is that of Cason [16], namely

> "Theory showing that depletion drive gas reservoirs will exhibit a straight line p/Z plot has been developed but the corollary, that a straight line p/Z plot proves the existence of depletion drive has not been proven".

And, it might be added, never will be. This is an important clarifying statement for it means that any perturbation, such as a change in the production rate, which results in a distortion in the linearity of the p/Z plot, implies that the reservoir is not of the volumetric depletion type.

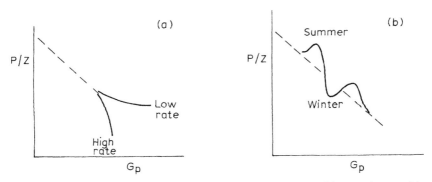

Fig. 6.9. Non-linear perturbations of p/Z plots for waterdrive fields: (a) rate change, (b) cyclic production.

There is a definite rate dependence in waterdrive gas fields. Increasing the rate will evacuate the much more mobile gas before the sluggish water can catch-up, Fig. 6.9 (a), which results in a reduction in pressure; whereas, reducing the rate permits the water influx to be more substantial and pressure to be maintained to a greater degree. Figure 6.9 (b), illustrates the common occurrence of cyclic production in which, to satisfy the fluctuating market demand for gas, fields are produced at higher rates during the winter than the summer months. This leads to peaks and troughs in the p/Z plot, as illustrated. But the point is, that if what appears to be a linear trend can be distorted in any way, as shown above, then Cason's statement must be taken into account and it is most likely that the field is not of the volumetric depletion type.

Referring to the two waterdrive plots in Fig. 6.7, the backward sloping straight line is the material balance at abandonment and its intersection with the non-linear p/Z plots determines the maximum gas that can be recovered: allowing for the fact that some of the gas will be completely by-passed by the water at flood-out while an additional volume is trapped at residual saturation behind the advancing water. The equation may be expressed at standard conditions as

$$
\begin{array}{ccccc}
\text{Abandonment} \\ \text{production}
& = & \text{GIIP} & - & \begin{array}{c}\text{Trapped} \\ \text{residual gas}\end{array} & - & \text{By-passed gas}
\end{array}
$$

$$
G_p \quad = \quad G \quad - \quad \left[\alpha \frac{G}{E_i} \frac{S_{gr}}{1 - S_{wc}} + (1 - \alpha)\frac{G}{E_i} \right] E_{ab} \qquad (6.21)
$$

in which α is the volumetric sweep efficiency at abandonment and E_{ab} the gas expansion factor at the same condition. The equation may be more conveniently expressed in terms of p/Z as,

$$
\frac{p}{Z_{ab}} = \frac{p_i}{Z_i} \frac{\left(1 - \dfrac{G_p}{G} \right)}{\alpha \left[\dfrac{S_{gr}}{1 - S_{wc}} + \dfrac{1 - \alpha}{\alpha} \right]} \qquad (6.22)
$$

and it is this linear function which is plotted in Fig. 6.7 dictating the maximum waterdrive recoveries at points *A* and *B* for moderate and strong waterdrives. Typically, in applying this equation the engineer varies α and S_{gr} (if it has not been determined experimentally) to test their effect on recovery. Reasonable values being $\alpha = 0.70$, $S_{gr} = 0.35$ PV.

Ultimate gas recovery is also dictated by the reservoir pressure at abandonment, as illustrated in Fig. 6.7. Gas contracts are usually stipulated in terms of a fixed production rate (sometimes expressed as a calorific rate) for a fixed period at a minimum wellhead pressure which is set by the operating pressure of the gas transmission network. Offshore, this can be high on account of the lack of booster stations. Working backwards to determine the abandonment reservoir pressure corresponding to final rate and wellhead pressure permits the equivalent, horizontal p/Z line to be plotted (Fig. 6.7). Where this intersects the material balance plot gives the ultimate recovery: provided the "material balance at abandonment" line does not constrain recovery first, as in the case of the strong waterdrive field, where the intersection occurs at point *B*. As can be seen, however, the moderate waterdrive reservoir, *A*, will, at high abandonment pressure, provide a greater recovery than a depletion type reservoir, in which there is no aquifer support. There is always the alternative of installing surface compression (Fig. 6.7). By this means, both the surface and reservoir pressures can be reduced while compressing the wellhead gas to the requisite input pressure of the delivery system. In this way, the gas recovery can be enhanced and this particularly favours depletion type reservoirs.

The impression is sometimes conveyed in the literature, however, that volumetric depletion must necessarily yield the highest recovery but Fig. 6.7 demonstrates the situation to be much more complex than that. It depends on the interplay between the aquifer flooding characteristics: aquifer size, residual gas saturation, volumetric sweep and the practical level of abandonment pressure. There are simply no rules of thumb for predicting ultimate gas recovery, it is necessary instead for the engineer to treat each field individually on its merits. Furthermore, in a field in which the condensate yield is high, liquid hydrocarbon recovery may be the primary aim; in which case the high degree of pressure support provided by a strong aquifer (case *B*, Fig. 6.7) may prove beneficial. The following example of a gas field development will illustrate the different techniques described in the section so far.

(d) Example field

A gas field and its radial aquifer are "invented" and its performance is then predicted exactly for a prescribed offtake rate. The results, in terms of cumulative gas production and pressure, are then analyzed, as though they represented actual field data, using the conventional p/Z plot and also the method of Havlena-Odeh. The intention is to illustrate the remarkable difference in sensitivity between the two in defining the drive mechanism and establishing the GIIP. The system is described below; the gas PVT properties are listed in Table 6.1 and the gas offtake-rate schedule in Table 6.2.

TABLE 6.1

PVT data, example field

p (psia)	Z	p/Z (psia)	B_g (rcf/scf)	p (psia)	Z	p/Z (psia)	B_g (rcf/scf)
3200	0.9135	3503.0	5.408×10^{-3}	1750	0.9010	1942.3	9.753×10^{-3}
3000	0.9070	3307.6	5.727×10^{-3}	1500	0.9080	1652.0	11.466×10^{-3}
2750	0.9008	3052.8	6.205×10^{-3}	1250	0.9178	1362.0	13.908×10^{-3}
2500	0.8968	2787.7	6.795×10^{-3}	1000	0.9302	1075.0	17.621×10^{-3}
2250	0.8955	2512.6	7.539×10^{-3}	750	0.9449	793.7	23.866×10^{-3}
2000	0.8968	2230.2	8.494×10^{-3}	500	0.9616	520.0	36.428×10^{-3}

GIIP = 823 Bscf k = 120 mD (effective, aquifer)

p_i = 3200 psia c_w = 3×10^{-6}/psi

h = 120 ft (reservoir/aquifer) c_f = 8×10^{-6}/psi

ϕ = 0.22 (reservoir/aquifer) f = 1 (360° radial encroachment)

S_{wc} = 0.23 PV r_o = 8350 ft (reservoir radius)

S_{gr} = 0.30 PV r_{eD} = 10 (aquifer/reservoir radius)

γ_g = 0.670 (air = 1) T = 210°F = 670°R

μ_w = 0.40 cp (water viscosity) B_w = 1.0 rb/stb; W_p = 0

Performance prediction

The water influx is calculated using the method of Hurst and van Everdingen. This has been described in detail and illustrated with examples in Chapter 9 of reference 2. Water influx is determined by superposing constant terminal pressure solutions of the diffusivity equation, W_D functions, to give

$$W_e = U \sum_{j=0}^{n-1} \Delta p_j W_D(T_D - t_{D_j})$$

in which a pressure drop, Δp_j, occurs at the start of the jth time step, and is maintained throughout it, and this is multiplied by the dimensionless influx, W_D, evaluated from the dimensionless time at which the influx is being calculated, T_D, back to the dimensionless time at which the pressure change occurred. The sum of such products over all the time steps gives the total influx, W_e, at time, T. In this equation, U is the aquifer constant

$$U = 1.119 \, f\phi hcr_o^2 \quad \text{(bbl/psi)}$$

in which $c = c_w + c_f$, the total aquifer compressibility, and for the above parameters.

$$U = 1.119 \times 1 \times 0.22 \times 120 \times 11 \times 10^{-6} \times 8350^2/10^6 = 0.02266 \text{ MMbbl/psi}$$

The dimensionless time argument, $T_D - t_D$, for which the W_D functions are evaluated, is calculated as

$$T_D - t_D = 2.309 \frac{k(T - t)}{\phi \mu c r_o^2} \quad (t, \text{years})$$

which for the example reservoir data gives

$$T_D - t_D = \frac{2.309 \times 120 \times (T - t)}{0.22 \times 0.4 \times 11 \times 10^{-6} \times 8350^2} = 4.11 \, (T - t) \tag{6.23}$$

In predicting the performance of the field, the water influx equation can be used in conjunction with either the p/Z expression of the material balance, equation 6.19, or the basic Havlena-Odeh material balance, equation 6.11, and a little bit of algebra on the latter will show that it amounts to precisely the same thing. It should be noted that using the p/Z formulation, which is conventional, is perfectly safe in this predictive application since both the reservoir and its aquifer are defined. In simultaneously solving the material balance and water influx equations it is necessary to determine both the pressure, p_n, and influx, W_{en}, at the end of the nth or current time step. Unfortunately, since the influx is also a function of p_n the process is necessarily iterative. At the start of calculations for the nth time step, the influx equation can be split into two components as

$$W_{e_n}(T) = U \sum_{j=0}^{n-2} \Delta p_j W_D(T_D - t_{D_j}) + U \frac{(p_{n-2} - p_n)}{2} W_D(T_D - t_{D_{n-1}}) \tag{6.24}$$

in which, as T is increased to correspond with the end of the nth time step, the first term is a calculable constant over all the previous pressure drops except the last [2]:

$$\Delta p_{n-1} = \frac{p_{n-2} - p_n}{2}$$

The influx resulting from this last pressure drop is separated out as the second term in the equation in which p_n is the unknown current pressure. The method of iterative solution to determine p_n and W_{en} is illustrated in Fig. 6.10 which has been extracted from reference 2. Usually two or three iterations per time step are sufficient to get convergence of values of p/Z. The corresponding values of the pressure can be obtained by interpolation using the PVT data in Table 6.1. Applying equation 6.2, E_i can be evaluated as 184.93 scf/rcf and therefore $G/E_i = 823/184.93 = 4.450$ Brcf = 792.58 MMrb. Consequently, the material balance, equation 6.19, can be expressed as

$$\frac{p}{Z} = 3503 \left(1 - \frac{G_p}{G}\right) / \left(1 - \frac{W_e}{792.58}\right) \quad (\text{psia})$$

in which W_e is in MMrb. It should be noted that in applying this equation the pore/water compressibility term has been omitted as a second-order effect. Using the input data, the water influx equation 6.24 is reduced to

$$W_e = 0.2266 \sum_{j=0}^{n-2} \Delta p_j W_D(T_D - t_{D_j}) + 0.01133(p_{n-2} - p_n) W_D(T_{D_n} - t_{D_{n-1}})$$

$$(\text{MMrb})$$

TABLE 6.2

Material balance prediction of the example field performance. Base case: rate increase towards the plateau level

Time (years)	t_D	Q (MMscf/d)	G_p (Bscf)	G_p/G	p/Z (psia)	p (psia)	Δp (psi)	W_D	W_e (MMrb)	F/E_g (Bscf)	α	Trapped gas (Bscf)
0					3503	3200						
0.25	1.028	59.0	5.384	0.00654	3482	3173	13.5	1.594	0.487	898	0.001	0.32
0.50	2.055	59.0	10.768	0.01308	3464	3156	22.0	2.489	1.556	967	0.003	0.96
0.75	3.083	59.0	16.152	0.01963	3447	3138	17.5	3.259	2.870	1010	0.006	1.90
0					3503	3200						
1	4.11	59.0	21.535	0.026	3430	3110	45	3.96	4	1033	0.008	2.52
2	8.22	115.5	63.693	0.077	3300	2990	105	6.44	16	1099	0.033	10.0
3	12.33	203.0	137.788	0.167	3070	2760	175	8.62	40	1115	0.083	23.4
4	16.44	225.0	219.913	0.267	2845	2540	225	10.64	77	1171	0.159	41.6
5	20.55	217.0	299.118	0.363	2645	2360	200	12.57	123	1221	0.254	61.8
6	24.66	225.0	381.243	0.463	2420	2170	185	14.37	176	1233	0.364	81.0
7	28.77	195.0	452.418	0.550	2245	2020	170	16.09	236	1260	0.488	100
8	32.88	180.0	518.118	0.630	2080	1875	148	17.72	300	1275	0.620	119
9	36.99	175.0	581.993	0.707	1915	1730	145	19.28	368	1284	0.760	134
10	41.10	160.0	640.393	0.778	1755	1590	143	20.76	441	1283	0.911	147

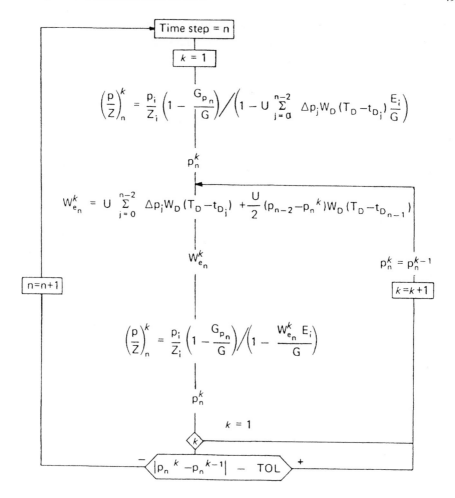

Fig. 6.10. Prediction of gas reservoir pressures resulting from fluid withdrawal and water influx (Hurst and van Everdingen).

The material balance predictions are presented in Table 6.2 for the prescribed wet gas offtake, Q (MMscf/d), which includes the relatively small amount of condensate ($r_{si} = 10$ stb/MMscf). In this base case, it will be noted that the rate increases during the first two years to a plateau level of >200 MMscf/d, implying production while drilling the development wells. Values of t_D have been calculated using equation 6.23 and the corresponding values of the dimensionless water influx, W_D, taken from the original Hurst and van Everdingen tables [6,17]. Calculation of p/Z, p, Δp and the cumulative water influx, W_e, is as described above. The value of F/E_g in column

11 is the required ordinate of the Havlena-Odeh plot (section 6.3b) in which the pore/water compressibility component, E_{fw}, has again been neglected. F/E_g may be calculated as

$$\frac{F}{E_g} = \frac{G_p B_g}{B_g - B_{gi}} \quad \text{(Bscf)}$$

in which B_g can be determined by interpolation of the values in Table 6.1, or by calculating it directly using equation 6.2 as

$$B_g = \frac{1}{E} = \frac{ZT}{35.37\,p} = \frac{18.943}{p/Z} \quad \text{(rcf/scf)}$$

The fractional volumetric sweep efficiency, α (column 12), is calculated from equation 6.17 as

$$\alpha = \frac{W_e}{\dfrac{G\,(1 - S_{gr} - S_{wc})}{E_i \quad (1 - S_{wc})}} = \frac{(1 - 0.7)W_e}{792.58(1 - 0.30 - 0.23)} = \frac{W_e}{484} \tag{6.25}$$

Finally, the "Trapped gas" (column 13) is determined using equation 6.18 as

$$\alpha\frac{G}{p_i/Z_i}\frac{S_{gr}}{(1 - S_{wc})}\frac{p}{Z} = \alpha\frac{823}{3503}\frac{0.30}{(1 - 0.23)}\frac{p}{Z} = 0.09154\alpha p/Z \quad \text{(Bscf)}$$

In evaluating the cumulative water influx in Table 6.2, it is of great assistance if equal time steps are selected. This means that the complex first term in equation 6.24 can be simply evaluated as the scaler or dot product:

$$[\Delta p_{n-2}, \quad \Delta p_{n-3}, \quad \ldots, \quad \Delta p_1, \quad p_0]\begin{bmatrix} W_D(T_D - 0) \\ W_D(T_D - t_{D_1}) \\ \vdots \\ W_D(T_D - t_{D_{n-3}}) \\ W_D(T_D - t_{D_{n-2}}) \end{bmatrix}$$

for which the values of Δp and W_D are available in the table. In this respect, above the dividing line in Table 6.2, equal quarter year time steps have been selected over the first nine months of production, whereas below the line, the calculations have been repeated from the start of production using annual time steps.

The p/Z plot across the full range of depletion (0–3503 psia) is shown in Fig. 6.11. As can be seen, this trend could be and is frequently interpreted as being linear, with the slight deviations from exact linearity being attributed to errors in the pressure measurements made in the field. If, however, the plot is restricted to the depletion range of interest (1600–3503 psia: Fig. 6.14, solid line) then it is at once evident that the trend is far from linear and, in fact, displays the typical shape (Fig. 6.8b) expected from a fairly strong waterdrive gas reservoir. Referring again to Fig. 6.11, it can be seen that the extrapolation of the "apparent" linear trend to the

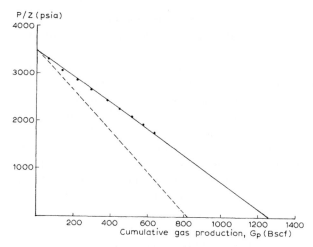

Fig. 6.11. p/Z plot for the example field, illustrating apparent linearity over a ten year period.

abscissa ($p/Z = 0$) would yield an estimate of the GIIP of 1260 Bscf which is 53% in excess of the correct value of 823 Bscf indicated by the dashed line. This example substantiates the claim made by Cason [16] that in using the conventional p/Z plot, it is possible to overestimate the GIIP by 30–50%.

Alternatively, a mere glance at the Havlena-Odeh plot of F/E_g versus G_p, Fig. 6.12, reveals immediately that the reservoir is being strongly energised by natural water influx. That is, the term $W_e B_w/E_g$ in equation 6.12 is both positive and increases as the depletion continues. If the field were strictly of the volumetric depletion type, then values of F/E_g would plot as the horizontal straight line (dashed line, Fig. 6.12) parallel to the abscissa — which is clearly not the case. It

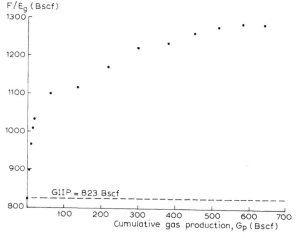

Fig. 6.12. Havlena-Odeh plot for the example field, illustrating the strong energising influence of natural water influx.

should also be noted that the linear extrapolation of the p/Z plot to determine the GIIP is valid "if and only if" the Havlena-Odeh plot is horizontal with a value, in this case, of $G = 823$ Bscf. Figure 6.12 also demonstrates that while the Havlena-Odeh interpretation method is far more sensitive than the p/Z plot in distinguishing the reservoir drive mechanism, the backward extrapolation of the F/E_g trend to $G_p = 0$ to establish the GIIP is hazardous. In this particular case, all that could be said is that the gas in place was less than 900 Bscf (first value, Table 6.2) but at least this is only some 9% in excess of the actual value of 823 Bscf. Part of the difficulty in performing the back extrapolation results from the manner of field operation, in that the gas production rate was gradually increased to the plateau level of 220 Bscf over the first three years. The initial production, at lower rates, permits the edge water to invade and significantly energise the reservoir thus causing the steep initial rise in the values of F/E_g.

To study the effect of rate dependence on the shape of the F/E_g plot, the water influx calculations have been repeated for two constant offtake rates of 168.5 and 270.6 MMscf/d, representing withdrawals of 7.5 and 12.0% of the GIIP per annum. The results of these calculations are presented in Table 6.3, in which, again, three monthly time steps have been taken over the initial phase of production; following which the calculations have been performed using annual time steps for the entire project lifetime. The Havlena-Odeh plots for the two constant-rate production cases are shown in Fig. 6.13 in comparison to the "base case" with its variable-rate history. As can be seen, producing at constant rate throughout would ease the backward extrapolation to determine the GIIP and the higher the rate the more straightforward the extrapolation. The shape of the Havlena-Odeh plots

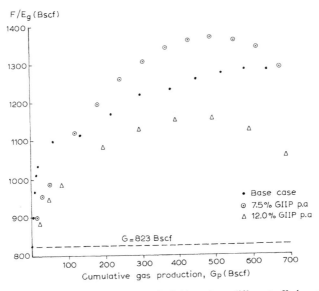

Fig. 6.13. Havlena-Odeh plots, example field, at three different offtake rates.

is dependent both on the offtake rate and the strength of the natural water influx. In the case of an infinite acting aquifer with favourable rock properties, values of F/E_g will increase continuously until abandonment. But for the two constant-rate cases shown in Fig. 6.13, for which the aquifer size is limited ($r_{eD} = 10$), the plots demonstrate strong, downward curvature towards the end of the project lifetime. This results from the reduction in water influx rate as the aquifer boundaries are "felt" and it begins to deplete, thus reducing the energising influence of the aquifer. For the "base case" the downward trend in values of F/E_g is not evident since there is a reduction in the gas offtake rate later in the project which increases the effect of the water influx. Altogether, the shapes of the F/E_g plots are very complex and it would be unwise to place too much reliance on their interpretation: other than to observe whether there is a waterdrive or not — which is usually quite obvious.

What is demonstrated in Fig. 6.13, for the constant-rate cases, is the significance of the offtake rate in reducing the influence of the aquifer. At 12.0% of the GIIP per annum the highly mobile gas is withdrawn before the less mobile water can "catch-up" and trap a large quantity of gas at high pressure, as described in section 6.3b. At 7.5% of the GIIP per annum, however, the aquifer is so much more effective in energising and therefore pressurising the reservoir which results in the loss of more gas behind the advancing waterfront. Considering an areal sweep at abandonment of $\alpha = 85\%$, which is quite reasonable for a central well cluster development, then the recovery statistics for the three cases considered may be determined by interpolation of the data in Tables 6.2 and 6.3 to give the results listed in Table 6.4. The figures demonstrate that producing at a constant rate of 12.0% compared to 7.5% per annum, would increase the recovery by 73 Bscf (8.9% of the GIIP) and this incremental gain would be achieved 2.9 years earlier. It will be noted that in each case the influx amounts to 411 MMrb, which is simply 85% of the movable gas volume of 484 MMrb (equation 6.25). The volume of by-passed gas may be calculated independently as the last term in equation 6.21, or simply by solving the overall material balance for this unknown. While the statistics look favourable for high rate offtake, it must be appreciated that the results relate solely to the aquifer/reservoir mechanics and may be influenced by such practical considerations as:

- Is there a market available to accommodate the higher rate?
- Would the lower pressures for high-rate production necessitate the drilling of more wells or the installation of surface compression?

The p/Z plots for the three cases are shown in Fig. 6.14 across an enlarged pressure scale from 1600 to 3600 psia. Such plots emphasize the non-linearity of the trends in comparison to that shown for the "base case" (Fig. 6.11), which is plotted across the entire range of p/Z. Nevertheless, the "early warning" of natural water influx as an active drive mechanism is much less pronounced than in the equivalent Havlena-Odeh plots (Fig. 6.12). Similarly, while the rate dependence of water influx, and resulting pressure support, is evident in the three p/Z plots, it is much clearer in Fig. 6.13. The dashed line in Fig. 6.14 represents the extrapolation of the early trend in the decline which intersects the abscissa at the value of $G_p = 485$ Bscf. Extrapolation of this line to $p/Z = 0$ would yield a value of $G = 894$ Bscf, which is about the

TABLE 6.3

Material balance predictions for the example field at constant offtake rates of 7.5 and 12.0% of the GIIP per annum

Time (years)	t_D	Q (MMscf/d)	G_p	G_p/G	p/Z (psia)	p (psia)	Δp (psi)	W_D	W_e (MMrb)	F/E_g (Bscf)	α	Trapped gas (Bscf)
Q = 168.5 MMscf/d: 7.5% GIIP p.a.												
0					3503	3200						
0.25	1.028	168.5	15.38	0.0187	3443	3130	35	1.594	1.263	898	0.003	0.95
0.50	2.055	168.5	30.75	0.0374	3390	3080	60	2.487	4.137	953	0.009	2.81
0.75	3.083	168.5	46.13	0.0561	3340	3022	54	3.259	7.912	991	0.016	4.92
0					3503	3200						
1	4.11	168.5	61.5	0.075	3284	2962	119	3.96	10.7	984	0.022	6.6
2	8.22	168.5	123.0	0.149	3119	2804	198	6.44	35.1	1122	0.073	20.9
3	12.33	168.5	184.5	0.224	2963	2658	152	8.62	65.8	1197	0.136	37.1
4	16.44	168.5	246.0	0.299	2818	2520	142	10.64	102	1258	0.211	54.7
5	20.55	168.5	307.5	0.374	2678	2396	131	12.57	144	1306	0.298	73.4
6	24.66	168.5	369.0	0.448	2540	2280	120	14.37	189	1342	0.390	91.1
7	28.77	168.5	430.5	0.523	2394	2148	124	16.09	240	1360	0.496	109
8	32.88	168.5	492.0	0.598	2241	2016	132	17.72	295	1366	0.610	126
9	36.99	168.5	553.5	0.673	2076	1870	139	19.28	355	1359	0.733	140
10	41.10	168.5	615.0	0.747	1895	1712	176	20.76	422	1340	0.872	152
Q = 270.6 MMscf/d: 12.0% GIIP p.a.												
0					3503	3200						
0.25	1.028	270.6	24.69	0.030	3405	3100	50	1.594	1.806	882	0.004	1.25
0.50	2.055	270.6	49.38	0.060	3320	3004	98	2.489	6.630	945	0.014	4.28
0.75	3.083	270.6	74.08	0.090	3240	2920	90	3.259	12.471	987	0.026	7.75
0					3503	3200						
1	4.11	270.6	98.8	0.120	3148	2836	182	3.96	16.3	975	0.034	9.8
2	8.22	270.6	197.6	0.240	2862	2564	318	6.44	55.2	1080	0.114	30.0
3	12.33	270.6	296.4	0.360	2582	2320	258	8.62	105	1127	0.217	51.5
4	16.44	270.6	395.2	0.480	2300	2068	248	10.64	166	1151	0.343	72.6
5	20.55	270.6	494.0	0.600	2005	1800	260	12.57	238	1155	0.492	90.8
6	24.66	270.6	592.8	0.720	1655	1500	284	14.37	324	1124	0.669	102
7	28.77	270.6	691.6	0.840	1210	1120	340	16.09	426	1056	0.880	98

TABLE 6.4

Recovery statistics, example field, for an abandonment sweep efficiency of 85%

	G_p (Bscf)	G_p/G (%)	Trapped gas (Bscf)	By-passed gas (Bscf)	W_e (MMrb)	$\bar{p}$ (psia)	p/Z (psia)	Time (years)
Base case	617	75.0	142	64	411	1647	1820	9.6
7.5% GIIP p.a.	605	73.5	150	68	411	1737	1923	9.8
12.0% GIIP p.a.	678	82.4	99	46	411	1174	1273	6.9

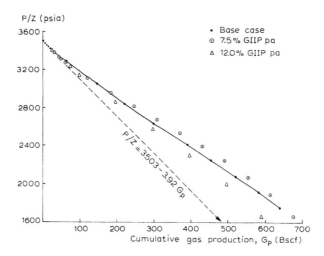

Fig. 6.14. Enlarged p/Z plots, example field.

same accuracy as attained using the method of Havlena-Odeh, and illustrates that although the water influx is small during the initial period of production — it is still finite. The recovery statistics presented in Table 6.4 can also be determined from the intersection of the equation of the material balance at abandonment (equation 6.22) with the p/Z plots, which will be left as an exercise for the reader.

The study of this fairly strong waterdrive "example" field clearly illustrates the superiority of the Havlena-Odeh over the p/Z plot in analyzing the production–pressure performance of gas fields. The former is much more sensitive in deciding on the nature of the drive mechanism but both suffer from deficiencies when applied to estimate the GIIP which worsens as the aquifer strength increases. For the p/Z plot, the determination requires a lengthy extrapolation of any early linear trend outside the range of the data points; whereas for the Havlena-Odeh plot there is a shorter extrapolation within the data points but it can be distinctly non-linear which reduces confidence in the resulting GIIP so determined.

The calculations for the example field were purely of a predictive nature for a defined reservoir/aquifer system. In reality, of course, the engineer is confronted by the converse problem in which the system requires definition by applying material balance to the basic data of production, pressure and PVT. The recommended

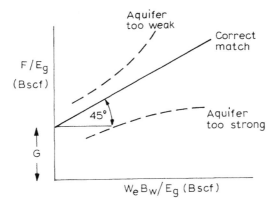

Fig. 6.15. Application of the Havlena-Odeh plot in history matching reservoir–aquifer performance.

method in history matching performance, with a view to subsequent prediction, is first of all to plot F/E_g versus G_p (backed-up by the p/Z plot), as described above. If the monitoring of the production/pressure history has been sound, as described in the following section, then this should lead to the definition of the GIIP and the occurrence of water influx. The next step, with the advice of exploration/development geologists, is to decide on the nature of the aquifer: its shape, size and rock properties and, for these assumptions, calculate the water influx using the method of Hurst and van Everdingen, for the observed pressure drop. Since the pressures are known, this process does not require the complexity of iteration, as in the example field in which the pressures are being predicted. The full Havlena-Odeh equation 6.12 is then applied graphically as illustrated in Fig. 6.15, in the same manner as described for oil fields described in Chapter 3, section 3.8. A correct aquifer model will simply provide a straight line of unit slope whose intercept on the ordinate gives the GIIP = G. If the selected aquifer model is ill-fitting, however, the trend will deviate above or below this line dependent upon whether the aquifer is too weak or too strong in providing water. Success in history matching means that the aquifer model selected can be used in performance prediction to determine the pressure decline as a function of the cumulative offtake and the ultimate gas recovery dependent on the abandonment conditions. The method is precisely the same as described in this section for the example field. For the reasons described in Chapter 3, section 3.8c, it is considered that this method of history matching is more accurate than that of attempting to match directly on the pressure decline which is very popular but which tends to treat the pressure as an unknown — which underrates it in importance.

(e) Gas field development

This section explores some of the broader implications of the application of material balance in gas field development studies. The method is necessarily investigative and may be described as "the minimum assumption route" through the subject, in comparison to numerical simulation in which the mere construction of a model, with geological maps and petrophysical input, assumes a GIIP rather than determines one and the simulation results obtained are then merely a reflection of the input assumption.

Matching production–pressure history

The *brutal insensitivity* in applying the traditional p/Z plot in isolation to assess the GIIP and drive mechanism has been illustrated at length in the previous section. But, without wishing to labour the point, it is worthwhile examining the reasons why Operators are so ready to believe the apparent linearity of the p/Z plot with its resulting over-assessment of the GIIP and failure to diagnose what can amount to a strong natural waterdrive. The reason for such attention is because it happens to be one of the more serious and unnecessary mistakes made in the oil and gas industry. Figure 6.16, for instance, illustrates the difference between the p/Z and F/E_g plots for a large offshore gas field with a strong natural waterdrive; the points represent annual pressure–production data. In spite of the fact that the average permeability of the reservoir was in excess of 1000 mD, and it was surrounded by an almost infinite acting aquifer, it was believed for the first five years, based solely on the observation of the p/Z decline, that the field was producing under volumetric depletion conditions; whereas a glance at the F/E_g plot after just one year of production would have confirmed otherwise. The actual GIIP (230 Bsm^3: 8.1 Tscf) was therefore overestimated by a very large amount but, in addition, decision making on raising the market rate to avoid gas wastage in the reservoir was deferred until it was no longer a practical proposition.

One reason why Operators are prepared to accept inflated GIIP figures resulting from p/Z plots is because if a field has a strong natural waterdrive, then it must

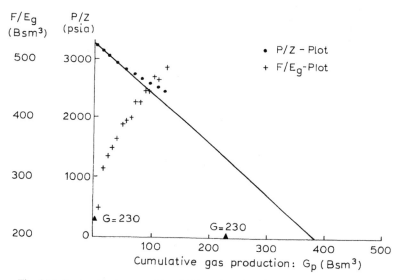

Fig. 6.16. Contrast between p/Z and F/E_g plots for a large waterdrive gas field.

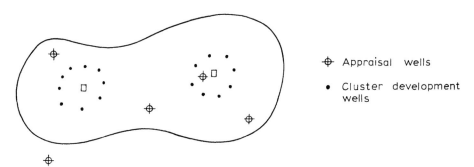

Fig. 6.17. Cluster development of an edge waterdrive gas field.

necessarily also have a reasonable *kh*-product. In such cases, it is both possible and preferable to drill the production wells in fairly confined clusters towards the centre of the accumulation and safely away from any edge water influx, as shown in Fig. 6.17. Unfortunately, in adopting this practice, little extra is learned from the development wells to improve on the initial GIIP estimate based on data acquired in the original, more widely spaced appraisal wells. Consequently, when the p/Z gas in place exceeds the early volumetric estimate, the Operator is more inclined to believe the former. Using the p/Z plot, the calculated GIIP will tend to increase annually [2,14] which is also a favourable state of affairs that seems readily acceptable. Of course, such optimism is usually confined to edge rather than basal waterdrive fields in which the vertical rise in the water is readily discernable.

It is in edge waterdrive fields where the risks are greatest since premature water breakthrough is rare and Operators are comforted, by the lack of water production, into believing that the reservoir is of the volumetric depletion type. The reason for this is because, on account of the low viscosity of gas, the end-point mobility ratio for water displacing gas is so small (typically 1/100) that it predominates over the influences of heterogeneity and gravity in making the displacement efficiency almost unconditionally stable. To illustrate this effect consider the case of waterdrive in a hypothetical reservoir that has been subdivided into seven layers of equal thickness. Porosity and irreducible water saturations are the same in all layers but the permeability of each doubles moving from layer to layer in the downward direction. That is, starting with $k = 10$ mD in the top layer, the permeability in basal layer is 640 mD (Fig. 6.18a, Table 6.5). In analogy with water–oil displacement (Chapter 5, section 5.5b) pseudo-relative permeabilities may be generated for water–gas displacement, under the assumed vertical equilibrium condition (VE-segregated displacement), using the following averaging procedures:

$$\overline{S}_{w_n} = \frac{\displaystyle\sum_{i=1}^{n} h_i \phi_i (1 - S_{gr_i}) + \sum_{i=n+1}^{N} h_i \phi_i S_{wc_i}}{\displaystyle\sum_{i=1}^{N} h_i \phi_i}$$

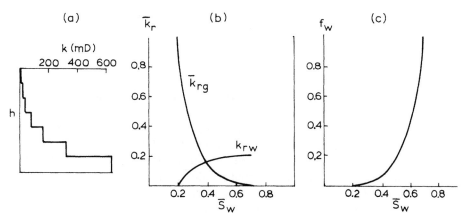

Fig. 6.18. Hypothetical edge waterdrive gas field: (a) permeability distribution, (b) pseudos (c) water fractional flow.

TABLE 6.5

Generation of pseudo-relative permeabilities/fractional flow: edge waterdrive gas field

Flood order	k (mD)	$\bar{S}_w$ (PV)	$\bar{k}_{rw}$	$\bar{k}_{rg}$	f_w
7	10	0.700	0.200	0	1
6	20	0.629	0.198	0.008	0.553
5	40	0.557	0.195	0.024	0.289
4	80	0.486	0.189	0.055	0.147
3	160	0.414	0.176	0.118	0.069
2	320	0.343	0.151	0.244	0.030
1	640	0.271	0.101	0.496	0.010

$$\bar{k}_{rw_n} = \frac{\displaystyle\sum_{i=1}^{n} h_i k_i k'_{rw_i}}{\displaystyle\sum_{i=1}^{N} h_i k_i}, \qquad \bar{k}_{rg_n} = \frac{\displaystyle\sum_{i=n+1}^{N} h_i k_i k'_{rg_i}}{\displaystyle\sum_{i=1}^{N} h_i k_i}$$

in which:

N = total number of layers (7)

n = the number of the layer which has just flooded with water (counting from the base to the top under the VE condition)

S_{gr_i} = residual gas saturation (0.30 PV — same in all layers)

k'_{rw_i} = end-point relative permeability to water (0.20 — same in all layers)

k'_{rg_i} = end-point relative permeability to gas (1.0 — same in all layers)

Finally, the fractional flow of water, for assumed horizontal displacement, may be calculated as

$$f_{w_n} = \cfrac{1}{1 + \cfrac{\mu_w}{\mu_g} \cfrac{\overline{k}_{rgn}}{\overline{k}_{rwn}}}$$

in which $\mu_w = 0.4$ cp and $\mu_g = 0.02$ cp. The results of these calculations are listed in Table 6.5 and plots of the pseudo-functions and fractional flow are shown in Fig. 6.18

Intuitively, the flooding efficiency in such a reservoir section might be anticipated to be very poor, with accelerated movement of the heavier water through the higher permeability, basal part of the section. And, inspection of the pseudo-relative permeabilities would tend to confirm this opinion: the water curve being strongly concave downward. As stressed in connection with the subject of water-drive, however, viewing the relative permeabilities alone is insufficient to decide upon the efficiency of the displacement. Instead, it is necessary to generate and plot the fractional flow, which incorporates the viscosities of the *in situ* reservoir fluids. Inspection of this function (Fig. 6.18c) reveals complete piston-like displacement of the gas by water, as manifest by the concave upward shape of the function across the entire moveable gas saturation range: $1 - S_{gr} - S_{wc}$ (Chapter 5, section 5.4). The dominant factor dictating this perfect sweep efficiency is the extremely low gas viscosity, in the present case 0.02 cp, which in turn results in a mobility ratio (equation 6.13) of $M = 0.01$. This value is so low and favourable that it completely dominates over the combined effects of heterogeneity and gravity and, as a consequence, perfect, piston like displacement occurs almost irrespective of the degree of reservoir heterogeneity. Because of this, premature water breakthrough is seldom observed in edge waterdrive fields and in a cluster type development (Fig. 6.17) the true severity of the water influx will not be directly observable until the centrally located wells suddenly and almost simultaneously water-out, just as in the example field described in the previous section.

In the development of suspect, edge waterdrive fields every effort should be made to measure DST/RFT pressures in any exploration/appraisal wells drilled in the area — even should they prove to be dry holes. The observation of pressure depletion in the aquifer can be very useful in quantifying the degree of water influx.

Development planning

It is imperative before production start-up of a reservoir that the engineer makes quite certain that the initial reservoir pressure has been accurately measured and that reliable fluid samples have been collected for PVT analysis. Failure to do so means that the material balance estimate of the GIIP can be greatly reduced in accuracy. This is because both the Havlena-Odeh and p/Z formulations of equations 6.12 and 6.19, always relate the current reservoir condition back to the initial state: the former equation contains the term $B_g - B_{gi}$ while the latter has the ratio p_i/Z_i. If there is error in the measurement of the initial pressure, even though slight, the effect it has on the Havlena-Odeh plot will be as illustrated in Fig. 6.19.

After the start of continuous production frequent pressure measurements should be made during the first few years again with the intention of facilitating the back

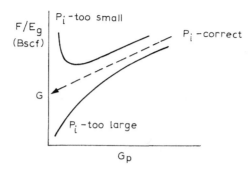

Fig. 6.19. Effect of an error in the measurement of the initial pressure on the F/E_g plot.

extrapolation of the F/E_g plot, or the forward extrapolation of the p/Z plot to establish the GIIP.

Consideration of the example field in this section would seem to suggest that it ought to be a sound basic policy to always produce gas fields at as high a rate as practically possible. Then, if there is no aquifer support, no harm will be done but if there is then there could be a significant gain in recovery due to the rapid evacuation of the gas. In addition the higher production rate eases the F/E_g extrapolation to determine the GIIP. In this respect, if a gas contract has to be satisfied by the production of a group of fields, then it would be preferable if they were developed sequentially at high rate rather than simultaneously at low average rates.

Unfortunately, there can be no certainty about whether a gas reservoir is influenced by an active aquifer or not until some time after the start of continuous production when material balance can be applied. By that time, however, the gas contract will have been negotiated, perhaps leaving the Operator with little flexibility for increasing the offtake rate. Nevertheless, if a rate increase is possible it should be considered and there are cases in the literature [18,19] where significant gains (20–30%) in recovery by the "accelerated" withdrawal of gas from moderate waterdrive fields have been reported. Even if there is not the opportunity to increase the rate, at least every effort should be made to maintain it at as high a constant level as possible. Since gas contracts are often specified for cyclic production with the peak occurring in the winter months and a reduced level during the summer, then if the Operator has the choice, the waterdrive fields should be produced at a high, continuous rate and any depletion type fields, which are not rate sensitive, should have their production reduced during the off season.

Another method of enhancing the recovery of waterdrive gas fields is by "blowing down" the pressure in the flooded-out regions. That is, gas production wells that have been flooded by the encroaching water should be completed with gas lift strings and converted to high-rate water producers. The aim in doing so is to reduce the pressure in the water invaded zone allowing the trapped residual gas to expand so that some portion of it can percolate updip into the gas column where it can be produced. As depletion occurs, the trapped gas volume (saturation) remains unaltered but, in accordance with equation 6.14, the quantity of trapped gas (n, lb-moles) is reduced. In one interesting case history in the North Alazan Field in Texas [20], four high-rate water producers were drilled and completed in the aquifer of the field while a further three wells in the water invaded zone were converted as water producers. By withdrawing water at a rate of 30,000 b/d the abandonment pressure was expected to be reduced from its natural waterdrive level of 2200 psia to 500 psia releasing 22 Bscf of trapped gas which raised the recovery factor by almost 30%. This technique is at its most effective in tight reservoirs in which it has been noted that there can be very significant pressure differentials across the water invaded zone caused, in part, by the low end-point relative permeability to water when displacing gas. The residual gas is therefore trapped at a higher pressure than prevailing in the gas zone thus enhancing the prospect of incremental recovery by the blowdown process. The lack of pressure equilibrium across the gas accumulation and aquifer invalidates the straightforward application of material balance to the entire system and a method has therefore been suggested [20] for modifying the equation to suit this condition. Unfortunately, it is more difficult to apply the blowdown technique offshore due largely to space constraints. There is the need to install bulky compression and power generation equipment to either gas lift the water producers or alternatively install downhole pumps; also there is the requirement to clean-up the large volumes of produced water, containing some liquid condensate, before dumping it in the sea. If the platform was not originally designed with sufficient deck space to cater for the blow-down phase, which occurs late in the lifetime of the field, then it is often difficult to accommodate the necessary additional equipment in the confined space.

6.4. THE DYNAMICS OF THE IMMISCIBLE GAS–OIL DISPLACEMENT

This section concentrates on the mechanics of gas–oil displacement (gas drive) in which there is no mass transfer between the gas and oil phases and the displacement is immiscible, as for waterdrive. Under these circumstances there will be a finite surface tension between the phases; meaning that a residual oil saturation will eventually be trapped in each pore space contacted by the injected gas. Miscible gas drive, with complex phase behaviour effects between the gas and oil at the flood front, is a subject which is not readily amenable to description using simple analytical techniques, although excellent attempts have been made to do so [21]. Instead, compositional simulation modelling may be required to account for the complex physical/chemical effects occurring in the reservoir. Nevertheless, the current description is not to be thought of as restrictive since many gas drive projects occur under immiscible conditions. Furthermore, in concentrating on displacement efficiency in macroscopic reservoir sections, many of the techniques are equally relevant to miscible flooding, as mentioned in this and the following section, which is devoted to dry gas recycling, regarded as a purely miscible type of flood. In fact, before embarking on the intricacies of modelling miscible displacement, the engineer would be well advised to apply the techniques described in this section to gain a sound perspective of the gas drive process and its efficiency. As for waterdrive, the point is stressed that there are three physical effects which control the efficiency of displacement: mobility ratio, heterogeneity and gravity and all must be fully accounted for in the description of this secondary recovery method.

(a) Mobility ratio

Included in this term, as in the description of waterdrive, are the results of all microscopic flooding experiments performed in the laboratory on thin core plugs — which restricts the description to the level of one dimension. The mobility ratio itself for gas–oil displacement is highly unfavourable: to use typical figures

$$M = \frac{k'_{rg}}{\mu_g} \bigg/ \frac{k'_{ro}}{\mu_o} = \frac{0.5}{0.025} \bigg/ \frac{1}{1} = 20$$

meaning that even when displacing a low-viscosity oil ($\mu_o = 1$ cp) the gas is capable of moving 20 times faster under a given pressure differential — and it does!

In Chapter 5, section 5.4f, it was asserted that in many cases water–oil relative permeabilities determined by laboratory experiment using high oil viscosity ($M \gg 1$) bore no relation to the flooding condition in the reservoir where, by choice, the oil viscosity is usually low and the mobility ratio less than unity. This is not the case for gas drive, however, for which both the laboratory experiments and the *in situ* displacement in the reservoir occur under unfavourable (unstable) conditions for which $M \gg 1$. Therefore, if a set of gas–oil relative permeabilities is used to generate a horizontal fractional flow relationship for gas which, in analogy with equation 5.18 for waterdrive, will have the form

$$f_{gh} = \frac{1}{1 + \dfrac{\mu_g}{\mu_o} \dfrac{k_{ro}}{k_{rg}}} \tag{6.26}$$

The result will be as depicted in Fig. 6.20. That is, on account of the unfavourable mobility ratio, the fractional flow will be concave downwards across the entire movable saturation range indicating that all gas saturations can be seen to be moving independently in this unstable form of displacement.

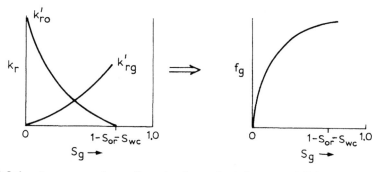

Fig. 6.20. Laboratory-measured, one-dimensional gas–oil relative permeabilities and fractional flow.

It will be noted that the relative permeabilities and fractional flow are plotted as functions of the increasing displacing phase saturation — the gas, S_g. For gas drive, the movable saturation range is, $\text{MOV} = 1 - S_{or} - S_{wc}$. Therefore, primarily on account of the abnormally low viscosity of gas, which gives rise to the high mobility ratio, the direct use of gas–oil relative permeability curves to describe gas drive displacement is usually quite accurate — at least on the scale of a core flooding experiment. But what is required to describe displacement in macroscopic reservoir sections is usually only the end-point values, k'_{rg} and k'_{ro}, as described in section 6.4b.

A debatable point in connection with gas drive relative permeabilities is the value of the residual oil saturation to gas, which many believe to be lower than for waterdrive. It has been demonstrated [22] in experiments modelling gas–oil displacement in the vertical direction (the gravity drainage process) that extremely low residual oil saturations (5% PV) can be achieved. In a practical sense, however, this matter is somewhat academic since in predominantly edge rather than ideal vertical displacement, the shape of both laboratory and pseudo-fractional flow curves is frequently as depicted in Fig. 6.20. The shape implies that it would require the circulation of so many pore volumes of gas, at high producing GOR, to approach the value of $\text{MOV} = 1 - S_{or} - S_{wc}$, that in practice the goal is unattainable.

The method of describing oil recovery in a one-dimensional, immiscible gas drive is precisely the same as for waterdrive: apply the theory of Buckley-Leverett using the practical graphical technique of Welge. In the latter, the fractional flow curve is plotted using the relative permeabilities and incorporating the gas/oil viscosity ratio. A tangent is drawn to the fractional flow from $S_g = 0$, the point of contact with the curve giving the gas saturation and fractional flow at breakthrough, at which the oil recovery is equal to the amount of gas that has been injected. Following this, gas saturations between breakthrough and flood-out, $S_g = 1 - S_{or} - S_{wc}$, are selected from the curve, S_{ge}, and for each, the corresponding value of the fractional flow of gas, f_{ge}, is read from the ordinate of the function. Then, in direct analogy with Welge's one-dimensional oil recovery equation for waterdrive (5.25), the gas drive recovery can be calculated as

$$N_{pd} = S_{ge} + (1 - f_{ge})G_{id} \quad (\text{PV}) \tag{6.27}$$

where

$$G_{id} = \frac{1}{\dfrac{\Delta f_{ge}}{\Delta S_{ge}}} \quad (\text{PV}) \tag{6.28}$$

which is the cumulative gas injected in pore volumes and is equal to the reciprocal of the derivative of the fractional flow curve at point S_{ge}, f_{ge}.

(b) Heterogeneity/gravity

These are strongly interrelated and must be treated in conjunction. Vertical gas drive (Fig. 6.21a), where it can be applied, is a stable and highly efficient form of

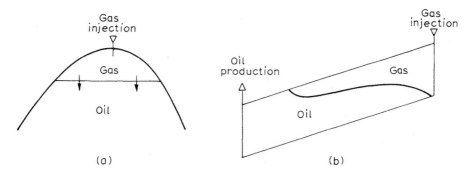

Fig. 6.21. Vertical and parallel gas injection.

recovery mechanism. It receives the maximum stabilizing influence from gravity, is little affected by heterogeneity and it is under this condition of gravity drainage, with slow frontal movement, that very low residual oil saturations have been observed. On the other hand, crestal gas injection with displacement in the down dip direction, parallel to the bedding planes, can be much more difficult to control (Fig. 6.21b) and is the type of gas drive focused upon in this section. It has been practised in several large fields, or parts of fields, in the Norwegian sector of the North Sea where there was initially too much gas available to supply the market requirement. Sometimes it is practised simply as a means of temporary gas storage until the market and delivery system are available but also because, in some cases, it is believed to provide a more efficient recovery of oil than downdip waterdrive.

The method of describing the vertical sweep efficiency in the two-dimensional cross-section follows the same pattern as for waterdrive: generate pseudo-relative permeabilities which reduce the description to one dimension and use the pseudos in the Buckley-Leverett displacement theory, applying the technique of Welge, to calculate oil recovery (vertical sweep) and the GOR development as a function of the fractional oil recovery. The latter is the equivalent of calculating the watercut development trend in waterdrive fields.

By direct analogy with equations 5.30 to 5.32 for waterdrive, the pseudo-relative permeabilities for gas drive may be calculated as

$$\overline{S}_{g_n} = \frac{\displaystyle\sum_{i=1}^{n} h_i \phi_i (1 - S_{or_i} - S_{wc_i})}{\displaystyle\sum_{i=1}^{N} h_i \phi_i} \tag{6.29}$$

$$\overline{k}_{rg_n} = \frac{\displaystyle\sum_{i=1}^{n} h_i k_i k'_{rg_i}}{\displaystyle\sum_{i=1}^{N} h_i k_i} \tag{6.30}$$

$$\overline{k}_{\text{ro}_n} = \frac{\displaystyle\sum_{i=n+1}^{N} h_i k_i k'_{\text{ro}_i}}{\displaystyle\sum_{i=1}^{N} h_i k_i} \tag{6.31}$$

which determine the thickness averaged gas saturation and corresponding pseudo-relative permeabilities to gas and oil after the nth layer out of a total of N has flooded with gas. The flooding order is dictated by the degree of pressure communication and therefore cross-flow, which is again best determined by running RFT pressure surveys in each new development well. The extreme conditions of vertical equilibrium and a total lack of it are described below. It will also be noticed, that in spite of the unfavourable mobility ratio for gas drive ($M \gg 1$) only the end-point relative permeabilities and saturations (k'_{rg_i}, k'_{ro_i}, S_{or_i}, S_{wc_i}) are used in the averaging procedures implying the either–or situation characteristic of piston-like displacement: either a layer is filled with gas to the flood out saturation, $S_{g_i} = 1 - S_{\text{or}_i} - S_{\text{wc}_i}$, or it is not — no in-between situation is catered for. Obviously, this assumption requires careful justification, which is provided below for the different flooding conditions. Apart from that, all the remarks made in Chapter 5, section 5.5. concerning the manner of evaluation of the individual parameters in the averaging procedures for waterdrive are equally relevant to the application of equations 6.29 to 6.31, for gas drive.

The pseudos incorporate all the reservoir heterogeneity in the vertical cross-section in reducing the description of displacement to one dimension. Gravity is introduced in the next important step of generating a fractional flow relationship for gas which, in analogy with equation 5.19 for waterdrive may be stated as

$$\overline{f}_g = \frac{1 - G}{1 + \dfrac{\mu_g \, \overline{k}_{\text{ro}}}{\mu_o \, \overline{k}_{\text{rg}}}}$$

or

$$\overline{f}_g = \overline{f}_{\text{gh}}(1 - G) \tag{6.32}$$

in which $\overline{f}_{\text{gh}}$ is the pseudo-fractional flow for horizontal displacement: as expressed by equation 6.26. The gravity term, G, has the form

$$G = 2.743 \times 10^{-3} \frac{\overline{k}k_{\text{ro}}}{v\mu_o} \Delta\rho \sin\theta \tag{6.33}$$

in which $\Delta\rho = \rho_o - \rho_g$ the density difference between oil and gas at reservoir conditions (water = 1). In waterdrive calculations, $\Delta\rho = \rho_w - \rho_o$: displacing minus displaced phase gravity difference which is the opposite way around and it is this difference that preserves the negative sign in equation 6.32 for gas–oil displacement in the down dip direction. For gas drive, however, the gravity difference is typically twice as large as for waterdrive which enhances the significance of the gravity term and means that it can play an important role in stabilizing the frontal advance of the gas. Consequently, evaluation of the gravity term, G, is mandatory for gas drive efficiency calculations. The stability will also improve by selecting reservoirs with high average permeability and dip angle and by reducing the average Darcy velocity of frontal advance, v (ft/day), which is controlled by the gas injection rate. The rate dependence of the stability of gas drive is illustrated in Exercise 6.1.

Having generated the pseudo-fractional flow, it is used in the Welge, one-dimensional equation

$$N_{pD} = \frac{\overline{S}_{ge} + (1 - \overline{f}_{ge})G_{id}}{1 - S_{wc}} \quad \text{(HCPV)} \tag{6.34}$$

which is analogous to equation 5.33 for waterdrive and in which

N_{pD} = oil recovery (vertical sweep) in HCPV
G_{id} = cumulative gas injected in PV ($= 1/\Delta \overline{S}_{ge}/\Delta \overline{f}_{ge}$)
$\overline{S}_{ge}, \overline{f}_{ge}$ = average gas saturation and fractional flow for which G_{id} has been evaluated. They are read from the fractional flow curve, after breakthrough, and represent the flooding at the end of the reservoir section being considered (producing wellbore).

The final step is to evaluate the relationship between the GOR increase and the fractional oil recovery, which is equivalent to the f_{ws} versus N_{pD} relationship for waterdrive and of equal importance. What is required is

$$\text{GOR} = \frac{q_g}{q_o} + R_S \quad \text{(scf/stb)}$$

whereas, what is evaluated in the displacement calculations is the reservoir fractional flow of gas

$$\overline{f}_g = \frac{q_g/E}{q_g/E + q_o B_o 5.615}$$

in which the surface rates have been expressed at reservoir conditions in rcf/day. Combining the two equations gives

$$\text{GOR} = \frac{5.615 B_o E}{\left(\dfrac{1}{\overline{f}_g} - 1\right)} + R_s \quad \text{(scf/stb)} \tag{6.35}$$

in which R_s (scf/stb) is the solution GOR at the flooding pressure.

In analogy with the simple waterdrive material balance at constant flooding pressure, equation 5.8, the gas drive material balance is

$$q_{gi} B_{gI} = q_o B_o + q_{gp} B_{gp} \quad \text{(rb/d)}$$

which, incorporating the GOR equation, becomes

$$q_{gi} = q_o \left[\frac{B_o}{B_{gI}} + (GOR - R_s) \frac{B_{gp}}{B_{gI}} \right] \quad (scf/d) \tag{6.36}$$

which appreciates that after processing the produced gas, before reinjection, there is usually a difference between the injected and produced gas FVFs, B_{gI} and B_{gp}, respectively.

This equation can be applied per producer–injector well pair, per reservoir or for the field as a whole, provided the average GOR versus N_{pD} relationship has been established for each situation. Of course, what has been determined by applying the above procedures to reservoir sections is $N_{pD} = E_v$, the vertical sweep for use in the overall recovery equation

$$\frac{N_p}{N} = E_v \times E_A$$

the areal sweep, E_A, is still unknown. But, as described for waterdrive, if E_v has been accurately determined, as manifest in the generation of pseudo-relative permeabilities, then inputting these to a reduced layered, numerical simulation model, will permit solution of the equation, for E_A, over the history matching period. Analytically, however, there is no sensible method of evaluating E_A for irregular well patterns and none is suggested in this text — this is the role and purpose of simulation modelling.

Nevertheless, the engineer need not be helpless, in this respect, in attempting to apply the recovery equation in a useful manner. The initial areal sweep efficiency is poor for gas drive on account of its high mobility ratio. Therefore, it may be assessed to be, say 10–20% at breakthrough rising to 70–80% at an abandonment GOR of 20–25000 scf/d. The figures are, of course, highly dependent on the flood pattern of the wells and must be reasonably assumed on that basis. These areal sweeps can then be incorporated in the GOR development trends of wells as shown in Fig. 6.22

The adjusted GORs allow for premature gas breakthrough while still preserving the effects of the vertical heterogeneity which often tends to predominate. Naturally, if there is any production history to match then the GOR correction for areal sweep becomes more meaningful for application in the equation. The purpose in doing so

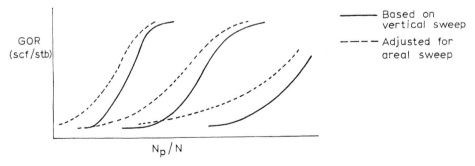

Fig. 6.22. GOR developments for different groups of wells.

is, as with waterdrive, to determine whether the capacities of the surface equipment for gas injection and production have been correctly sized — which is especially important for offshore developments considering the confined deck space available on platforms.

The surface equipment for gas drive projects consists of the produced gas processing equipment for extraction of the gas liquids and the injection plant which reinjects the dry gas perhaps supplemented by make-up gas from other reservoirs or fields. Inadequate capacities will act as a constraint on the circulation of gas and hence the oil production. Equation 6.36 can then be applied to the GOR trends of individual or groups of wells (Fig. 6.22) to generate production profiles. Wells are closed-in, as necessary, to meet the overall GOR or injection equipment constraints or alternatively the plant can be upgraded to maintain production at high GOR.

(c) Displacement condition

Whether for vertical equilibrium (VE), or a complete lack of it or some in-between case, the displacement condition in the reservoir dictates the order in which the selected layers in the section will flood and therefore the order in which the thickness averaging procedures for generating pseudos, equations 6.29 to 31, must be evaluated.

Vertical equilibrium

This condition relies on there being a lack of barriers or restrictions to vertical fluid movement across the reservoir. The gas and oil are then considered to segreg-ate instantaneously with the lighter fluid, the gas, ascending to the top of the section. The flooding order will therefore be from the top to the base of the layers selected to define the heterogeneity. It is also assumed that there will be a sharp interface between the gas and oil with virtually no capillary transition zone between them in comparison to the reservoir thickness. The VE condition is more readily satisfied for gas drive than waterdrive, in a given reservoir section, because the increased gravity difference between the fluids and the extremely low gas viscosity promote more rapid segregation. Furthermore the larger gravity difference reduces the extent of any capillary transition zone. This conforms with field observations that, when drilling through partially gas flooded zones, a sharp interface between the gas and oil is usually seen on petrophysical logs run in such wells. Consequently, the mixing of gas and oil that occurs in core flooding experiments, in which there is no gravity effect, does not occur in the field under the VE condition. Instead, the either–or situation prevails in which a layer contains gas at flood-out saturation, $S_g = 1 - S_{or} - S_{wc}$, or it does not, $S_g = 0$, there is no intermediate state.

Since the displacing phase, gas, is lighter than the oil, gas drive can be regarded as the inverse of waterdrive. Consequently, the influence of permeability distributions on the stability of displacement is also the inverse of waterdrive, as depicted in Fig. 6.23, which should be compared with Fig. 5.39. A coarsening upward sequence, so favourable for waterdrive, represents the worst possible condition for gas drive. The injected gas preferentially enters the upper high-permeability layers and, being the

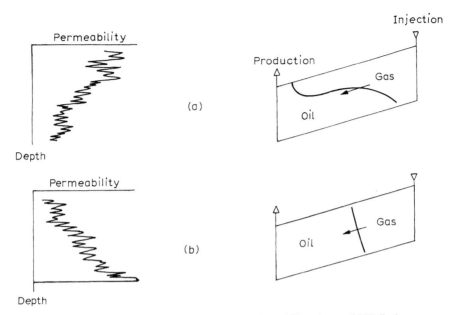

Fig. 6.23. Influence of permeability distribution on the stability of gas–oil VE displacement.

lighter phase, it stays there leading to gas overriding the oil and its breakthrough to the production wells prematurely, resulting in early abandonment at high GOR and attendant loss of oil recovery. Coarsening downward, Fig. 6.23b has precisely the opposite effect. Most of the injected gas enters at the base of the section but after travelling only a short distance into the reservoir gravity becomes more significant than the viscous, driving force and the gas rises to the top of the reservoir providing a highly efficient form of piston-like displacement across the entire section.

Because of the increased significance of the gravity term, G, in the fractional flow, however, matters are a bit more complicated than depicted in Fig. 6.23 and, as demonstrated in Exercise 6.1, a reasonable degree of stability can be achieved even in a most unlikely looking reservoir having high-permeability layers at the top of the formation. This is achieved by reducing the injection rate and therefore the average Darcy velocity of frontal advance of the flood (ft/day). Stability is also encouraged, in this instance, by favourable values of k, θ, μ_o and $\Delta\rho$ in the gravity term, equation 6.33.

Neglecting the gravity term in the fractional flow equation results in a highly unfavourable curve, case A, Fig. 6.24, whereas reducing the velocity to 1 ft/day, which represents a practical rate for this field gives a degree of inflexion in the fractional flow such that the recovery after breakthrough, bt-case B, proves quite satisfactory up to an abandonment GOR of 10,000 scf/stb. Reducing the rate further to 0.25 ft/day would lead to the condition that the tangent to the fractional flow would intersect the curve at the flood-out saturation (case C), implying complete

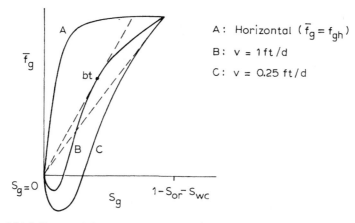

Fig. 6.24. Influence of the velocity of the gas flood on the stability of frontal advance.

piston-like displacement across the section. The fact that for small gas saturations the fractional flow is negative is not significant because, as described in Chapter 5, section 5.4f, until breakthrough, the curve is "virtual" with all gas saturations caught up in the shock front development.

Total lack of vertical equilibrium

This implies dealing with the deltaic type of depositional environment described in Chapter 5, section 5.7, in which the layers defined are separated from each other by impermeable barriers to vertical fluid movement: a condition confirmed by the observation of a lack of pressure equilibrium across the reservoir section in RFT surveys conducted in development wells. It might be thought that for this type of flood the individual layers are of similar dimensions to core plugs and therefore considering the high mobility for gas drive ($M \gg 1$) it would be quite inappropriate to assume piston-like displacement in each separate sand — but this is not the case. In Exercise 5.4, for instance, the 8 sands have thicknesses ranging from 2 to 14 ft., considerably greater than used for conventional core flooding experiments. Consequently, within each layer, there will be segregation of the gas and oil which, dependent on the permeability distribution, will have extremes of configuration shown in Fig. 6.23a or b. Whichever it is, however, is not of great importance for even if the worst case, a, in a thin sand, the distance between the leading edge and the flood-out saturation will be slight, in comparison to the distance between injection and production wells, which would justify the assumption of piston-like displacement. If greater accuracy is required, however, the thicker sands can be subdivided into more discrete layers, according to their permeability distributions, as described in Chapter 5, section 5.7b, and illustrated in Fig. 6.25.

Assuming VE flooding within each sand, the coarsening upward sequence, a, must be represented as three layers while for coarsening downward, b, one layer will suffice. It will be noted that Fig. 6.25 demonstrates the inverse situation to that

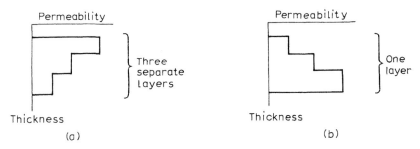

Fig. 6.25. Subdividing individual sands in a deltaic environment into discrete layers for gas drive.

depicted in Fig. 5.60. The total number of discrete layers for the entire section is then re-ordered in the sequence in which they will flood and, since the mobility ratio is significantly greater than unity, displacement efficiency calculations must be performed applying the method of Dykstra-Parsons. The techniques are precisely the same as described in Chapter 5, sections 5.7b and d.

It must be admitted however, that the whole concept of gas drive in this type of reservoir environment is somewhat academic and most operators would not even consider the likelihood of such a flood. The reason is because there is little possibility of any assistance from gravity in stabilizing the frontal advance across the macroscopic sand section: the gravity term in the pseudo-fractional flow, equation 6.32, is redundant. That is, within each separate layer the gravity term is effective in the individual fractional flows but, as described above, considering the length of the section in comparison to the layer thicknesses, the effect is small and piston-like displacement is approximated within each. Considering the section as a whole, however, there is no cross-flow between the layers and therefore the displacement efficiency is governed by mobility ratio and heterogeneity with an absence of gravity. Consequently, the macroscopic fractional flow is approximately the same as for horizontal displacement which is usually so unfavourable as to preclude the application of gas drive altogether in this type of reservoir environment, as demonstrated in Exercise 6.1.

In-between cases

This refers to sand sections, as described in Chapter 5, section 5.10, in which there may be cross-flow within some of the N layers included in the completion interval but a lack of vertical fluid movement between some of the layers in the section. Then, unless detailed RFT surveys are available in development wells revealing the state of equilibrium, the engineer, whether performing analytical or numerical simulation studies of recovery efficiency, is obliged to make an intelligent guess — hopefully based on experience in the area, of the flooding order of the layers. Within each of the more significant sand intervals, in which the VE condition pertains, the layering may be undertaken as shown in Fig. 6.25. Again, however, this must be considered as a risky reservoir environment in which to consider a lateral gas drive.

Exercise 6.1: Immiscible gas drive in a heterogeneous reservoir under the VE condition

Introduction

The lateral displacement of oil by gas is studied in a heterogeneous reservoir in which it might be intuitively expected that the process would be inefficient. Yet, as demonstrated, the significance of the gravity term in the fractional flow relationship is such as to make the situation quite tolerable. The exercise illustrates just how sensitive lateral gas drive efficiency is to variation of the many parameters in the fractional flow.

Question

Gas drive in the downdip direction is being considered in the reservoir whose permeability distribution across the 47 m thick sand section is plotted in Fig. 6.26. The severe coarsening upward in permeabilities would suggest that this may prove a most unsatisfactory reservoir in which to conduct a gas drive. The reservoir section has been divided into 18 layers, whose properties are listed in Table 6.6. Other reservoir and fluid property data at initial pressure, close to which the gas drive will be conducted, are as follows

p_i = 3630 psia; μ_g = 0.02 cp
T = 180ºF = 640ºR; Oil gravity = 28.3ºAPI
θ = 15º (uniform dip); B_{oi} = 1.283 rb/stb
γ_g = 0.640 (air = 1); R_{si} = 520 scf/stb
Z_i = 0.909; μ_o = 0.75

The reservoir is characterised by a single set of rock relative permeability curves for which the end-points are k'_{rg} = 0.47, k'_{ro} = 1.0 and the residual oil saturation is 0.25 PV.

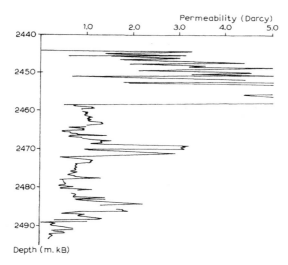

Fig. 6.26. Permeability distribution across the proposed gas-drive reservoir.

TABLE 6.6

Layer data corresponding to the permeability distribution plotted in Fig. 6.26

Layer No.	h_i (m)	ϕ_i	k_i (mD)	S_{wc_i} (PV)	Layer No.	h_i (m)	ϕ_i	k_i (mD)	S_{wc_i} (PV)
1	3.00	0.221	2030	0.120	10	2.30	0.314	1000	0.125
2	4.35	0.243	3340	0.106	11	3.70	0.300	760	0.130
3	6.65	0.264	7560	0.095	12	1.30	0.305	990	0.136
4	2.00	0.319	735	0.130	13	1.70	0.294	435	0.157
5	1.00	0.339	1285	0.133	14	3.65	0.301	830	0.146
6	2.35	0.341	710	0.139	15	3.00	0.329	1550	0.127
7	2.30	0.339	1220	0.127	16	2.00	0.310	880	0.146
8	1.00	0.350	3160	0.107	17	2.65	0.315	390	0.163
9	1.70	0.338	1980	0.105	18	2.30	0.329	290	0.174

- Generate pseudo-relative permeabilities for gas drive in this formation and fractional flow relationships at average frontal velocities of $v = 1$, 0.5 and 0.25 ft/d.
- Calculate the vertical sweep efficiency, N_{pD} and GOR development at the flooding velocity of 1 ft/d. The individual well abandonment criterion is a GOR of 10,000 scf/stb.

From observation of the flooding behaviour of neighbouring reservoirs, it may be assumed that the VE condition will prevail.

Solution

VE pseudo-relative permeabilities for gas drive are presented in Table 6.7, in which equations 6.29 to 6.31 have been evaluated for the gas flooding order of the layers from the top to the base of the section. It will be noted that the fact that the depths are in metres, rather than the usual feet, is of no consequence because the values of h_i merely act as weighting factors in the calculations. The pseudos are plotted in Fig. 6.27 as the solid lines, from which it can be seen the extent to which their shapes are initially influenced by the flooding of the first three high-permeability layers at the top of the section. The curves have been smoothed (dashed lines), without losing their characteristic shapes, to facilitate the generation of fractional flow relationships required for displacement efficiency calculations.

The fractional flow of gas is calculated as

$$f_g = f_{gh}(1 - G) \tag{6.32}$$

in which

$$G = 2.743 \times 10^{-3} \frac{\overline{kk_{ro}}}{v\mu_o} \Delta\rho \sin\theta \tag{6.33}$$

In evaluating this it is necessary to calculate $\Delta\rho$ the gas–oil gravity difference (water = 1) in the reservoir at initial conditions, in which state the flood is being conducted. That is [2]

TABLE 6.7

Generation of pseudo-relative permeabilities and averaged formation properties

Layer No.	h_i (m)	ϕ_i	k_i (mD)	S_{wc_i} (PV)	$\overline{S}_g$ (PV)	$\overline{k}_{rg}$	$\overline{k}_{ro}$
1	3.00	0.221	2030	0.120	0.030	0.028	0.941
2	4.35	0.243	3340	0.106	0.079	0.094	0.800
3	6.65	0.264	7560	0.095	0.161	0.324	0.311
4	2.00	0.319	735	0.130	0.189	0.330	0.297
5	1.00	0.339	1285	0.133	0.204	0.336	0.284
6	2.35	0.341	710	0.139	0.239	0.344	0.268
7	2.30	0.339	1220	0.127	0.274	0.357	0.241
8	1.00	0.350	3160	0.107	0.290	0.371	0.210
9	1.70	0.338	1980	0.105	0.316	0.387	0.178
10	2.30	0.314	1000	0.125	0.349	0.397	0.155
11	3.70	0.300	760	0.130	0.398	0.410	0.128
12	1.30	0.305	990	0.136	0.415	0.416	0.115
13	1.70	0.294	435	0.157	0.436	0.419	0.108
14	3.65	0.301	830	0.146	0.484	0.433	0.079
15	3.00	0.329	1550	0.127	0.528	0.454	0.034
16	2.00	0.310	880	0.146	0.555	0.462	0.017
17	2.65	0.315	390	0.163	0.590	0.467	0.006
18	2.30	0.329	290	0.174	0.621	0.470	0

$\sum h_i$ = 46.95 m; $\overline{S}_{or}$ = 0.25 PV
$\sum h_i \phi_i$ = 13.984 m; $\overline{\phi}$ = 0.298   MOV = $(1 - 0.25 - 0.129)$ = 0.621 PV
$\sum h_i k_i$ = 102927 mD·m; $\overline{k}$ = 2192 mD k'_{rg} = 0.47; k'_{ro} = 1.0
$\sum h_i \phi_i S_{wc_i}$ = 1.806 m; $\overline{S}_{wc}$ = 0.129 PV M = 17.6

Gas:

$$\rho_g = \rho_{g_{sc}} E$$

$$= 0.0763 \frac{\gamma_g \times 35.37}{ZT} p = \frac{0.0763 \times 640 \times 35.37 \times 3630}{0.909 \times 640} = 10.777 \text{ lb/cu.ft}$$

$$\rho_g = 10.777/62.43 = 0.173 \quad (\text{water} = 1)$$

In this calculation 0.0763 lb/cu.ft is the density of air at s.c. and the density of pure water is 62.43 lb/cu.ft.

Oil:

$$\rho_{o_{sc}} = \frac{141.5}{^\circ API + 131.5} = 0.885: \ 0.885 \times 62.43 = 55.25 \text{ lb/cu.ft}$$

$$\rho_{or} = \frac{[5.615 \times \rho_{osc}] + [R_{si} \times 0.0763 \times \gamma_g]}{[B_o \times 5.615]} \text{ lb/cu.ft}$$

$$= \frac{[5.615 \times 55.25] + [520 \times 0.0763 \times 0.64]}{[1.283 \times 5.615]} = 46.6 \text{ lb/cu.ft}$$

$$\rho_{or} = 46.6/62.43 = 0.746 \quad (\text{water} = 1)$$

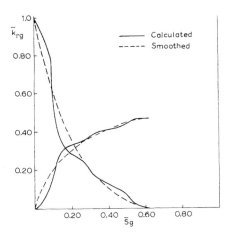

Fig. 6.27. Calculated and smoothed pseudo-relative permeabilities for VE gas drive.

Therefore, $\Delta\rho = 0.746 - 0.173 = 0.573$ (water $= 1$), and using the data provided in the question and Table 6.7.

$$G = 2.743 \times 10^{-3} \times \frac{2192}{0.75} \times \frac{\overline{k}_{ro}}{v} \times 0.573 \times \sin 15^{\circ} = 1.190 \frac{\overline{k}_{ro}}{v}$$

and consequently

$$f_g = f_{gh}\left(1 - 1.19\frac{\overline{k}_{ro}}{v}\right)$$

in which the horizontal fractional flow, f_{gh}, is evaluated using equation 6.26. This, together with the fractional flow relationships at average flooding velocities of $v = 1, 0.5$ and 0.25 ft/day are listed in Table 6.8 and plotted in Fig. 6.28. As can be seen, for horizontal flow, $G = 0$, the extremely harsh nature of the fractional flow would preclude any consideration of gas drive in this reservoir but including the dip of $\theta = 15^{\circ}$ and reducing the average Darcy velocity to $v = 1$ ft/day, or less has a very favourable influence on the displacement efficiency until at $v = 0.25$ ft/day a tangent to the fractional flow would intersect at $S_g = (1 - \overline{S}_{or} - \overline{S}_{wc})$, the flood-out saturation indicating perfect, piston-like displacement. At a rate of 1 ft/day, the tangent from $S_g = 0$ to the fractional flow indicates gas breakthrough (bt — Fig. 6.28) at $\overline{S}_g = 0.175$ PV, $\overline{f}_g = 0.50$. Welge calculations have been performed for gas saturations in excess of this value, using equation 6.34 for the oil recovery and equation 6.35 to calculate the GOR from the reservoir fractional flow, that is

$$\text{GOR} = \frac{E B_o \times 5.615}{\left(\dfrac{1}{\overline{f}_g} - 1\right)} + R_{si} = \frac{1590}{\left(\dfrac{1}{\overline{f}_g} - 1\right)} + 520 \text{ scf/stb}$$

TABLE 6.8

Fractional flow relationships for different velocities of frontal advance

$\overline{S}_g$	$\overline{k}_{rg}$	$\overline{k}_{ro}$	$f_{g\,(horiz)}$	$f_g\,(v = 1\text{ ft/d})$	$f_g\,(v = 0.5\text{ ft/d})$	$f_g\,(v = 0.25\text{ ft/d})$
0.025	0.082	0.885	0.777	−0.041		
0.050	0.130	0.785	0.861	0.057		
0.075	0.172	0.695	0.903	0.156		
0.100	0.207	0.618	0.926	0.245		
0.125	0.238	0.530	0.944	0.349		
0.15	0.260	0.460	0.955	0.432	−0.091	
0.20	0.303	0.354	0.970	0.561	0.153	
0.25	0.335	0.272	0.979	0.662	0.345	−0.289
0.30	0.364	0.200	0.986	0.751	0.517	0.047
0.35	0.390	0.150	0.990	0.831	0.637	0.283
0.40	0.411	0.109	0.993	0.864	0.735	0.478
0.45	0.432	0.070	0.996	0.913	0.830	0.664
0.50	0.444	0.041	0.998	0.949	0.901	0.803
0.55	0.458	0.021	0.999	0.974	0.949	0.899
0.621	0.470	0	1	1	1	1

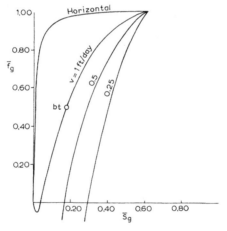

Fig. 6.28. Gas-drive fractional flow relationships for velocities of $v = 1$, 0.5 and 0.25 ft/day.

Results of the calculations are listed in Table 6.9 and indicate an oil recovery of 0.614 (HCPV) or 0.614(1 − 0.129) = 0.535 PV which is 86% of the MOV of 0.713 (HCPV)/0.621 PV, abandoning at a GOR of 10,000 scf/stb. Considering the highly unfavourable nature of the permeability distribution this is a very satisfactory result. It is influenced by several of the parameters in the gravity term of the fractional flow: the reasonable dip angle, $\theta = 15^\circ$, low oil viscosity, $\mu_o = 0.75$ cp, large gravity difference, $\Delta\rho = 0.573$ but, in particular by the high average permeability across the sand section of $k = 2192$ mD. This produces an almost tank-like effect when injecting gas into such a reservoir making the displacement reasonably safe. The

TABLE 6.9

Welge gas drive displacement calculations to determine vertical sweep and GOR

$\overline{S}_{ge}$	$\overline{f}_{ge}$	$\Delta \overline{S}_{ge}$	Δf_{ge}	G_{id} (PV)	$\overline{S}_{ge}$	$\overline{f}_{ge}$	N_{pD} (HCPV)	GOR (scf/stb)
0.175 (bt)	0.500	0.175	0.500	0.350	0.088	0.250	0.402	1050
0.20	0.561	0.025	0.061	0.410	0.188	0.531	0.437	2320
0.25	0.662	0.050	0.101	0.495	0.225	0.612	0.479	3030
0.30	0.751	0.050	0.089	0.562	0.275	0.707	0.505	4360
0.35	0.813	0.050	0.062	0.806	0.325	0.782	0.575	6230
0.40	0.864	0.050	0.051	0.980	0.375	0.839	0.612	8800
0.45	0.913	0.050	0.049	1.020	0.425	0.889	0.618	13250
0.50	0.949	0.050	0.036	1.389	0.475	0.931	0.655	21970
0.55	0.974	0.050	0.025	2.000	0.525	0.962	0.690	40770
0.621	1	0.071	0.026	2.731	0.586	0.987	0.714	121240

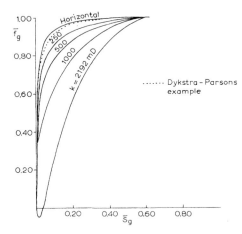

Fig. 6.29. Gas-drive fractional flow relationships for $v = 1$ ft/day for different average permeabilities.

influence of the permeability on the fractional flow is illustrated in Fig. 6.29, in which the value has been successively reduced from $k = 2192$ mD while adhering to the flooding velocity of $v = 1$ ft/d. As can be seen, for anything less than 1000 mD, still a very high average for a reservoir, the fractional flow relationships become so unfavourable as to preclude consideration of lateral gas drive.

The importance of the gravity term is stabilizing the frontal advance of gas drive cannot be over emphasised. Its complete absence is also demonstrated by the dotted line in Fig. 6.29. This function has been generated for the sand section shown in Fig. 5.63 in which the sands are physically separated from each other. Gas drive has been considered in this environment using the mobility ratio in the present exercise ($M = 17.6$) and for the same movable saturation (MOV = 0.621 PV). The method is precisely the same as described in Exercise 5.5 for Dykstra-Parsons displacement in which, as in this section, the overall influence of gravity is neglected ($G = 0$). As can be seen, the fractional flow is so unfavourable that gas drive could not be considered in this type of situation. Unless the gravity term in the fractional is significant, then gas drive in the downdip direction is not a practical proposition.

6.5. DRY GAS RECYCLING IN RETROGRADE GAS-CONDENSATE RESERVOIRS

The aim in dry gas recycling is to maintain the pressure at a high level and thus minimize the deposition and loss of retrograde liquid condensate in the reservoir. At the same time the injected dry gas displaces the wet towards the producing wells in what is generally regarded as a highly efficient manner. Two types of recycling can be distinguished:

Full recycling
All the separated dry gas, save that used for power generation at the surface, is re-injected into the reservoir with the aim of maintaining the pressure above the dew point so that all the wet gas produced contains its maximum condensate yield, r_{si} (stb/MMscf) (Fig. 6.1b), none is deposited in the reservoir. Naturally, during such recycling there is a material balance deficiency so that there is the risk that the pressure may slowly decline — even below the dew point. To prevent this happening, dry "make-up" gas can be imported from elsewhere to supplement the injection. Not only does the high-pressure recycling reduce the loss of condensate resulting from deposition in the reservoir but also, if there is an aquifer, the pressure maintenance inhibits water influx and therefore the trapping of wet gas with its contained condensate behind the advancing water front (section 6.3).

The process and its typical production profile are illustrated in Fig. 6.30. During the initial years of exclusive recycling only the liquid condensate is available for sale. Consequently, the most important parameter in evaluating the economic viability

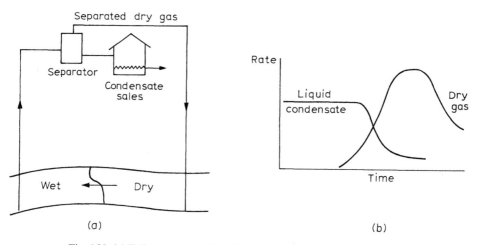

Fig. 6.30. (a) Full dry gas recycling. (b) Liquid and gas production profiles.

of the project is the value of the condensate yield above the dew point, r_{si} — stb/MMscf (Fig. 6.1b) and this is especially the case for costly offshore projects. The recycling is terminated when dry gas breaks through to the production wells and the condensate production rate is inadequate to sustain the required cash flow. Then, there is a gradual change-over from recycling to dry gas production and sales as the blow-down period commences.

Partial recycling

This amounts to anything that provides less than full pressure maintenance above the dew point. Usually, a gas sales contract is being satisfied but the facilities exist (injection wells/compressors) to inject any gas that is produced in excess of the contract requirement. During the off-peak summer months, for instance, the field can still be produced at its maximum gross rate, commensurate with the number of production wells, and any excess gas produced is recycled. If the pressure declines continuously during the operation then below the dew point the condensate yield will steadily decrease below its initial, maximum value of r_{si} (stb/MMscf) yet it will remain above the value for straightforward depletion thus increasing the ultimate recovery of condensate and hopefully justifying the added expense associated with the partial recycling. Alternatively, if there is a strong degree of natural water influx and pressure maintenance, the limited gas injection will reduce water encroachment and the trapping of residual gas with its contained condensate at high pressure.

As with waterdrive or gas drive, described in the previous section, dry gas recycling in reservoirs must be viewed in terms of the three factors which most influence the recovery efficiency of the process: mobility ratio, heterogeneity and gravity.

(a) Mobility ratio

Dry gas displacing wet is a miscible process in which the phases physically mix to the extent that there is no sharp interface between them and therefore no surface tension. As a result, in a laboratory core flooding experiment in a homogeneous core plug, there will be no residual wet gas saturation remaining in the pores contacted by the dry gas. One-dimensional, rock relative permeabilities to describe this ideal form of displacement are simply linear functions across the movable saturation range [23] as depicted in Fig. 6.31. These represent complete mixing such that the sum of the relative permeabilities is always unity. They are plotted as functions of the increasing displacing phase saturation of the dry gas, S_{gd}, and since there is no residual wet gas saturation, the flood-out saturation is $1 - S_{wc}$ which is also the total movable saturation range. The viscosity of the lighter dry gas is less than that of the wet gas and therefore the mobility ratio — using typical fluid property data, is

$$M = \frac{k'_{rgd}}{\mu_{gd}} \bigg/ \frac{k'_{rgw}}{\mu_{gw}} = \frac{1}{0.02} \bigg/ \frac{1}{0.03} = 1.5$$

which, although greater than unity, should give no cause for concern: the process is reasonably stable on the microscopic scale.

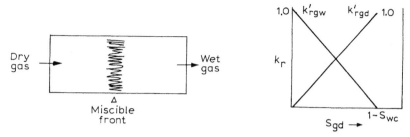

Fig. 6.31. Dry gas recycling flooding experiment and resulting one-dimensional relative permeabilities.

Because recycling is miscible, attention has been focused in the literature on modelling the displacement using compositional simulation with the PVT data processed using equations of state calibrated against the results of CVD experiments. In this, the numerous hydrocarbon and non-hydrocarbon constituents are grouped into a reduced number of pseudo-components which are used in the simulator with no significant loss of accuracy, in terms of the amounts of hydrocarbon in each component in the liquid or vapour phase, at each stage of depletion. Yet, as pointed out by Coats [10], such sophistication is unnecessary in two of the principal methods of hydrocarbon recovery from gas-condensate fields, namely, depletion and full, high-pressure recycling above the dew point pressure. For both these processes a very reliable match on the results of full compositional modelling can be achieved using a much simpler, two-phase modified black oil simulator. The normal black oil model has solution gas dissolved in the oil (R_{si} — scf/stb), the modification caters for the vaporization of oil into the gas phase. Effectively, this is a two-component (stock tank oil/separator gas) model for which the PVT input consists of tables of B_o, R_s, B_g and r_s as functions of pressure. The reason such a simple approach is acceptable for depletion and high-pressure recycling is because for the former the fluid behaviour in the reservoir should match that in the CVD experiment, provided the reservoir fluids have been sampled correctly during testing; while recycling above the dew point is a miscible two-phase displacement with no phase equilibrium effects. Full compositional modelling is required, however, in partial recycling in which the pressure may decline below the dew point. It will then be necessary to account for the individual hydrocarbon components being deposited or vapourized during depletion. In an example of recycling below the dew point of a near critical condensate, Coats has demonstrated [10] the need to use a minimum of seven pseudo-components for acceptable accuracy.

In this author's experience, the results of material balance calculations (section 6.3) accurately match those of compositional simulation for depletion fields while for high-pressure recycling the dominant factors dictating hydrocarbon recovery efficiency are the influence of reservoir heterogeneity and gravity, as described below.

TABLE 6.10

Wet and dry gas compositions for dry gas recycling

Component	Mole (%)		Component	Mole (%)	
	wet	dry		wet	dry
N_2	0.24	0.24	$i\text{-}C_5$	0.45	0.23
CO_2	1.00	1.27	$n\text{-}C_5$	0.49	0.22
C_1	81.15	89.24	C_6	0.75	0.41
C_2	5.39	4.23	C_7	1.14	1.05
C_3	2.21	2.01	C_8	1.10	
$i\text{-}C_4$	0.49	0.39	C_9	0.74	
$n\text{-}C_4$	1.00	0.71	C_{10+}	3.85	

(b) Heterogeneity/gravity

In examining the combined influence of these, it is first necessary to consider the gravity difference between the wet and dry gases. A compositional analysis up to C_{10+} is provided in Table 6.10 for a wet gas at an initial reservoir pressure of 4250 psia, which is close to dew point, and injected dry gas, which has been analysed as far as the C_{7+} fraction.

Applying standard correlations [3] at a reservoir temperature of 170°F (630°R), the following PVT properties have been calculated for the gases.

	Wet gas	Dry gas
Gravity, γ_g (air = 1)	0.937	0.672
Pseudo-critical temperature [2], T_c (°R)	426	374
Pseudo-critical pressure [2], p_c (psia)	641	663
Z-factor at initial presure	0.894	0.936
Gas expansion factor at p_i, E_i (scf/rcf)	267	255
Gas FVF at p_i, B_{gi} (rcf/scf)	0.00375	0.00392
Gas density in the reservoir (lb/cu.ft)	19.1	13.1

In evaluating these properties, the gas density in the reservoir has been evaluated [2] as

$$\rho = 0.0763\gamma_g E \quad \text{(lb/cu.ft)}$$

The first point to notice in comparing the gas properties is that the FVF of the dry gas is greater than for the wet. Therefore, considering 1 scf of wet and dry gas at the surface; if these were returned to the reservoir at high pressure and temperature, the less dense dry gas would occupy a larger volume, which is quite normal. Typical plots of wet and dry gas FVFs are shown in Fig. 6.32; there is usually a cross-over between the functions at intermediate pressures. Therefore, although there is a material balance deficiency in recycling, due to the removal of liquid condensate from the wet gas, this is partially compensated for by the relatively larger volume occupied by the dry gas when returned to the reservoir for high-pressure recycling.

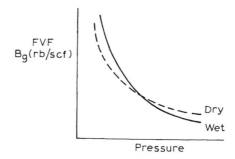

Fig. 6.32. FVFs of wet and dry gas in recycling operations.

Also of importance is the density difference between the gases in the reservoir. In absolute value it may not amount to much (6 lb/cu.ft) but in relative terms it is significant, the wet gas being 46% denser than the dry. This percentage difference is almost as large as that between water and oil in a waterflood, yet it is frequently overlooked in displacement calculations. The combination of slightly unfavourable mobility ratio and significant gravity difference means that the stability of dry gas recycling in the vertical cross-section will be dependent on the nature of the heterogeneity and particularly the permeability distribution across the reservoir. The manner in which the vertical sweep efficiency may be accounted for both for the VE condition and a total lack of it is described below.

Vertical equilibrium

If there are no restrictions to vertical fluid movement across the reservoir, as established in RFT surveys in each new development well, then there will be a tendency for segregation to occur with the lighter dry gas rising to the top of the section. The situations depicted in Fig. 6.33 are the same as for gas–oil displacement, Fig. 6.23. If there is a coarsening upward in rock properties the lighter dry gas will override the wet leading to its premature breakthrough at the production well and the requirement to circulate large volumes of dry gas to recover all of the movable wet gas, $(1 - \bar{S}_{wc})$ PV, in the section. Alternatively, if the better permeabilities are towards the base of the reservoir, the dry gas will rise producing the effect of piston-like displacement across the entire reservoir. In this case, the flood will resemble that of a large core flooding experiment and the beneficial effects of the miscible displacement will be fully realized. If there is override, however, (Fig. 6.33a), then although the displacement is miscible at the point of contact of the gases all along the front, the overall effect of miscibility can be seriously downgraded. That is, if the override is so severe that the recycling has to be terminated for practical reasons before all the movable wet gas has been recovered, then there will be an average residual wet gas saturation, $\bar{S}_{gw}$, remaining in the reservoir at the termination of the flood, meaning that on the macroscopic scale the full effect of miscibility has not been achieved.

Even though recycling is a miscible process on the microscopic scale, it is still perfectly valid to generate pseudo-relative permeabilities to describe the displacement

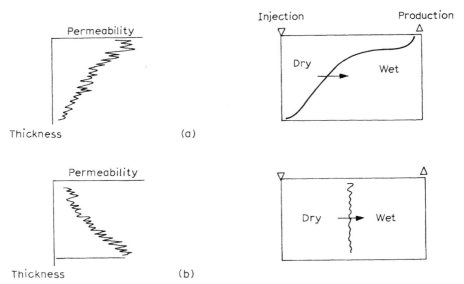

Fig. 6.33. Influence of permeability distribution on the efficiency of dry gas recycling under the VE condition.

across the entire reservoir section. If the VE condition pertains then the flooding order of the N layers selected will be from the top to the base. The thickness averaged dry gas saturation, $\bar{S}_{gd}$, and pseudo-relative permeabilities for the flow of dry and wet gas, after the nth layer has flooded may be calculated as

$$\bar{S}_{gd_n} = \frac{\sum\limits_{i=1}^{n} h_i \phi_i (1 - S_{wc_i})}{\sum\limits_{i=1}^{N} h_i \phi_i} \tag{6.37}$$

$$\bar{k}_{rgd_n} = \frac{\sum\limits_{i=1}^{n} h_i k_i k'_{rgd}}{\sum\limits_{i=1}^{N} h_i k_i} \tag{6.38}$$

$$\bar{k}_{rgw_n} = \frac{\sum\limits_{i=n+1}^{N} h_i k_i k'_{rgw}}{\sum\limits_{i=1}^{N} h_i k_i} \tag{6.39}$$

in which it is assumed that an individual layer has been completely miscibly flooded to a dry gas saturation of $1 - S_{wc_i}$, or it has not: no in-between state is catered for. Consequently, only the end-point relative permeabilities are required in averaging procedures and these both have the value of unity (Fig. 6.31) for miscible displacement. Justification for the use of only the end-points is, as described for waterdrive (Chapter 5, section 5.6) and gas oil displacement (section 6.4b), on account of the gravity segregation that occurs and the relatively small mixing zone between the gases when the displacement is viewed on the macroscopic scale. The generation of VE pseudos for dry gas recycling is illustrated in Exercise 6.2, for both favourable and unfavourable permeability distributions, in which analytical pseudos are validated against those derived by cross-sectional numerical simulation modelling.

Total lack of vertical equilibrium

If the recycling occurs in a reservoir environment in which there are barriers to vertical fluid movement, then the influence of gravity on the displacement is excluded. In this case, a dry gas fractional flow may be generated directly using the method of Dykstra-Parsons or, since the mobility ratio is only slightly greater than unity, applying the method of Stiles to generate pseudo-relative permeability using equations 6.37 to 6.39. Both methods, which yield almost identical results in displacement calculations, are applied in the same manner as described for waterdrive in Chapter 5, section 5.7.

(c) Vertical sweep efficiency

The generation of pseudos effectively reduces the description of the dry–wet gas displacement to one dimension in which it is suitable for the application of the Buckley-Leverett theory using the application method of Welge. In this, the first step is to generate a fractional flow relationship for the dry gas which, including the gravity term, may be expressed

$$\overline{f}_{gd} = \frac{1 - 2.743 \times 10^{-3} \dfrac{\overline{kk}_{rgw}}{v\mu_{gw}} \Delta\rho \sin\theta}{1 + \dfrac{\mu_{gd}}{\mu_{gw}} \dfrac{\overline{k}_{rgw}}{\overline{k}_{rgd}}} \tag{6.40}$$

which is in direct analogy with equation 5.19 for waterdrive and equation 6.32 for gas drive. In the equation $\Delta\rho = \rho_{gw} - \rho_{gd}$, (water = 1) at reservoir conditions and it is assumed that the displacement by the dry gas is in the more stable, downdip direction for which the negative sign is appropriate. In fact, as for waterdrive, the gravity term in the fractional flow for recycling is not usually very significant, unless the dip angle is large, and can normally be omitted from the equation — after first checking its magnitude. This is because although the percentage gravity difference between the wet and dry gas is quite large, in absolute terms it is small being only 6 lb/cu.ft in the example quoted in section 6.5b.

Having generated the pseudo-fractional flow, the equation of Welge may be applied in precisely the same manner as described for water and gas drive. In the present case, the recovery equation is

$$G_{pD} = \frac{\overline{S}_{gde} + (1 - \overline{f}_{gde})G_{id}}{1 - S_{wc}}$$ (6.41)

in which

G_{pD} = cumulative recovery of wet gas (HCPV)
G_{id} = cumulative dry gas injected (PV) = $\Delta \overline{S}_{gde}/\Delta f_{gde}$.

Exercise 6.2: Generation of pseudo-relative permeabilities for dry gas recycling

Introduction
    There is little in the literature relating to the need or method of generating pseudo-relative permeabilities for evaluating the vertical sweep efficiency of recycling. One example is to be found, however, in an SPE paper from 1970 concerning an edge injection scheme in the giant Kaybob Field in Alberta, Canada [25]. In this, pseudo-relative permeabilities were derived by detailed cross-sectional, numerical simulation modelling and these are compared in this exercise to the equivalent pseudos generated analytically.

Question
    In the reported simulation modelling the total sand section was divided into two reservoirs, the Upper and Lower, separated from each other by an impermeable barrier. The only formation properties supplied were the horizontal and vertical permeability distributions across the sands; these are listed in Table 6.11 and the

TABLE 6.11

Permeability distributions across the upper and lower reservoirs of the Kaybob Field

Thickness	Permeabilities		Thickness	Permeabilities	
h (ft)	k (mD)	k_v (mD)	h (ft)	k (mD)	k_v (mD)
Upper reservoir			*Lower reservoir*		
13.1	133.1	21.3	12.5	59.5	59.6
15.7	242.8	11.3	12.9	39.7	1.0
15.0	174.8	25.8	9.5	27.5	4.8
14.7	108.8	49.1	18.0	195.5	76.2
			12.6	228.1	5.4
			9.7	528.7	13.4
			5.2	87.8	7.9
			5.1	162.3	17.3
			5.3	558.0	65.1
			4.8	335.9	4.4
			3.7	417.3	23.4
$\sum h_i$ = 58.5 ft			$\sum h_i$ = 99.3 ft		
$\sum h_i k_i$ = 9777 mD·ft			$\sum h_i k_i$ = 20437 mD·ft		

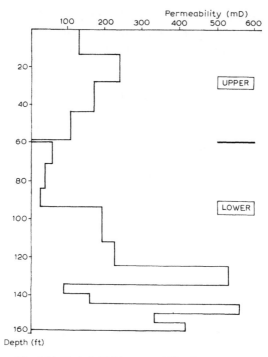

Fig. 6.34. Kaybob Field: permeability distributions.

former is plotted in Fig. 6.34. It must be assumed that in comparison to the permeability the variation in both porosity and irreducible water saturation, which were not detailed, are slight.

- Generate pseudos and fractional flow relationships for dry gas recycling and calculate the vertical sweep efficiency across the reservoirs.

It may be assumed that the rock relative permeabilities are linear with unit end points (Fig. 6.31) representing fully miscible displacement on the microscopic scale. The average, irreducible water saturations are $S_{wc} = 0.14$ PV for the Upper and 0.15 PV for the Lower reservoir and the gas viscosities are $\mu_{gw} = 0.03$ cp and $\mu_{gd} = 0.02$ cp at the average flooding pressure.

Solution
Merely by inspection of the horizontal permeability distributions across the reservoirs (Fig. 6.34) it may be predicted that the vertical sweep efficiency in the Upper will be poor on account of the tendency for dry gas override encouraged by the coarsening upward sequence in permeability. Furthermore, because of the relatively high vertical permeabilities (Table 6.11) the engineer will have little difficulty in accepting that the VE condition will apply and that the flooding order will be from the top to the base of the reservoir in the four layers defined. Conversely, in the Lower reservoir there is a very definite coarsening downwards in

the permeability which intuitively suggests that there should be a stable, piston-like displacement of the wet gas by the dry. It may be more difficult to accept, however, that the VE condition, implying flooding from top to base, will be appropriate in the eleven layers of the Lower reservoir, especially since there are parts of the section in which the vertical permeabilities are low. Nevertheless, pseudos will be generated for both reservoirs assuming that VE pertains and the results compared with those derived by numerical simulation modelling. In performing the calculations, neither the vertical permeabilities nor the gravity difference between the gases are used explicitly as in the simulation. Instead, it is assumed that both are sufficiently large to permit the instantaneous segregation of the gas phases. The pseudos have been generated using equations 6.37 to 6.39 but, since the porosity is taken as constant, the expression to calculate the thickness averaged dry gas saturation is simplified as

$$\overline{S}_{gd} = \frac{\sum\limits_{i=1}^{n} h_i(1 - S_{wc})}{\sum\limits_{i=1}^{N} h_i}$$

in which the irreducible water saturations are also constant in the two reservoirs. The pseudos are listed in Table 6.12 and plotted in Fig. 6.35 together with the functions derived from the cross-sectional simulation study. As can be seen, the close correspondence between the two validates the assumption of the VE condition, in which the reservoirs flood from the top to the base, and particularly in the less likely case of Lower reservoir with its coarsening downward sequence of permeabilities. The next step is the incorporation of the dry/wet gas viscosity ratio, $\mu_{gd}/\mu_{gw} = 0.02/0.03 = 0.667$, which has been used in equation 6.40 to calculate the fractional

TABLE 6.12

Generation of pseudo-relative permeabilities and fractional flow relationships for dry gas recycling in the upper and lower reservoirs of the Kaybob Field (VE condition, flooding order from top to base)

h (ft)	k (mD)	$\overline{S}_{gd}$ (PV)	$\overline{k}_{rgd}$	$\overline{k}_{rgw}$	f_{gd}	h (ft)	k (mD)	$\overline{S}_{gd}$ (PV)	$\overline{k}_{rgd}$	$\overline{k}_{rgw}$	f_{gd}
Upper reservoir						*Lower reservoir*					
13.1	133.1	0.193	0.178	0.822	0.245	12.5	59.5	0.107	0.036	0.964	0.053
15.7	242.8	0.423	0.568	0.432	0.664	12.9	39.7	0.217	0.061	0.939	0.089
15.0	174.8	0.644	0.836	0.164	0.884	9.5	27.5	0.299	0.074	0.926	0.107
14.7	108.8	0.860	1.000	0	1	18.0	195.5	0.453	0.246	0.754	0.329
						12.6	228.1	0.561	0.387	0.613	0.486
						9.7	528.7	0.644	0.638	0.362	0.726
						5.2	87.8	0.688	0.660	0.340	0.744
						5.1	162.3	0.732	0.701	0.299	0.779
						5.3	558.0	0.777	0.846	0.154	0.892
						4.8	335.9	0.818	0.924	0.076	0.948
						3.7	417.3	0.850	1.000	0	1

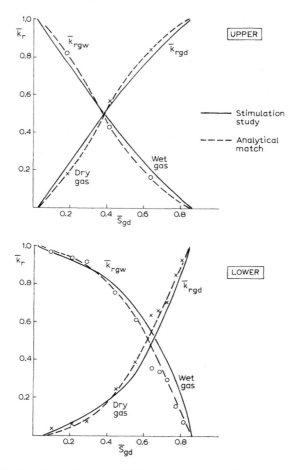

Fig. 6.35. Pseudo-relative permeabilities for dry gas recycling (Kaybob Field).

flow of dry gas for both reservoirs. In this instance, the gravity term is negligible and the equation is therefore reduced to the form

$$\overline{f}_{gd} = \cfrac{1}{1 + \cfrac{\mu_{gd}}{\mu_{gw}} \cfrac{\overline{k}_{rgw}}{\overline{k}_{rgd}}}$$

These calculations (which were not reported in reference 25) are listed in Table 6.12 and plotted in Fig. 6.36 from which it can be seen that they confirm intuitive judgement concerning the displacement efficiency in the two reservoirs. In the Upper, the plot is concave downwards across a significant portion of the movable wet gas saturation range (breakthrough occurs at $\overline{S}_{gd} = 0.415$ PV: $\overline{f}_{gd} = 0.650$) indicating a degree of dry gas override. Conversely, in the Lower, the fractional flow is concave upwards across the entire movable saturation range implying, as

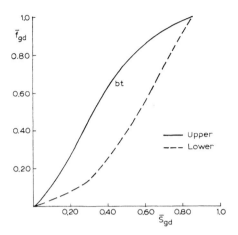

Fig. 6.36. Dry gas pseudo-fractional flow relationships: upper and lower reservoirs, Kaybob Field.

described for waterdrive, a complete piston-like displacement of the wet gas by the dry in which the full benefit of miscibility will be attained.

At this point, great care must be exercised in deciding how the pseudo-relative permeability curves should be used, especially if considering their input to a numerical simulation model. Suppose, for instance, the pseudos for the Lower reservoir (Fig. 6.35) were to be used in a model with the intention of reducing the original number of eleven layers to just one — which is the usual aim in generating pseudos. Inputting the curves, which are continuous and finite across the entire movable saturation range will confuse the coarse gridded, one layer model into thinking that all dry gas saturations are separately mobile and it will therefore disperse the dry gas. Then even though the shapes of the two pseudos are basically favourable, it will produce a pessimistic result which is akin to dry gas override. Yet a glance at the fractional flow for the Lower reservoir (Fig. 6.36) confirms that no dry gas saturations should have independent mobility, they are all caught-up in the piston-like shock front. This is another example of the paradoxical situation described for waterdrive (Chapter 5, section 5.8): that for an excellent reservoir, in which the displacement is perfect, it is not possible to generate pseudos from a detailed layered model and use them in a one-layered coarse grid block model. Instead, to accurately model the piston-like effect, it will be necessary to construct a detailed layered model with rock relative permeabilities in each so that horizontal and vertical numerical dispersion of the dry gas will compensate each other leading to the development of the required shock front. The same is partially true for the upper reservoir for which the fractional flow (Fig. 6.36) indicates a one-dimensional (Buckley-Leverett) shock front of $\overline{S}_{gd} = 0.415$ PV. If using the pseudo-relative permeabilities for this reservoir in a one-layered model, some means must be incorporated of prohibiting movement of the lower dry gas saturations for values of $\overline{S}_{gd} < 0.415$ PV (refer Chapter 5, section 5.4f) otherwise the override will appear worse than it is in this reservoir. Once again it must be stressed that judgement in

TABLE 6.13

Calculation of the dry gas vertical sweep (Welge) for the upper reservoir of the Kaybob field

$\overline{S}_{gde}$	$\overline{f}_{gde}$	$\Delta \overline{S}_{gde}$	$\Delta \overline{f}_{gde}$	$\overline{S}_{gde}$	$\overline{f}_{gde}$	G_{id} (PV)	G_{pD} (HCPV)
0.415 (bt)	0.650						
0.45	0.703	0.035	0.053	0.433	0.677	0.660	0.751
0.50	0.760	0.05	0.057	0.475	0.732	0.877	0.826
0.55	0.810	0.05	0.050	0.525	0.785	1.000	0.860
0.60	0.855	0.05	0.045	0.575	0.833	1.111	0.884
0.65	0.893	0.05	0.038	0.625	0.874	1.316	0.920
0.70	0.927	0.05	0.034	0.675	0.910	1.471	0.939
0.75	0.956	0.05	0.029	0.725	0.942	1.724	0.959
0.80	0.980	0.05	0.024	0.775	0.968	2.083	0.979
0.86	1	0.06	0.020	0.830	0.990	3.000	1.000

the use of pseudos can only be exercised properly by taking the extra step of plotting and inspecting the fractional flows. In the Kaybob simulation, it appears that the pseudos were used directly in one-layered models since the results indicated dry gas override in both reservoirs and while this is to be expected in the Upper, it is difficult to imagine how it could happen in the Lower reservoir whose permeability distribution is perfect for recycling.

The final part of the exercise is to calculate the vertical sweep efficiency in the reservoirs. In the lower, as noted, it should be perfect with the movable wet gas volume $1 - S_{wc} = 0.85$ PV = 1 HCPV being recovered by the injection of the same volume of dry gas. In the upper reservoir, however, Welge displacement calculations (equation 6.41) have been performed after dry gas breakthrough using the less favourable fractional flow plotted in Fig. 6.36 and the results are listed in Table 6.13. In performing the calculations it is the values of $\overline{S}_{gde}$ and $\overline{f}_{gde}$ in columns 5 and 6 that have been used in Welge's equation. These are the average values of the figures in columns 1 and 2 at which the cumulative injected dry gas, G_{id} (PV), has been evaluated. As can be seen, the flooding efficiency is much poorer in the Upper than the Lower reservoir on account of dry gas override and it would take three pore volumes of injected gas to recover the movable wet gas. In a situation like this it would be appropriate to have a higher ratio of injection to production wells on the Upper reservoir than the Lower to accelerate the recovery from the former.

REFERENCES

[1] Standing, M.B. and Katz, D.L.: Density of Natural Gases, Trans. AIME, 1942, Vol. 146 (140).
[2] Dake, L.P.: Fundamentals of Reservoir Engineering, Elsevier, Amsterdam, 1978.
[3] Hewlett-Packard.: Petroleum Fluids Pac, Hewlett-Packard, Corvallis, Oreg., 1983.
[4] Benedict, M., Webb, G.B. and Rubin, L.C.: An Empirical Equation for Thermodynamic Properties of Light Hydrocarbons and their Mixtures, Chem. Eng. Prog., August 1951.
[5] Whitson, C.H. and Torp, S.B.: Evaluating Constant Volume Depletion Data, JPT, March 1983.

[6] Craft, B.C. and Hawkins, M.F.: Applied Petroleum Reservoir Engineering, Prentice Hall, Inc., N.J., 1959.

[7] Redlich, O. and Kwong, J.N.S.: On the Thermodynamics of Solutions, and Equation of State, Fugacities of Gaseous Solutions, Chem. Rev., 1949, Vol. 44, 233.

[8] Soave, G.: Equilibrium Constants from a Modified Redlich-Kwong Equation of State, Chem. Eng. Sci. 1972, Vol. 27, 1197.

[9] Peng, D.Y. and Robinson, D.B.: A New Two Constant Equation of State, Ind. Eng. Chem. Fundam. 1976, Vol. 15, 59.

[10] Coats, K.H.: Simulation of Gas Condensate Reservoir Performance, SPE Symposium Reservoir Simulation, New Orleans, 1982, SPE 10512.

[11] Moses, P.L.: Engineering Applications of Phase Behaviour of Crude Oil and Condensate Systems, JPT, July 1986, 715.

[12] Katz, D.L.: Overview of Phase Behaviour in Oil and Gas Production, JPT, June 1983, 1205.

[13] Fishlock, T.P., Smith, R.A., Soper, B.M. and Wood, R.W.: Experimental Studies on the Waterflood Residual Gas Saturation and its Production by Blowdown, SPE Reservoir Engineering, May 1988.

[14] Bruns, J.R., Fetkovich, M.J. and Meitzen, V.C.: The Effect of Water Influx on p/Z-Cumulative Gas Production Curves, JPT, March 1965, 287.

[15] Agarwal, R.G., Al-Hussainy, R. and Ramey, H.J., Jr.: The Importance of Water Influx in Gas Reservoirs, JPT, November 1965, 1336.

[16] Cason, L.D., Jr.: Waterflooding Increases Gas Recovery, JPT, October 1989, 1102.

[17] van Everdingen, A.F. and Hurst, W.: The Application of the Laplace Transformation to Flow Problems in Reservoirs, Trans. AIME, 1949, Vol. 186, 305.

[18] Brinkman, F.P.: Increased Gas Recovery from a Moderate Waterdrive Reservoir, JPT, December 1981, 2475.

[19] Chesney, T.P., Lewis, R.C. And Trice, M.L.: Secondary Gas Recovery from a Moderately Strong Waterdrive Reservoir: A Case History, JPT, September 1982, 2149.

[20] Lutes, J.L. et al.: Accelerated Blowdown of a Strong Waterdrive Gas Reservoir, JPT, December 1977, 1533.

[21] Sandrea, R. and Nielsen, R.F.: Dynamics of Petroleum Reservoirs Under Gas Injection, Gulf Publishing Company, Houston, Texas.

[22] Dumoré, J.M. and Schols, R.S.: Drainage Capillary-pressure Functions and the Influence of Connate Water, Soc. Pet. Eng. J., October 1974, 437.

[23] Williams, J.K. and Dawe, R.A.: Near-Critical Condensate Fluid Behaviour in Porous Media-A Modelling Approach, SPE-Reservoir Engineering, May 1989.

[24] Shank, G.D. and Vestal, C.R.: Practical Techniques in Two Pseudocomponent Black Oil Simulation, SPE-Reservoir Engineering, May 1989.

[25] Field, M.B., Wytrychowski, I.M. and Patterson, J.K.: A Numerical Simulation of Kaybob South-Gas Cycling Projects, SPE Fall Meeting, Houston, October 1970.

SUBJECT INDEX

Printed and bound by CPI Group (UK) Ltd, Croydon, CR0 4YY

08/05/2025

01864930-0002